SELOC'S PERSONAL WATERCRAFT

Volume I

Early Days thru 1991

Tune-Up
and
Repair Manual

SELOC
PUBLICATIONS

MARINE
MANUALS

Other books by Joan Lorton Coles and/or Clarence W. Coles

Glenn's Complete Bicycle Manual
*The Book of Harry (Limited Edition)
*Glenn's Flat Rate Manual -- American Cars
Seloc's OMC Stern Drive Tune-up and Repair Manual
Seloc's Marine Jet Drive Tune-up and Repair Manual
Seloc's Mercury Volume I Tune-up and Repair Manual
Seloc's Mercury Volume II Tune-up and Repair Manual
Seloc's Mercury Volume III Tune-up and Repair Manual
Seloc's Force Outboard Tune-up and Repair Manual
Seloc's Yamaha Volume I Tune-up and Repair Manual
Seloc's Yamaha Volume II Tune-up and Repair Manual
Seloc's Yamaha Volume III Tune-up and Repair Manual
Seloc's Mariner Volume I Tune-up and Repair Manual
Seloc's Mariner Volume II Tune-up and Repair Manual
Seloc's Chrysler Outboard Tune-up and Repair Manual
Seloc's OMC Cobra Stern Drive Tune-up and Repair Manual
Seloc's MerCruiser Stern Drive Volume I Tune-up and Repair Manual
Seloc's MerCruiser Stern Drive Volume II Tune-up and Repair Manual
Seloc's MerCruiser Stern Drive Volume III Tune-up and Repair Manual
Seloc's Mercury/Mariner Volume I Tune-up and Repair Manual
Seloc's Mercury/Mariner Volume II Tune-up and Repair Manual
Seloc's Mercury/Mariner Volume III Tune-up and Repair Manual
Seloc's Johnson/Evinrude Volume I Tune-up and Repair Manual
Seloc's Johnson/Evinrude Volume II Tune-up and Repair Manual
Seloc's Johnson/Evinrude Volume IIA Tune-up and Repair Manual
Seloc's Johnson/Evinrude Volume III Tune-up and Repair Manual
Seloc's Johnson/Evinrude Volume IV Tune-up and Repair Manual
Seloc's Johnson/Evinrude Volume V Tune-up and Repair Manual
Seloc's Personal Watercraft Volume I Tune-up and Repair Manual
Seloc's Personal Watercraft Volume IA Tune-up and Repair Manual
Seloc's Personal Watercraft Volume II Tune-up and Repair Manual
Seloc's Personal Watercraft Volume IIA Tune-up and Repair Manual
Seloc's Personal Watercraft Volume III Tune-up and Repair Manual
Seloc's Personal Watercraft Volume IIIA Tune-up and Repair Manual
Seloc's Volvo Penta Stern Drive Volume I Tune-up and Repair Manual
Seloc's Volvo Penta Stern Drive Volume II Tune-up and Repair Manual
Seloc's Volvo Penta Stern Drive Volume III Tune-up and Repair Manual
Briggs & Stratton Single Cylinder OHV Tune-up and Repair Manual
Briggs & Stratton Single Cylinder Horizontal Crankshaft Tune-up & Repair Manual
Briggs & Stratton Single Cylinder Vertical Crankshaft Tune-up and Repair Manual
*Out of print.

SELOC'S
PERSONAL WATERCRAFT

Volume I

Kawasaki

Early Days thru 1991

Tune-Up and Repair Manual

by

Joan and Clarence Coles

Cover
An original painting
by
Ray Goudey II
B.F.A. Art Center
Pasadena, California

Technical Director
Dani Fisk

Graphics
Barbara Flotow

Third Printing -- March, 1997

Inquiries should be addressed to:

Seloc Publications
10693 Civic Center Drive
Rancho Cucamonga, California 91730

ISBN 0-89330-032-2

Printed in the United States of America

FOREWORD

This is a comprehensive tune-up and repair manual for Personal Watercraft units from early days thru 1991, produced by:

Kawasaki
Heavy Industries Ltd.
Consumer Products & Components Group
Japan

The information contained in this manual has been designed and written to cover stock, factory "out the door" engines and jet pumps. Modifications for higher than manufacturer's rated performance are so extensive and so varied, no attempt has been made to include them in this volume.

All Seloc manuals, including this book, are directed toward the professional mechanic, the do-it-yourselfer, and the student developing her/his mechanical skills.

ACKNOWLEDGMENTS

The authors wish to express their thanks and appreciation to Kawasaki Motors Corp. U.S.A., Santa Ana, California for the courtesies and assistance extended to us during the preparation of material for this book.

A personal **"THANKS"** is extended to Patrick Kelly, Kawasaki Motors, for his patience and helpful suggestions in keeping us moving in the right direction during our photography sessions at their facility.

Permission by Kawasaki Motors Corp. to use certain exploded illustrations is a positive contribution to this manual for which we are most grateful.

Joan & Clarence Coles
Seloc Publications

TABLE OF CONTENTS

6 FUEL

7 IGNITION

8 ENGINE

9 ELECTRICAL

10 JET PUMP

11 CONTROL ADJUSTMENTS

APPENDIX

1
DESCRIPTION & OPERATION

1-1 BRIEF HISTORY

The jet drive system for propelling a craft through the water arrived on the scene in the early 1960's with the jet drive boat. In the early days the jet drive system was mated only with "high performance" power-plants -- engines in the 454 cu. in. class and larger. For this reason, during the "gas crunch" of the 1970's, the jet drives were labeled as "inefficient" and as "gas hogs". In addition to these two negative terms, they earned the reputation as "bad boy" boats due to their noisey "straight" exhaust, high rpm operation, and their almost unbelievable maneuverability. These combined factors did little to enhance their image and certainly restricted their popularity.

With "new and improved" technology, personal watercraft arrived on the scene around 1973. Personal watercraft, as we know them today, were developed using the same principles as the jet boats. Only single and twin cylinder engines are used with single stage pumps (one impeller) to make the craft highly efficient.

A personal watercraft, well maintained and given just a reasonable amount of "TLC", will reward the owner and his/her friends with hours of trouble free fun on the water.

Series Covered

The following Kawasaki Series are covered in this manual, from early model years thru 1991.

JS300 - 300 series
300SX - 300 series
JS440 - 440 series
JS550 - 550 series
550SX - 550 series
JS650 - 650 series
JF650 - 650 series
JB650 - 650 series
650SX - 650 series
JetMate - 650 series
X-2 - 650 series
TS - 650 series
SC - 650 series

Other Seloc Personal Watercraft Manuals

This manual is the first in a series of three volumes produced in 1991 covering personal watercraft. Volume II covers the Bombardier Sea-Doo models and Volume III covers all Yamaha personal watercraft and other craft powered with a Cuyuna engine.

The largest engine installed in a Kawasaki Personal Watercraft is the 635cc engine, classed as the "650 Series".

1-2 PRINCIPLES OF OPERATION

One of the first lessons to be learned in any elementary physics class is Newton's basic law: "for every force, there is an opposite and equal force". This statement is the basic principle of the jet pump. Water is "sucked" and "scooped" in from under the craft by a powerful pump rotating at incredible speed and then discharged, "blown" out, sternward in the opposite direction. In this manner the watercraft is propelled forward.

The personal watercraft covered in this manual are equipped with either a single or twin cylinder water cooled two-stroke engine, matched with a single stage (one impeller) jet pump.

On a very few models, a "gate" is swung down over the pump outlet nozzle forcing the exhausted water back in a forward direction thus moving the craft sternward.

Personal watercraft jet pumps may be classified as "axial flow" or "mixed flow".

ALL Kawasaki Personal Watercraft are equipped with an axial flow jet pump, with the exception of the JS550. The JS550 is equipped with a mixed flow jet pump.

Axial Flow

Water in an axial pump moves on a single axis, as depicted in the adjacent illustration, thus the word "axial". In simple terms, water is ingested and discharged parallel to the axis of impeller rotation. Some manufacturers classify their units as "axial flow" while others with virtually the same water movement label the unit as "mixed flow".

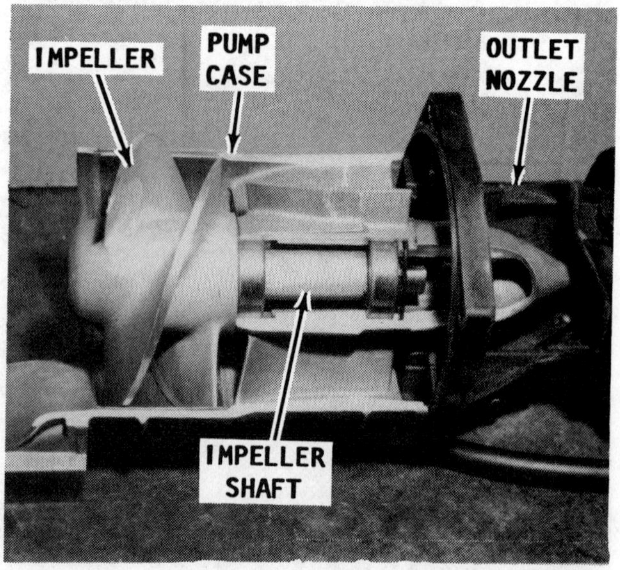

Photo of a cutaway "demonstration" type Kawasaki axial flow jet pump. A few major parts are identified.

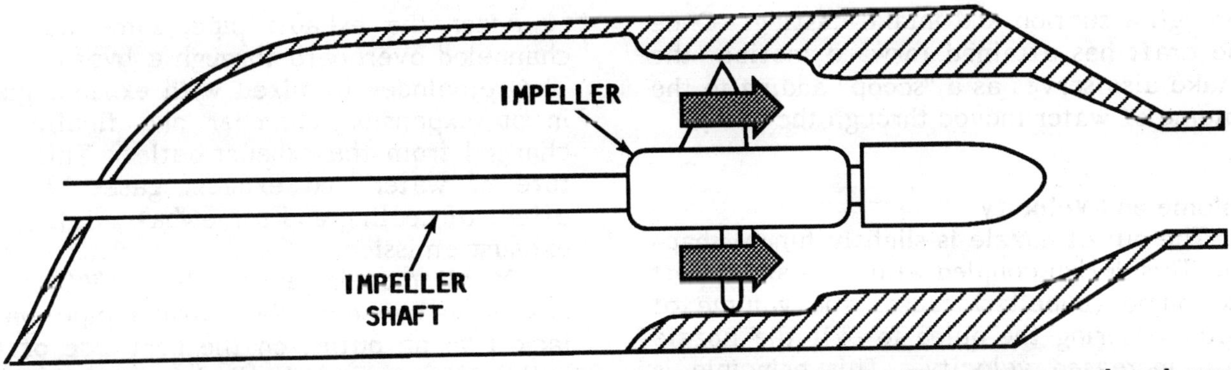

Cross-section line drawing to depict water flow through an axial flow jet pump. The water passes through parallel to the axis of impeller rotation.

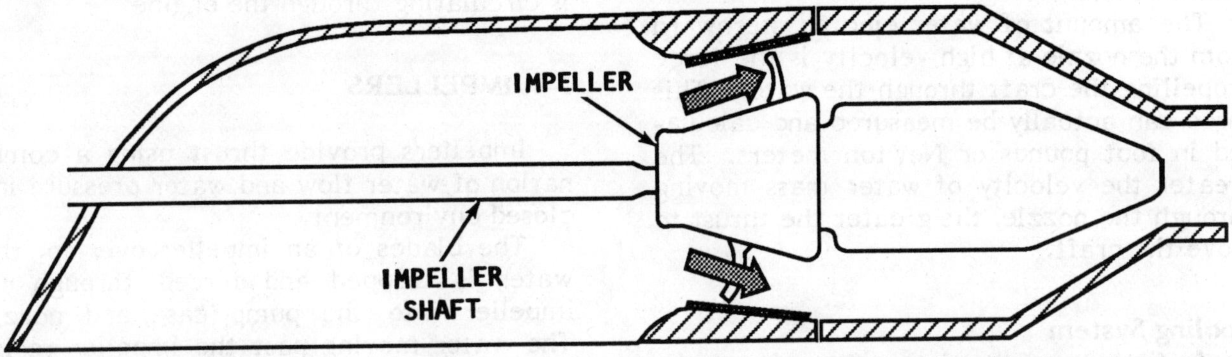

Line drawing to depict water flow through a mixed flow jet pump. Notice how the water is directed upward at an angle to the axis of impeller rotation.

The inside of an axial pump case is parallel to the water flow -- unlike the inside of a mixed flow pump which has a slightly tapered surface.

Mixed Flow

Many experts consider the mixed flow system to be the most efficient for personal watercraft application. The water flow is discharged at an inclined angle to the axis of impeller rotation, as indicated by the arrows in the accompanying illustration. The impeller blades are inclined in such a manner to cause the water to be thrown off at a 45 degree angle to the impeller shaft within the pump case. From the pump case the water is then forced out through the outlet nozzle and steering nozzle. The vanes in the pump case help straighten the water flow through the nozzle.

The inside of the pump case is inclined at an angle to the water flow. Therefore, this design pump incorporates shim material to space the impeller blades at a specified distance from the pump case. Mixed flow pump efficiency decreases by 30% as blade-to-pump case clearance increases by 0.04" (1mm).

Axial or Mixed?

Axial and mixed flow pumps are intended for different applications. One is not necessarily better than the other. The axial pump can be operated at higher engine rpm and is more suitable for racing. A mixed flow pump adds thrust at low engine rpm due to the pump design. Therefore, the mixed flow pump performs more efficiently at low engine rpm. This type pump is more suitable for a craft equipped to carry extra weight.

A jet pump seldom requires service except for the following:

Impeller-to-pump case clearance becomes excessive -- mixed flow only.

Impeller is damaged.

The oil seal around the impeller shaft fails -- allowing water to enter hull.

1-3 ENGINE AND JET DRIVE

The basic principle and arrangement of engine and pump are almost identical for all manufacturers.

The engine crankshaft is coupled either directly to the pump shaft or through a short driveshaft without any gear reduction. An impeller mounted on the pump shaft draws water into an opening in the hull through a suction and intake casting. Once the craft has attained forward motion the intake also serves as a "scoop" adding to the volume of water moved through the pump.

Volume and Velocity

The outlet nozzle is slightly funnel shaped. This design coupled with the capacity of the pump (impeller) causes the volume of water entering the pump to exit the nozzle with increased velocity. This principle is similar to air passing through the venturi of a carburetor.

The amount of water ejected sternward from the nozzle at high velocity is the force propelling the craft through the water. This force can actually be measured and calculated in foot pounds or Newton meters. The greater the velocity of water mass moving through the nozzle, the greater the thrust to move the craft.

Cooling System

A fitting on the jet pump, aft of the impeller, syphens off cooling water. This water is delivered first to the exhaust manifold -- the hottest part of the engine. The water is then channeled around the cylinder walls, cylinder head and exhaust pipe.

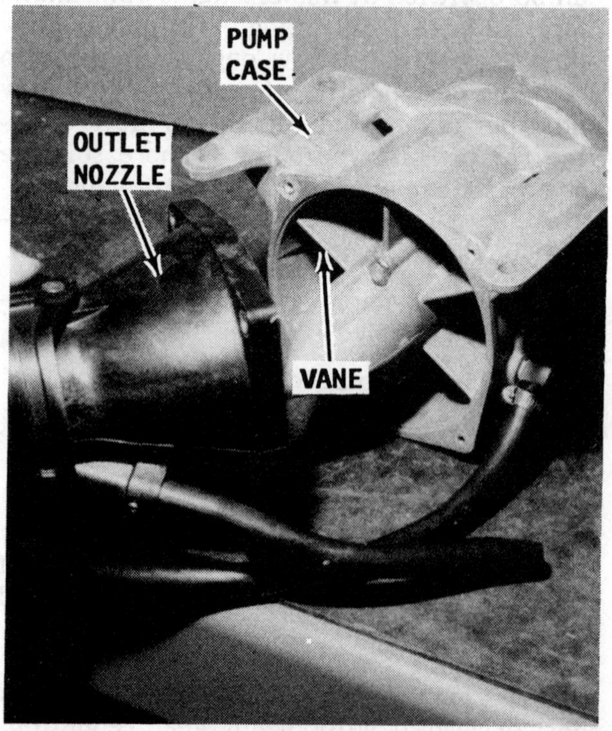

The vanes in the pump case straighten water flow and the conical shape of the outlet nozzle increases flow velocity.

After the exhaust pipe, some water is channeled overboard through a bypass hose. The remainder is mixed with exhaust gases in an expansion chamber and finally discharged from the exhaust outlet. This mixture of water and exhaust gases has the effect of cooling and somewhat uieting the exhaust emission.

On all models, except the JS440, a bypass hose connects the exhaust pipe water jacket to an outlet on the port side of the hull. When water is discharged from this outlet, the operator is assured cooling water is circulating through the engine.

1-4 IMPELLERS

Impellers provide thrust using a combination of water flow and water pressure in a closed environment.

The blades of an impeller overlap, thus water is trapped and forced through the impeller into the pump case and nozzle. The water moving past the impeller resists cavitation because it is under pressure. Blades in the pump case are angled exactly opposite to those of the impeller. These blades redirect the water to establish a concentrated flow through the nozzle.

The manufacturer recommends a radius of about 1/64" (0.3-0.5mm) for the leading edge of the impeller blade. A sharper radius will result in cavitation and a greater radius will reduce pump efficiency.

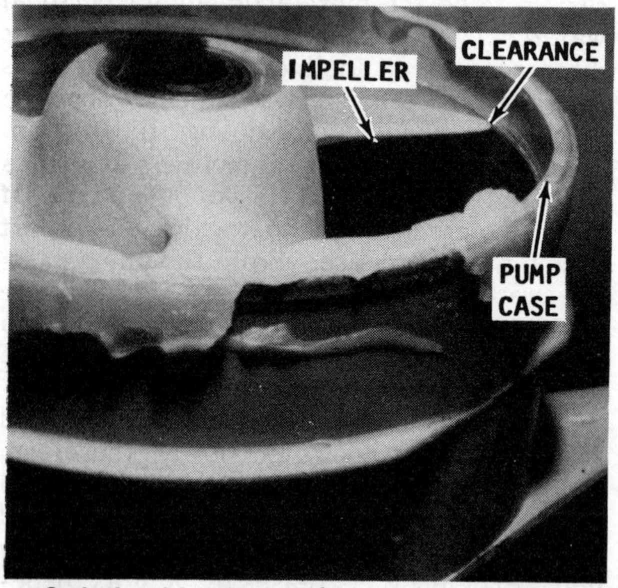

On both axial and mixed flow Kawasaki jet pumps, a feeler gauge is inserted between the impeller blade and the pump case to measure clearance.

Cavitation Burns

Cavitation burns are the worst enemy of an impeller. These burns literally "eat away" material from the impeller, leaving holes and weakening the impeller structure. In extreme cases entire blades have been known to "depart" from the impeller hub because the cavitation burns at the base of the blade were so severe.

Cavitation burns are the result of imperfections (damage) on the impeller blades or air mixed with the water flow.

Cavitation burns may be caused by:

- Wave jumping - air is sucked into the pump as the craft leaves the surface of the water.
- Bad seal between the pump case and outlet nozzle -- allowing air to enter.
- Leading edge of the impeller becoming damaged.

To explain exactly what causes cavitation burns, first, let us examine the last cause. If the impeller leading edge is damaged and "mushrooms" over the blade, a low pressure area will form under the mushroomed lip.

At atmospheric pressure, water boils at 212°F. Anytime there is an imperfection on the surface of an impeller blade and water passes over that imperfection, the water pressure is lowered. If the pressure is lowered, water will boil at a much lower temperature. Therefore, air bubbles will form in the water boiling under the mushroomed lip. The bubbles will creep down the surface of the blade and accumulate at the impeller hub. A high pressure area is formed at the base of the blade. Here, the air bubbles will collapse and reform back into water with a release of energy. This energy is absorbed by the impeller and results in material being eaten away.

The impeller should be dressed to a slightly rounded edge. It is impossible to achieve a perfectly straight sharp edge, as one side will always be convex and the other side concave. The concave side will form an area of low pressure and encourage cavitation as described above.

The other two causes of cavitation -- defective seals and wave jumping -- occurs when air is sucked in with the water flow. The same principles apply, as just described, in the previous paragraphs. Air bubbles will creep down the surface of the blade and accumulate at the impeller hub. A high pressure area is formed at the base of the blade. Here, the air bubbles will collapse and reform back into water with a release of energy. This energy is absorbed by the impeller and results in material being eaten away.

Cooling Water and Bilge Hoses

On most pumps two hoses are attached to the pump case and outlet nozzle. One hose channels some of the water flowing through the pump to the exhaust manifold -- the hottest part of the engine -- to cool the block during operation. The other hose siphons water out of the bottom of the hull, and is referred to as the bilge system. The cooling hose has water flowing **FROM** the pump. The bilge hose has water flowing **TO** the pump. Both hoses attach to the pump in a similar manner.

What determines the direction of water flow?

The answer to this question is in a simple explanation of high and low pressure areas along the inside surface of the pump.

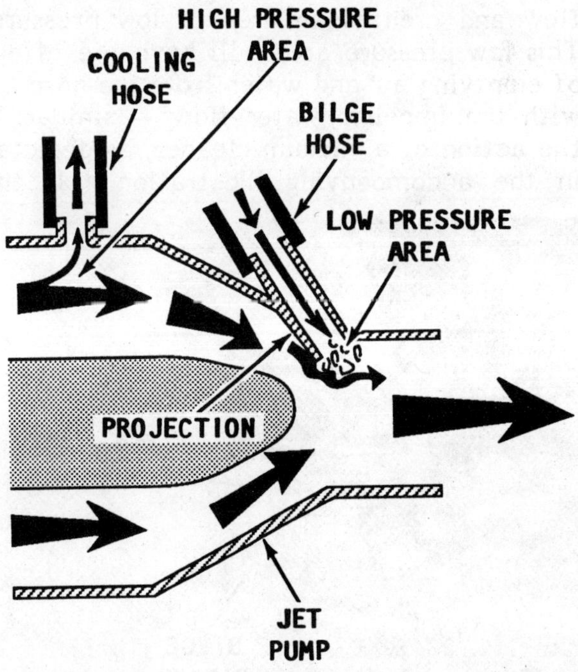

The direction of water flow through the cooling hose and the bilge hose is opposite. The cooling hose routes water from the pump to the engine. Water is "vacuumed" from the bilge through the bilge hose to the pump where it is forced out through the nozzle.

The bilge fitting protrusion extends into the water flow area, causing a low pressure area and a "vacuuming" effect.

Close inspection of the area inside the pump at the cooling water hose fitting reveals a smooth rounded shoulder with no obstruction or obstacle to impede the flow of water. A high pressure area develops here and draws the water down the hose attached to the fitting.

Further inspection of the area inside the hosing at the bilge hose fitting reveals a small protrusion around the opening. This protrusion causes disturbance to the water flow and creates an area of low pressure. This low pressure area will have the effect of emptying air and water from the hose aft with the impeller water flow -- similar to the action of a vacuum cleaner, as depicted in the accompanying illustration. If the other end of the hose is submerged in water inside the hull, the water will be "vacuumed" out. In this manner the bilge system drains the bilge.

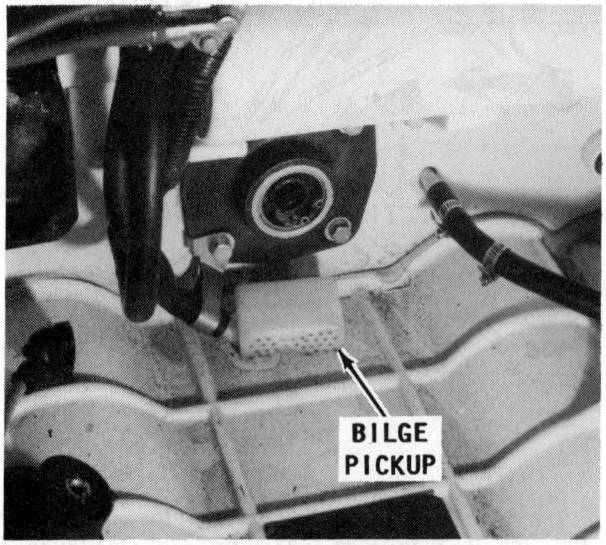

Location of the bilge pickup in the lowest part of the engine compartment. (Photograph taken with the engine removed for clarity.)

Many an owner -- with good intentions -- thinking his action would smooth water flow and increase pump performance -- has filed the protrusion described in the previous paragraph smooth with the surrounding area of the pump. During operation of the water craft, the bilge system worked in **REVERSE.** Water was pumped from the impeller housing into the bilge area, quickly filling the bilge and possibly sinking the craft before the rider had time to realize what was happening.

Bilge Breather Fitting

A bilge breather fitting is installed at the highest point of the bilge line. This fitting has a small breather hole to prevent siphoning water from the pump into the bilge. If the engine was shut down with the bilge hose filled with water and the jet pump fitting was positioned higher then the end of the bilge line, water would flow back and fill the bilge area. The small breather hole in the fitting prevents this back-siphoning.

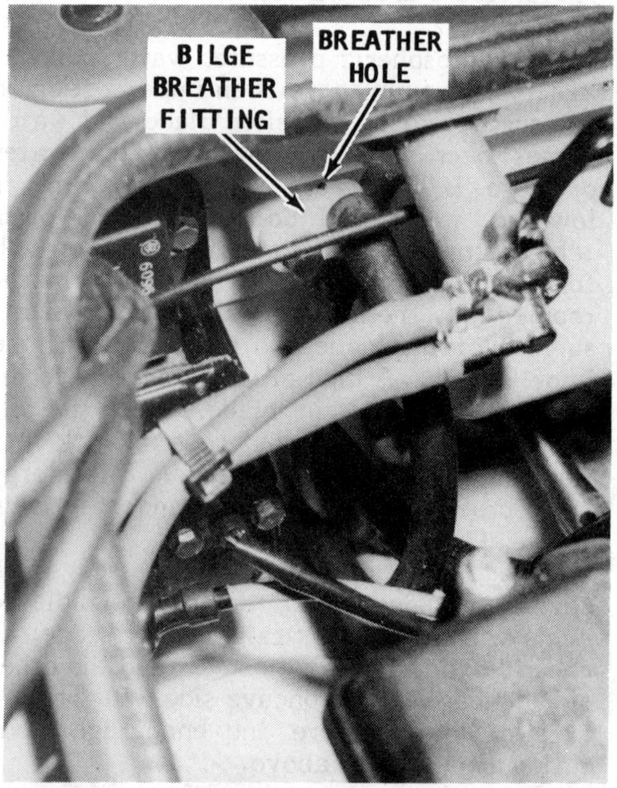

*A bilge line breather fitting is installed at the highest point in the bilge hose. This fitting has a small hole which **MUST** be kept clean and unobstructed.*

1-5 REPLACEABLE PARTS

Any time the engine or jet pump is disassembled, all manufacturers and the authors heartily recommend all parts included in an overhaul kit be installed. An engine kit usually includes all gaskets, seals, O-rings, and other parts required to rebuild the engine. Pistons, rings, bearings, crankshafts and even blocks should be inspected to determine if they are fit for further service. The pump kits usually include all gaskets, seals, O-rings, and other parts required to restore full efficiency and "like new" condition to the jet pump unit.

1-6 DEBRIS REMOVAL AND ENGINE OVERHEATING

As may be expected, a clogged impeller will slow the craft. The volume of water passing through the nozzle and water velocity will be reduced. A clogged impeller may cause restricted cooling water through the engine.

Many times the cooling water supply hose, if not properly secured with a hose clamp, may "blow off". The engine loses its supply of cooling water and internal damage quickly occurs.

Some units are equipped with an overheating warning horn. If the horn sounds, **SHUT DOWN** the engine immediately and remove the restriction or check the cooling hose connection.

Caution must be exercised when moving the craft off a beach to avoid sucking in mud, sand, pebbles, or rocks through the intake. Such ingestion could cause damage to the impeller, the vanes in the impeller duct, or clog engine cooling passages.

The engine cooling passages must be flushed with clear water on a regular basis. This is accomplished by simply connecting a garden hose to the cooling water hose or fitting on the manifold, if so equipped. Such an easy maintenance practice should become a habit and will prevent accumulation of debris which could restrict engine coolant water flow and cause overheating.

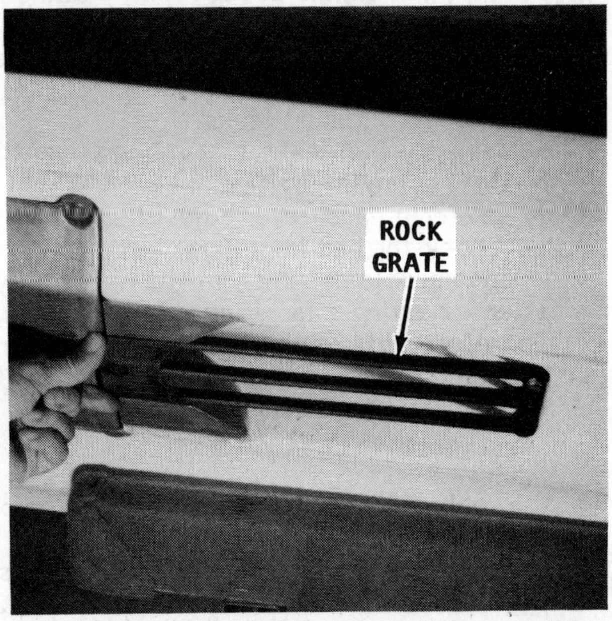

Any obstruction, such as a plastic bag, sucked up against the rock grate will drastically restrict water volume through the pump. A reduction in water flow affects craft performance and may quickly lead to a serious engine overheating problem.

1-7 REVERSE CAPABILITY

The Kawasaki JetMate and SC are equipped with reverse capability. As mentioned earlier, "reverse" is accomplished by a gate installed in such a manner to swing down over the steering nozzle. The gate is controlled with a cable.

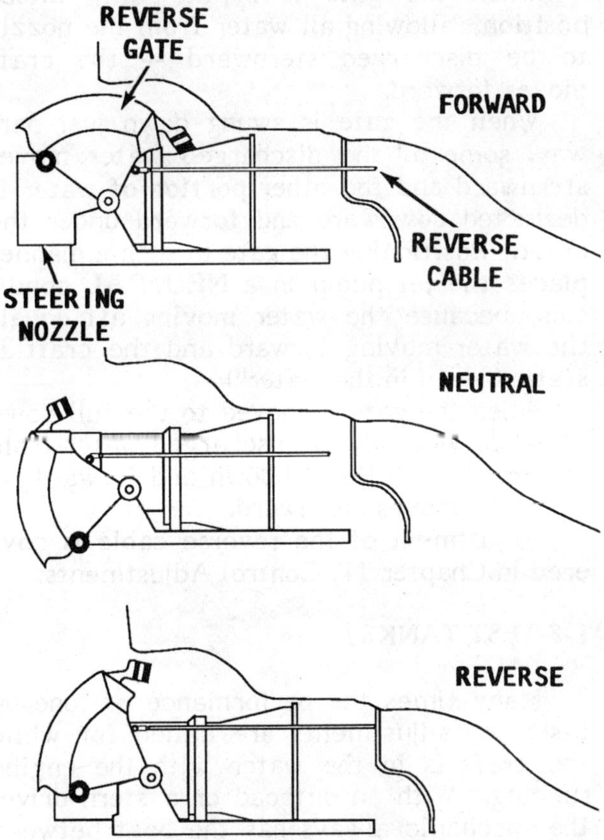

*Simple line drawings to illustrate water flow from the nozzle with the reverse gate in the full **FORWARD** position (top), in the **NEUTRAL** position (center), and in full **REVERSE** (bottom).*

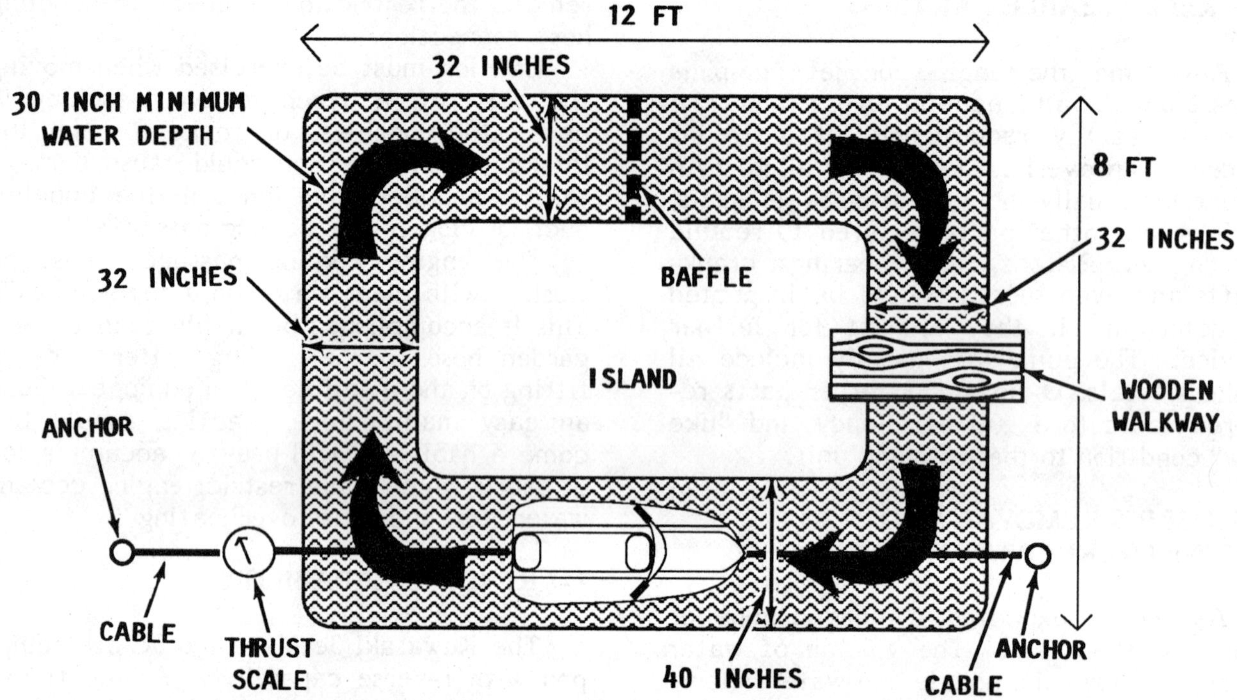

Layout of a typical test tank for testing personal watercraft. Such a tank is most valuable in verifying the completed work following service on the engine or the pump — especially when making a timing adjustment. The tank is described in detail on this and the following page.

When the gate is in the fully raised position, allowing all water from the nozzle to be discharged sternward -- the craft moves forward.

When the gate is swung down just part way, some of the discharged water moves sternward and the other portion of water is deflected downward and forward under the craft. Positioning the gate in such a manner places the jet pump in a **NEUTRAL** condition, because the water moving aft equals the water moving forward and the craft is static ("dead in the water").

When the gate is moved to the full down position, virtually all discharged water from the nozzle is deflected down and forward -- the craft moves sternward.

Adjustment of the reverse cable is covered in Chapter 11, Control Adjustments.

1-8 TEST TANKS

Many times the performance of tune-up tasks, or adjustments are called for while the craft is in the water with the engine running. With an outboad or a stern drive, the mechanic always has the boat between him/her and the wet stuff. On a personal watercraft, this is impossible without getting into the water, unless a test tank is used.

The advantages of using a test tank might be listed as:

a- Simulate actual water conditions. The craft may be operated without an exhaust extraction pipe and without using an auxiliary cooling water supply.
b- Test the engine at high rpm, placing a load on the impeller.
c- Reveal water leaks in the engine compartment and hull.
d- Uncover exhaust leaks.
e- Identify fuel leaks.
f- Detect engine noises.
g- Perform carburetor adjustments.
h- Check for bad impellers -- slippage.
i- Spot problems which might occur only after running for say 30 minutes. Therefore, prevent being stranded in the middle of a lake or far from shore if a serious malfunction should occur.

The only area a test tank cannot help diagnose is a craft handling problem.

The accompanying illustration depicts the type of test tank manufacturers use, with **approximate** dimensions. This type of tank is partially sunk into the ground. The craft sits in a body of circulating water and the mechanic has access to both sides of the craft without getting wet and has a dry place for tools etc.

The tank design with an island in the center allows water to circulate around the tank. With no circulation the water would foam and the jet pump would cavitate and lose thrust. A series of baffle plates installed on the opposite side to the craft, slow and straighten the water flow. The walls must be strong enough to hold the weight of the water and to contain the thrust of the pump, but coated to seal the walls and not scratch the sides of the craft.

The bow is secured with a stout line anchored to the edge of the tank to prevent the craft from "wandering". The stern is secured to the opposite edge of the tank to prevent the craft from moving forward. A thrust scale is sometimes attached to the stern line to determine the thrust exerted by the pump.

Seek out dealers with a test tank and attempt to use/rent their facilities to test the completed work.

1-9 SPECIAL FEATURES

The Kawasaki Series covered in this manual have a couple of unique features worth mentioning at this time. One is an rpm limiter to prevent an almost "runaway" engine condition when the operator is wave jumping, using a ramp, or performing any number of other gyrations for fun or during competion.

A minor, but important feature --labeled "Throttle Opening Limiter" -- is a simple arrangement to limit engine rpm entirely. Such a device is important to outfits renting personal watercraft to novice users.

Another feature is a self-circling system enabling an operator who may be thrown from the craft to get back aboard and regain control of the craft.

RPM Limiter

Jumping waves can cause extensive internal engine stresses. As soon as the craft leaves the water, engine speed -- if left unchecked -- would rapidly increase to as much as 12,000 rpm. As soon as the craft re-enters the water, engine speed instantly decreases to about 6,000 rpm. This sudden increase and decrease in engine speed is dramatically hard on engine life. Repeated jumps have been known to cause bent crankshafts and/or broken connecting rods.

Therefore, the need certainly existed for an rpm limiter which would automatically reduce engine rpm under such conditions.

Prior to 1990, a mechanical rpm limiter connected to a relay inside the electrical box was used. The return fuel line from the carburetor passed through the rev limiter which was actually a solenoid. When engine rpm reached 7,800 rpm, a signal from the electrical box was immediately sent to the solenoid. This solenoid, instantly and mechanically blocked off the returning fuel from the carburetor to the fuel tank and diverted it back to the carburetor. This arrangement resulted in an overrich mixture which slowed engine rpm. Operators discovered this design could be easily bypassed, by simply unplugging the solenoid. As a result, in 1990 the manufacturer changed design.

After 1990, an electronic rev limiter was used on all models. This type limiter was, as the factory intended, much more difficult to bypass. No! We are not going to tell you exactly how it functions or how it can be bypassed.

Throttle Opening Limiter

Some late model personal watercraft are equipped with a device to limit throttle opening. If the owner wishes to limit craft speed while the unit is on loan -- or renting -- to a novice rider, a throttle limiter, next to the twist grip on the handlebar, may be adjusted for this purpose.

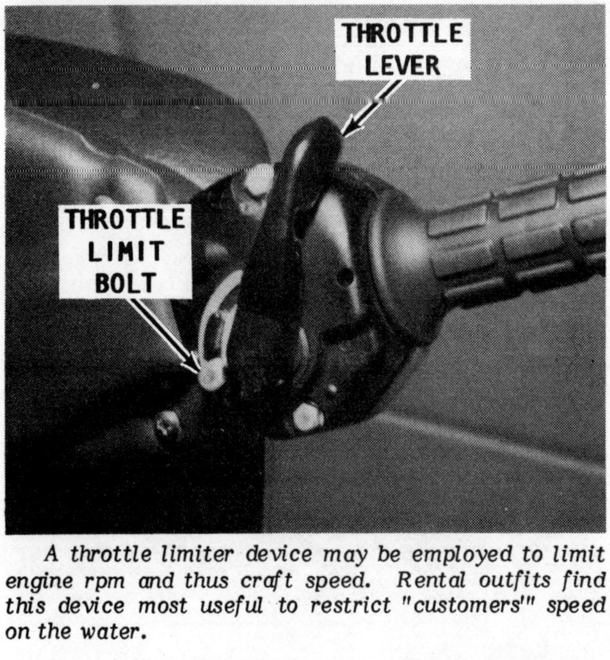

A throttle limiter device may be employed to limit engine rpm and thus craft speed. Rental outfits find this device most useful to restrict "customers'" speed on the water.

The owner of the craft simply loosens the bolt -- slides the lever on the throttle limiter to the desired position -- tightens the bolt to hold the adjustment -- and thus, engine rpm is limited and the craft's speed is restricted.

Self-Circling Mode

The JetMate and the SC are the only personal watercraft equipped with an emergency tether behind the engine "kill" switch. On all other craft, Kawasaki engineers incorporated a self-circling feature causing the craft to automatically follow a large circular course if the operator should fall or be thrown from the craft.

The theory behind this feature is similar to that of a free rolling automobile tire. When a single tire is set rolling freely, it will eventually slow down -- begin to wobble -- finally circle on its edge on one side -- eventually coming to rest on the ground.

With the personal watercraft -- once the operator leaves the craft, engine rpm is automatically reduced to an idle speed. Because the center of gravity for the craft shifts dramatically forward due to the loss of the rider, the bow drops and the craft begins to "plow" through the water instead of "planning" on an even keel.

Now, the action of waves against the hull will act to provide a continuous force preventing the craft from following a straight course. This wave action constantly pushing on the craft will cause it to follow a circular course. Without steering from an operator, the craft slowly makes a circular sweep to port -- or to starboard -- giving the operator the chance to catch the craft as it comes around -- climb aboard -- and continue his fun ride.

This self-circling feature will **ONLY** work if the engine idle speed is set correctly. If idle speed is set too high -- the craft will continue in a straight course.

2
SAFETY

2-1 INTRODUCTION

Personal watercrafting is one of the fastest growing leisure activites in the world. Riding a personal watercraft can be an exciting, exhilarating, and safe experience provided the operator exercises prudent behavior and remains in constant control of the vehicle.

Personal watercraft may be likened to a firearm. Weapons do not injure people, but the individual handling them is the true culprit. In a similar manner, if the personal watercraft is operated in an unresponsible manner the driver, passengers, and others using the same waterway may be in danger.

The greatest hazard in operating a personal watercraft stems not from the craft itself, but from the behavior of the operator. The Personal Watercraft Industry Association (PWIA) and the National Marine Manufacturers Association (NMMA) recognize there are a number of problems due to irresponsible craft operators contributing to tarnishment of the sport. Specifically, these problems range from alcohol and substance abuse to excessive noise, speed, and reckless maneuvering. Insensitive operators can disrupt an otherwise enjoyable day for other craft users through misconduct on the water. These individuals often endanger their own lives as well as the lives of others by losing full control of their craft.

Organized events, races, and exhibitions, properly planned and controlled have proven to be a spectactular event appealing to a wide range and number of participants and spectators.

Craft Classification

Any craft equipped with propulsion machinery is classified as a motorboat.

Personal watercraft are classified as "Class A inboard boats" by the Coast Guard and therefore are subject to most of the same laws and requirements as more conventional larger craft. A decal, usually affixed somewhere in the stern area, lists exceptions the Coast Guard has granted for this type of water vehicle.

Information

In 1989, several states recognized the need to impose specific regulations governing operation of personal watercraft. Rules and regulations differ from state to state and are constantly being revised. Therefore, it would be an impossible task to present current legislation on personal watercraft activites. Before the printing ink for this publication was dry, additions, deletions, and revisions would already be in effect.

For current information, the reader may call 1-800-336-BOAT for the name and address of the local Boating Law Adminis-

This Kawasaki Personal Watercraft reflects its owner's pride and interest in keeping it "squared away". With his/her pride may also come an adult responsibility to exercise safe and sensible practices while on the water.

trator. The reader may also contact the PWIA Government Relations Representative, Washington, D.C.

Regulation Enforcement

The U.S. Coast Guard and state enforcement officers have the authority to stop any craft to check for compliance with federal or state law. They can order a craft to return to the closest dock and remain out of the water until the hazardous condition of the craft is corrected.

Examples are:

Inadequate number of Personal Flotation Devices.

Missing/Inoperative fire extinguisher.

Overloading.

Operation at night without proper lights.

Fuel leakage.

Fuel transportation (other than approved fuel tank).

Failure to meet ventilation requirements.

Failure to meet carburetor backfire flame arrestor requirements.

Excessive leakage or accumulation of water in the bilge.

The U.S. Coast Guard and state enforcement officers have the authority to stop any craft if they determine the craft is being operated in a negligent manner. Negligent operation is defined as "the failure to excercise the degree of care necessary to prevent the endangering of life, limb, or property of any person".

Examples are:

Operating while under the influence of drugs or alcohol.

Excessive speed in a congested area.

Excessive speed in stormy or foggy conditions.

Coming too close to another vessel.

Operating in a swimming area where bathers are present.

Towing water skiers where obstructions exist or where a fall might cause skier to be struck by another vessel.

Operating in the vicinity of a dam.

Cutting through a marine parade or regatta.

2-2 MINIMUM LEGAL REQUIREMENTS FOR EQUIPMENT ON BOARD CRAFT

All personal watercraft owners **MUST** provide the following equipment:

One approved Type I, II, III or IV PFD (personal flotation device) for each person on board or being towed on water skis, etc.

An efficient sound producing device.

A B-I approved hand portable fire extinquisher.

Visual distress signals for nighttime use, if operated on coastal waters.

Personal Flotation Devices (PFDs)

The Coast Guard requires an approved life-saving device be worn by each person on board. Devices approved are identified by a tag indicating Coast Guard approval. Such devices may be life preservers, or buoyant vests. Ring buoys, or buoyant cushions are not acceptable.

Life preservers have been classified by the Coast Guard into five type categories. All PFDs presently acceptable on recreational craft fall into one of these five designations. Only four of the five catagories are stipulated by legal requirements. Type V is not acceptable for use on a personal watercraft.

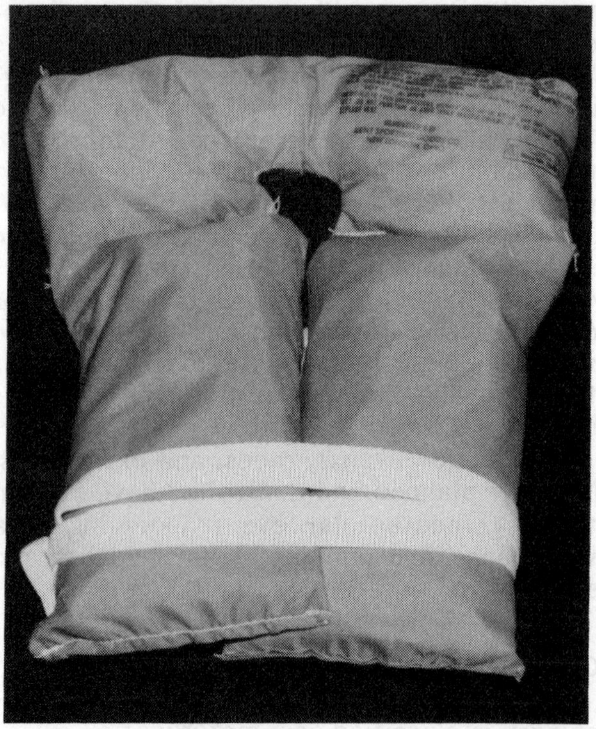

Type I PFD Coast Guard approved life jacket. This type flotation device provides the greatest amount of buoyancy.

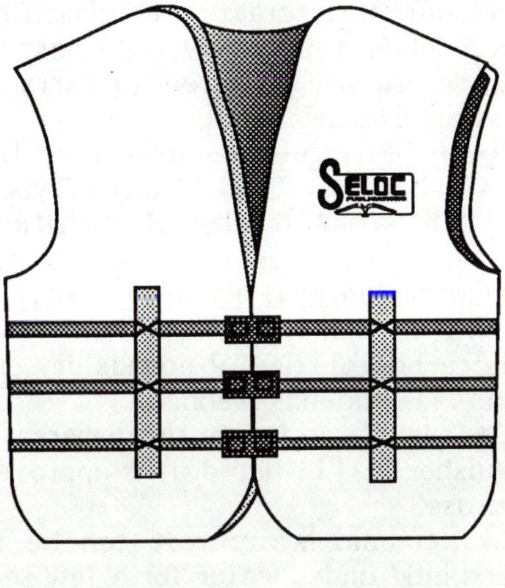

A Type III PFD requires the wearer to be active to remain upright in the water. This type device is comfortable and allows the wearer to actually swim while wearing it in the water.

A PFD **MUST** be U.S. Coast Guard approved, in good and serviceable condition, and of an appropriate size for the person intending to wear it.

It is an accepted fact that most boating people own life preservers, but too few actually wear them. There is little or no excuse for not wearing one because the modern comfortable designs available today do not subtract from an individual's boating pleasure.

PFDs may have a long serviceable life, if proper care is exercised and the device is not abused. Wipe them clean and dry before storing in a well ventilated area during the off-season. **NEVER** use a PFD as a fender. A crushed PFD loses its buoyancy and becomes useless.

Type I PFD has the greatest required buoyancy and is designed to turn most **UNCONSCIOUS** persons in the water from a face down position to a vertical or slightly backward position. The adult size device provides a minimum buoyancy of 22 pounds and the child size provides a minimum buoyancy of 11 pounds. The Type I PFD provides the greatest protection to its wearer and is most effective for all waters and conditions.

Type II PFD is designed to turn its wearer in a vertical or slightly backward position in the water. The turning action is not as pronounced as with a Type I. The device will not turn as many different type persons under the same conditions as the Type I. An adult size device provides a minimum buoyancy of $15\frac{1}{2}$ pounds, the medium child size provides a minimum of 11 pounds, and the infant and small child sizes provide a minimum buoyancy of 7 pounds.

Type III PFD is designed to permit the wearer to place himself (herself) in a vertical or slightly backward position. The Type III device has the same buoyancy as the Type II PFD but it has little or no turning ability. Many of the Type III PFD are designed to be particularly useful when water skiing, sailing, hunting, fishing, or engaging in other water sports. Several of this type will also provide increased hypothermia protection.

Type IV PFD is designed to be thrown to a person in the water and grasped and held by the user until rescued. It is **NOT** designed to be worn. The most common Type IV PFD is a ring buoy or a buoyant cushion.

ONE LAST WORD

Common sense dictates the "cheap and cheerful" approach does not pay when applied to a life preserving device. Select and buy the very best. The life you save may be your own.

NOW, THESE WORDS

Due to the nature and maneuvering characteristics of personal watercraft, and from a "practical" standpoint, the following paragraph on bells, horns, and whistles hardly apply. The information is included to make the manual "complete" and because according to the U.S. Coast Guard they are required. Also, the authors feel the reader may "skipper" or "crew" on a larger craft and therefore, might be required to use the information presented here.

Bells, Horns, and Whistles

All craft must carry some means of producing an efficient sound signal and the sound must be audible for a specific distance and duration, depending on the size of the craft. For personal watercraft, an audible sound must carry for one half mile and be of four to six second duration.

Law enforcement craft are the **ONLY** boats allowed to use sirens.

Personal watercraft operators are required to sound fog and maneuvering signals when underway in fog or rain as a means of communication between craft which are not visible. This system was devised to alert nearby craft of the presence of others in the water and as a means of communicating maneuvering intentions. Signals indicating maneuvering tactics are short one second blasts.

One blast indicates the craft is turning to starboard.
Two blasts indicate the craft is turning to port.
Three blasts indicate the craft is in reverse.
Five or more blasts indicate - stay away - danger.

A prolonged blast lasting 4 to 6 seconds duration every two minutes indicates a straight line course in poor visibility.

These signals should be memorized by every personal watercraft operator, not necessarily because the operator would give them, but in the event of a potential head on collision -- with say a large powerboat -- the skipper may signal his intention to turn either to port or starboard using this system of communication.

Personal watercraft are required to carry a Type B-1 fire extinguisher aboard. The compartment is almost always in the stern, as shown, where it is readily available.

Fire Extinguishers

Personal watercraft are classified as "Class A inboard boats" by the Coast Guard and therefore are required to carry a B-I fire extinguisher.

All fire extinguishers must bear Underwriters Laboratory (UL) "Marine Type" approved labels. With the UL certification, the extinguisher does not have to have a Coast Guard approval number.

A B-I contains 1-1/4 gallons foam, 4 pounds carbon dioxide, 2 pounds dry chemical, and 2-1/2 pounds freon.

READ labels on fire extinguishers. If the extinguisher is U.L. listed, it is approved for marine use.

Any personal watercraft can be rolled over and held under water for a few seconds to extinguish the fire. Turning the craft over in the water will prevent oxygen from entering the engine compartment. Without oxygen, the fire will soon exhaust the remaining oxygen in the compartment and then go out. **DO NOT** open the engine compartment until the smoke has completely stopped, then open with great caution and have a fire extinguisher at hand.

Visual Distress Signals

Coast Guard Regulations require personal watercraft to carry night visual distress signals when used on coastal waters - even though it is illegal to operate a personal watercraft at night.

Coastal waters include: the Great Lakes, the territorial seas and those waters directly connected to the Great Lakes and the territorial seas, up to a point where the waters are less than two miles wide.

These signals assist rescuers in locating and aiding people in distress. This information is useful to every personal watercraft owner, to know how to respond to others in distress.

For night use: Pyrotechnic visual distress signaling devices (red flares, hand held or aerial) **MUST** be Coast Guard Approved, in serviceable condition and stowed to be readily accessible. If they are marked with a date showing the serviceable life, this date must not have passed.

Coast Guard Approved pyrotechnic devices carry an expiration date. This date can **NOT** exceed 42 months from the date of manufacture and at such time the device can no longer be counted toward the minimum requirements.

The only non-pyrotechnic visual distress signaling device approved by the coast guard for use at night is an electric distress light -- not a flashlight but an approved electric distress light which **MUST** automatically flash the international **SOS** distress signal (. . . - - - - - - . . .)four to six times each minute.

2-3 MINIMUM LEGAL REQUIREMENTS FOR REGISTRATION OF CRAFT

Federal regulations require all personal watercraft to be issued a registration number and have the number displayed in a specific manner on the craft.

The "Certificate of Number" is issued by the state and must be carried on board at all times. The Hull Identification Number (HIN) is necessary to register the craft with the state.

The registration number must be displayed as follows:

The figures must be read from left to right.

The figures must be displayed on the forward half of each side of the bow of the craft.

The figures must be in bold, block letters of good proportion and must not be less than 3 inches high.

The figures must be of contrasting color to the craft hull, or background.

The figures must be as high above the waterline as practical.

No figures other than those assigned to the craft can be displayed on the forward half of the craft.

Letters must be separated from numbers by spaces or hyphens.

Validation decals (if required by state) must be displayed within 6 inches of the registration number.

Elaborate color schemes and custom detailing are very much a part of the fun of owning a personal watercraft. **HOWEVER,** the paint design must not obstruct the registration numbers or the validation decal. Like an automobile without a license plate, sooner or later a craft with no visible registration numbers will attract the attention of the law.

Hull Identification Number

Federal regulations require all personal watercraft to have a Hull Identification Number (HIN) permanently attached to the craft in two separate locations: on the starboard side of the transom, above the waterline. The HIN must also be displayed in an unexposed location, which is usually left to the discretion of the manufacturer.

2-4 SAFETY PRACTICES

Riding Techniques

Of all watercraft, large and small, the personal watercraft are the most responsive to steering and maneuverability. Therefore, a personal watercraft operator should take it upon herself/himself to stay clear of other boaters and especially swimmers.

A beginner should operate the craft at a low speed until the operator feels comfortable and is aware of its limitations and responses to steering.

While riding the craft, keep feet, hands, hair, and clothing away from the pump intake.

Full Throttle Operation

Prolonged operation at full throttle will raise the cylinder head temperature by about 100° above normal. Some units are equipped with a thermo-switch engine temperature sensor. This sensor is connected directly to the CDI unit and grounded to the engine kill switch. If the engine temperature exceeds a maximum value specified by

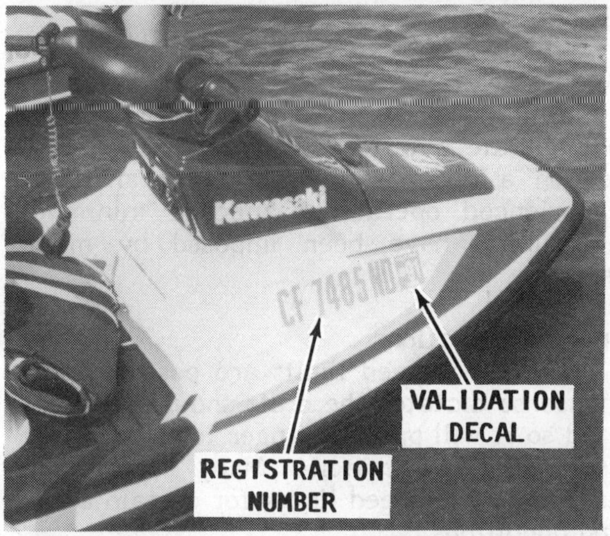

The validation decal, required by most states, and the registration number, required by all states, must be prominently displayed -- port and starboard -- on the forward portion of the hull.

the manufacturer, the ignition will be grounded and the engine shut down to prevent damage through overheating.

Jumping Waves

In the very early days of personal watercraft, jumping could cause extensive internal engine damage. When the craft left the water, unchecked engine rpm rapidly increased to as high as 12,000 rpm. As soon as the craft re-entered the water, engine rpm instantly decreased to about 6,000 rpm. This sudden increase and decrease in engine speed dramatically shortened engine life. In those days, repeated jumps were known to cause bent crankshafts, sheared flywheel keys, and broken connecting rods.

Kawasaki engineers developed an rpm limiter system to prevent high engine rpm and an almost "runaway" condition with the design of a rpm limiter. This limiter restricts engine speed to about 8,000 rpm. Thus, wave jumping, using a ramp, and a host of other almost unbelievable manuevers could be performed without undue stress on the engine. Today, with an rpm limiter system installed, a bent crankshaft or broken connecting rod is almost non-existent.

Alcohol and Substance Use

Operating a personal watercraft while under the influence of either alcohol or drugs is both illegal and life threatening. This fact cannot be stated too strongly. Such substances affect vision and slow reaction times. The danger is quadrupled in a craft on the water compared to on the shore.

Age Restrictions

Children under 14 years of age should not operate a watercraft unless accompanied on a two passenger craft by an adult experienced operator. Specific minimum age limits have been imposed by many states.

Speed Restrictions

Where no speed limits are posted, commonsense dictates the craft should be operated so it will pose no danger to other water users or the operator. Many states now have enforced speed limits for certain areas and conditions.

When entering "no wake" or 5 mph zones, such as fishing areas, swimming areas or marinas, the operator must slow to headway speed, the slowest speed at which appropriate steering capability can be maintained.

Individuals responsible for law enforcement on the water are keeping an ever vigilant eye on personal watercraft in many areas. Horseplay such as "wetting down" or "buzzing" other water users may result in a "ticket" for negligent operation.

Operation at Night

Personal watercraft are not equipped with lights and therefore, it is illegal to operate such a craft on the water after sunset.

The law requires any craft operated between sunset and sunrise to display certain colored lights mounted in specific locations, depending on the size and class of the craft. These lights will indicate the position and intended direction of the craft to operators of other watercraft in the area.

Personal watercraft are extremely difficult to see at night because of their small size. Even with lights, because of their almost unbelievable maneuverability, confusing signals would be sent to other traffic.

Anchor Light

Personal watercraft are not required to be equipped with an anchor light **UNLESS** anchored in or near a narrow channel, fairway, anchorage, or where other boats normally navigate. Since this situation is probably unlikely to occur, personal watercraft may be considered exempt from this requirement. However, if special circumstances exist, an anchor light should be affixed to the craft. An anchor light is a white light, with a 360° arc, mounted in a prominent position and must be visible for two miles.

Shallow Water Operation

Under normal conditions, personal watercraft may be operated in very shallow waters, and it is possible to maneuver the craft directly onto a slopping beach.

Caution must be exercised when moving the craft off a beach to avoid sucking in mud, sand, pebbles, or rocks through the intake. Such ingestion could cause damage to the impeller, the vanes in the pump bowl, or clog engine cooling passages.

WARNING
Expelled sand and small pebbles may cause INJURY to persons in the area.

A grated section set into the hull bottom helps prevent the ingestion of small rocks and pebbles.

Carry about a foot of stiff wire aboard for use in probing into the impeller area to dislodge debris where fingers cannot reach.

Engine Compartment Ventilation

All motorboats (personal watercraft included) built after April 25, 1940 and before August 1, 1980, powered by a gasoline engine or by fuels having a flashpoint of 110° F or less **MUST** have the following, which is quoted from a recent Coast Guard publication:

> At least two ventilation ducts fitted with cowls or their equivalent for the purpose of properly and efficiently ventilating the bilges of every engine and fuel tank compartment. Each duct must be at least two inches in internal diameter.
>
> There shall be at least one exhaust duct installed so as to extend from the lower portion of the bilge to cowls in the open air, and at least one intake duct installed so as to extend to a point at least midway to the bilge or at least below the level of the carburetor air intake.

Flame Arrestors

A flame arrestor, as the name suggests, safeguards against flame caused by engine backfire.

A gasoline engine installed in a motorboat or motor vessel after April 25, 1940, except outboard motors, must have a Coast Guard Approved flame arrestor fitted to the carburetor. This requirement applies to personal watercraft because the engine is enclosed.

Fuel System

All parts of the fuel system should be selected and installed to provide maximum service and protection against leakage.

The capacity of the fuel filter must be large enough to handle the demands of the engine as specified by the engine manufacturer.

All fittings and outlets must come out the top of the tank.

In order to obtain maximum circulation of air around fuel tanks, the tank should not come in contact with the hull except through the necessary supports. The supporting surfaces and hold-downs must fasten the tank firmly and they should be insulated from the tank surfaces. This insulation material should be non-abrasive and non-absorbent material.

Taking On Fuel

The fuel tank should be kept almost full to allow for expansion, but still prevent water from entering the system through condensation caused by temperature changes. Water droplets forming are one of the greatest enemies of the fuel system. By keeping the tank almost full, the air space in the tank is kept to an absolute minimum and there is less room for moisture to form. It is a good practice not to store fuel in the tank over an extended period, say for six months. Today, fuels contain ingredients that change into gums when stored for any length of time. These gums and varnish products will cause carburetor problems and poor spark plug performance. An additive (Sta-Bil) is available and can be used to prevent gums and varnish from forming.

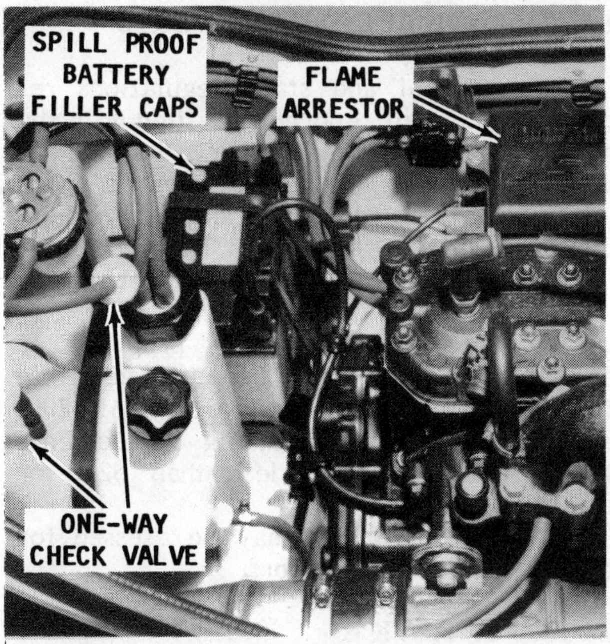

Carburetors on personal watercraft are required to be fitted with a flame arrestor because the engine compartment is an enclosed area. Other safety features include spill-proof battery filler caps and one-way check valves on the fuel and oil tanks, to prevent spillage under all conditions.

Excessive Noise

Every manufacturer designs a stock exhaust system to keep noise to a minimum. A poorly maintained or modified exhaust system will produce excessive and disturbing noise. This is a form of air polution and is illegal in most states.

Automotive Replacement Parts

When replacing fuel, electrical, and other parts, check to be sure they are marine type and Coast Guard Approved. Automotive parts are not made to the high standards of marine parts. The carburetors must **NOT** leak fuel; electrical equipment **MUST** be able to operate in a gas fume enclosed area without exploding; etc. Automotive parts could cause a fire endangering the operator, passengers, and the craft. The part may look the same and even have a similar number, but if it is not **MARINE** it is not safe to use and will not pass Coast Guard inspection.

Overloading

Each personal watercraft is designed to carry a specific number of persons. Attempting to board more passangers than the craft is designed to carry, will upset weight distribution causing a marked decrease in performance and possible loss of control.

2-5 BOATING ACCIDENT REPORTS

New federal and state regulations require an accident report to be filed with the nearest state boating authority within 48 hours if a person is lost, disappears, or is injured to the degree of needing medical treatment beyond first aid. The time limit is reduced to 24 hours if the accident results in a loss of life.

Accidents involving only property or equipment damage **MUST** be reported within 10 days, if the damage is in excess of $200. Some states require reporting of accidents with property damage less than $200 or total loss.

A **$500 PENALTY** may be asessed for failure to submit the report.

WORDS OF ADVICE

Take time to make a copy of the report to keep for your records or for the insurance company. Once the report is filed, the Coast Guard will not give out a copy, even to the person who filed the report.

The report must give details of the accident and include:

1- The date, time, and exact location of the occurrence.

2- The name of each person who died, was lost, or injured.

3- The number and name of the vessel.

4- The names and addresses of the owner and operator.

If the operator cannot file the report for any reason, each person on board **MUST** notify the authorities, or determine that the report has been filed.

In nautical terms, the front of the craft is the **bow;** the rear is the **stern;** the right side, when facing forward, is the **starboard** side; and the left side is the **port** side. One easy way to remember this basic fundamental is to consider the words "port" and "left" both have four letters and go together.

All directional references in this manual use this terminology. Therefore, the direction from which an item is viewed is of no consequence, because **starboard** and **port** **NEVER** change no matter where the individual is located, or standing, even on his or her head.

2-6 SECURITY

As mentioned in the opening paragraphs of this chapter, personal watercrafting is one of the fastest growing leisure activites in the world. Like most personal possessions, personal watercraft have value and valuables are subject to be lost to unscrupulous characters -- thieves. The best advice for a personal watercraft owner is "lock it or lose it".

Do not leave the craft unattended on the beach or dock. **ALWAYS** remove the kill switch tether from the switch and of course, the ignition key, if the unit is equipped with these features. Certain aftermarket security devices are available for installation to prevent theft of personal watercraft. These include: A "secret" ignition grounding switch, which grounds one side of the ignition coil, and a hidden fuel shut-off valve.

Most owners transport their craft on trailers or in the bed of a pickup truck. If loaded on a trailer, chain and lock the craft to the trailer. Invest in a hitch lock so the trailer cannot be detached from the towing vehicle. If the craft is loaded on a pickup truck, chain and lock it to the bed, or to a large and heavy object.

3
TUNING

3-1 INTRODUCTION

The efficiency, reliability, fuel economy and enjoyment available from engine performance are all directly dependent on having the engine tuned properly. The importance of performing service work in the sequence detailed in this chapter cannot be overemphasized. Before making any adjustments, check the specifications in the Appendix. **NEVER** rely on memory when making critical adjustments.

Before beginning to tune any engine, check to be sure the engine has satisfactory compression. An engine with worn or broken piston rings, burned pistons, or scored cylinder walls, cannot be made to perform properly no matter how much time and expense is spent on the tune-up. Poor compression must be corrected or the tune-up will not give the desired results.

A practical maintenance program that is followed throughout the year, is one of the best methods of ensuring the engine will give satisfactory performance at any time.

The extent of the engine tune-up is usually dependent on the time lapse since the last service. A complete tune-up of the entire engine would entail almost all of the work outlined in this manual. A logical sequence of steps will be presented in general terms. If additional information or detailed service work is required, the chapter containing the instructions will be referenced.

Each year higher compression ratios are built into modern marine engines and the electrical systems become more complex, especially with electronic (capacitor discharge) units. Therefore, the need for reliable, authoritative, and detailed instructions becomes more critical. The information in this chapter and the referenced chapters fulfill that requirement.

3-2 TUNE-UP SEQUENCE

During a major tune-up, a definite sequence of service work should be followed to return the engine to the maximum performance desired. This type of work should not be confused with attempting to locate problem areas of "why" the engine is not performing satisfactorily. This work is classified as "troubleshooting". In many cases, these two areas will overlap, because many times a minor or major tune-up will correct the malfunction and return the system to normal operation.

The following list is a suggested sequence of tasks to perform during the tune-up service work. The tasks are merely listed here. In most cases, procedures are

A clean, properly tuned, and adequately maintained power unit can multiply by many fold the enjoyment derived from owning and operating a personal watercraft.

given in subsequent sections of this chapter. For more detailed instructions, see the referenced chapter, for the unit being seviced.

1- Perform a compression check of each cylinder. See Ignition Chapter.
2- Inspect the spark plugs to determine their condition. Test for adequate spark at the plug. See Ignition Chapter.
3- Start the engine in a body of water or connect a "Flushing Kit" with a garden hose and check the water flow through the engine, See Engine Chapter.
4- Check the carburetor adjustments and the need for an overhaul. See Fuel Chapter.
5- Check the fuel pump for adequate performance and delivery. See Fuel Chapter.
6- Make a general inspection of the ignition system. See Ignition Chapter.
7- Test the cranking motor and the solenoid. See Electrical Chapter.
8- Check the internal wiring.

Performing a compression test on a two cylinder engine. Both spark plug leads have been grounded to prevent placing an unnecessary extra load on the ignition coil.

3-3 COMPRESSION CHECK

A compression check is extremely important, because an engine with low or uneven compression between cylinders **CANNOT** be tuned to operate satisfactorily. Therefore, it is essential that any compression problem be corrected before proceeding with the tune-up procedure.

If the engine shows any indication of overheating, such as discolored or scorched paint, inspect the piston skirts visually thru the transfer ports, using a mirror and flashlight, for possible scoring. It is possible for a cylinder with satisfactory compression to be scored slightly.

Checking Compression

Remove the spark plug wires. **ALWAYS** grasp the molded cap and pull it loose with a twisting motion to prevent damage to the connection. Remove both spark plugs and remember from which cylinder they came for evaluation later. Ground both spark plug leads to the engine to render the ignition system inoperative while performing the compression check.

Insert a compression gauge into the forward spark plug opening. Crank the engine with the cranking motor, thru at least four complete revolutions of the crankshaft, with the throttle at the wide-open position, to obtain the highest possible reading. Record the highest reading. Repeat the test and record the compression for the other cylinder. The manufacturer specifies the compression pressure for a new engine should be 142.2 psi (981 kpa). A variation between cylinders is far more important than the actual readings. A variation of more than 15 psi between the two cylinders indicates the lower compression cylinder is defective. The problem may be worn, broken, or sticking piston rings, scored pistons or worn cylinders.

Use of an engine cleaner will help to free stuck rings and to dissolve accumulated carbon. Follow the directions on the can.

3-4 SPARK PLUG INSPECTION

Inspect both spark plugs for badly worn electrodes, glazed, broken, blistered, or lead fouled insulators. Replace both the plugs, if either shows signs of excessive wear.

Make an evaluation of the cylinder performance by comparing the spark condition with those shown in the Ignition Chapter. Check each spark plug to be sure they are both of the same manufacturer and have the same heat range rating.

Inspect the threads in the spark plug opening in the head and clean the threads before installing the plug.

When purchasing new spark plugs, **ALWAYS** ask the dealer if there has been a spark plug change for the engine being serviced. Set the spark plug gap to 0.030" for all series engines covered in this manual.

Crank the engine through several revolutions to blow out any material which might have become dislodged during cleaning.

Install the spark plugs and tighten them to a torque value of 14ft lbs (20Nm). **ALWAYS** use a new gasket and wipe the seats in the head clean. The gasket must be fully compressed on clean seats to complete the heat transfer process and to provide a gas tight seal in the cylinder. If the torque value is too high, the heat will dissipate too rapidly. Conversely, if the torque value is too low, heat will not dissipate fast enough.

After a compression check or inspection of the spark plugs, install the high tension leads firmly over the spark plugs.

Broken Reed

A broken reed is usually caused by metal fatigue over a long period of time. The failure may also be due to the reed flexing too far because the reed stop has not been adjusted properly or the stop has become distorted.

If the reed is broken, the loose piece **MUST** be located and removed, before the engine is returned to service. The piece of reed may have found its way into the crankcase, behind the by-pass cover. If the broken piece cannot be located, the engine must be completely disassembled until it is located and removed.

An excellent check for a broken reed on an operating engine is to hold an ordinary business card in front of the carburetor. Under normal operating conditions, a very small amount of fine mist will be noticeable, but if fuel begins to appear rapidly on the card from the carburetor, one of the reeds is broken and causing the backflow through the carburetor onto the card.

A broken reed will cause the engine to operate roughly and with a "pop" back through the carburetor.

The reeds must **NEVER** be turned over in an attempt to correct a problem. Such action would cause the reed to flex in the opposite direction and the reed would break in a very short time.

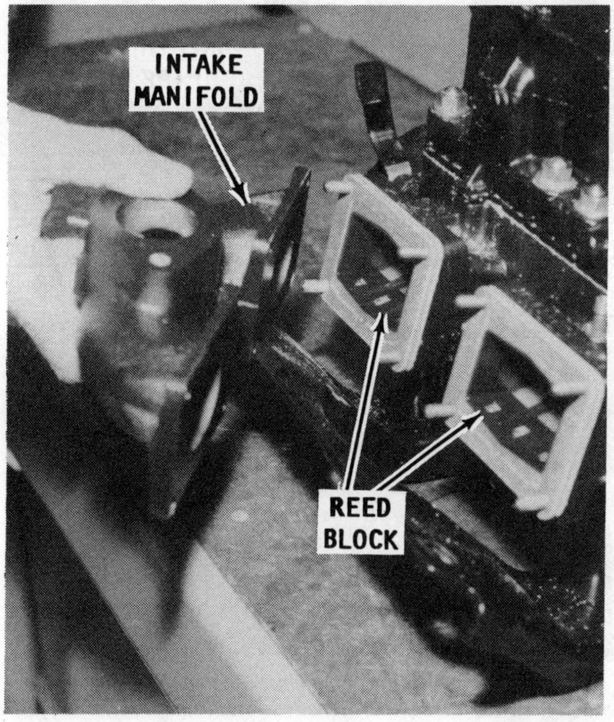

Removing the intake manifold in preparation to servicing the reed blocks.

3-5 ELECTRICAL POWER SUPPLY

Remove the battery from the engine compartment for service. Inspect and service the battery, cables and connections. Check for signs of corrosion. Inspect the battery case for cracks or bulges, dirt, acid, and electrolyte leakage.

Make sure the breather pipe is attached to the battery and not pinched by any part of the engine compartment.

Clean the top of the battery. The top of a 12-volt battery should be kept especially clean of acid film and dirt, because of the high voltage between the battery terminals and prevent contact between the acid and the two rubber hold down bands. For best results, first wash the battery with diluted ammonia or baking soda solution to neutralize any acid present. Flush the solution off the battery with clean water. Keep the vent plugs tight to prevent the neutralizing solution or water from entering the cells.

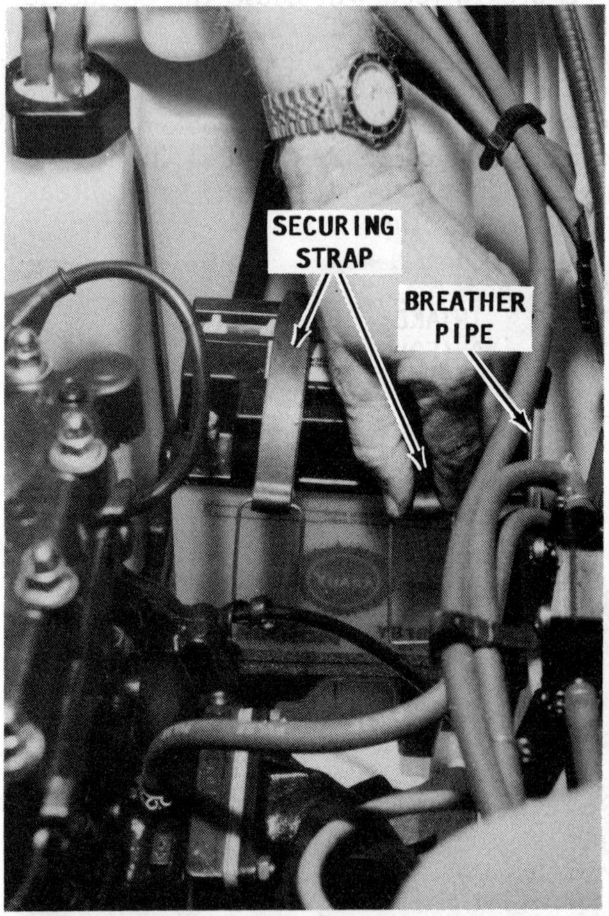

*Due to the violent maneuvers experienced by a personal watercraft — more so than any other type craft -- the battery **MUST** be extremely well-secured with **NO** movement of any kind possible.*

Filling the Battery

Remove each filler cap using a pair of pliers. Fill each cell to the proper level with distilled water. The level of electrolyte should be between the upper and lower level marks during operation. When filling the battery, fill to the upper line and allow the battery to stand for twenty minutes. Check the level again and replenish as necessary.

Use a temperature corrected hydrometer to test the specific gravity of the electrolyte. At 20°C (68°F) the hydrometer reading should be 1.26 on the scale. If it is necessary to charge the battery, leave the filler caps lightly resting on the cell openings to allow the gases to escape. The charging current should not exceed 1.9 amps for 10 hours.

After charging, push each filler cap into place and wipe the top of the battery clean before installation.

Clean the battery posts and battery cable ends with a wire brush to ensure good, clean connections. Clean the top surface of the battery and identify the "POS" or "+" and "NEG" or "-" embossed symbols. Correctly connect the battery cables to the battery observing polarity. If the cables are connected backwards, the ignition system **WILL** be destroyed the first time the engine is started.

Check to be sure the battery is fastened securely in position. The hold-down rubber straps should be tight enough to prevent any movement of the battery in the holder under the most violent maneuvers of the watercraft.

If the battery posts or cable terminals are corroded, the cables should be cleaned separately with a baking soda solution and a wire brush. Apply a thin coating of Multipurpose Lubricant to the posts and cable clamps before making the connections. The lubricant will help to prevent corrosion.

Jumper Cables

If booster batteries are used for starting an engine the jumper cables must be connected correctly and in the proper sequence to prevent damage to either battery, or the rectifier diodes.

ALWAYS connect a cable from the positive terminals of the dead battery to the positive terminal of the good battery **FIRST. NEXT,** connect one end of the other cable to the negative terminal of the good battery. Finally, connect the other end of the cable to the far side of the **ENGINE** for a good ground. By making the ground connection on the engine, any spark created will not be near the battery. An arc near the battery could cause an explosion, destroying the battery and causing serious personal injury.

If a trickle charger is used on a dead battery installed in the craft, one battery lead **MUST** be disconnected prior to connecting the charger, to prevent destroying the diodes in the rectifier.

NEVER use a trickle charger as a booster to start the engine because the diodes in the rectifier will be **DAMAGED.**

3-6 CARBURETOR ADJUSTMENT

Fuel and Fuel Tanks

Take time to check the fuel tank and all of the fuel lines, fittings, couplings, valves, flexible tank fill and vent. Turn on the fuel supply valve. If gas was not drained at the end of the previous season, make a careful inspection for gum formation. When gasoline is allowed to stand for long periods of time, particularly in the presence of copper, gummy deposits form. This gum can clog the filters, lines, and passageways in the carburetor.

If the condition of the fuel is in doubt, drain, clean, and fill the tank with fresh fuel, with the proper fuel/oil mixture recommended by the manufacturer.

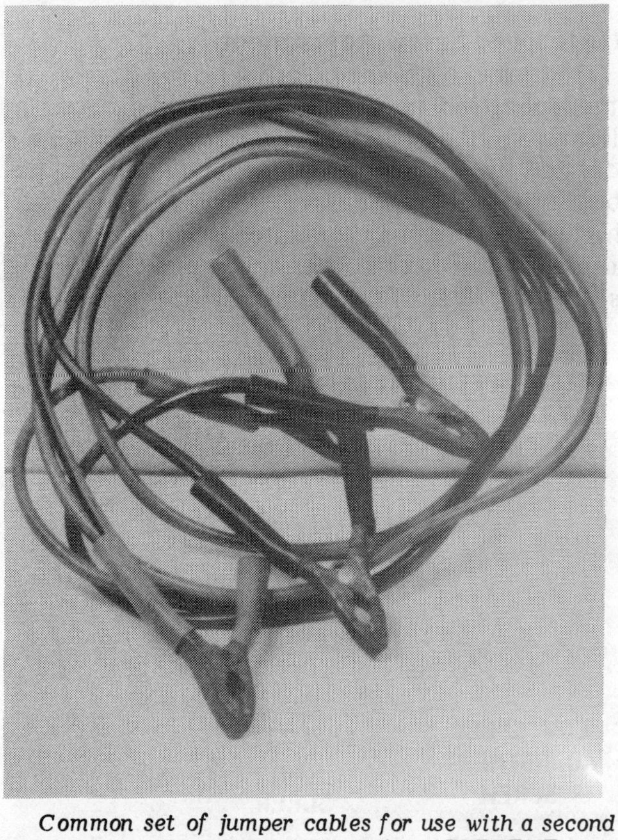

*Common set of jumper cables for use with a second battery to crank and start the engine. **EXTREME** care should be exercised when using a second battery, as explained in the text.*

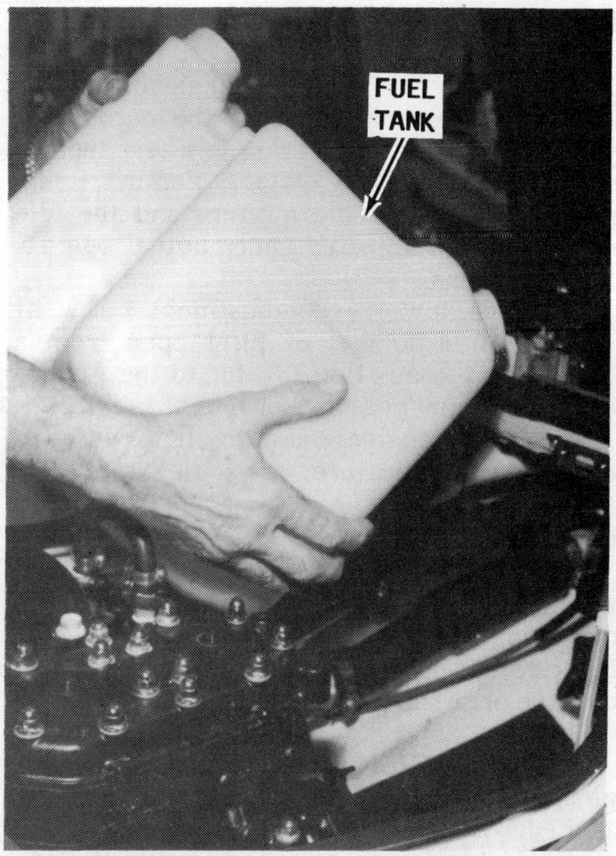

Removing the fuel tank in order to drain old fuel which may have aged and "soured".

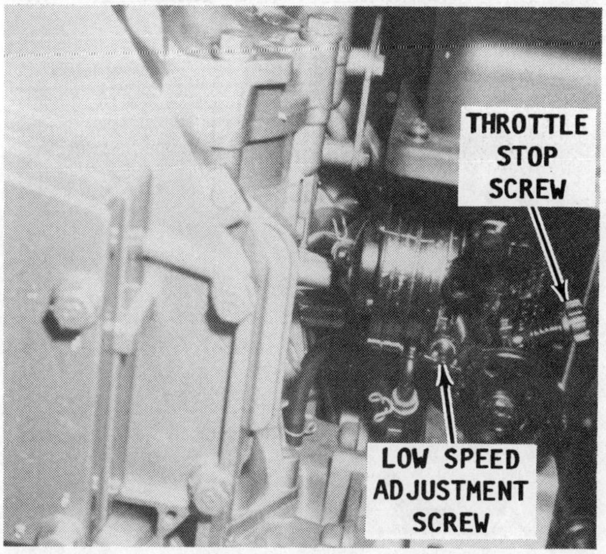

Location of the throttle stop screw and the low speed adjustment screw on a Mikuni carburetor.

Low Speed and Idle Mixture Adjustment

The idle mixture and idle speed are set at the factory. Due to local conditions, it may be necessary to adjust the carburetor while the craft is operating in a test tank or secured in a body of water. For maximum performance, the idle mixture and the idle rpm should be adjusted under actual operating conditions.

Set the low speed adjustment screw at the specified number of turns open from a lightly seated position. Refer to the Appendix for the model being serviced.

Start the engine and allow it to warm to operating temperature.

CAUTION

Water must circulate from the jet pump to and from the engine anytime the engine is operating. Just a few minutes without cooling circulating water may cause extensive damage or seizure of the engine.

NEVER, AGAIN NEVER, operate the engine at high speed with a flush device attached. An engine operating at high speed with such a device attached, would **RUNAWAY** from lack of a load on the impeller shaft, causing extensive damage.

Connect a tachometer to the engine.

Tachometer Connections

Each Kawasaki model has different color coded wiring connections. Therefore, it is impossible to give general instructions on tachometer connections. Obtain a Kawasaki electronic tachometer, and follow the in-

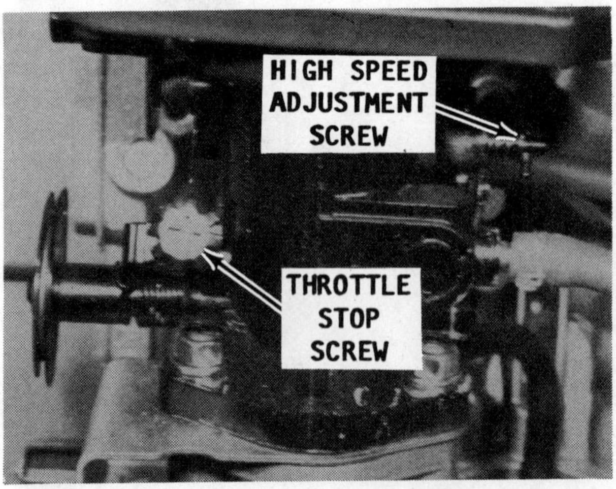

A different view of the throttle stop screw with the high speed adjustment screw clearly visible.

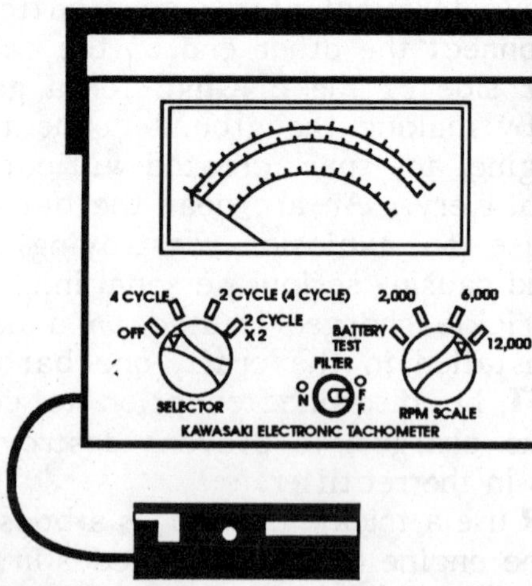

Maximum engine performance can only be obtained through proper tuning using a tachometer.

structions to make the necessary connections.

With the engine running, slowly turn the throttle stop screw until the engine idles at 1,500 rpm. Then rotate the low speed screw in or out 1/8th turn until the cylinders fire evenly and engine rpm increases. Readjust the throttle stop scew to reduce the rpm to idle speed.

High Speed Screw Adjustment

Set the high speed adjustment screw at the specified number of turns open from a lightly seated position. Refer to the Appendix for the model being serviced. This is the best setting for sea level. For operating at higher elevations, a leaner mixture is obtained by rotating the screw **CLOCKWISE** slightly, 1/8 turn at a time.

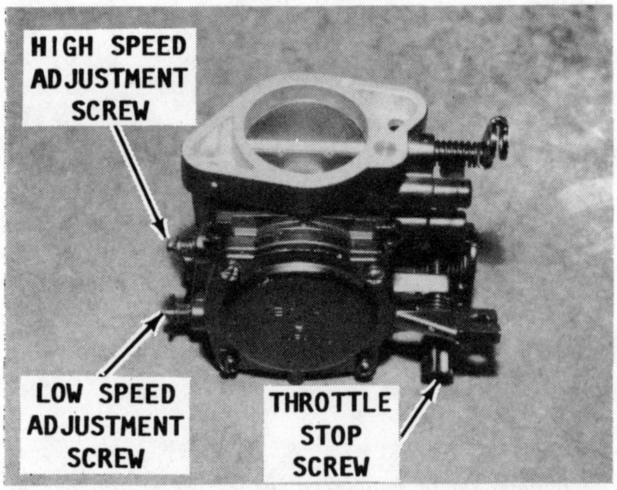

Location of the high speed and low speed adjustment screws and the throttle stop screw on a Keihin carburetor.

3-7 FUEL PUMPS

Many times, a defective fuel pump diaphragm is mistakenly diagnosed as a problem in the ignition system. The most common problem is a tiny pin-hole in the diaphragm. Such a small hole will permit gas to enter the crankcase and wet foul both spark plugs at idle-speed. During high-speed operation, gas quantity is limited, the plug is not fouled and will therefore fire in a satisfactory manner.

If the fuel pump fails to perform properly, an insufficient fuel supply will be delivered to the carburetor. This lack of fuel will cause the engine to run lean, lose rpm or cause piston scoring.

NEVER use liquid Neoprene on fuel line fittings. Always use Permatex when making fuel line connections. Permatex is available at almost all marine and hardware stores.

To service the fuel pump, see the Fuel Chapter.

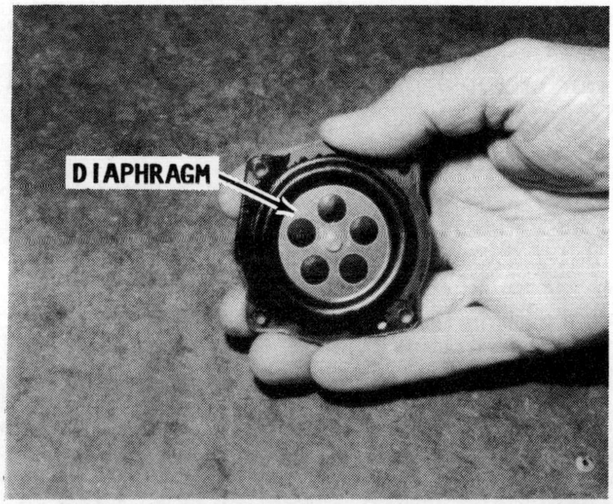

Close view of the fuel pump diaphragm from a Mikuni carburetor with integral fuel pump. The diaphragm must be carefully inspected for the tiniest hole or tear. Such damage will "wet foul" the spark plug at idle speed.

3-8 CRANKING MOTOR AND SOLENOID

Cranking Motor Test

Check to be sure the battery is fully charged. Would you believe, many cranking motors are needlessly disassembled, when the battery is actually the culprit.

Lubricate the pinion gear and screw shaft with No. 10 oil.

Connect one lead of a voltmeter to the positive terminal of the cranking motor. Connect the other meter lead to a good ground on the engine. Check the battery voltage under load by depressing the ignition switch and observing the voltmeter reading.

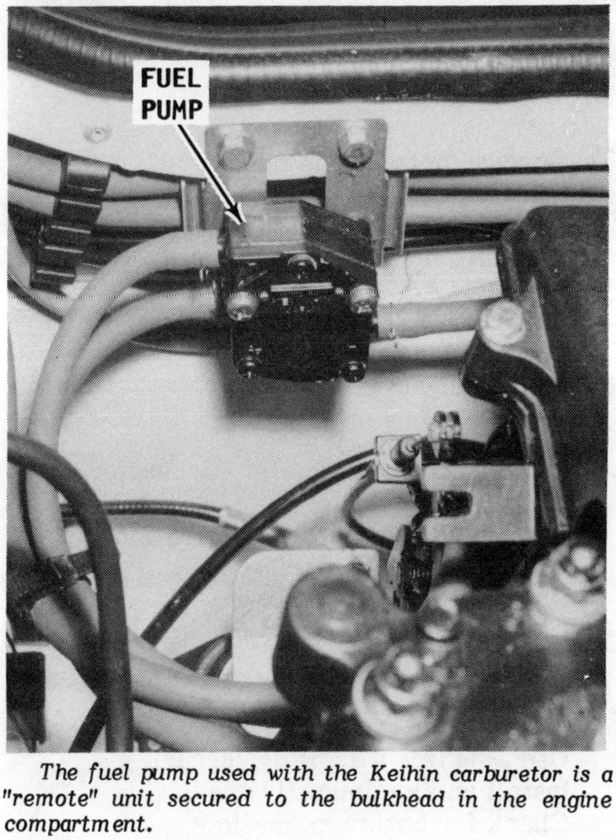

The fuel pump used with the Keihin carburetor is a "remote" unit secured to the bulkhead in the engine compartment.

Removing the cranking motor from the engine for service work and/or testing.

If the reading is 9-1/2 volts or greater, and the cranking motor fails to operate, repair or replace the unit. See the Electrical Chapter.

Solenoid Test

An ohmmeter is required for this test. Select the Rx1000 ohm scale. Calibrate the meter by making contact with the meter leads to each other, and then rotating the calibration knob on the meter until the needle indicates zero ohms. Separate the meter leads.

Make contact with the Red meter lead to one of the large terminals on the solenoid. Make contact with the Black meter lead to the other large solenoid terminal. **NEVER** connect the battery leads to the large terminals of the solenoid, or the meter will be damaged.

Using battery jumper leads, connect the positive battery lead to the small terminal of the solenoid. Connect the negative battery lead to the small terminal of the solenoid. If the meter needle indicates continuity, the solenoid is serviceable. If the meter fails to indicate continuity, the solenoid is defective and **MUST** be replaced.

3-9 JET PUMP

Very few personal watercraft are equipped with a reverse gate. However, if the unit being serviced has a reverse gate, the gate **MUST** be properly adjusted to obtain maximum performance from the craft. When properly adjusted, the gate will permit the pump to deliver its full potential of thrust with no drag.

Gate Position

The shift rod lever adjustment should be checked from time to time to ensure the gate is firmly against the pump housing, when the unit is in the **FORWARD** position (wide open).

With the gate against the housing, any rattle noises will be avoided as the craft moves through the water. Proper positioning of the gate in forward gear will prevent wave action from accidently shifting the gate into reverse as the craft is operated through violent maneuvers.

A simple "tuning" task with the jet pump is to make a thorough check of the linkage and gate movement. Such a check will ensure maximum thrust and use of the horsepower developed by the engine.

Impeller

Excessively rounded jet impeller edges will reduce the efficiency of the jet drive.

The term cavitation "burn" is a common expression used throughout the world among people working with pumps, impeller blades, and forceful water movement.

"Burns" on the impeller blades are caused by cavitation air bubbles exploding with considerable force against the impeller blades. The edges of the blades may develop small "dime size" areas resembling a porus sponge, as the aluminum material is actually "eaten" by the condition just described.

3-10 MODIFYING FOR HIGHER PERFORMANCE

Normally Seloc manuals only cover stock engine and jet pump components. However, personal watercraft enthusiasts are notorious for modifying their equipment to gain higher performance. Therefore, a large aftermarket industry has grown to accommodate this need for speed. A few well chosen words on modifying for performance seem to be in order.

The following general guidelines will provide improved two stroke performance. The tasks listed below, in order of effectiveness, or possibly a combination of modifications and components, may give the desired results.

To increase bottom end power:
- Increase the compression.
- Advance the timing.
- Use a smaller diameter carburetor.
- Use a long chamber exhaust.
- Use a low pitch impeller.
- Use a larger nozzle bore.
- Use improved reed petals.

To increase top end power:
- Install a larger diameter carburetor.
- Use a short exhaust chamber.
- Go to higher porting specifications.
- Use high-flow reeds.
- Use a higher pitched impeller.
- Use a ride plate that generates lift.

To increase acceleration:
- Increase the compression.
- Use a lighter weight driveshaft.
- Install an aluminum flywheel.
- Check for optimum air flow.

4
MAINTENANCE

4-1 INTRODUCTION

The watercraft being prepared for service probably represents a sizeable investment In order to protect this investment and for the owner to receive the maximum amount of enjoyment from the craft it must be cared for properly while being used and when it is out of the water. Always store the watercraft with the bow higher than the stern and be sure to remove the inner hull drain plug/s. If you use any type of cover to protect craft, plastic, canvas, whatever, be sure to allow for some movement of air through the hull. Proper ventilation will assure evaporation of any condensation that may form due to changes in temperature and humidity.

GOOD WORDS

The authors estimate 75% of engine repair work can be directly or indirectly attributed to lack of proper care for the engine. This is especially true of care during the off-season period. There is no way on this green earth for a mechanical engine, particularly a marine engine, to be left sitting idle for an extended period of time, say for six months and then be ready for instant satisfactory service.

Imagine, if you will, leaving your automobile for six months, and then expecting to turn the key, have it roar to life, and be able to drive off in the same manner as a daily occurrence.

It is critical for a marine engine to be run at least once a month, preferably, in the water, but if this is not possible, then a flush attachment **MUST** be connected either directly to the engine block or in the hose line from the jet pump to the engine.

Move the craft to a body of water or connect a garden hose to the engine cooling water supply fitting or flush fitting on the cylinder head.

Start the engine and allow the rpm's to stabilize at idle speed **FOR JUST A FEW SECONDS,** and then turn the water on.

Adjust the water flow until a small trickle is discharged from the bypass outlet on the port side of the hull.

When the engine is to be shut down turn the water off **FIRST** -- raise the aft portion of the hull -- **WHILE THE ENGINE IS OPERATING AT IDLE** -- "rev" the engine just a **COUPLE** times to clear water from the exhaust system -- and then shut it down.

NEVER allow the engine to operate without cooling water for more than 15 seconds.

There is no substitute and few valid excuses for not developing good habits leading to regular maintenance practices. The results of a good program are always evident in top performance, maximum enjoyment, and "pride of ownership", all at a minimum cost.

NEVER, AGAIN NEVER, operate the engine at high speed with a flush device attached. The engine, operating at high speed with such a device attached, would **RUNAWAY** from lack of load on the impeller, causing extensive damage.

At the same time, the shift mechanism, if the unit is so equipped, should be operated through the full range several times. The steering should also be operated from hard-over to hard-over.

Only through a regular maintenance program can the owner expect to receive long life and satisfactory performance at minimum cost.

The material presented in this chapter is divided into four general areas.

1- General information every owner should know.

2- Maintenance tasks that should be performed periodically to keep the craft operating at minimum cost.

3- Care necessary to maintain the appearance of the craft and to give the owner that "Pride of Ownership" look.

4- Winter storage practices to minimize deterioration during the off-season when the craft is not in use.

In nautical terms, the front of the craft is the **bow;** the rear is the **stern;** the right side, when facing forward, is the **starboard** side; and the left side is the **port** side. All directional references in this manual use this terminology. Therefore, the direction from which an item is viewed is of no consequence, because **starboard** and **port** **NEVER** change no matter where the individual is located or his position -- even standing on his/her head.

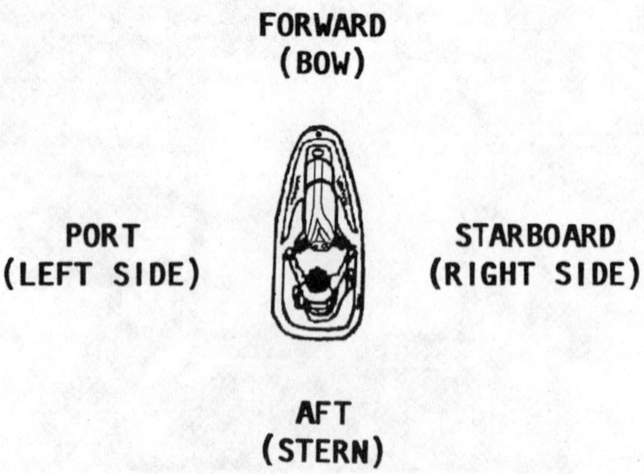

*Common terminology used throughout the world and extensively in this manual for reference designation on water craft of **ALL** sizes. "Port", "Starboard", "Forward", and "Aft", never change, even if standing on your head.*

*After a fun day on the water and before moving the craft out of the water, **SLOWLY** open the fuel tank cap to release built-up pressure in the tank.*

After Use Tasks

At the end of a day's use, make a habit of opening the engine compartment to allow fumes to escape and sponging out all water in the bottom of the hull before trailering the craft. Leave the engine cover off when the craft is stored overnight to allow any moisture in the compartment to evaporate. Slowly open the fuel tank cap to release built up pressure. Pressure in the tank will increase considerably during hot weather. Releasing the pressure reduces the chance of an explosion and ensures smooth fuel flow when the engine is next started.

If the craft will remain unused for a week, spray storage oil inside the engine and a light film of lubricant, such as WD-40, Super Lube or Justice Bros. 80, on the engine exterior. Keeping the engine exterior lightly oiled at all times will result in water sliding off the engine and draining away rather than accumulating in casting webs -- resulting in corrosion.

Cooling

The engine cooling passages must be flushed with clear water on a regular basis. This is accomplished by simply connecting a "Flushing Kit" per the manufacturer's instructions **OR** by disconnecting the cooling hose from the pump to the engine and connecting a garden hose to the disconnected hose. Such an easy maintenance practice should become a habit and will prevent accumulation of debris which could restrict engine coolant water flow and cause overheating.

Connect a garden hose to the engine cooling water supply fitting or flush fitting on the cylinder head.

Start the engine and allow the rpm's to stabilize at idle speed **FOR JUST A FEW SECONDS,** and then turn the water on.

Adjust the water flow until a small trickle is discharged from the bypass outlet on the port side of the hull.

When the engine is to be shut down, turn the water off **FIRST** -- raise the aft portion of the hull -- **WHILE THE ENGINE IS OPERATING AT IDLE** -- "rev" the engine just a **COUPLE** times to clear water from the exhaust system -- and then shut it down.

NEVER allow the engine to operate without cooling water for more than 15 seconds.

Cooling System Flushing

Connect the garden hose to the engine cooling water supply fitting or flush fitting on the cylinder head.

The manufacturer recommends the engine be started and engine rpm allowed to stabilize at idle speed **FOR A FEW SECONDS,** before turning on the water. If the water is turned on before the engine is started, the water may flow back through the exhaust pipe into the engine. Should this occur the cylinder would fill with water and if the engine was cranking at the time -- trying to start -- the piston would be attempting to compress water and the connecting rod would buckle under the stress.

As soon as the engine stabilizes at idle rpm, turn on the cooling water and adjust the water flow until a small trickle is discharged from the bypass outlet on the portside of the hull.

Perform whatever adjustments or flushing is necessary and then turn off the water **WHILE THE ENGINE IS OPERATING AT IDLE.**

Work **QUICKLY** -- raise the aft section of the craft and "rev" the engine a few times to clear water from the exhaust system.

DO NOT allow the engine to operate without cooling water for more than 15 seconds.

Shut down the engine.

Controlling Corrosion

Since man first started out on the water, corrosion has been his enemy. The first form was merely rot in the wood and then it was rust, followed by other forms of destructive corrosion in the more modern materials.

The higher the concentration of salt in the water, the worse the corrosion which will take place.

One defense against corrosion is to use similar metals in the manufacture of the parts exposed to the water.

A second defense against corrosion is to insulate dissimilar metals. This can be done by using an exterior coating of Sea Skin or by insulating them with plastic or rubber gaskets.

Pump Impeller

The pump impeller is a precisely machined and dynamically balanced aluminum spiral. Observe the drilled recesses at exact locations to achieve the delicate balancing.

Excessive vibration of the jet pump may be attributed to an out-of-balance condition caused by the impeller being struck excessively by rocks, gravel or cavitation "burn".

The term cavitation "burn" is a common expression used throughout the world among people working with pumps, impeller blades, and forceful water movement.

"Burns" on the impeller blades are caused by cavitation air bubbles exploding with considerable force against the impeller

Location of the flush fitting on a 650 Series engine. A garden hose connected to this fitting is used to flush the engine -- a very important task, especially following operation in salt water.

blades. The edges of the blades may develop small "dime size" areas resembling a porous sponge, as the aluminum is actually "eaten" by the condition just described.

Excessive rounding of the impeller edges will reduce efficiency and performance. Therefore, the impeller should be inspected at regular intervals. If rounding is detected, the impeller should be removed from the housing, placed on a work bench and the edges restored to as sharp a condition as possible using a file. Refer to the Jet Pump chapter for detailed procedures on removal and installation of the impeller.

4-2 SERIAL NUMBERS

Most manufacturers use three different identification numbers stamped, embossed or decaled somewhere on the craft.

The engine serial number is the manufacturer's key to engine changes. This number identifies the year of manufacture, the qualified horsepower rating, and the parts book identification. If any correspondence or parts are required, the engine serial number **MUST** be used or proper identification is not possible.

Federal regulations require all personal watercraft to have a Hull Identification Number (HIN) permanently attached to the craft in two separate locations: one on the transom on the starboard side, above the waterline and the other in an unexposed location, which is usually left to the discretion of the manufacturer This "HIN" number can be most helpful when attempting to retrieve a stolen craft. The "HIN"

The state license identification number must be prominently displayed on the forward portion of the hull -- port and starboard.

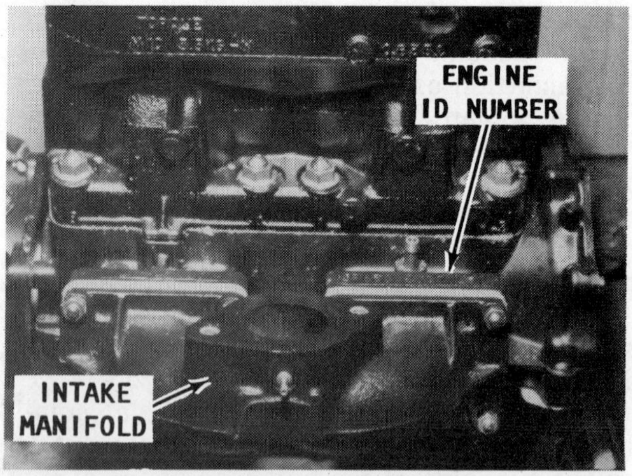

The engine ID number is embossed on the block casting just above the intake manifold.

number is required in most states in order to properly register the craft.

Some manufacturers also use a Primary I.D. Number stamped on a plate attached to the inside of the engine compartment.

ONE MORE WORD

As a theft prevention measure, all metal plates with any type of identification number are especially made wafer thin. Any attempt to remove these plates will result in cracks across the serial number.

4-3 LUBRICATION - COMPLETE UNIT

As with every type mechanical invention with moving parts, lubrication plays a prominent role in operation, enjoyment, and longevity of the unit.

If a personal watercraft is operated in salt water the frequency of applying lubricant to fittings is usually cut in half as compared with the unit being used in fresh water. The few minutes involved in moving around the craft applying lubricant and at the same time making a visual inspection of its general condition will pay in rich rewards with years of continued service.

The first personal watercraft arrived on the scene in the United States in the early 1970s. It is not uncommon to see outboard units well over 20 years of age moving a boat through the water as if the unit had recently been purchased from the current line of models. An inquiry with the proud owner will undoubtedly reveal his main credit for its performance to be regular periodic maintenance. There is no reason this same longevity of performance cannot be true for personal watercraft.

The accompanying chart can be used as a guide to the frequency of maintenance while the craft is being used during the season. Unless otherwise stated use a good brand name multi-purpose water resistant lubricant. Detailed lubrication instructions are given below for locations requiring special attention.

In addition to the normal lubrication listed in the lubrication chart, the prudent owner will inspect and make checks on a regular basis as listed in the accompanying chart.

Throttle Cable Lubrication

Hold the throttle lever in the full open position against the grip. Use a small screwdriver and pry out the small seal from the groove of the throttle stop and a little way along the cable. Spray rust inhibitor along the exposed length of cable. Fill the groove with multi-purpose water resistant lubricant and then pull the seal down along the cable and back into place inside the groove. Wipe away any excess lubricant.

Steering Cable Lubrication

Pry the barrel of the steering cable free of the ball joint on the steering post arm. Coat the ball joint with lubricant and then snap the barrel back onto the ball joint.

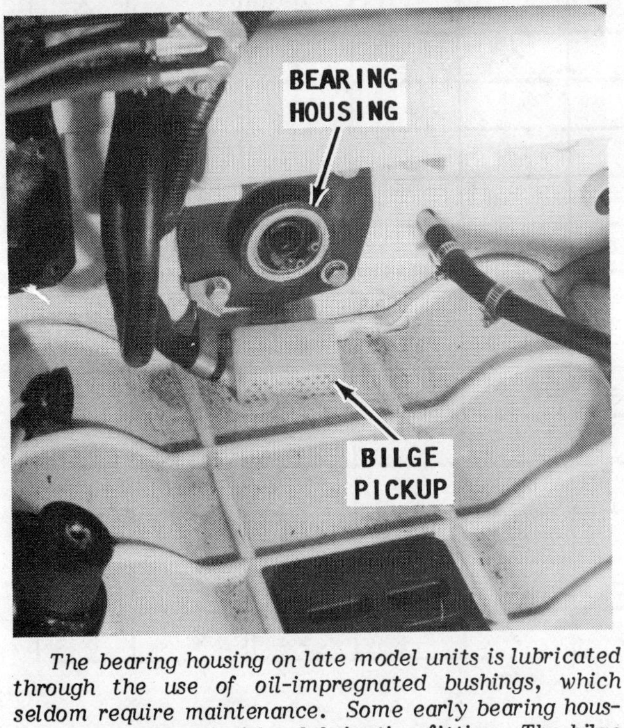

The bearing housing on late model units is lubricated through the use of oil-impregnated bushings, which seldom require maintenance. Some early bearing housings were equipped with a lubrication fitting. The bilge pickup holes should be checked regularly to ensure they are open, ready to pass bilge water when required.

The steering shaft ball joint (300 Series shown), requires periodic lubrication.

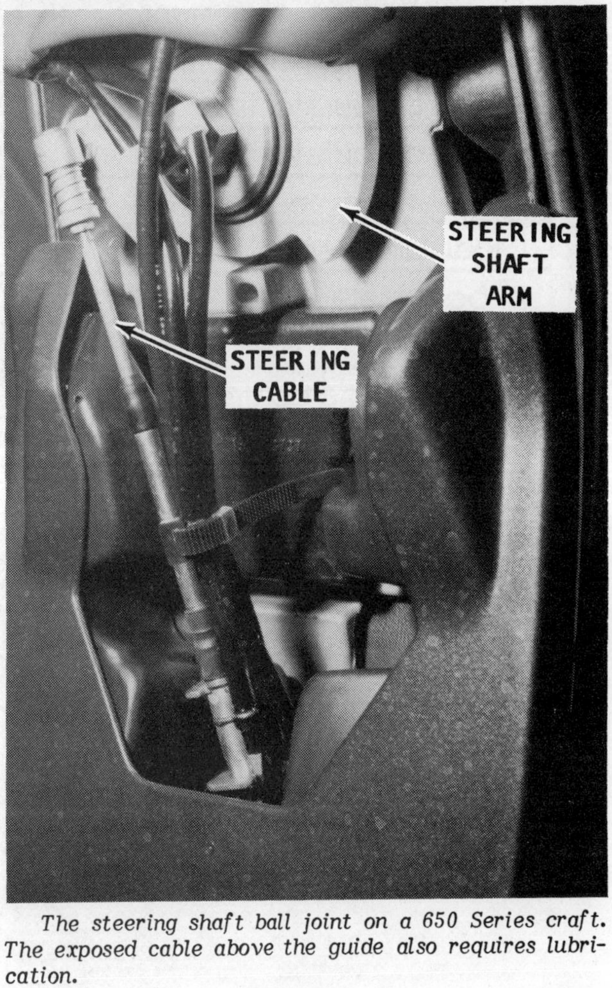

The steering shaft ball joint on a 650 Series craft. The exposed cable above the guide also requires lubrication.

Lubrication Points:	Initially every 10hrs	Initially every 50hrs	Initially every 100hrs	Then every 100hrs	Then every 200hrs
Carburetor throttle shaft			❖	❖	
Throttle cable - engine end			❖	❖	
Throttle cable - handlebar stop			❖	❖	
Choke cable - engine end			❖	❖	
Choke knob shaft			❖	❖	
Steering nozzle pivot shaft			❖	❖	
Steering post arm ball joint			❖	❖	
Reverse gate pivot pins			❖	❖	
Jet pump			❖	❖	
Seat latch			❖	❖	

Maintenance Tasks:	Initially every 10hrs	Initially every 50hrs	Initially every 100hrs	Then every 100hrs	Then every 200hrs
Check engine alignment					❖
Clean and gap spark plugs	❖	❖	❖	❖	
Adjust carburetor idle	❖		❖	❖	
Adjust carburetor mixture	❖		❖	❖	
Drain and inspect fuel tank					❖
Clean and inspect fuel filter	❖	❖	❖	❖	
Inspect fuel line			❖	❖	
Inspect check valve			❖	❖	
Adjust throttle control			❖	❖	
Adjust choke system			❖	❖	
Adjust steering system			❖	❖	
Inspect impeller		❖	❖	❖	
Check impeller to pump case clearance			❖	❖	
Inspect rubber coupling					❖
Inspect electrical connections	❖	❖	❖	❖	
Check battery electrolyte level	❖	❖	❖	❖	
Clean bilge strainer			❖	❖	
Inspect rubber water seal	❖		❖	❖	

Fuel/Oil Mixture

All two-stroke engines, except those equipped with oil injection - such as the 300 Series and 650 Series - must be operated on an exact mixture of fuel and oil.

A new or rebuilt engine must be properly "broken-in" if it is to develop maximum performance. An engine which is incorrectly broken-in will never be able to achieve the maximum rated horsepower. During the initial breaking-in period, the fuel/oil ratio should be very rich, about 20:1 for the first two to four tank fulls. Twenty parts of fuel to one part oil. Two-stroke oil be added to the fuel tank at the ratio of 25:1 for the first ten hours of operation after the unit is brought out of storage. This ratio will **ENSURE** adequate lubrication of moving parts which have been drained of oil during the storage period. Operation of an engine without the proper amount of oil will quickly lead to a significant loss of horsepower.

After the first ten hours of operation, a lower ratio of fuel/oil mixture may be used. In the past, manufacturers of two-stroke engines gave a minimum of 100:1. Kawasaki recommends a minimum of 50:1. However, experienced two-stroke mechanics run their own units between 16:1 and 32:1 on the premise that it is cheaper to replace a fouled spark plug than to have a complete engine overhaul with rebored cylinders, oversize pistons and rings. These same mechanics who own oil injected units add an extra ounce of oil per each gallon of fuel. The addition of this extra oil may prevent engine seizure in the event of an oil pump failure or clog in an oil delivery line.

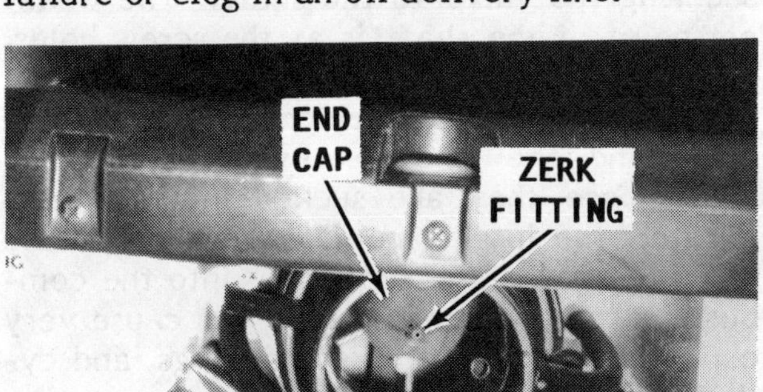

Most Kawasaki jet pumps are equipped with a Zerk lubrication fitting. This fitting is accessible through the nozzle opening. If no Zerk fitting is provided, remove and fill the end cap with lubricant.

Use only outboard marine oil in the mixture, never automotive oils. Four stroke automotive engine oil is not formulated to burn completely -- only to lubricate. Therefore, automotive oil, if used in a two stroke engine will leave an undesirable residue.

Jet Pump Lubrication

On most models, the bearing and seal housing installed on the aft end of the drive shaft is lubricated through an externally mounted zerk fitting. This fitting is located on the end cap, easily accessible through the nozzle opening. Some older units were not equipped with a zerk fitting. To service these older units, remove the three bolts securing the end cap through the nozzle opening. Clean out the end cap and refill it with fresh lubricant. Use water resistant multi-purpose lubricant at this location every 50 hours of operation.

The forward end of the pump shaft is supported by the bearing housing at the bulkhead. Older craft are equipped with a lubrication fitting on the bearing housing to lubricate the forward driveshaft bearing.

4-4 INSPECTION AND SERVICE

Fuel Tank Check Valve and Oil Tank Check Valve 300 Series and 650 Series

Before disturbing any hose be sure to have a new hose tie wrap or a small hose clamp to replace the one removed. Original equipment tie wraps must be cut through to remove them from a hose.

Disconnect the ventilation hose from the fitting on the fuel tank. Remove the two tie wraps, one on each end of the check valve.

Blow through the fitting centered on one face of the valve, in the direction of the embossed arrow. The air should pass through the valve **ONLY** in this direction. If the check valve will not allow air to pass in this direction, the fuel will remain in the tank and not be drawn by the fuel pump. Now, blow through the fitting which is off-center, on the other face of the valve. The valve should prevent air flow in this direction.

If the valve allows air to flow in the wrong direction, it will also allow fuel to flow out of the tank, through the valve and spill into the hull as the craft is subjected to violent maneuvers. Therefore, replace the valve if it fails to function correctly.

Install the check valve with new tie wraps or hose clamps with the arrow embossed on the valve pointing **TOWARD THE TANK.**

If the arrows are missing, the fitting in the center of the valve face is the air "in" fitting -- air should pass through in only one direction. The off-center fitting should have the hose connected to the fuel tank.

If working on a 300 or 650 Series engine perform the same check for the oil tank check valve. Both the oil and fuel valves are identical.

A faulty fuel tank check valve will starve the engine of fuel. If the condition becomes really serious, the engine will falter and die.

A faulty check valve will also reduce critical lubrication to the engine. Such a condition will quickly cause extensive and expensive damage.

Fuel Filler Cap

Remove the fuel filler cap and inspect the condition of the rubber seal. Replace the seal if it is worn or cracked.

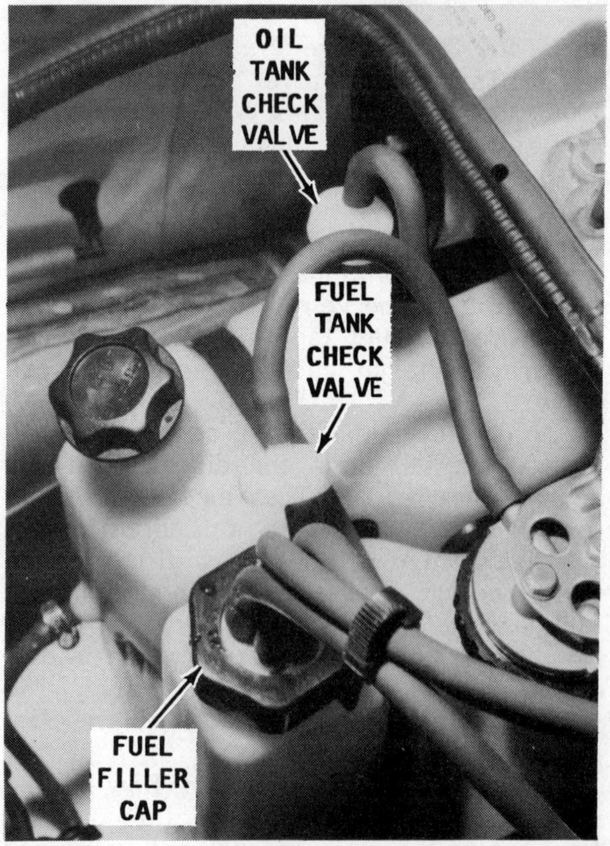

Location of the oil check valve and the fuel check valve on a 300 Series and 650 Series engine.

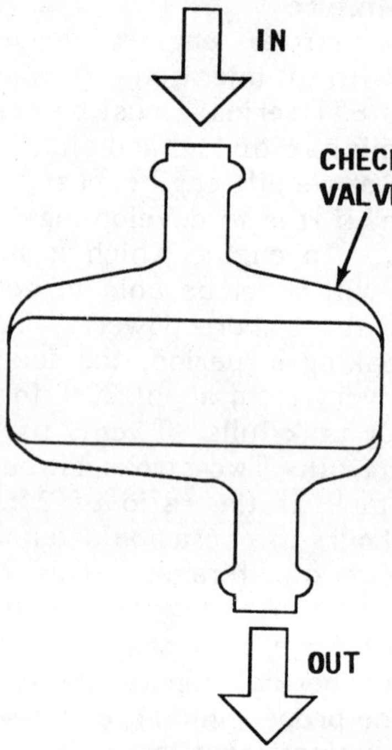

Very simple line drawing to depict the one-way check in the oil tank. This valve allows air to pass into the tank to replace oil used, but will prevent oil from leaving the tank, even if the craft is rotated 360°.

Check the tightness of the large hose clamp around the fuel tube to the tank.

Loose Bolts Down the Carburetor!

Periodically remove the flame arrestor and inspect the condition of both the choke and throttle shafts. Most shaft designs incorporate a cutout portion to enable the butterfly valve to lay flat against the shaft. Two small holes are drilled in this cutout portion to accommodate the small screws securing the valve to the shaft. The weakest point of the shaft is at the screw holes. Small cracks form at the holes due to metal fatigue. These cracks elongate the threaded holes and the small screws shake loose. Once loose they are sucked into the carburetor, the intake manifold, the crankcase and eventually find their way into the combustion chamber, where they can cause very expensive damage to pistons, rings, and cylinder walls.

Other bolts in danger of being vibrated loose and sucked down into the carburetor, are the three Phillips head screws securing the lower flame arrestor to the carburetor.

Check the tightness of all bolts likely to be a problem in or above the carburetor, particularly the bolts of the choke and throttle shafts. Should any of these bolts come free, they could cause extensive and expensive damage.

Drain Plug

Remove the drain plug from the transom. Inspect the condition of the O-ring. Replace as necessary. Clean off any dirt or sand from the plug threads. Install the plug and tighten securely.

4-5 IMPELLER-TO-PUMP CASE CLEARANCE — ALL AXIAL AND MIXED FLOW JET PUMPS

SPECIAL WORDS

The following procedure may be performed with the watercraft elevated on saw horses enabling a person to work underneath, or with the watercraft on its side allowing access to the rock grate and impeller.

If the impeller-to-pump case clearance is to be determined with the watercraft raised on saw horses, the battery may be left secured in the craft.

If the watercraft is to be positioned on its side, first disconnect the electrical leads at both battery terminals, and then remove the battery from the watercraft.

Remove the three bolts and washers securing the rock grate to the jet pump. If the seal between the hull and the pump housing is broken, or the foam packing is damaged, reseal the housing and intake area with silicone sealant to prevent impeller cavitation.

Insert a feeler gauge between each impeller blade and the pump case and determine the clearance. Make a note of all three clearances, one for each blade. Calculate the average of the three clearances.

On a new craft the clearance will be what the manufacturer refers to as the "standard" clearance -- the first column. The second column is the "service limit". This value is the maximum allowable clearance.

IMPELLER-TO-PUMP CASE CLEARANCE

Series	Standard Clearance	Service Limit
300	0.0138-0.0177" (0.35-0.45mm)	0.0350" (0.87mm)
440	0.0138-0.0177" (0.35-0.45mm)	0.0350" (0.87mm)
550	0.0079-0.0118" (0.20-0.30mm)	0.0236" 0.060mm
All 650 Including JetMate	0.0079-0.0118" (0.20-0.30mm)	0.0236" (0.60mm)

Axial Flow Pump
300 Series, 440 Series, and 650 Series

If the clearance is less then the maximum valve listed in the previous table, no action is necessary.

If the clearance is more than the service limit, inspect the condition of the pump case. If the pump case has scratches deeper

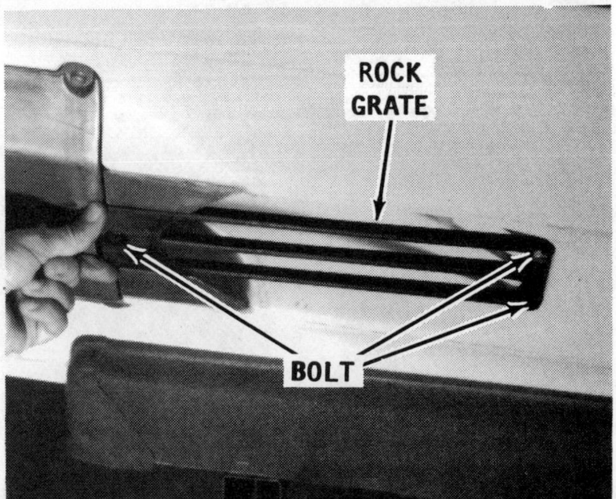

Removing the bolts securing the rock grate to the hull to gain access to the impeller.

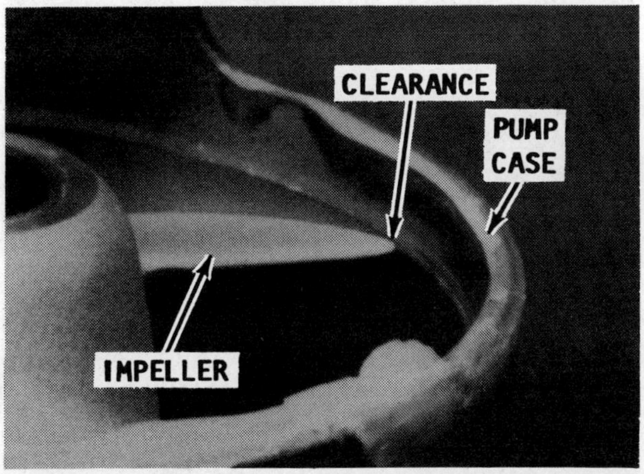

Close view to show clearance area between the impeller blades and the pump case, as explained in the text.

than 0.04" (1mm), replace the pump case. If the pump case is satisfactory, the problem must be with the impeller. Visually inspect the impeller for nicks, scratches, pitting or a "mushroomed" edge. If the cause of the excessive clearance cannot be determined visually, the pump must be disassembled and the pump case measured and compared to specifications.

The jet pump must be removed from the craft to perform this work, see Chapter 10.

Mixed Flow Pump
550 Series

If the clearance is less then the maximum valve listed in the table, no action is necessary.

If the clearance is greater than the maximum value, and the pump case is in good condition, the impeller clearance must be adjusted. Shim material must be added between the pump case and the impeller. The addition of 0.04" (1mm) of shim material will decrease the impeller clearance by 0.005" (0.144mm).

Mixed flow pump efficiency decreases as much as 30%, if the blade-to-pump case clearance increases by 0.04" (1mm).

If the clearance is greater than the maximum value, but the pump case has scratches deeper than 0.04" (1mm), replace the pump case.

The jet pump must be removed from the craft to add shim material behind the impeller. Procedures for pump removal are presented in Chapter 10, Section 10-4. The impeller-to-pump case clearance is **NOT** used to determine the total amount or the additional amount of shim material used behind the impeller. The thickness of shim material used depends entirely on the height of the end cap shoulder and the depth to which the bearings seat inside the guide vane casting.

A sample calculation and a bearing shim material chart are provided in the assembling procedures, in Chapter 10, to correctly space the impeller on the pump shaft for the specified impeller-to-pump case clearance.

If the clearance is not kept to a minimum, the jet pump will lose efficiency by allowing excess water to pass between the blades and the housing. This water has no thrust and will actually slow the stream of water under pressure from the impeller.

Install the rock grate and tighten the attaching bolts securely.

Rubber Coupler Inspection

If the engine and jet pump are correctly aligned, the rubber coupler will show little wear for many seasons. If the alignment is incorrect, the rubber coupler will wear quickly and possibly even fail. If the rubber coupler requires replacement, the engine must be removed. The task of engine alignment -- both horizontal and vertical -- must always be the final step of coupler replacement. Refer to the Engine chapter for engine removal, installation and alignment.

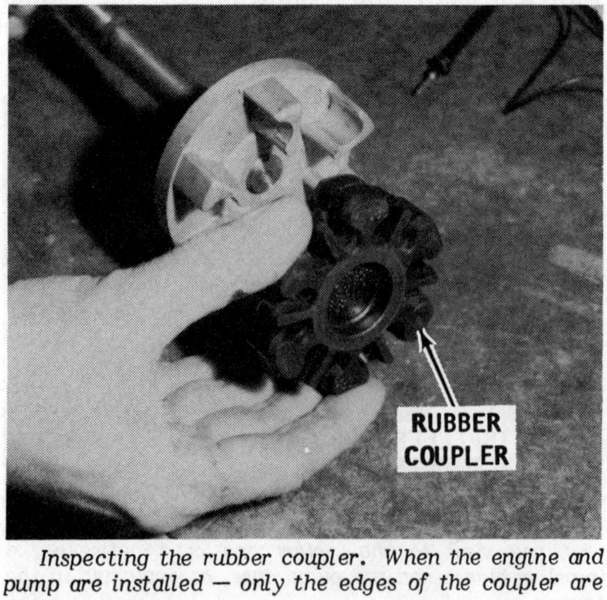

Inspecting the rubber coupler. When the engine and pump are installed — only the edges of the coupler are visible.

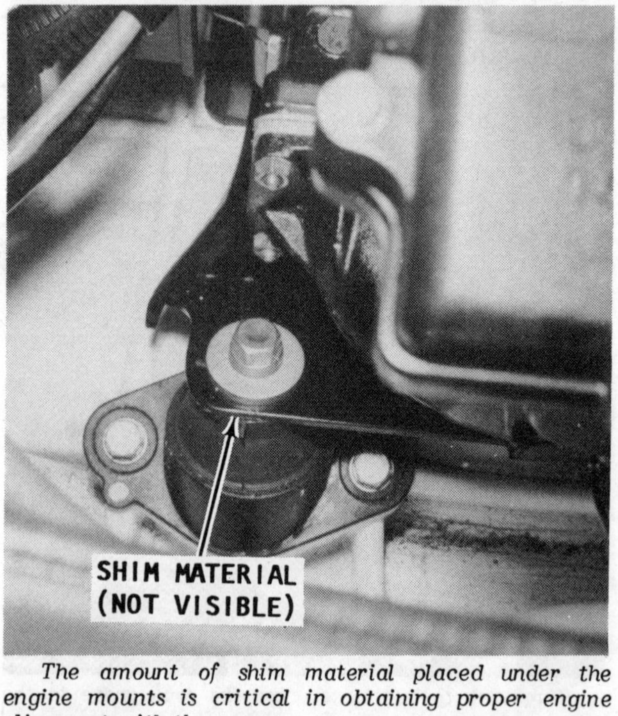

The amount of shim material placed under the engine mounts is critical in obtaining proper engine alignment with the pump.

However, if the rubber coupler is only slightly worn, then engine alignment procedures may be performed to reduce further coupler wear. Refer to the Engine chaper for detailed engine alignment procedures.

4-6 PRE-SEASON PREPARATION

Satisfactory performance and maximum enjoyment can be realized if a little time is spent in preparing the unit for service at the beginning of the season. Assuming the unit has been properly stored, as outlined in Section 4-10, a minimum amount of work is required to prepare the unit for use.

The following steps outline an adequate and logical sequence of tasks to be performed before using the craft the first time in a new season.

1- Lubricate the craft according to the manufacturer's recommendations. Refer to the lubrication chart. Remove, clean, inspect, adjust, and install the spark plugs with new gaskets (if they require gaskets). Make a thorough check of the ignition system. This check should include: the CDI igniter, magneto assembly, condition of the wiring, and the battery electrolyte level and charge.

If the exhaust outlet was plugged at the end of the last season, remove the plug.

2- Take time to check the gasoline tank and all fuel lines, fittings, couplings, valves, including the flexible tank fill and vent. Turn on the fuel supply valve at the tank. If the fuel was not drained at the end of the previous season, make a careful inspection for gum formation. When gasoline is allowed to stand for long periods of time, gummy deposits form. This gum can clog the filters, lines, and passageways in the carburetor.

Just a small amount of time spent in pre-season preparation tasks will reward the owner with satisfactory performance and maximum enjoyment.

GOOD WORDS

All manufacturers recommend the fuel filter be replaced at the start of each season or at least once a year. There may be as many as three fuel filters on any engine. One at the main fuel tank pickup, another at the reserve tank pickup, and a third in-line filter, mounted between the tank and the fuel pump. Some model carburetors may have an internal filter which can only be serviced by partially dismantling the carburetor.

Location of the electrical box on a 650 Series engine. This box contains the rectifier/regulator and the starter relay.

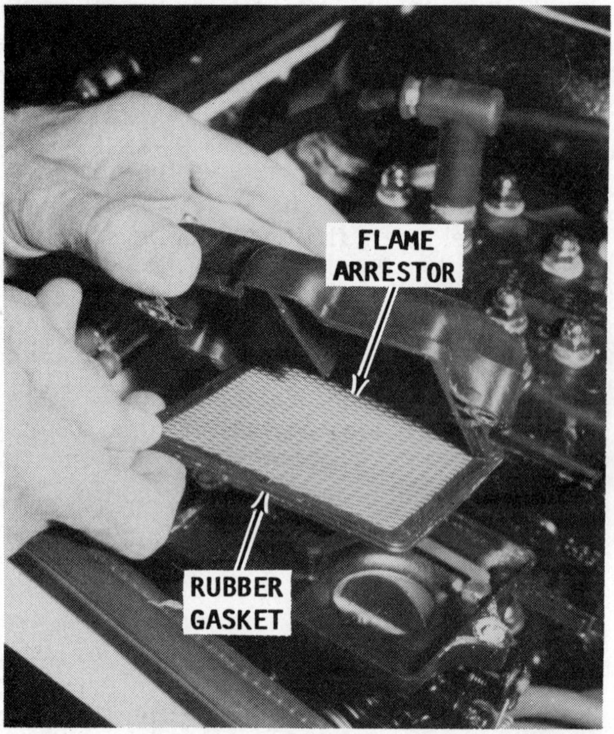

The flame arrestor should be kept fairly clean and the rubber gasket around the arrestor must remain in good condition.

3- Close all water drains. Check and replace any defective water hoses. Check to be sure the connections do not leak. Replace any spring-type hose clamps, if they have lost their tension, or if they have distorted the water hose, with band-type clamps.

The exhaust system hose clamps should be tightened periodically as a prevention against possible exhaust leaks.

4- The engine can be run with the jet pump in water to flush it. If this is not practical, a flush attachment may be used. This device is attached to the engine water supply hose, between the engine and the pump.

Connect a garden hose to the engine cooling water supply fitting or flush fitting on the cylinder head.

Start the engine and allow the rpm's to stabilize at idle speed **FOR JUST A FEW SECONDS,** and then turn the water on.

Adjust the water flow until a small trickle is discharged from the bypass outlet on the port side of the hull.

When the engine is to be shut down, turn the water off **FIRST** -- raise the aft portion of the hull -- **WHILE THE ENGINE IS OPERATING AT IDLE** -- "rev" the engine just a **COUPLE** times to clear water from the exhaust system -- and then shut it down.

NEVER allow the engine to operate without cooling water for more than 15 seconds.

Check the exhaust outlet for water discharge. Check for leaks.

5- Check the electrolyte level in the battery and the voltage for a full charge. Clean and inspect the battery terminals and cable connections. **TAKE TIME** to check the polarity, if a new battery is being installed. Cover the cable connections with lubricant or special protective compound as a prevention to corrosion formation. Check all electrical wiring and grounding circuits.

The engine cooling hose and bilge hose connections on the jet pump should be checked regularly to ensure they are tight.

6- Check all electrical parts on the engine and lower portions of the hull to be sure they are of a type to prevent ignition of an explosive atmosphere. Rubber caps help keep spark insulators clean and reduce the possibility of arcing. Electric cranking motors and high-tension wiring harnesses should be of a marine type to prevent an explosive mixture from igniting.

4-7 SEALANTS, ADHESIVES, LUBRICANTS, AND FUEL STABILIZERS

It is common practice for the larger manufacturers of personal watercraft to market their own line of products for use on their craft. Kawasaki chemical engineers have developed such a line available for use with Kawasaki outboard units, personal watercraft, snowmobiles, and motorcycles. Throughout this manual, the authors recommend the application of the manufacturer's products as a first choice. All products listed are alternatives of equal value, and may be used with confidence if the manufacturer's line is not available.

Sealants and Adhesives

Four sealants are recommended and they are **NOT** interchangeable. Each is designed to perform under a different set of conditions. Follow the directions on the package for cleaning and preparing surfaces. Sealants and adhesives **MUST** be applied **ONLY** to clean and dry parts. Apply sparingly -- excessive amounts may block oil passageways and cause serious damage.

Check the electrolyte level and hydrometer reading as part of the pre-season maintenance program. When installed, the battery should be secured to prevent even the slightest amount of movement in any direction during the most violent maneuvers the craft may be performing.

Loctite Lock N' Seal is a non-hardening, non-permanent locking compound. This material is recommended for application to the threads of load bearing fasteners. Loctite helps prevent loosening of the bolt due to vibration, thread wear, and corrosion.

Loctite Stud N' Bearing Mount compound is also a non-hardening, non-permanent locking compound. This material is recommended for application to the threads of load bearing fasteners submerged under water.

Loctite Superflex is a water resistant silicone sealer which provides a very effective flexible seal and is able to withstand high temperatures. This sealer is recommended for use where gaskets are required next to a metal surface at high temperature, for example, on the exhaust manifold cover gaskets. Loctite Superflex is blue in color.

Kawasaki Bond is a non-hardening material and is recommended for metal-to-metal joints such as the crankcase halves. This substance is highly resistant to oil and gasoline.

Lubricants

Different lubricants are recommended in the lubrication procedures presented in this manual. These lubricants are **NOT** interchangeable, each is designed to perform under varying conditions.

Shell Alvania EP1 is a general marine lubricant, chemically formulated to resist salt water. This lubricant is recommended for application to bearings, bushings, and oil seals.

Kawasaki Lubricant is a two-stroke engine oil. It has a petroleum base, and is considered a clean burning lubricant. Yamalube reduces carbon deposits and ensures maximum protection against engine wear. This lubricant also contains an ashless detergent to keep piston rings "free". Oil additives are usually not recommended by the manufacturer, and in some cases the use of such a substance may invalidate the warranty.

Fuel Stabilizer

"Sta-Bil" Fuel Conditioner and Stabilizer is recommended during engine operation and during the storage period. This fluid absorbs water in the fuel system and protects against corrosion.

If used during operation, this fuel additive will prevent the formation of gum and varnish deposits and greatly extend the period between required carburetor overhauls.

When added to the fuel during storage, the additive will prevent the fuel from "souring" for up to twelve full months.

4-8 FIBERGLASS HULLS

Fiberglass reinforced plastic hulls are tough, durable, and highly resistant to impact. However, like any other material they can be damaged. One of the advantages of this type of construction is the relative ease with which it may be repaired. Because of its break characteristics, and the simple techniques used in restoration, these hulls have gained popularity throughout the world.

A fiberglass hull has almost no internal stresses. Therefore, when the hull is broken or stove-in, it retains its true form. It will not dent to take an out-of-shape set. When the hull sustains a severe blow, the impact will be either absorbed by deflection of the laminated panel or the blow will result in a definite, localized break. Repairs are usually confined to the general area of the rupture.

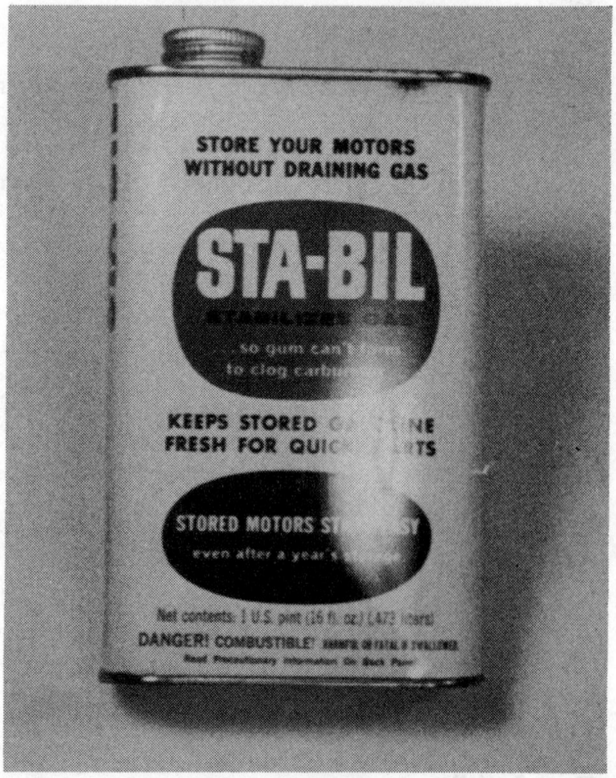

A gasoline stabilizer and conditioner may be used to prevent fuel from "souring" for up to twelve full months.

Cleaning, Waxing, and Polishing

Any craft should be washed with clear water after each use to remove surface dirt and any salt deposits from use in salt water. Regular rinsing will extend the time between waxing and polishing. It will also give "pride of ownership", by having a sharp looking piece of equipment. Elbow grease, a mild detergent, and a brush will be required to remove stubborn dirt, oil, and other unsightly deposits.

Avoid harsh abrasives or strong chemical cleaners. A white buffing compound can be used to restore the original gloss to a scratched, dull, or faded area. The finish should be thoroughly cleaned, buffed, and polished at least once each season. Take care when buffing or polishing with a marine cleaner not to overheat the surface being worked, because the area will become "burned".

4-9 SUBMERGED ENGINE SERVICE

A submerged engine is always the result of an unforeseen accident. Once the craft is recovered, special care and service procedures **MUST** be closely followed in order to return the unit to satisfactory performance.

NEVER, again we say **NEVER** allow an engine that has been submerged to stand more than a couple hours before following the procedures outlined in this section and making every effort to get it running. Such delay will result in serious internal damage. If all efforts fail and the engine cannot be started after the following procedures have been performed, the engine should be disassembled, cleaned, assembled, using new gaskets, seals, and O-rings, and then started as soon as possible.

Submerged engine treatment is divided into three unique problem areas: submersion in salt water; submerged while powerhead was running; and a submerged unit in fresh water.

The most critical of these three circumstances is the engine submerged in salt water, with submersion while running, a close second.

Salt Water Submersion

NEVER attempt to start the engine after it has been recovered. This action will only result in additional parts being damaged and the cost of restoring the engine increased considerably. If the engine was submerged in salt water the complete unit **MUST** be disassembled, cleaned, and assembled with new gaskets, O-rings, and seals. The corrosive effect of salt water can only be eliminated by the complete job being properly performed.

Submerged While Running
Special Instructions

If the engine was running when it was submerged, the chances of internal engine damage is greatly increased. Remove the spark plugs to prevent compression in the cylinders. Use a socket wrench on the flywheel nut to rotate the crankshaft. If the attempt fails, the chances of serious internal damage, such as: bent connecting rod, bent crankshaft, or damaged cylinder, is greatly increased. If the crankshaft cannot be rotated, the engine must be completely disassembled.

CRITICAL WORDS

Never attempt to start an engine that has been submerged. If there is water in the cylinder, the piston will not be able to compress the liquid. The result will most likely be a bent connecting rod.

Submerged Engine — Fresh Water
SPECIAL WORDS

As an aid to performing the restoration work, the following steps are numbered and should be followed in sequence. However, illustrations are not included with the procedural steps because the work involved is general in nature.

1- Recover the craft as quickly as possible.

2- Remove the engine cover and the spark plugs.

3- Remove the carburetor float bowl cover, or the bowl.

4- Flush the outside of the engine with fresh water to remove silt, mud, sand, weeds, and other debris. **DO NOT** attempt to start the engine if sand has entered any part of the engine. Such action will only result in serious damage to engine components. Sand in the engine means the unit must be disassembled.

CRITICAL WORDS

Never attempt to start an engine that has been submerged. If there is water in the cylinder, the piston will not be able to compress the liquid. The result will most likely be a bent connecting rod.

5- Remove as much water as possible from the engine. Most of the water can be eliminated by first holding the engine in a horizontal position with the spark plug holes **DOWN**, and then the flywheel with a socket wrench on the flywheel nut. Rotate the crankshaft through at least 10 complete revolutions. If you are satisfied there is no water in the cylinders, proceed with Step 6 to remove moisture.

6- Alcohol will absorb moisture. Therefore, pour alcohol into the carburetor throat and again rotate the crankshaft.

7- Pour alcohol into the spark plug openings and again rotate the crankshaft.

8- With assistance, rotate the craft into the horizontal position until the spark plug openings are facing **DOWN**. Pour engine oil into the carburetor throat and, at the same time, rotate the crankshaft to distribute oil throughout the crankcase.

9- With the craft upright, pour approximately one teaspoon of engine oil into each spark plug opening. Rotate the crankshaft to distribute the oil in the cylinders.

10- Install and connect the spark plugs.

11- Install the carburetor float bowl cover, or the bowl.

Move the craft to a body of water or connect a garden hose to the engine cooling water supply fitting or flush fitting on the cylinder head.

12- Obtain **FRESH** fuel and attempt to start the engine.

If using a garden hose, after the engine starts, allow the rpm's to stabilize at idle speed **FOR JUST A FEW SECONDS,** and then turn the water on.

Adjust the water flow until a small trickle is discharged from the bypass outlet on the port side of the hull.

Allow the engine to run for approximately an hour to eliminate any unwanted moisture remaining in the engine.

When the engine is to be shut down, turn the water off **FIRST** -- raise the aft portion of the hull -- **WHILE THE ENGINE IS OPER-**

ATING AT IDLE -- "rev" the engine just a **COUPLE** times to clear water from the exhaust system -- and then shut it down.

NEVER allow the engine to operate without cooling water for more than 15 seconds.

CAUTION

Water must circulate through the jet pump -- to and from the engine, anytime the engine is operating. Circulating water will prevent overheating -- which could cause damage to moving engine parts and possible engine seizure.

13- If the engine fails to start, determine the cause, electrical or fuel, correct the problem, and again attempt to get it running. **NEVER** allow an engine to remain unstarted for more than a couple hours without following the procedures in this section and attempting to start it. If attempts to start the engine fail, the unit should be disassembled, cleaned, assembled, using new gaskets, seals, and O-rings, as **SOON** as possible.

4-10 WINTER STORAGE

Taking extra time to store the craft properly at the end of each season, will increase the chances of satisfactory service at the next season. **REMEMBER**, idleness is the greatest enemy of a marine engine. The unit should be run on a monthly basis. The steering and shifting mechanism should also be worked through complete cycles several times each month. The owner who spends a small amount of time involved in such maintenance will be rewarded by satisfactory performance, and greatly reduced maintenance expense for parts and labor.

Lubricate the ball joint on the steering post as part of the winter storage program.

Proper storage involves adequate protection of the unit from physical damage, rust, corrosion, and dirt.

The following steps provide an adequate maintenance program for storing the unit at the end of a season.

1- Empty all fuel from the carburetor.

For many years there has been the widespread belief simply shutting off the fuel at the tank and then running the engine until it stops is the proper procedure before storing the engine for any length of time. Right? **WRONG!**

First, it is **NOT** possible to remove all fuel in the carburetor by operating the engine until it stops. Considerable fuel is trapped in the float chamber and other passages and in the line leading to the carburetor. The **ONLY** guaranteed method of removing **ALL** fuel is to take the time to remove the carburetor, and drain the fuel.

Secondly, if the engine is operated with the fuel supply shut off until it stops, the fuel and oil mixture inside the block is removed, leaving bearings, pistons, rings, and other parts without any protective lubricant.

Remove the spark plugs and pour about one ounce of two-cycle engine oil into each spark plug hole. Wait a minute or two and then slowly rotate the flywheel/crankshaft through two complete revolutions to evenly distribute the oil. Install the spark plugs.

2- Drain the fuel tank and the fuel lines. Pour approximately one quart (0.96 liters) of benzol (benzine) into the fuel tank, and then rinse the tank and pickup filter with the benzol. Drain the tank. Store the craft with the engine compartment door open to allow air to circulate around the tank and with the fuel vent **OPEN** to allow air to circulate through the tank.

3- Clean the in-line fuel filter with benzol, see the Fuel chapter.

4- Lubricate the throttle and shift linkage. Lubricate the steering pivot shaft with water resistant multi-purpose lubricant.

5- Plug the exhaust outlet with a shop towel or similar item to prevent any type of contaminant from entering the exhaust system.

Battery Storage

Remove the battery from the craft and keep it charged during the storage period. Clean the battery thoroughly of any dirt or corrosion, and then charge the battery to the full specific gravity reading. After charging, store the battery in a clean cool dry place where it will not be damaged or knocked over.

NEVER store the battery with anything on top of it or cover the battery in such a manner as to prevent air from circulating around the filler caps. All batteries, both new and old, will discharge during periods of storage, more so if they are hot than if they remain cool. Therefore, the electrolyte level and the specific gravity should be checked at regular intervals. A drop in the specific gravity reading is cause to charge the battery back to a full reading.

One of the largest personal watercraft battery manufacturers recommends that a battery in storage should be checked every two weeks and charged if necessary. The electrolyte level can also be replenished at this time to prevent sulfation occurring.

In cold climates, **EXERCISE CARE** in selecting the battery storage area. A fully-charged battery will freeze at about 60° below zero. A discharged battery, almost dead, will have ice forming at about 19° above zero.

4-11 PRE-SEASON CHECK

Before attempting to start the craft after winter storage, check the steering and throttle action for smooth operation, without binding. If the battery was removed for storage, check to make sure it has a full charge.

Filling the Battery

Remove each filler cap using a pair of pliers. Fill each cell to the proper level with distilled water. The level of electrolyte should be between the upper and lower level marks during operation. When filling the battery, fill to the upper line and allow the battery to stand for twenty minutes. Check the level again and replenish as necessary.

Use a temperature corrected hydrometer to test the specific gravity of the electrolyte. At 20°C (68°F) the hydrometer reading should be 1.26 on the scale. If it is necessary to charge the battery, leave the filler caps lightly resting on the cell openings to allow the gases to escape. The charging current should not exceed 1.9 amps for 10 hours.

After charging, push each filler cap into place and wipe the top of the battery clean before installation.

Clean the battery posts and battery cable ends with a wire brush to ensure good, clean connections. Clean the top surface of the battery and identify the "POS" or "+" and "NEG" or "-" embossed symbols. Correctly connect the battery cables to the battery observing polarity. If the cables are connected backwards, the ignition system **WILL** be destroyed the first time the engine is started.

Make sure the breather pipe is in place on the battery and the pipe is not pinched when the battery is installed.

Battery life may be extended considerably if the battery is removed from the craft and stored in a warm dry area with proper air circulation during the off-season period.

First Time Start

If the proper maintenance was observed at the end of the previous season, both the carburetor and fuel tank should be completely empty. Fill the gas tank with the proper mixture and open the fuel valve, if equipped. Open the choke. **DO NOT** attempt to crank the engine until it starts, **AT THIS TIME.** Considerable cranking time will be spent before the oil/fuel mixture reaches engine bearings and other moving parts. Expensive parts may be damaged due to lack of lubrication and the cranking motor will become excessively hot, which is also **BAD NEWS.**

If the craft cannot be moved to a body of water, connect a garden hose to the engine cooling water supply fitting or flush fitting on the cylinder head.

Start the engine and allow the rpm's to stabilize at idle speed **FOR JUST A FEW SECONDS,** and then turn the water on.

Adjust the water flow until a small trickle is discharged from the bypass outlet on the port side of the hull.

When the engine is to be shut down, turn the water off **FIRST** -- raise the aft portion of the hull -- **WHILE THE ENGINE IS OPERATING AT IDLE** -- "rev" the engine just a **COUPLE** times to clear water from the exhaust system -- and then shut it down.

NEVER allow the engine to operate without cooling water for more than 15 seconds.

CAUTION

Water must circulate through the jet pump -- to and from the engine, anytime the engine is operating. Circulating water will prevent overheating -- which could cause damage to moving engine parts and possible engine seizure.

The correct procedure for starting is to obtain an **AEROSOL** can of WD-40. It must be an aerosol canister because the liquid must be atomized, a squirt can will not do the job. **DO NOT** use ether or a commercial starting fluid because they will evaporate almost instantly. These products work well with four-stroke engines equipped with an oil pump, but they do not contain a lubricant essential for a two-stroke engine. WD-40 is an extremely flammable oil.

Make sure both spark plugs are tight, so the mist from the canister will not ignite. Spray the WD-40 through the flame arrestor and continue spraying with **NO** throttle until the engine starts. As soon as the engine is operating, cease spraying the WD-40. The engine may emit some smoke for a few seconds after it starts. Watch the sediment bowl (if so equipped). As soon as it fills, the engine should operate on the regular oil/fuel mixture. Work the throttle to keep the engine at idle speed until it reaches operating temperature, then close the choke. Advance the throttle a few times and the engine should then idle smoothly.

5
TROUBLESHOOTING

5-1 INTRODUCTION

This chapter is divided into five main sections as follows:

5-2 Mechanical engine problems.
5-3 Fuel system problems.
5-4 Ignition system faults.
5-5 Cranking system failures.
5-6 Charging system malfunctions.
5-7 Troubleshooting Charts

If a change in engine rpm is experienced at normal operating speed, such a change is usually a symptom of an engine problem or trouble in the jet pump. An increase or a decrease in rpm may be an indication of the area to be checked. The following generalizations may be helpful in isolating the problem.

A noticeable change in craft performance, as evidenced by one of the following symptoms, may be attributed to one or more of the areas listed:

Lower than Normal Engine Rpm

Mechanical problems in the engine:
- Plugged flame arrestor.
- Fouled spark plug/s, incorrect heat range or gap.
- Shorted or weak ignition coil.
- Contaminated fuel.
- Restricted exhaust system.
- Inadequate fuel pump pressure.
- Loss of compression due to blown head gasket.

Mechanical problems in the jet drive:
- Worn or damaged impeller.
- Main bearing failure.
- Rubber coupler failure.

Higher than Normal Engine Rpm

Mechanical problem in the engine:
- None.

Mechanical problems in the jet drive:
- Clogged intake grate -- possibly plastic bag or other debris.
- Line, sea weed, etc. entangled in impeller.
- Ingested sand.

If a reduction in speed is experienced, accompanied with no loss or gain of engine rpm, the loss of performance can fairly accurately be attributed to some type of drag on the craft or restriction of the water ejected from the jet pump.

Drag on the hull is usually caused by damage to the craft.

Many times, drag from the jet pump is caused by a partially lowered reverse gate, if so equipped, or from dragging debris caught in the intake grate.

A partially lowered reverse gate may be caused by incorrect cable adjustment or by binding around the gate pivot pins.

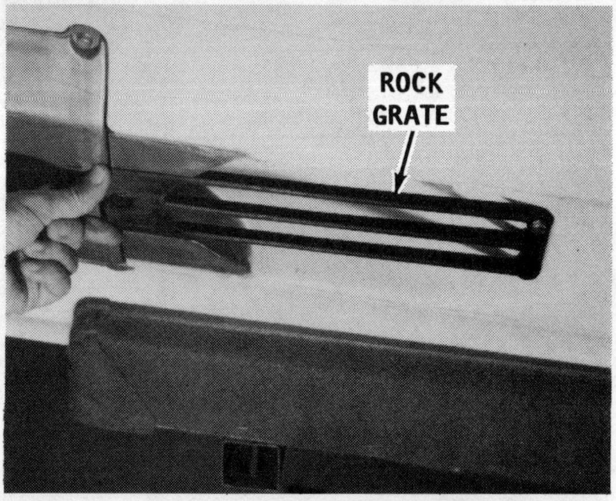

One of the most hazardous conditions for a personal watercraft is restriction of water through the rock grate caused by a plastic bag, plastic wrap, or by some type of sea weed. Even the smallest reduction in water volume will affect jet pump performance and could quickly lead to engine overheating.

Engine Troubleshooting

Troubleshooting must be a well thought-out procedure. Always attempt to proceed with the troubleshooting in an orderly manner. The "shot in the dark" approach will only result in wasted time, incorrect diagnosis, replacement of unnecessary parts, and frustration.

Obviously, if the instructions are to be of maximum benefit as a guide, they must be fully understood and followed in the proper sequence.

When an engine fails to start, the trouble must be localized to one of four general areas: cranking system; ignition system; fuel system; or compression. Each of these areas must be systematically inspected until the trouble is isolated. At that time, detailed tests of the system or area must be made to determine the part causing the problem. The last section of this chapter describes charging system troubleshooting.

Troubleshooting Check

When using this Troubleshooting Check, proceed sequentially through each test until the defect is uncovered. Then skip to the detailed testing procedure and check for that system. For example, if, when using the Troubleshooting Check procedure, the first two systems, cranking and ignition tests OK, but the third test shows there is trouble in the fuel system, then proceed to the detailed test under the Fuel System Troubleshooting Check in Section 5-3.

Cranking System Test

1- Turn the key switch to the **START** position or press the ignition button. The cranking motor should crank the engine at a normal rate of speed.

If the cranking motor cranks the engine slowly or fails to crank it at all, the trouble is either in the battery or in the cranking system. Test the battery first, then if the battery checks out, proceed directly to Troubleshooting the Cranking System in Section 5-7 for detailed testing procedure to isolate the problem.

Ignition System Test

2- Connect a standard spark tester (available at almost any automotive parts store at very nominal cost), between the spark plug and the spark plug lead. Hold the spark plug lead with an insulated holder (to

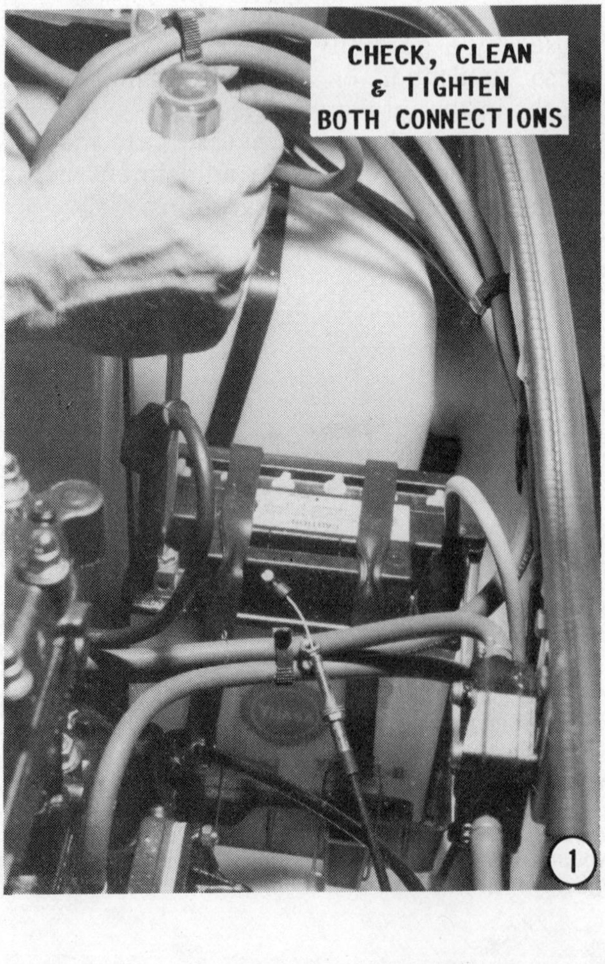

avoid receiving a "shock" when the current passes through). Crank the engine with the cranking motor. If there is **NO** spark, or if the spark is very weak, the trouble is in the ignition system. Proceed directly to the Ignition Troubleshooting in Section 5-4 for detailed testing procedures to uncover the problem.

3- Remove the flame arrestor. Look down into the throat of the carburetor and at the same time open and close the throttle several times. Observe the throttle valve to verify it moves from the fully closed to the fully open position. Perform the same check with the choke valve.

Attempt to determine the age of the fuel in the fuel tank. Many fuels tend to "sour" in three to four months, especially if an additive such as Sta-Bil has not been added. Other fuels take longer to "sour". In no case should an attempt be made to start the engine if the fuel in the tank is more than 12 months old.

Compression Test

4- Good compression is the key to proper engine performance. An engine with worn piston rings or a blown gasket cannot be made to perform satisfactorily until the mechanical defects contributing to low compression are corrected. Generally, a compression gauge is used to determine the cranking pressure within each cylinder.

To make a compression test, remove both spark plugs and remember or identify from which cylinder they were taken. Evaluation of the spark plug firing end can be most useful in determining how the cylinder is functioning. After the spark plugs have been removed, install a compression gauge into one cylinder and crank the engine.

The throttle valve and choke **MUST** be in the **WIDE OPEN** position in order to obtain maximum readings. Crank the engine through several revolutions and observe the highest reading on the compression gauge.

The cranking compression for a new engine is 142 psi. The significance in a compression test is the variation in pressure readings between the cylinders. As long as this variation is between 20-30 psi, cylinder compression is considered as "satisfactory". If a greater variation exists, then the lower reading cylinder should be checked by mak-

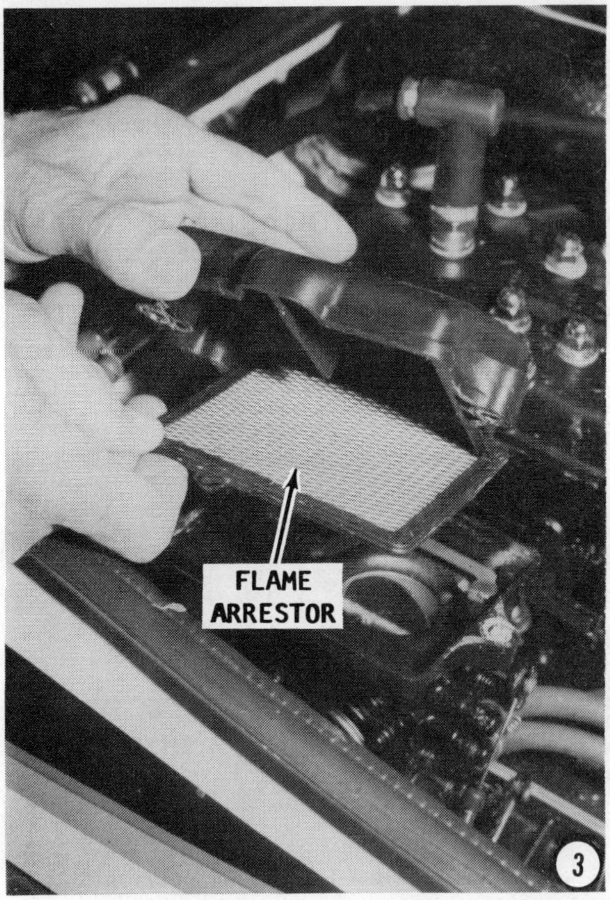

A gasoline stabilizer and conditioner may be used to prevent fuel from "souring" for up to twelve full months.

ing a cylinder leak test. A simple leak test can be made by first inserting a teaspoonful of oil into the spark plug opening of the lower reading cylinder, and then cranking the engine a few times to distribute the oil.

Check the compression again to see if inserting the oil caused a change. The oil helps make a temporary seal around the rings and increases the compression. If the reading increased, the compression loss was probably due to worn rings. If the reading did not change appreciably, the loss may be due to a blown head gasket.

5-2 MECHANICAL ENGINE PROBLEMS

Jumping Waves

Even though "jumping waves" with a personal watercraft may be considered **GREAT** sport, consider the extensive internal engine damage which will surely occur if this type action is consistently practiced.

During wave jumping or using a wet ramp, when the craft clears the water, the engine RPM approaches the limit set by the RPM limiter in an unloaded condition.

Once the craft returns to the water surface, the sudden load on the impeller is tranferred along the driveline. Speed of the rotating pump shaft and the crankshaft is

suddenly reduced while the flywheel continues to spin rapidly due to its mass. This condition, especially when repeated over a period of time, often shears the Woodruff key and the flywheel becomes repositioned on the crankshaft.

A very small misalignment will result in a change in ignition timing. The craft operator will notice a drastic loss of power. A worse condition would be the flywheel shifting to the degree engine start would not be possible.

This sudden increase and decrease in engine speed dramatically shortens engine life. Bent crankshafts and broken connecting rods can usually be directly attributed to this type of "fun play".

Easy Check for Twisted Crankshaft

A twisted crankshaft may be detected by first removing the spark plugs and setting up a dial indicator in one of the spark plug openings. If a dial indicator is not available, a pencil, straw, or other small diameter straight object may be used.

With the indicator, or other device, in place, determine the TDC (top dead center) of the No. 1 piston. Now, make two matching marks -- one on the flywheel and the other adjacent on the stator. Next, rotate the flywheel **EXACTLY** 180° -- preferably using a degree wheel on the flywheel. Make a second mark on the stator adjacent to the mark on the flywheel.

NOW, check for TDC of the other piston. The dial indicator should indicate the same reading as for the first piston **AND** the two marks on the stator should be **EXACTLY** opposite each other through the center of the crankshaft. If the indicator is off or if the marks on the stator are not aligned, as

Removal of the crankshaft is not necessary to check for a "twisted" condition, as explained in the text.

described -- the crankshaft is twisted and **MUST** be replaced.

Engine Noises

Engine noises can be generally classified as knocks, slaps, clicks, or squeaks. These noises are usually caused by loose bearings, sloppy pistons, or other moving parts of the engine.

A main bearing knock is usually identified by a dull thud which is noticeable under engine load. Attempting to move the craft under power with it tied to the dock will bring out a main bearing knock. If you pull the spark plug wire from one plug at a time and the noise disappears when one of the wires is removed, then the noise is probably coming from that cylinder. The cause may be either the rod bearing, piston pin, or the piston, which has quieted down because the load has been removed from the part. If a rod bearing has failed, the noise will be loudest when the engine is decelerating. Piston pin noise and piston pin slap are generally louder when a cold engine is first started.

Carbon build up in the combustion chamber can cause interference with a piston.

Many times the source of an unusual sound may be isolated by using a stethoscope or other listening device. One such device is a long shank screwdriver. Allow your hand to extend over the end of the handle, press an ear to your hand, and then probe with the other end of the screwdriver by touching it firmly against the block at each cylinder or noise area. Take care and use good judgment when using this method of attempting to detect the cause of a problem because the noise will travel through other metallic parts of the engine and could lead to a false interpretation of what you are hearing.

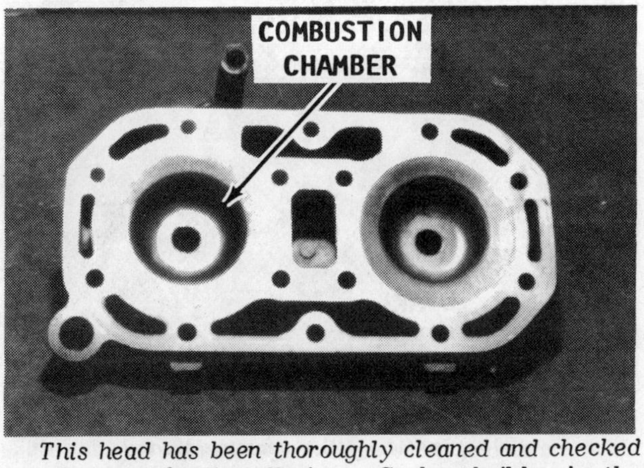

This head has been thoroughly cleaned and checked and is ready for installation. Carbon buildup in the combustion chamber can cause interference with a piston.

5-3 FUEL SYSTEM PROBLEMS

The following paragraphs provide an orderly sequence of tests to pinpoint problems in the system. It is very rare for the carburetor by itself to cause failure of the engine to start.

Many times fuel system troubles are caused by a plugged fuel filter, a defective fuel pump, or by a leak in the fuel line from the fuel tank to the fuel pump or by a vacuum leak in the line between the pump and the crankcase.

Fuel will begin to sour in three to four months and will cause engine starting problems. A fuel additive such as Sta-Bil may be used to prevent gum from forming during storage or prolonged idle periods.

When the engine is hot, the fuel system can cause starting problems. After a hot engine is shut down, the temperature inside the fuel bowl may rise to 200°F and cause the fuel to actually boil. All carburetors are vented to allow this pressure to escape to the atmosphere. However, some of the fuel may percolate over the high-speed nozzle and overflow into the intake manifold.

In order for this raw fuel to vaporize enough to burn, considerable air must be added to lean out the mixture. Therefore,

The tiniest "pin" hole in the carburetor diaphragm will seriously affect carburetor performance.

The throttle stop screw must be set properly or the self-circling feature of the craft will be lost, if the operator is thrown into the water.

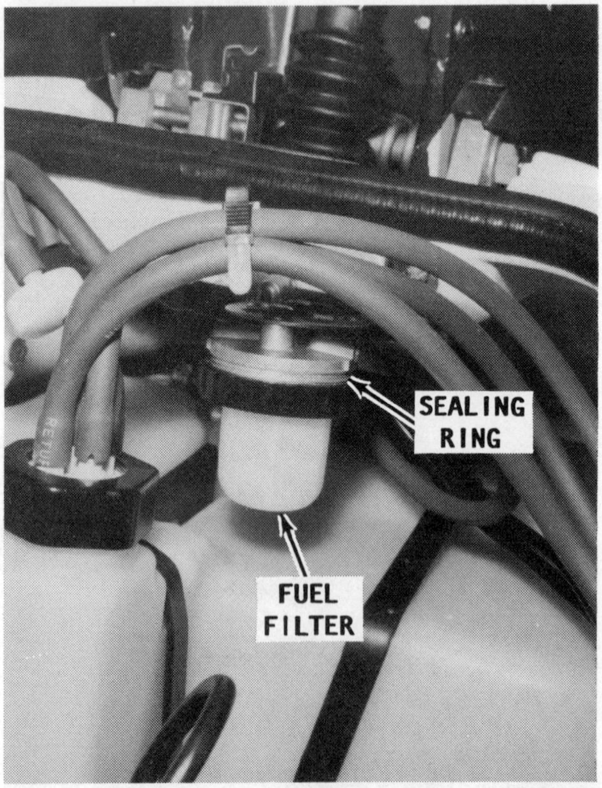

A leaking sealing ring at the fuel filter will draw air into the system. The leak may not be evident until the engine is shut down -- fuel will then leak out past the damaged sealing ring.

the only remedy is to open the throttle as wide as possible and to crank the engine until enough air is drawn in to provide the proper mixture for the engine to start. **NEVER** move the throttle lever back-and-forth in an attempt to start a hot engine. This action will only compound the problem by adding more fuel to an already too-rich mixture.

A leak between the fuel tank and the pump many times will not appear when the engine is operating because the suction created by the pump sucking fuel will not allow the fuel to leak. Once the engine is turned off and the suction no longer exists, fuel may begin to leak.

Engine Surge

If the engine operates as if the load on the craft was being constantly increased and decreased, even though the operator attempts to maintain a constant engine speed, the problem can most likely be attributed to the fuel pump.

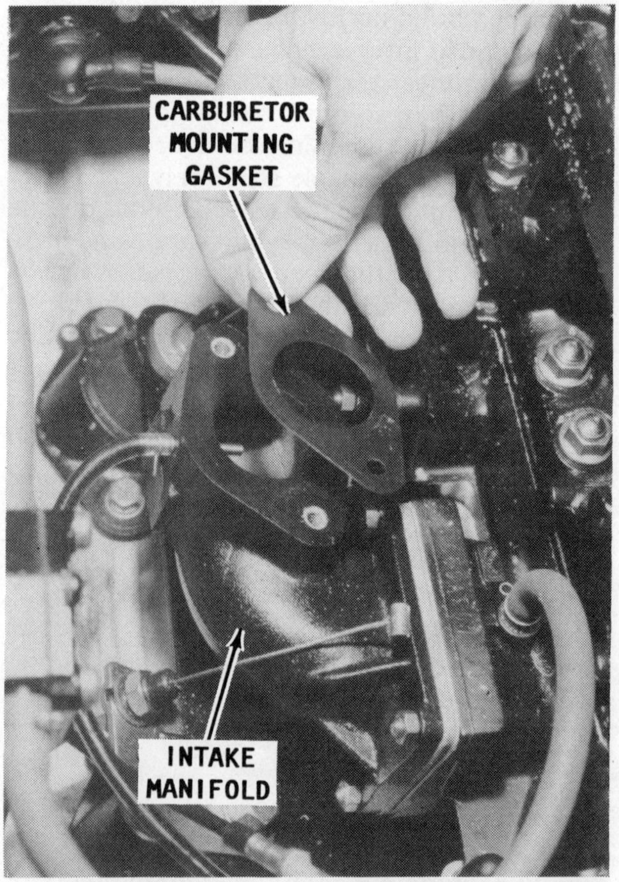

A small leak between the carburetor and the intake manifold, caused by a defective carburetor mounting gasket, will result in rough engine performance at idle speed.

Rough Engine Idle

Problems preventing an engine from running smoothly include: a fouled spark plug; an air leak in the intake manifold; and uneven compression between the cylinders.

Of course any problem in the carburetor affecting the air/fuel mixture will also prevent the engine from operating smoothly at idle speed. These problems usually include: leaking needle valve and seat; a dirty flame arrestor; defective choke; and improper adjustments for idle mixture or idle speed.

5-4 IGNITION SYSTEM FAULTS

WARNING

Check to be sure the engine compartment is well ventilated and free of any gasoline vapors before starting any of the following tests. A spark will be generated creating a potential fire hazard if fuel vapors are present.

1- Check to be sure the battery has a full charge. If not, correct the condition by charging the battery or making a substitution.

2- Check all the terminal connections inside the electrical box. Check all primary

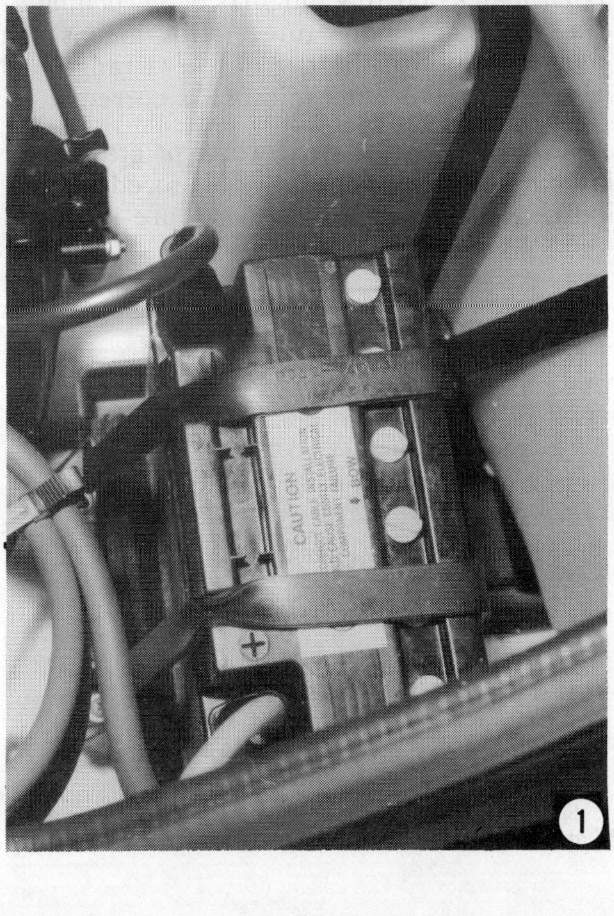

1

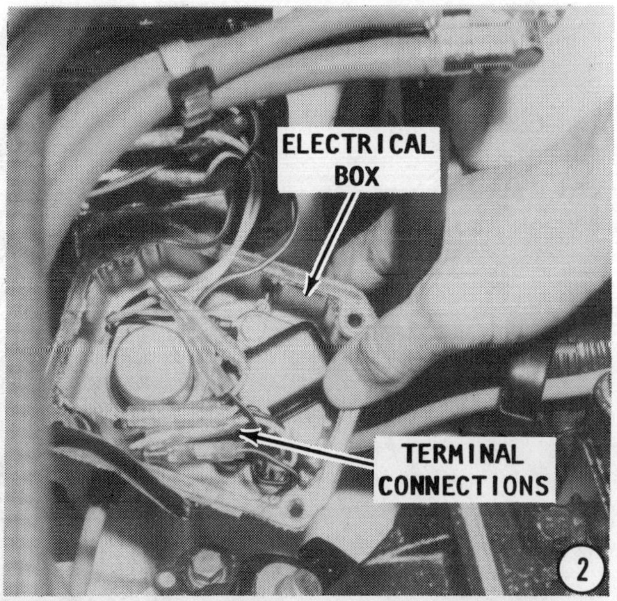

2

secondary ignition wiring, the engine "kill" switch and the ignition switch.

3- Use a spark tester and check for spark. If a spark tester is not available, hold the plug wire about 1/4" (6.4mm) from the engine. Crank the engine through a few revolutions using the cranking motor and check for spark. A strong spark over a wide gap must be observed when testing in this manner, because under compression a strong spark is necessary in order to ignite the air/fuel mixture in the cylinder. This means it is possible to think a strong spark is present, when in reality the spark will be too weak when the plug is installed. If

3

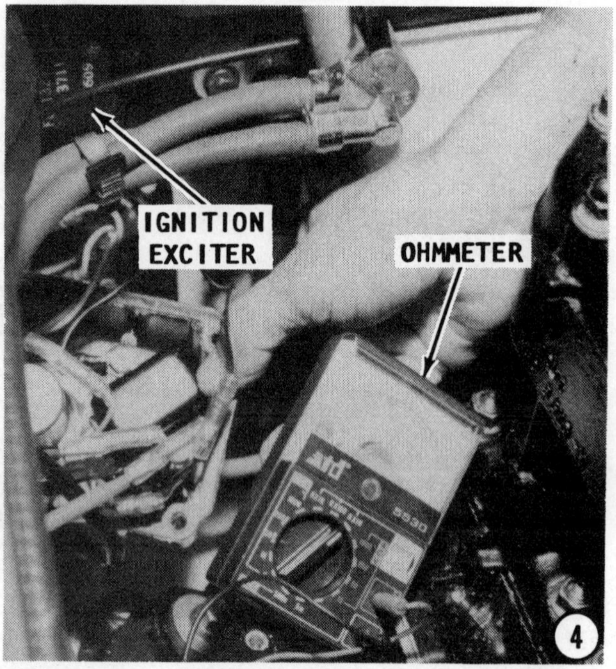

there is no spark, or if the spark is weak, the trouble is most likely under the flywheel in the stator assembly.

4- Perform the resistance test procedures given in the Ignition chapter to identify the faulty component.

Intermittent or Multiple Problems

Many ignition problems occur only during engine operation, when the component is subject to vibration and/or an increase in temperature. Many of these defective components return to normal once the engine is shut down and allowed to cool.

There is also the possibility of more than one defective component in the ignition system. One component failure may have caused a surge of power and "fried" another component.

Make every attempt to avoid the **BUY** and **TRY** method of troubleshooting. In **MOST** cases electrical components are not returnable, once the item leaves the store.

SPARK PLUG EVALUATION

Removal: Remove the spark plug wires by pulling and twisting on only the molded cap. **NEVER** pull on the wire or the connection inside the cap may become separated or the boot damaged. Remove the spark plugs and identify from which cylinder they were taken. **TAKE CARE** not to tilt the socket as you remove the plug or the insulator may be cracked.

Examine: Carefully examine the firing end to determine the firing condition in each cylinder.

Correct Color: A proper firing plug should be dry and powdery. Hard deposits inside the shell indicate the engine is starting to use some oil, but not enough to cause concern. The most important evidence is the light gray color of the porcelain, which is an indication this plug has been running at the correct temperature. This means the plug is one with the correct heat range and also that the air/fuel mixture is correct.

Overheating: A dead white or gray insulator, which is generally blistered, is an indication of overheating and pre-ignition.

Example of overheating, heavy load. Use a spark plug with a lower heat rating.

Spark plug fouled by oil. The engine may be in need of an overhaul.

The electrode gap wear rate will be more than normal and in the case of pre-ignition, will actually cause the electrodes to melt. Overheating and pre-ignition are usually caused by overadvanced timing, detonation from using too-low an octane rating fuel, an excessively lean air/fuel mixture, or problems in the cooling system.

Rich Mixture: A black, sooty condition on both the spark plug shell and the porcelain is caused by an excessively rich air/fuel mixture, both at low and high speeds. The rich mixture lowers the combustion temperature so the spark plug does not run hot enough to burn off the deposits.

Deposits formed only on the shell is an indication the low-speed air/fuel mixture is too rich. At high speeds with the correct mixture, the temperature in the combustion chamber is high enough to burn off the deposits on the insulator.

Too Cool: A dark insulator, with very few deposits, indicates the plug is running too cool. This condition can be caused by low compression or by using a spark plug of an incorrect heat range. If this condition shows on only one plug it is most usually caused by low compression in that cylinder. If both plugs have this appearance, then it is probably due to the plugs having a too-low heat range.

Fouled: A fouled spark plug may be caused by the wet oily deposits on the insulator shorting the high-tension current to ground inside the shell. The condition may also be caused by ignition problems which prevent a high-tension pulse to be delivered to the spark plug.

Carbon Deposits: Heavy carbon-like deposits are an indication of excessive oil consumption. This condition may be the result of worn piston rings.

Electrode Wear: Electrode wear results in a wide gap and if the electrode becomes carbonized it will form a high-resistance path for the spark to jump across. Such a condition will cause the engine to misfire during acceleration. If both plugs are in this condition, it can cause an increase in fuel consumption and very poor performance at high-speed operation. The solution is to replace the spark plugs with a rating in the proper heat range and gapped to specification.

Red rust-colored deposits on the entire firing end of a spark plug can be caused by water in the cylinder combustion chamber. This can be the first evidence of water entering the cylinders through the exhaust manifold. This condition **MUST** be corrected at the first opportunity.

5-5 CRANKING SYSTEM FAILURES

Regardless of how or where the solenoid is mounted, the basic circuits of the starting system on all makes of cranking motors are the same and similar tests apply. In the

Powdery deposits have melted and shorted-out this spark plug.

Red, brown, or yellow deposits, are by-products of combustion from fuel and lubricating oil.

following testing and troubleshooting procedures, the differences are noted.

WARNING

ALWAYS TAKE TIME TO VENT THE BILGE WHEN MAKING ANY OF THE TESTS AS A PREVENTION AGAINST IGNITING ANY FUMES ACCUMULATED IN THAT AREA.

Faulty Symptoms

If the cranking motor spins, but fails to crank the engine, the cause is usually a corroded or gummy Bendix drive. The drive should be removed, cleaned, and given an inspection.

If the cranking motor cranks the engine too slowly, the following are possible causes and the corrective actions that may be taken:

a- Battery charge is low. Charge the battery to full capacity.
b- High resistance connections at the battery, solenoid, or motor. Clean and tighten all connections.
c- Undersize battery cables. Replace cables with sufficient size.

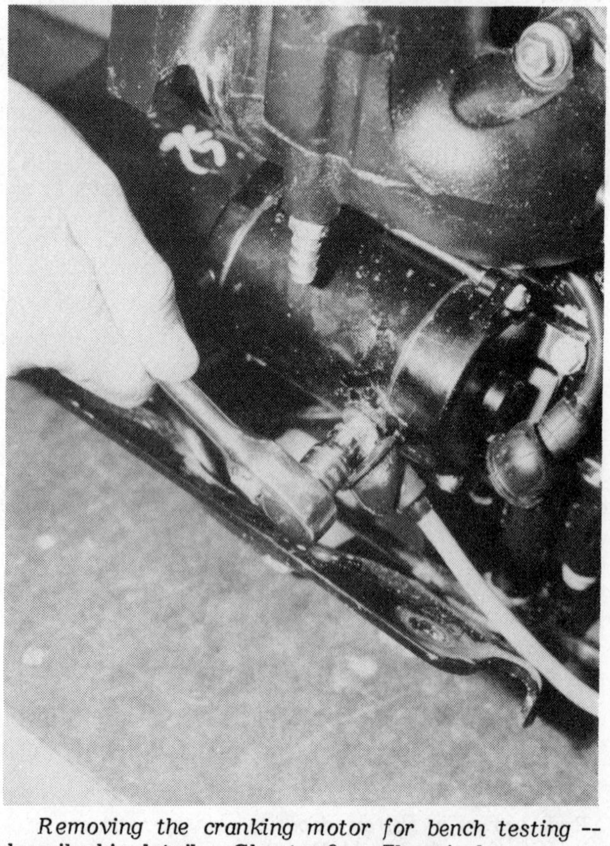

Removing the cranking motor for bench testing -- described in detail — Chapter 9 —, Electrical.

All engine cranking problems fall into one of three problem areas:

1- The cranking motor fails to turn.
2- The cranking motor spins rapidly, but does not crank the engine.
3- The cranking motor cranks the engine, but too slowly.

The following paragraphs provide a logical sequence of tests designed to isolate a problem in the cranking system.

Before wasting too much time troubleshooting the cranking motor circuit, the following checks should be made. Many times, the problem will be corrected.

a- Battery fully charged.
b- Main fuse is "good" (not blown).
c- All electrical connections clean and tight.
d- Wiring in good condition -- insulation not worn or frayed.

The following troubleshooting procedures are presented in a logical sequence.

Do not operate the cranking motor for more than 15 seconds. Prolonged cranking motor operation will cause overheating and damage the motor.

After each test, allow the cranking motor to cool for a minute or so.

Never depress the **START** button to activate the cranking motor while the engine is operating. Such action will damage the pinion and/or flywheel gears.

Cranking Circuit Tests

1- Remove both spark plug leads from the plugs and ground the ends to the engine

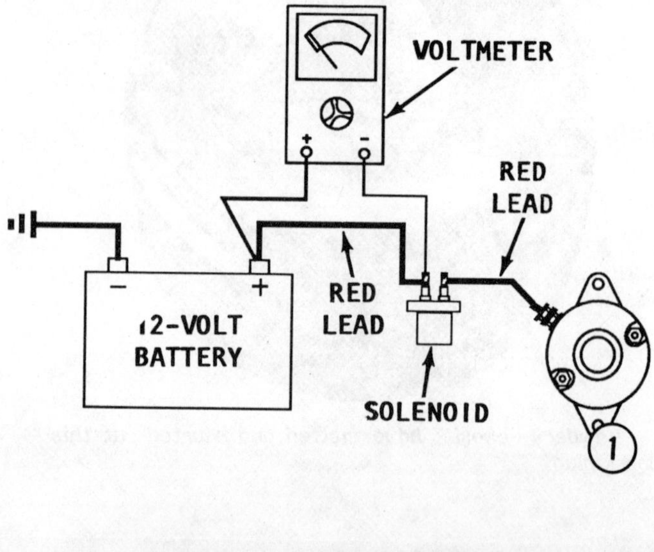

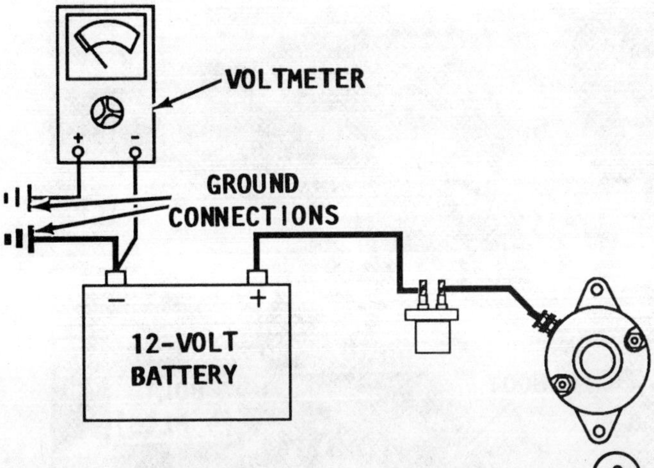

2

to prevent accidental engine start. Keep the leads grounded for the following tests.

Obtain a voltmeter and select the Vx1 DC scale. Observe the starter solenoid. Both large terminals have Red leads connected. One terminal has the battery cable connected and the other terminal has the lead from the cranking motor connected. Make contact with the negative Black meter lead to the terminal on the starter solenoid with the battery cable connected.

Make contact with the positive Red meter lead to the positive battery terminal. The voltmeter should register less then 0.25V DC. If the reading exceeds 0.25V DC, clean the connections at both ends of the positive battery cable and repeat the test. If the reading does not change, replace the cable.

2- Keep the meter on the same scale. Make contact with the negative Black voltmeter lead to the negative terminal on the battery. Make contact with the positive Red voltmeter lead to a good engine ground.

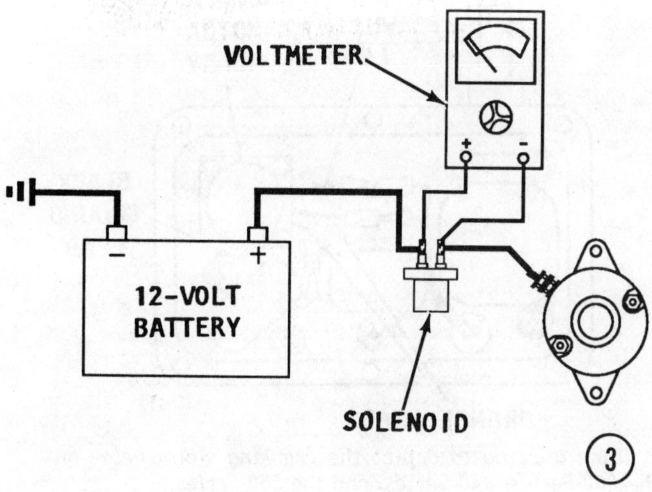

3

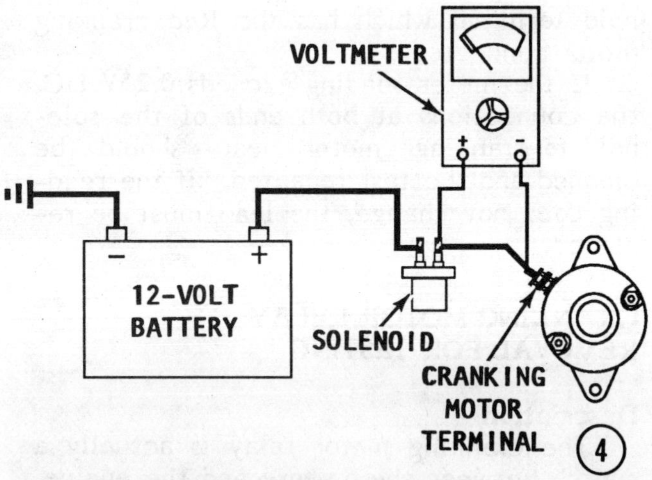

4

The voltmeter should register less then 0.25V DC. If the reading exceeds 0.25V DC, clean the connections at both ends of the negative battery cable and repeat the test. If the reading does not change, replace the cable.

CRITICAL WORDS

During the following tests -- Step 3 and Step 4, both meter leads must be placed on the solenoid terminals **WHILE** the engine is being cranked. **ALSO,** both meter leads must be removed from the solenoid terminals **WHILE** the engine is being cranked and **BEFORE** the engine has stopped cranking. If these precautions are not followed, the voltmeter may be damaged.

3- Keep the meter on the same scale. Obtain the services of an assistant.

Signal the assistant to start cranking the engine. While the engine is being cranked, make contact with the negative Black voltmeter lead to the solenoid terminal which has the lead from the cranking motor connected, and then quickly make contact with the positive Red voltmeter lead to the solenoid terminal which has the battery cable connected.

If the meter reading exceeds 0.25V DC, the starter solenoid must be replaced. Remove the meter leads and signal the assistant to cease cranking.

4- Keep the meter on the same scale and the assistant at the **START** button. Observe the same precautions as before in Step 3. While the engine is being cranked, make contact with the negative Black voltmeter lead to terminal on the cranking motor, and then quickly make contact with the positive Red voltmeter lead to the solenoid terminal which has the Red cranking motor connected.

If the meter reading exceeds 0.25V DC, the connections at both ends of the solenoid-to-cranking motor lead should be cleaned and the test repeated. If the reading does not change, the lead must be replaced.

CRANKING MOTOR RELAY REMOVAL FOR TESTING

Description

The cranking motor relay is actually a switch between the battery and the engine. The switch cannot be serviced. Therefore, if testing indicates the switch is faulty, it **MUST** be replaced. The relay is housed inside the electrical box. The box must first be removed from the engine, and then the relay removed from the box for testing purposes.

ELectrical Box Removal 300, 440, and 550 Series

1- Disconnect the battery ground cable from the battery. Remove the spark plug lead/s from the spark plug/s. Remove the two bolts securing the connector cover to

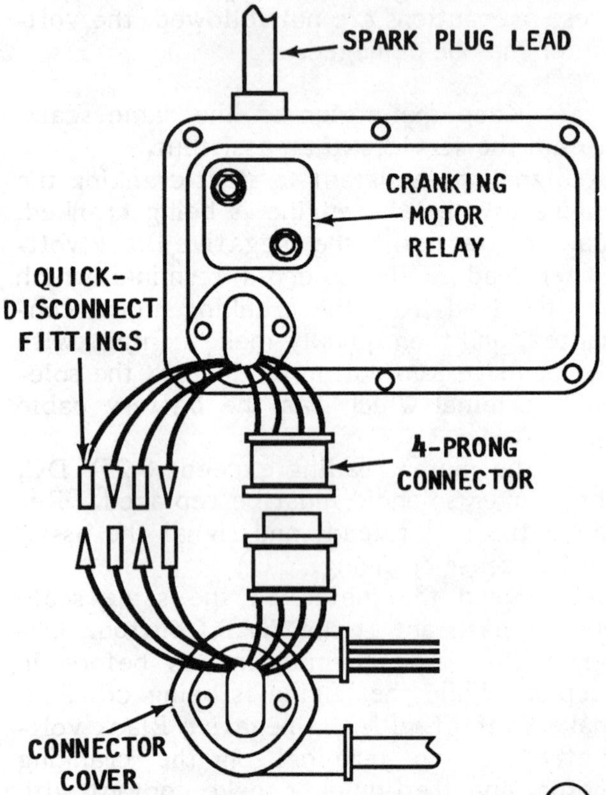

1

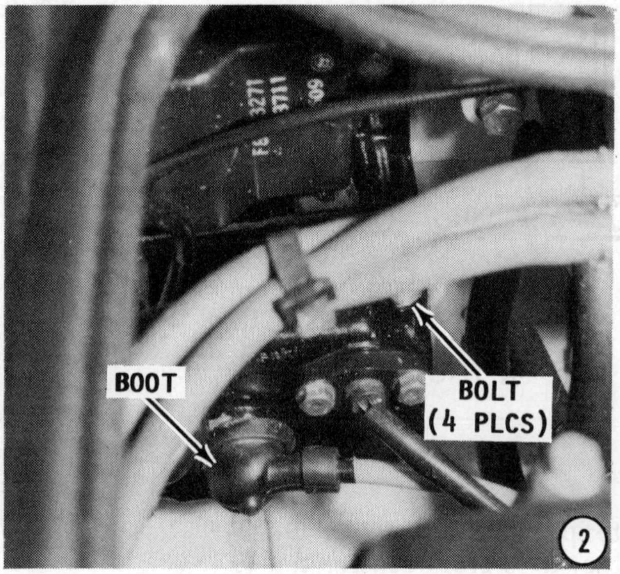

the electrical box. Pull off the cover. Disconnect the 4-pin connector and the four free leads at their quick disconnect fittings.

Pull back the two boots over the large cables at the cranking motor relay. Disconnect both large cables from the relay terminals. Remove the six bolts securing the electrical box to the engine compartment bulkhead. Remove the two bolts securing the two halves of the box together, and then open the box.

Electrical Box Removal 650 Series Only

2- Disconnect the battery ground cable from the battery. Remove the spark plug leads from the spark plugs. Pull back the large boot over both large cables at the cranking motor relay. Disconnect both cables from the relay terminals. Remove the four bolts securing the cover to the electrical box on the engine compartment

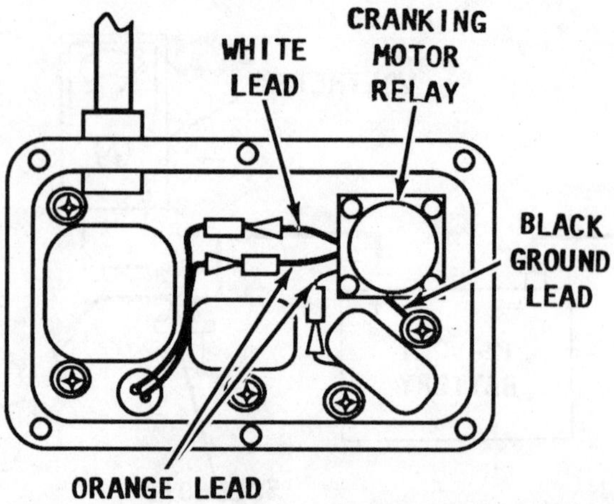

Line drawing to depict the cranking motor relay on the 300 Series, 440 Series, and the 550 Series.

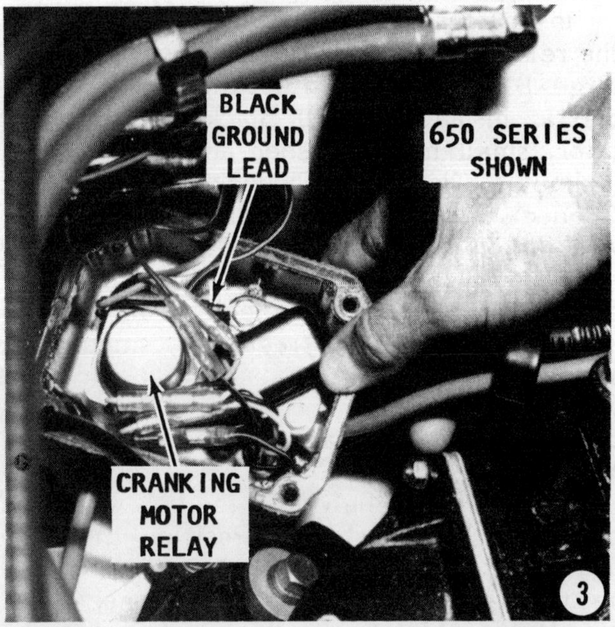

bulkhead. Note there are two different bolt sizes.

Cranking Motor Relay Removal All Models

3- Inside the box -- disconnect the ground switch from the relay at the rectifier/regulator. Disconnect the two Orange leads and the White lead at their quick-disconnect fittings -- if working on a Jet Mate, disconnect the Yellow/Red, Red and Black leads. Outside the box -- remove the nuts from the two large relay terminals. Pull the relay out of the box, taking care to observe the arrangement of washers and insulating grommets.

Cranking Relay Testing

4- Select the Rx1 ohm scale on the ohmmeter. Make contact with the two meter leads across the two large relay terminals. The meter should read an infinite resistance. If the meter shows a resistance of less than infinity, the relay is defective and must be replaced.

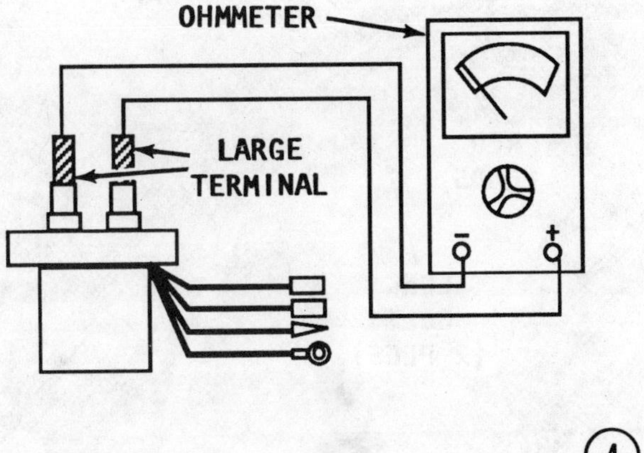

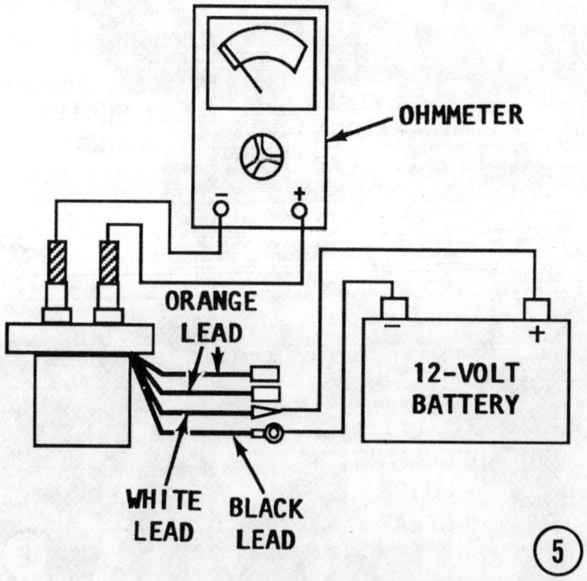

5- Keep the ohmmeter on the same scale. Obtain a 12V battery. Connect the negative battery terminal to the small Black ground lead from the relay. Connect the positive battery terminal to the small White lead from the relay -- if working on a JetMate, connect the positive battery terminal to the Yellow/Red relay lead. Make contact with the two meter leads across the two large relay terminals. If the relay emits a "click" and the ohmmeter registers zero ohms, the relay is good. If no "click" is heard, or if the meter registers a high or infinite resistance, the relay is defective and must be replaced.

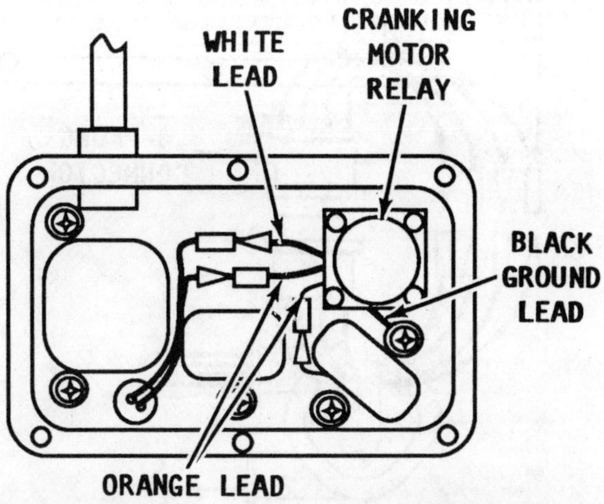

The cranking motor relay is housed inside the electrical box on the 300 Series, 440 Series, and the 550 Series.

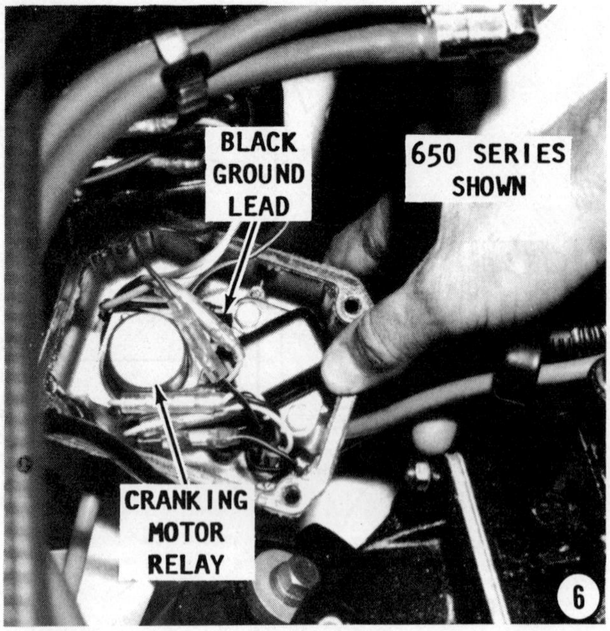

Cranking Relay Installation

6- Apply a coat of multi-purpose water restistant lubricant to all the insulating washers and grommets. Slide a large flat metal washer over each of the two large terminals, followed by a flat white plastic insulator. Then, insert the relay into the electrical box.

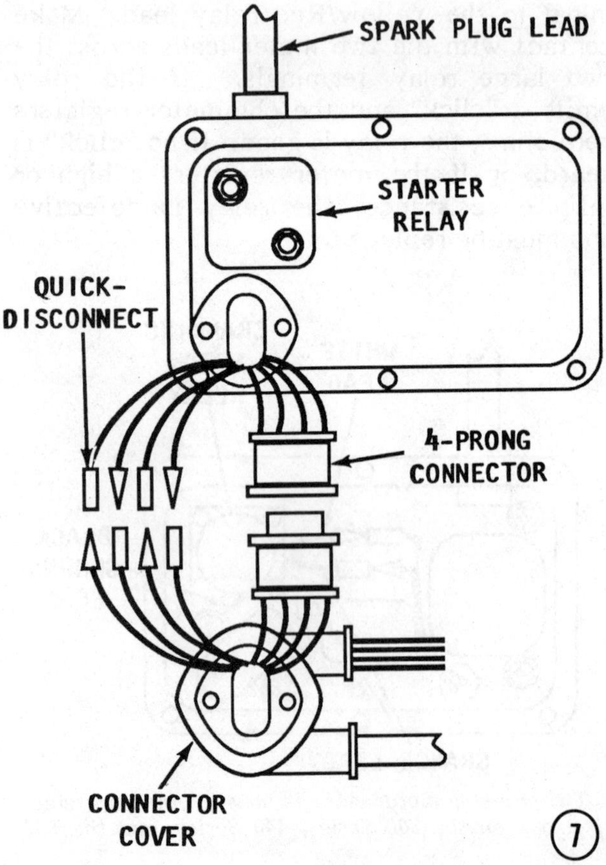

Install the following items over each of the relay terminals protruding from the box: large Black plastic insulator, large White plastic insulator, small Black plastic insulator, small metal washer and finally, the locknut. Tighten both locknuts securely. Connect the Orange and White leads -- or if working on a JetMate, the Yellow/Red, Red and Black leads, matching color-to-color and secure the relay Black ground lead under the rectifier/regulator mounting screw.

Electrical Box Installation
Models 300, 440 and 550

7- Secure the halves of the box with the two bolts and washers. Position the box against the engine compartment wall and secure it with the six bolts. Connect the two large cables to the relay terminals, one from the battery and one to the cranking motor. Cover the terminals with the boots, one on each cable.

Connect the two halves of the 4-prong connector. Connect the four free leads, matching color-to-color, at their quick-disconnect fittings. Install the connector cover over the electrical box and secure it in place with the two bolts.

Install the spark plug leads and connect the battery ground cable.

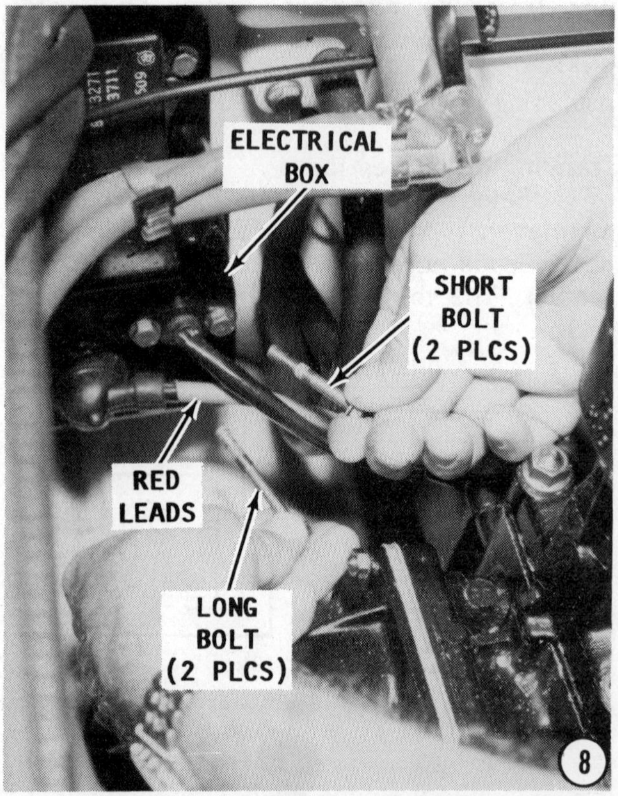

Electrical Box Installation
Model 650

8- Install the cover over the electrical box mounted on the engine compartment wall. Secure the cover to the box with the four bolts and washers. Notice there are two different length bolts, as shown. Connect the two large cables to the relay terminals, one from the battery and one to the cranking motor. Cover the terminals with the single large boot.

Install the spark plug leads and connect the battery ground cable.

Typical location for the electrical box which houses and protects the rectifier/regulator and the cranking motor relay. The box is watertight with grommets surrounding the harness leads where they enter the box.

5-6 CHARGING SYSTEM MALFUNCTIONS

The lighting coil, rectifier/regulator, battery, and the necessary wiring to connect it all together comprise the charging system.

Before the charging system is blamed for battery problems, consider other areas which may be the cause:

1- Operating at low speed for short periods.

2- Voltage losses due to high resistance.

3- Corroded battery cables, connectors, and terminals.

4- Low electrolyte level in the battery cells.

5- Prolonged disuse of the battery causing a self-discharged condition.

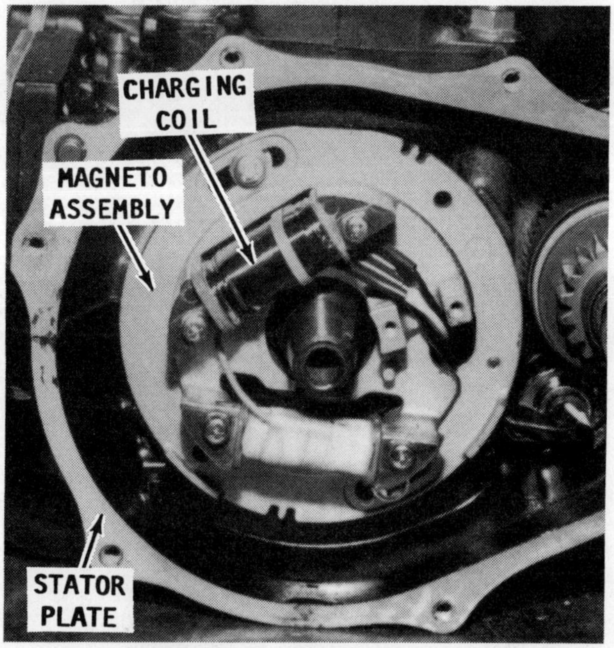

The charging coil is mounted on the magneto assembly. The magneto assembly is secured to the stator plate.

Service the battery, refer to Section 5-3. If, after charging, the battery fails to give the required specific gravity reading, then the battery will never hold a charge and must be replaced. This problem may have originated with the battery, a shorted cell perhaps, or may be attributed to the failure of the rectifier/regulator or charging coil. Refer to the Ignition chapter for resistance tests on these two components.

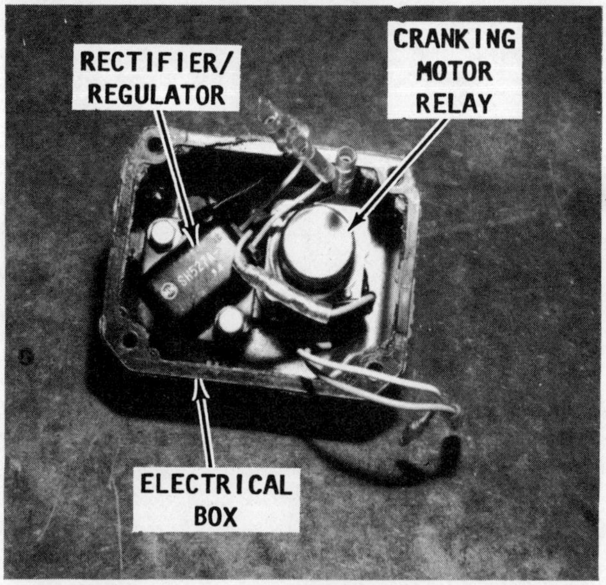

Opened electrical box exposing the rectifier/regulator and the cranking motor relay.

5-7 TROUBLESHOOTING CHARTS

The following six charts are presented as an aid to troubleshooting the engine and jet pump. Refer to the heading in the uppermost box which most closely identifies the problem, and then work downward from there. The probable causes are listed in columns under a specific system, for example: Fuel system, or Electrical system. Once a possible cause is identified, refer to the specific chapter for procedures on testing and service.

When using the charts -- proceed step-by-step through each of the tests or checks until the defect or cause of the problem is uncovered.

FIRST CHART -- Page 5-17 -- sets up the condition of the engine failing to start.

SECOND CHART -- Page 5-18 -- covers situation of engine starting, then stopping, without cause.

THIRD CHART -- Page 5-19 -- deals with an engine misfiring.

FOURTH CHART -- Page 5-20 -- most involved chart of the set -- covers loss of engine power.

FIFTH CHART -- Page 5-21 -- checks out strange unexplained engine noises.

SIXTH CHART -- Page 5-22 -- covers problems involving the jet pump.

ONE FINAL WORD

Before going out on the water, **TAKE TIME** to check and verify the drain plug is installed. Countless number of excursions have had a very sad beginning because the craft was eased into the water only to have the hull begin filling with water.

Keep your gas tank full, the fuel pump pumping, the spark plugs sparking, the charging circuit charging, and the pistons, well -- keep them working too.

Joan & Clarence Coles
and the crew
Seloc Publications

CHART ONE

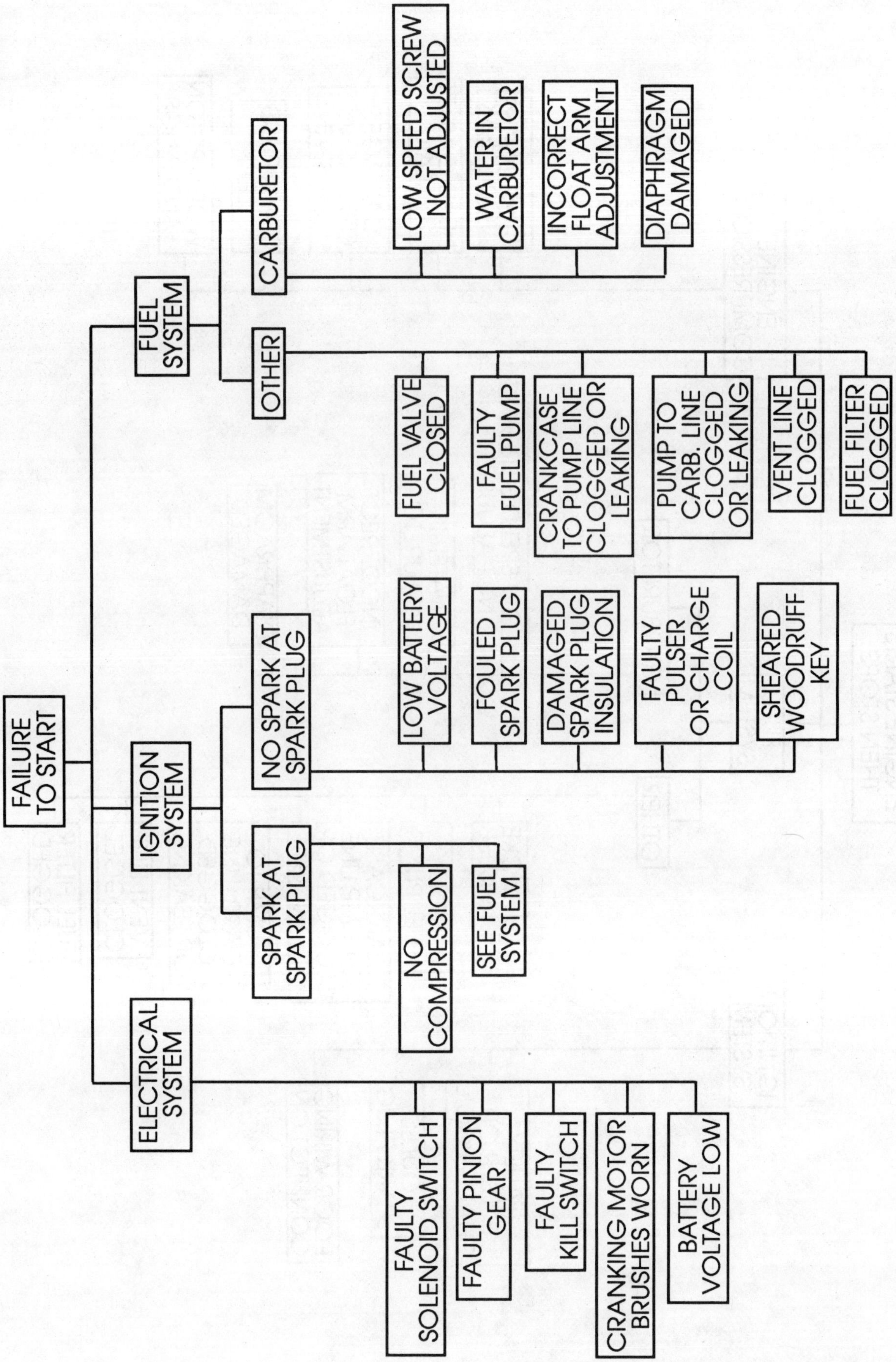
FAILURE TO START
FUEL SYSTEM
IGNITION SYSTEM
ELECTRICAL SYSTEM
CARBURETOR
OTHER
LOW SPEED SCREW NOT ADJUSTED
WATER IN CARBURETOR
INCORRECT FLOAT ARM ADJUSTMENT
DIAPHRAGM DAMAGED
FUEL VALVE CLOSED
FAULTY FUEL PUMP
CRANKCASE TO PUMP LINE CLOGGED OR LEAKING
PUMP TO CARB. LINE CLOGGED OR LEAKING
VENT LINE CLOGGED
FUEL FILTER CLOGGED
NO SPARK AT SPARK PLUG
LOW BATTERY VOLTAGE
FOULED SPARK PLUG
DAMAGED SPARK PLUG INSULATION
FAULTY PULSER OR CHARGE COIL
SHEARED WOODRUFF KEY
SPARK AT SPARK PLUG
NO COMPRESSION
SEE FUEL SYSTEM
FAULTY SOLENOID SWITCH
FAULTY PINION GEAR
FAULTY KILL SWITCH
CRANKING MOTOR BRUSHES WORN
BATTERY VOLTAGE LOW

CHART TWO

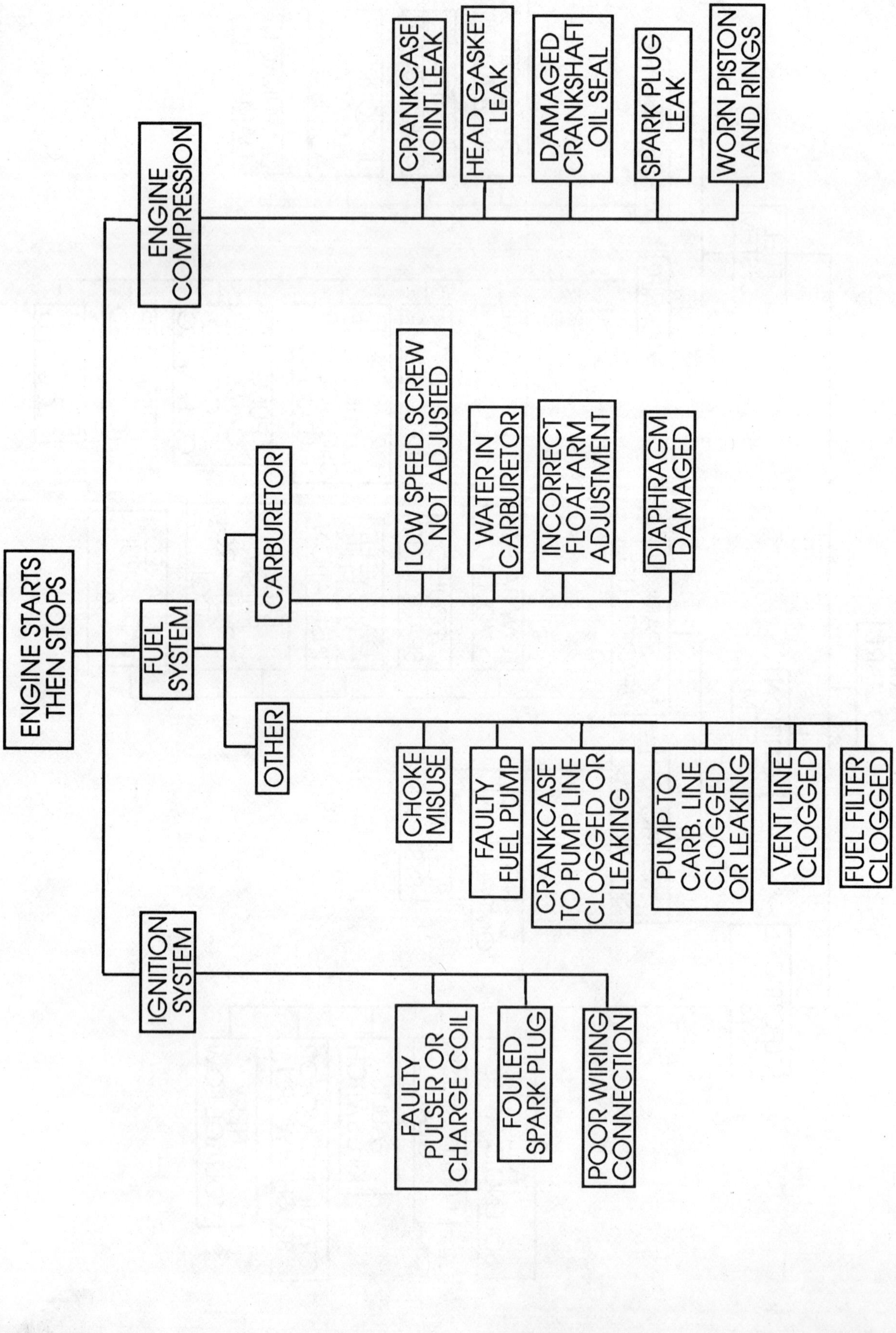
ENGINE STARTS THEN STOPS
ENGINE COMPRESSION
CRANKCASE JOINT LEAK
HEAD GASKET LEAK
DAMAGED CRANKSHAFT OIL SEAL
SPARK PLUG LEAK
WORN PISTON AND RINGS
FUEL SYSTEM
CARBURETOR
LOW SPEED SCREW NOT ADJUSTED
WATER IN CARBURETOR
INCORRECT FLOAT ARM ADJUSTMENT
DIAPHRAGM DAMAGED
OTHER
CHOKE MISUSE
FAULTY FUEL PUMP
CRANKCASE TO PUMP LINE CLOGGED OR LEAKING
PUMP TO CARB. LINE CLOGGED OR LEAKING
VENT LINE CLOGGED
FUEL FILTER CLOGGED
IGNITION SYSTEM
FAULTY PULSER OR CHARGE COIL
FOULED SPARK PLUG
POOR WIRING CONNECTION

CHART THREE

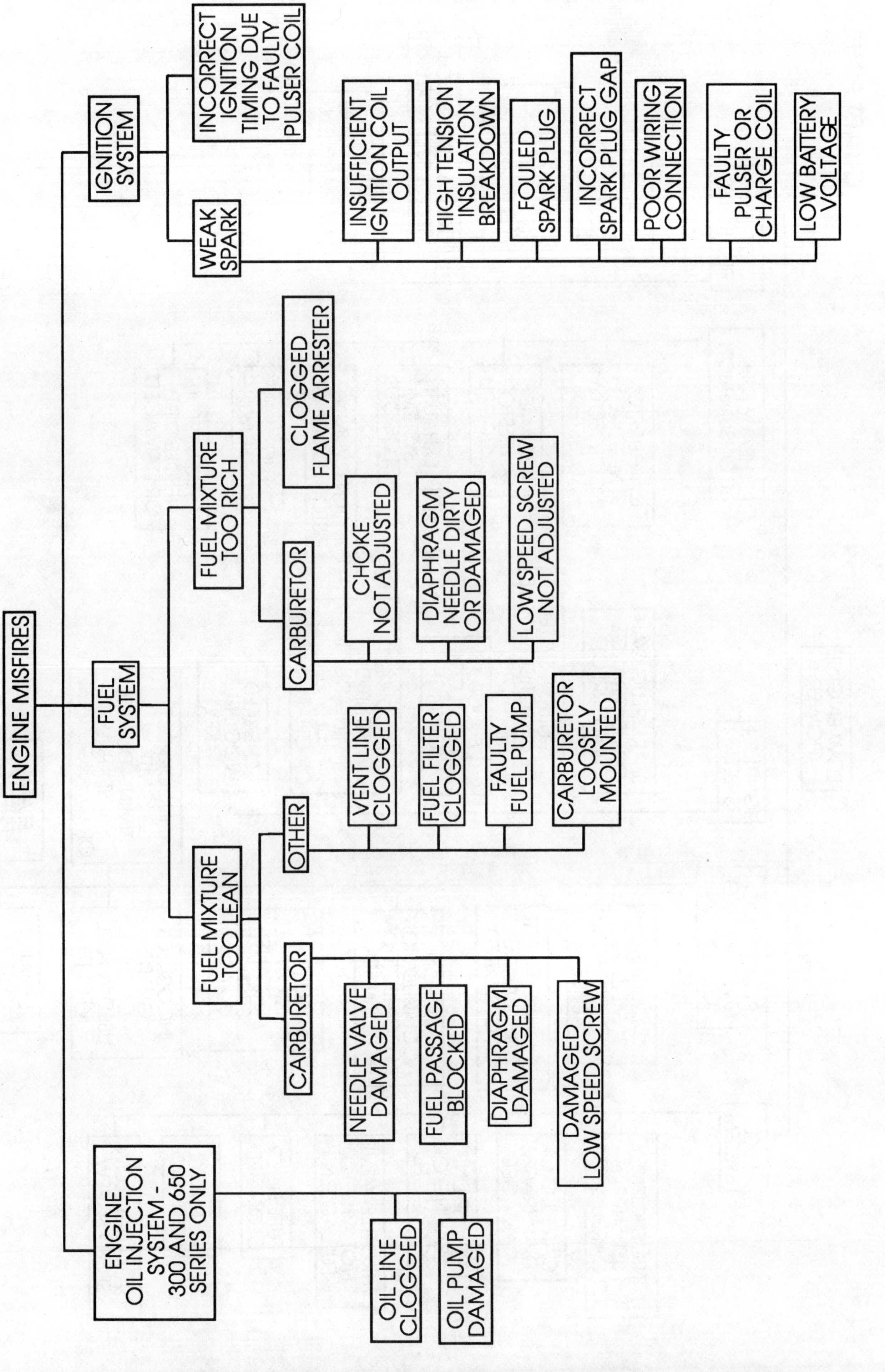
ENGINE MISFIRES
IGNITION SYSTEM
INCORRECT IGNITION TIMING DUE TO FAULTY PULSER COIL
WEAK SPARK
INSUFFICIENT IGNITION COIL OUTPUT
HIGH TENSION INSULATION BREAKDOWN
FOULED SPARK PLUG
INCORRECT SPARK PLUG GAP
POOR WIRING CONNECTION
FAULTY PULSER OR CHARGE COIL
LOW BATTERY VOLTAGE
FUEL SYSTEM
FUEL MIXTURE TOO RICH
CLOGGED FLAME ARRESTER
CARBURETOR
CHOKE NOT ADJUSTED
DIAPHRAGM NEEDLE DIRTY OR DAMAGED
LOW SPEED SCREW NOT ADJUSTED
FUEL MIXTURE TOO LEAN
OTHER
VENT LINE CLOGGED
FUEL FILTER CLOGGED
FAULTY FUEL PUMP
CARBURETOR LOOSELY MOUNTED
CARBURETOR
NEEDLE VALVE DAMAGED
FUEL PASSAGE BLOCKED
DIAPHRAGM DAMAGED
DAMAGED LOW SPEED SCREW
ENGINE OIL INJECTION SYSTEM - 300 AND 650 SERIES ONLY
OIL LINE CLOGGED
OIL PUMP DAMAGED

CHART FOUR

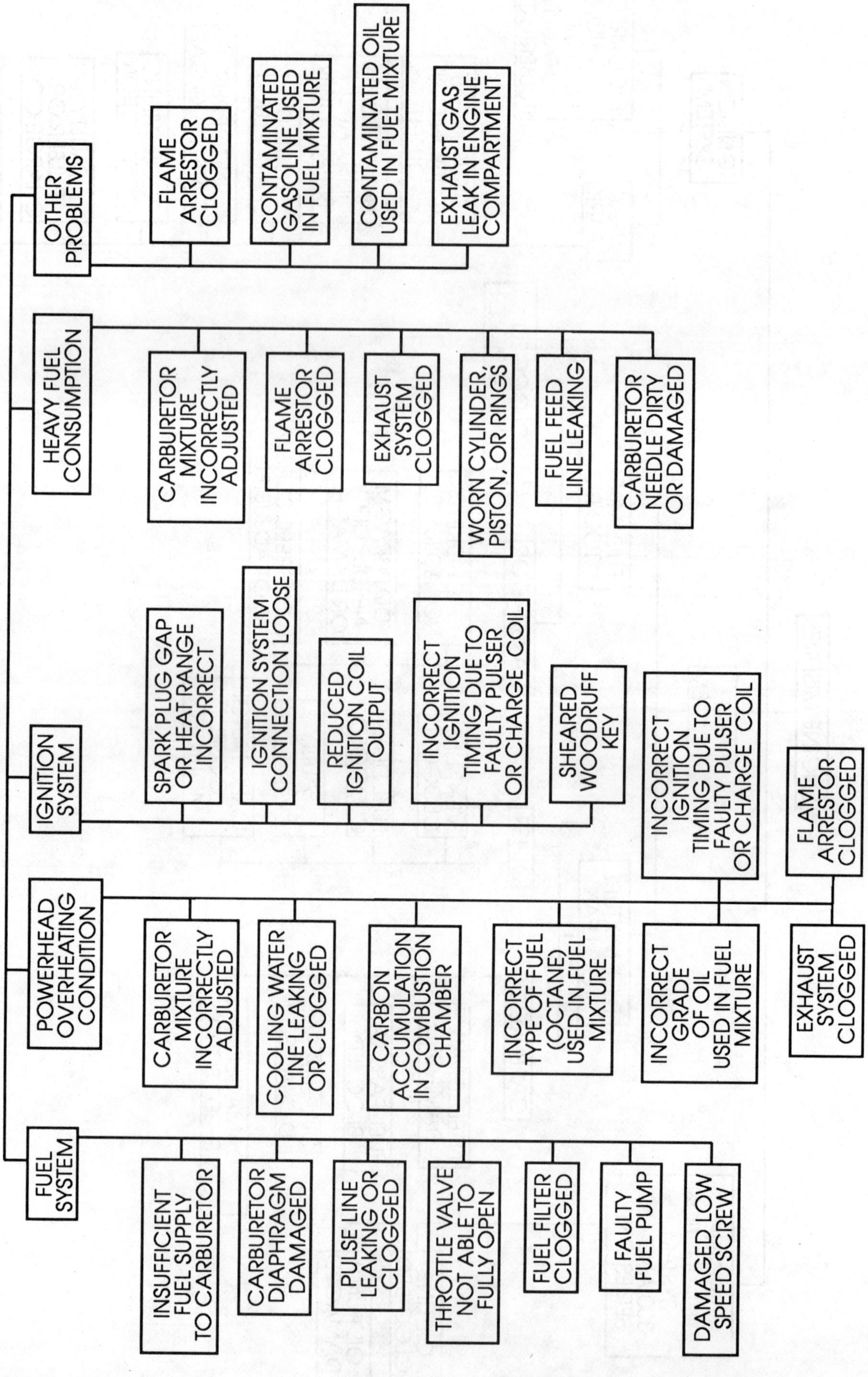
LOW ENGINE POWER
FUEL SYSTEM
INSUFFICIENT FUEL SUPPLY TO CARBURETOR
CARBURETOR DIAPHRAGM DAMAGED
PULSE LINE LEAKING OR CLOGGED
THROTTLE VALVE NOT ABLE TO FULLY OPEN
FUEL FILTER CLOGGED
FAULTY FUEL PUMP
DAMAGED LOW SPEED SCREW
POWERHEAD OVERHEATING CONDITION
CARBURETOR MIXTURE INCORRECTLY ADJUSTED
COOLING WATER LINE LEAKING OR CLOGGED
CARBON ACCUMULATION IN COMBUSTION CHAMBER
INCORRECT TYPE OF FUEL (OCTANE) USED IN FUEL MIXTURE
INCORRECT GRADE OF OIL USED IN FUEL MIXTURE
EXHAUST SYSTEM CLOGGED
INCORRECT IGNITION TIMING DUE TO FAULTY PULSER OR CHARGE COIL
FLAME ARRESTOR CLOGGED
IGNITION SYSTEM
SPARK PLUG GAP OR HEAT RANGE INCORRECT
IGNITION SYSTEM CONNECTION LOOSE
REDUCED IGNITION COIL OUTPUT
INCORRECT IGNITION TIMING DUE TO FAULTY PULSER OR CHARGE COIL
SHEARED WOODRUFF KEY
HEAVY FUEL CONSUMPTION
CARBURETOR MIXTURE INCORRECTLY ADJUSTED
FLAME ARRESTOR CLOGGED
EXHAUST SYSTEM CLOGGED
WORN CYLINDER, PISTON, OR RINGS
FUEL FEED LINE LEAKING
CARBURETOR NEEDLE DIRTY OR DAMAGED
OTHER PROBLEMS
FLAME ARRESTOR CLOGGED
CONTAMINATED GASOLINE USED IN FUEL MIXTURE
CONTAMINATED OIL USED IN FUEL MIXTURE
EXHAUST GAS LEAK IN ENGINE COMPARTMENT

CHART FIVE

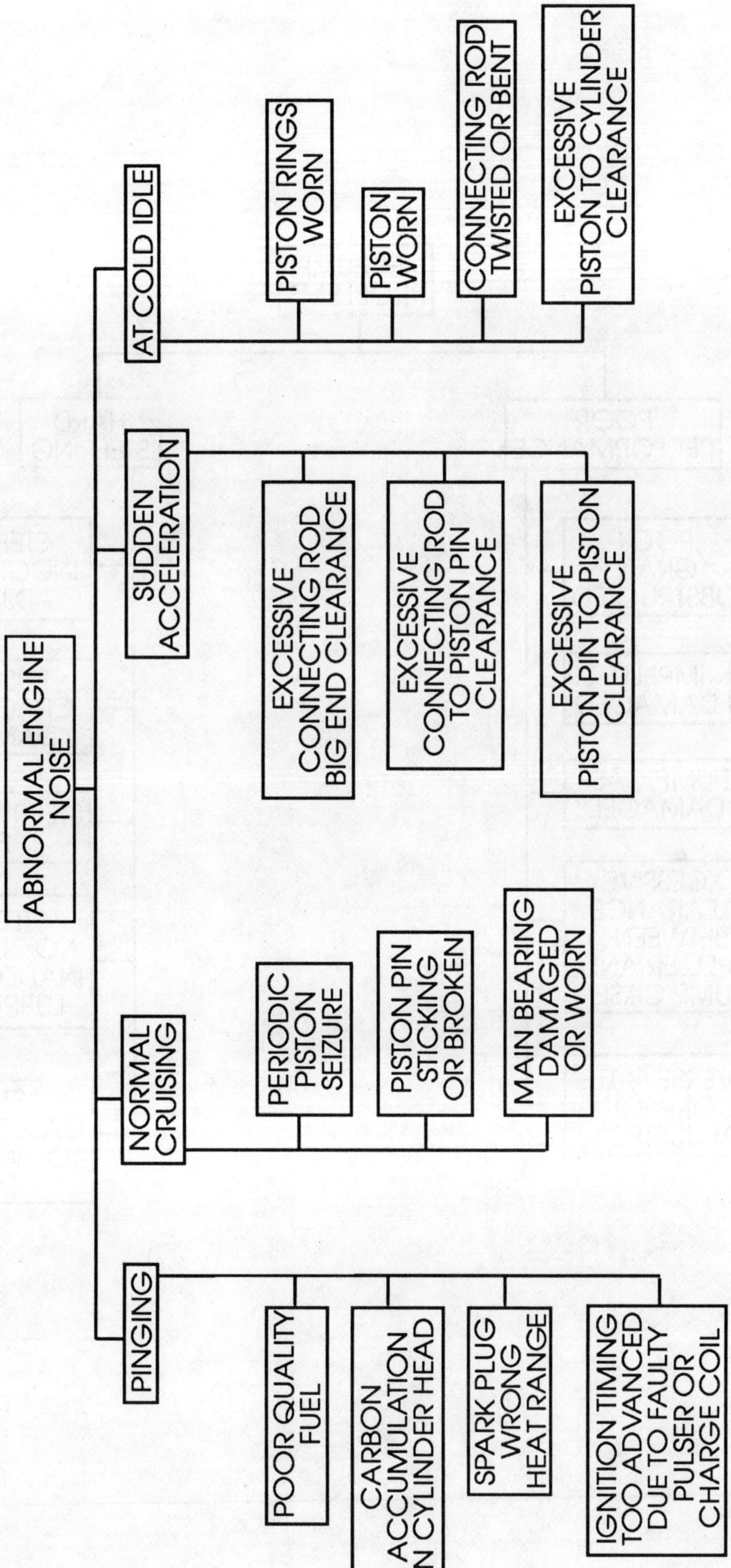
ABNORMAL ENGINE NOISE
AT COLD IDLE
PISTON RINGS WORN
PISTON WORN
CONNECTING ROD TWISTED OR BENT
EXCESSIVE PISTON TO CYLINDER CLEARANCE
SUDDEN ACCELERATION
EXCESSIVE CONNECTING ROD BIG END CLEARANCE
EXCESSIVE CONNECTING ROD TO PISTON PIN CLEARANCE
EXCESSIVE PISTON PIN TO PISTON CLEARANCE
NORMAL CRUISING
PERIODIC PISTON SEIZURE
PISTON PIN STICKING OR BROKEN
MAIN BEARING DAMAGED OR WORN
PINGING
POOR QUALITY FUEL
CARBON ACCUMULATION IN CYLINDER HEAD
SPARK PLUG WRONG HEAT RANGE
IGNITION TIMING TOO ADVANCED DUE TO FAULTY PULSER OR CHARGE COIL

CHART SIX

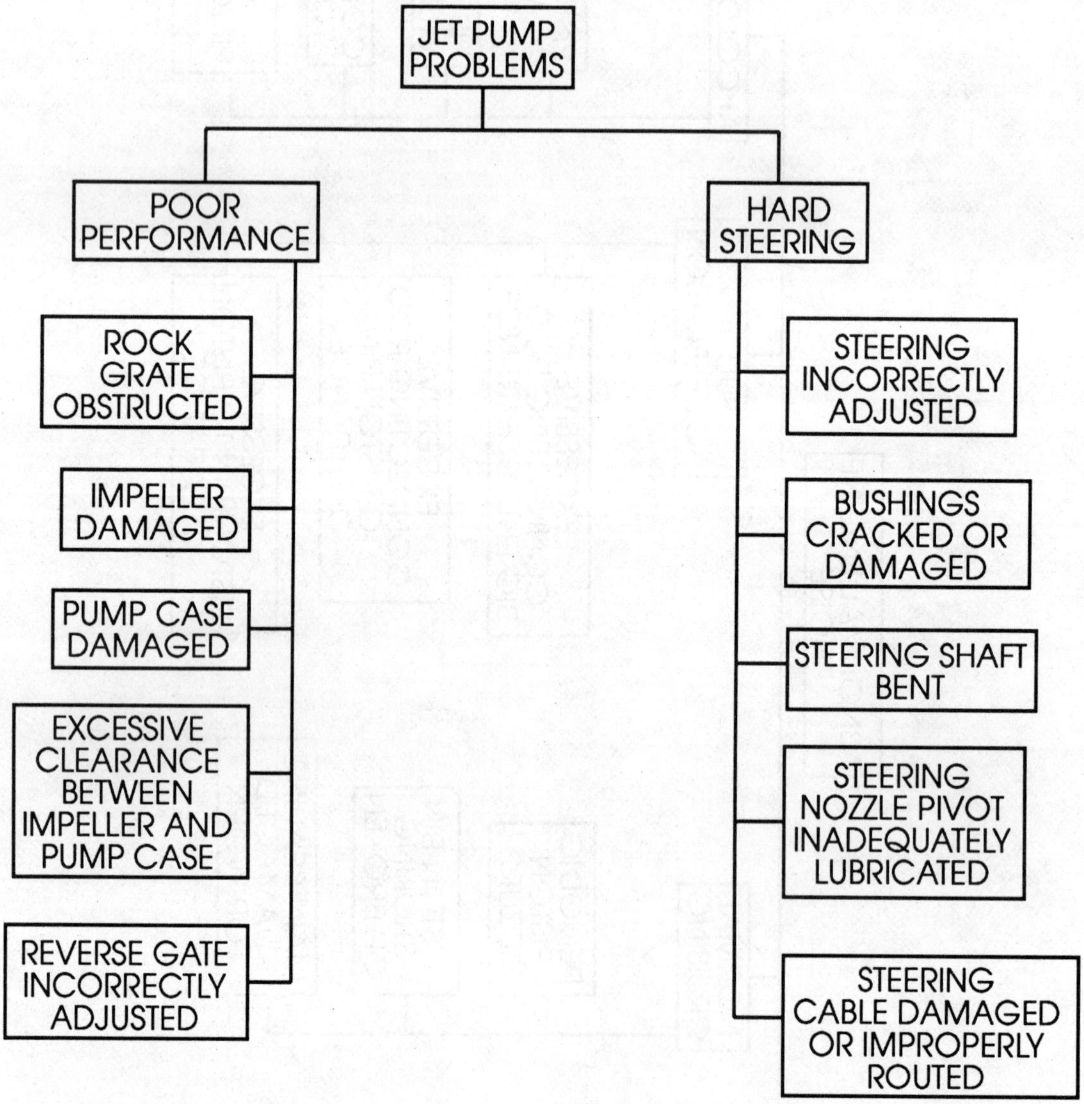
JET PUMP PROBLEMS
POOR PERFORMANCE
HARD STEERING
ROCK GRATE OBSTRUCTED
IMPELLER DAMAGED
PUMP CASE DAMAGED
EXCESSIVE CLEARANCE BETWEEN IMPELLER AND PUMP CASE
REVERSE GATE INCORRECTLY ADJUSTED
STEERING INCORRECTLY ADJUSTED
BUSHINGS CRACKED OR DAMAGED
STEERING SHAFT BENT
STEERING NOZZLE PIVOT INADEQUATELY LUBRICATED
STEERING CABLE DAMAGED OR IMPROPERLY ROUTED

6
FUEL

6-1 INTRODUCTION

The carburetion and ignition principles of two-cycle engine operation **MUST** be understood in order to perform a proper tune-up on a marine engine.

If you have any doubts concerning your understanding of two-cycle engine operation, it would be best to study the Introduction section in the first portion of Chapter 8, before tackling any work on the fuel system.

The fuel system includes the fuel tank, fuel pump, fuel filters, carburetor, and the associated parts to connect it all together. Regular maintenance of the fuel system to obtain maximum performance, is limited to changing the fuel filters at regular intervals and using fresh fuel.

If a sudden increase in gas consumption is noticed, or if the engine does not perform properly, a carburetor overhaul, including boil-out, or replacement of the fuel pump may be required.

6-2 GENERAL CARBURETION INFORMATION

The carburetor is merely a metering device for mixing fuel and air in the proper proportions for efficient engine operation. At idle speed, a marine engine requires a mixture of about 8 parts air to 1 part fuel. At high speed or under heavy duty service, the mixture may change to as much as 12 parts air to 1 part fuel.

DIAPHRAGM CARBURETORS

Description

The two carburetors used on Kawasaki personal watercraft are diaphragm type carburetors. One carburetor, used on Kawasaki 300, 440, 550 and 650 Series (1991 and on), is manufactured by Mikuni. This carburetor has an integral fuel pump. The second carburetor, used only on the 650 series -- prior to 1991 and all JetMates, was manufactured by Keihin and has a remote fuel pump.

A diaphragm carburetor has no float bowl or float. Instead, a diaphragm takes the place of the float.

A hinged arm, still called a float arm, rests against the diaphragm and opens and closes the needle and seat assembly.

Maintenance type service of the fuel system is limited to using fresh fuel; changing the fuel filter at regular intervals; and cleaning the flame arrestor screen.

The Mikuni carburetor, with an integral fuel pump, is used on the Kawasaki 300, 440, 550 Series and on the 650 Series after 1990.

The Keihin carburetor, with a remote fuel pump is used on the 650 Series prior to 1991 and on all Jet-Mates.

The remote fuel pump used with all Keihin carburetors on the series listed for the previous illustration.

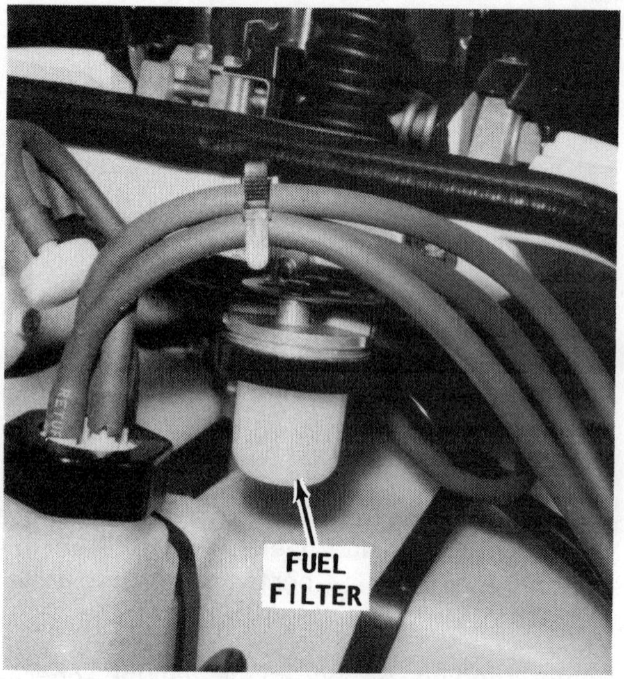

All Kawasaki Personal Watercraft series have some type of in-line fuel filter between the fuel tank and the fuel pump, as shown here.

Operation

Following the fuel through its course, from the fuel tank to the combustion chamber of the cylinder, will provide an appreciation of exactly what is taking place.

After the engine starts, the fuel passes through the pump to the carburetor. All systems have some type of filter installed somewhere in the line between the tank and the carburetor.

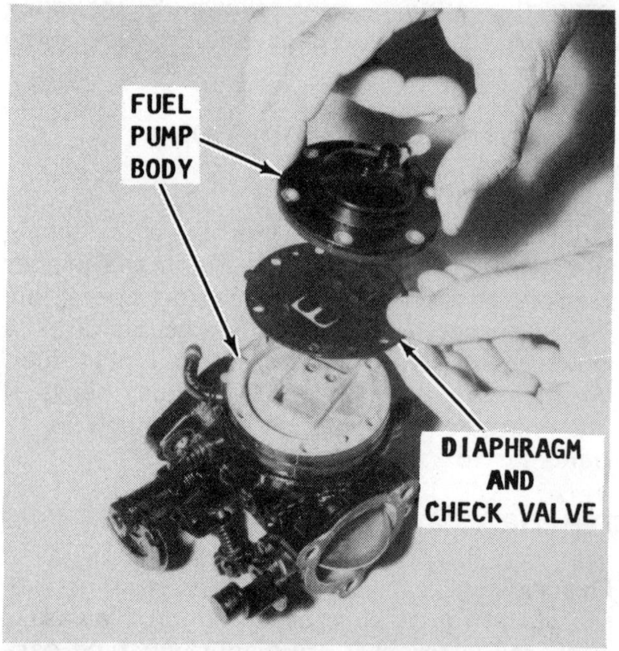

The integral fuel pump of a Mikuni carburetor opened to show some of the internal parts.

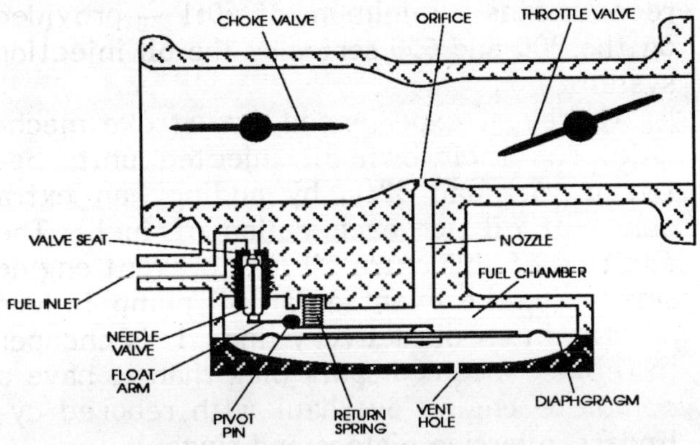

Cross-section drawing of a diaphragm type carburetor used on all series and model years covered in this manual.

Fuel is delivered to an inlet valve under pressure from the integral or remote fuel pump. When the choke valve is closed and the engine is cranked, or when the piston is on its upward stroke, the pressure in the carburetor throat orifice and the crankcase is less than atmospheric. The fuel chamber above the diaphragm is connected to the crankcase through a vacuum tube. A nozzle channel connecting the carburetor throat with the fuel chamber also allows the fuel chamber to be at less than atmospheric pressure.

The chamber on the other side of the diaphragm is vented to atmosphere, and therefore, is always at atmospheric pressure.

With the two pressure differentials, the diaphragm will flex toward the lowest pressure -- towards the fuel chamber. As the diaphragm flexes, it raises the float arm which, because of a return spring, is in constant contact with the diaphragm. The float arm is hinged. As one side goes up, the other side comes down and allows the inlet valve to open. At the same time, the needle valve falls away from the valve seat. As soon as the inlet valve opens, fuel flows under pressure from the fuel pump into the fuel chamber.

When the piston is on its downward stroke, the crankcase, the carburetor throat, and the fuel chamber are all at a higher pressure than atmosphere. The fuel chamber filled with fuel at a higher pressure now flexes toward the side with the lower pressure -- the side vented to atmosphere.

The return spring pushes the float arm down against the diaphragm. The other side of the float arm raises up and pushes the needle valve against its seat closing the inlet valve.

On the next piston downward stroke, the diaphragm flexes back again forcing the fuel in the fuel chamber to be ejected through a jet into the carburetor throat. On the same stroke the fuel chamber is replenished with fresh fuel through the open needle valve. This cycle repeats itself, sending a squirt of fuel into the carburetor throat each downward stroke of the piston.

In order to obtain the proper air/fuel mixture for all engine speeds, some models have high and low speed jets. These jets have adjustable needle valves which are used to compensate for changing atmospheric conditions. In almost all cases, the high-speed circuit has fixed high-speed jets and are not adjustable.

A throttle valve controls the flow of air/fuel mixture drawn into the combustion chambers. A cold engine requires a richer fuel mixture to start and during the brief period it is warming to normal operating temperature. A choke valve is placed ahead of the metering jets and venturi. As this valve begins to close, the volume of air intake is reduced, thus enriching the mixture entering the cylinders.

When this choke valve is fully closed, a very rich fuel mixture is drawn into the cylinders.

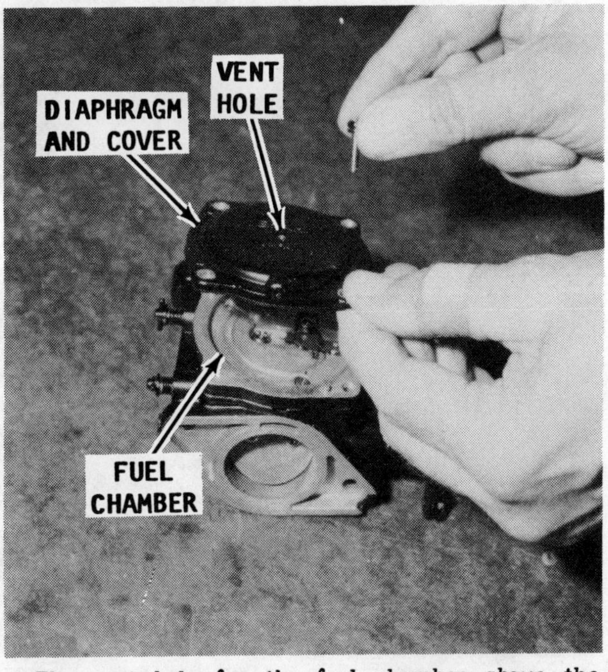

The vent hole for the fuel chamber above the diaphragm must be kept clear to the atmosphere at all times for the diaphragm to function efficiently.

Fuel/Oil Mixture

The 300 and 650 series Kawasaki personal watercraft are equipped with oil injection. The 440 and 550 series operate on a pre-mix of 40:1.

A new or rebuilt engine must be properly "broken-in" if it is to develop maximum performance. An engine which is incorrectly "broken-in" will never be able to achieve the maximum rated horsepower. During the initial "breaking-in" period, the fuel/oil ratio should be very rich, about 20:1 for the first two to four tank fulls -- twenty parts of fuel to one part oil.

Two-stroke oil must be added to the fuel tank at the ratio of 25:1 for the first ten hours of operation after the unit is brought out of storage. This ratio will **ENSURE** adequate lubrication of moving parts which have been drained of oil during the storage period. Operation of an engine without the proper amount of oil will quickly lead to a significant loss of horsepower.

After the first ten hours of operation, a lower ratio of fuel/oil mixture may be used. In the past, manufacturers of two-stroke engines gave a minimum of 100:1. Kawasaki recommends a minimum of 50:1 -- provided on the 300 and 650 series by the oil injection system.

However, experienced two-stroke mechanics run their own oil injected units between 16:1 and 32:1, by adding an extra ounce of oil per each gallon of fuel. The addition of this extra oil may prevent engine seizure in the event of an oil pump failure or clog in an oil delivery line. It is cheaper to replace a fouled spark plug than to have a complete engine overhaul with rebored cylinders, oversize pistons and rings.

Use only two-stroke outboard marine oil in the mixture -- never automotive oils. Four-stroke automotive engine oil is not designed to burn completely -- only to lubricate. Therefore, automotive oil, if used in a two-stroke engine, will leave an undesirable residue in the combustion chamber.

The oil injection system is covered in Section 6-10, beginning on Page 6-49.

Air/Fuel Mixture

A suction effect is created in the intake manifold each time the piston moves upward

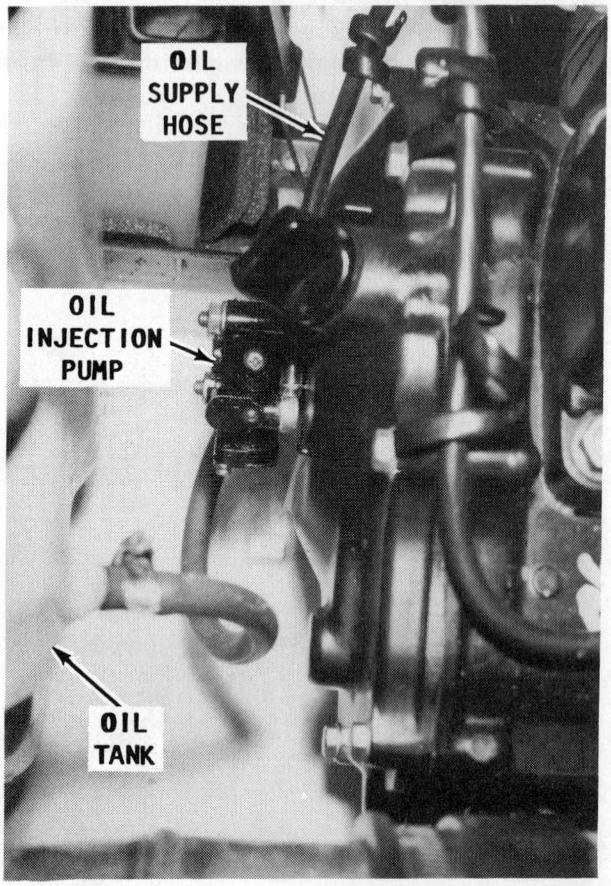

Close view of the oil injection system on a 650 Series engine. The system on a 300 Series engine is almost identical.

Oil is injected through a nozzle into the intake manifold under the carburetor. Here, the carburetor has been removed to show the oil injection nozzle.

in the cylinder. This suction draws air through the throat of the carburetor. A restriction in the throat, called a venturi, controls air velocity and has the effect of reducing air pressure at this point.

The difference in air pressures at the throat and in the fuel chamber, causes the fuel to be pushed out of a metering jet extending down into the fuel chamber. When the fuel leaves the jet, it mixes with the air passing through the venturi. This air/fuel mixture should then be in the proper proportion for burning in the cylinders for maximum engine performance.

Throttle and Choke Valves

A throttle valve controls the flow of air/fuel mixture drawn into the combustion chambers. A cold engine requires a richer fuel mixture to start and during the brief period it is warming to normal operating temperature. A choke valve is placed ahead of the metering jets and venturi. As this valve begins to close, the volume of air drawn in is reduced, thus enriching the mixture entering the cylinders.

When this choke valve is fully closed, a very rich fuel mixture is drawn into the cylinders aiding engine start.

Engine Revolution Limiter 550 Series Only Prior to 1990

Up to 1990, the manufacturer used a mechanical fuel revolution limiter. This device was connected to a relay inside the electrical box. The return fuel line from the carburetor passed through a solenoid activated diverter valve. At engine rpm above 7,800, a signal from the electrical box was sent to the solenoid which then activated the valve and mechanically blocked off the returning fuel from the carburetor to the fuel tank. This fuel was redirected back to the carburetor. This "excess" fuel resulted in an overrich mixture which slowed engine rpm. Such an arrangement was easy to bypass -- by simply unplugging the solenoid. Therefore, in 1990 engineers changed design and operating principle.

After 1990, an electronic rev limiter was used on all 550 series models. This type limiter was, as the factory intended, much more difficult to bypass. No!, we are not going to tell you how to do it.

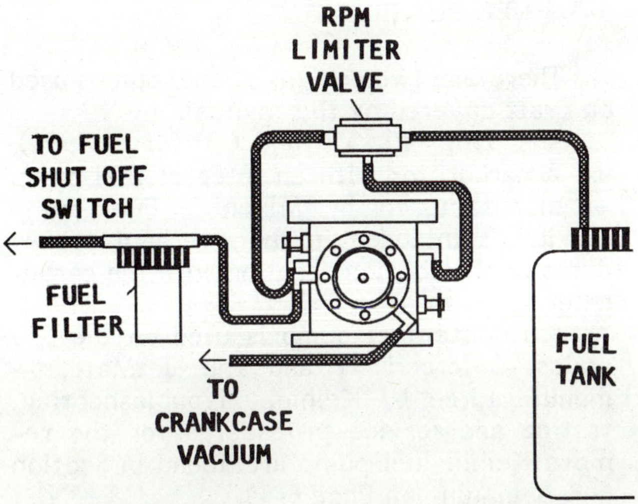

Functional diagram to indicate operation of the mechanical fuel revolution limiter described in the text. This arrangement was used on the 550 Series prior to 1991.

Engine Revolution Limiter 300, 440, and 650 Series — All Model Years 550 Series Only After 1990

The rev limiter on the 300, 440, and 650 Series — all model years, and the 550 Series after 1990 is an integral part of the CDI igniter. Any attempt to bypass the limiter could have serious and expensive consequences.

Fuel Filter and Sediment Bowl

Some models are equipped with a fuel filter and sediment bowl mounted inside the engine compartment.

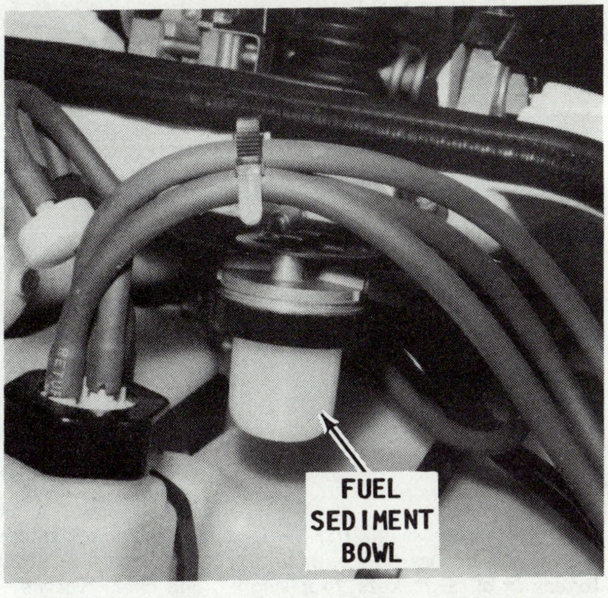

This fuel filter and sediment bowl installation on a 650 Series engine is typical for other series engines.

6-3 FUEL PUMP

There are two designs of fuel pumps used on craft covered by this manual.

The 300, 440, 550 and 650 (after 1991), use a carburetor with an integral fuel pump -- manufactured by Mikuni. Fuel pump service is included in the overhaul procedures in the sections dealing with the carburetor.

A remote fuel pump is used on the 650 series, before 1991 and the JetMate, -- manufactured by Keihin. Troubleshooting, testing and service procedures for the remote Keihin fuel pump are found in Section 6-9, beginning on Page 6-44.

Integral Fuel Pump Description and Operation

The fuel pump consists of a series of diaphragms and check valves operated by crankcase pressure. The pump feeds fuel directly into the fuel chamber through the inlet valve.

As the piston moves upward, a suction, or negative pressure, is created in the crankcase. This low pressure acts upon the inner diaphragm through the carburetor. The inner diaphragm is, therefore, pulled inward. When the diaphragm moves inward,

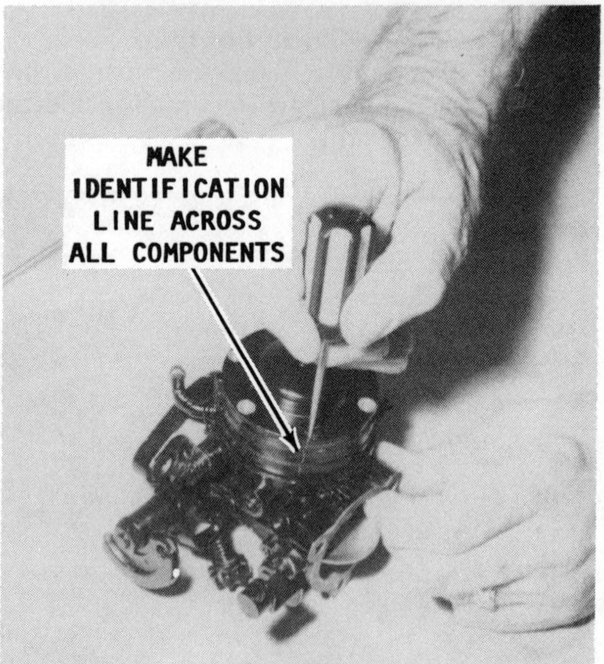

*Prior to disassembling the fuel pump, a mark should be scribed across all components of the pump -- eight layers in all --gaskets, diaphragms, and check valves -- to **ENSURE** all parts are assembled back in their original positions.*

the check valve opens and fuel is drawn from the bottom of the fuel pump exterior chamber into the fuel pump interior chamber.

As the piston moves downward, a positive pressure is created in the crankcase causing the inner diaphragm to be pushed outward. The delivery check valve opens under pressure and allows fuel to flow into the upper part of the exterior chamber and on to the inlet valve of the carburetor.

The function of the outer diaphragm is to absorb the pulsations of the fuel and allow a smooth uninterrupted fuel flow.

If the fuel pump diaphragm tears or is punctured, fuel will enter the crankcase through the pulse line causing an overrich mixture. This mixture will foul the plugs and in extreme cases flood the crankcase.

6-4 FUEL IN THE SYSTEM

The fuel system includes the fuel tank, fuel pump, fuel filters, carburetor, connecting fuel lines, and the associated parts to connect it all together. Regular maintenance of the fuel system to obtain maximum performance, is limited to changing the fuel filters at regular intervals and using fresh fuel.

Even with the high price of fuel, removing gasoline that has been standing unused over a long period of time, is still the easiest and least expensive preventive maintenance possible. In most cases this old gas, even with some oil mixed with it, can be used without harmful effects in an automobile using regular gasoline.

If a sudden increase in gas consumption is noticed, or if the engine does not perform properly, a carburetor overhaul, including boil-out, or replacement of the fuel pump may be required.

Leaded Gasoline

The manufacturers of the units covered in this manual recommends the engines be operated using either regular unleaded or regular leaded gasoline with a minimum octane rating of 84 or higher.

In the United States, the Environmental Protection Agency (EPA) has slated a proposed national phase-out of leaded fuel, "Regular" gasoline, by 1993. Lead in gasoline boosts the octane rating (energy). Therefore, if the lead is removed, it must be replaced with another agent. Unknown to the general public, many refineries are adding alcohol in an effort to hold the octane rating.

Alcohol in gasoline can have a deteriorating effect on certain fuel system parts. Seals can swell, pump check valves can swell, diaphragms distort, and other rubber or neoprene composition parts in the fuel system can be affected.

Since about 1981, every effort has been made to use materials that will resist the alcohol being added to fuels.

Fuels containing alcohol will slowly absorb moisture from the air. Once the moisture content in the fuel exceeds about 1%, it will separate from the fuel taking the alcohol with it. This water/alcohol mixture will settle to the bottom of the fuel tank. The engine will fail to operate. Therefore, storage of this type of gasoline for use in marine engines is not recommended for more than just a few days.

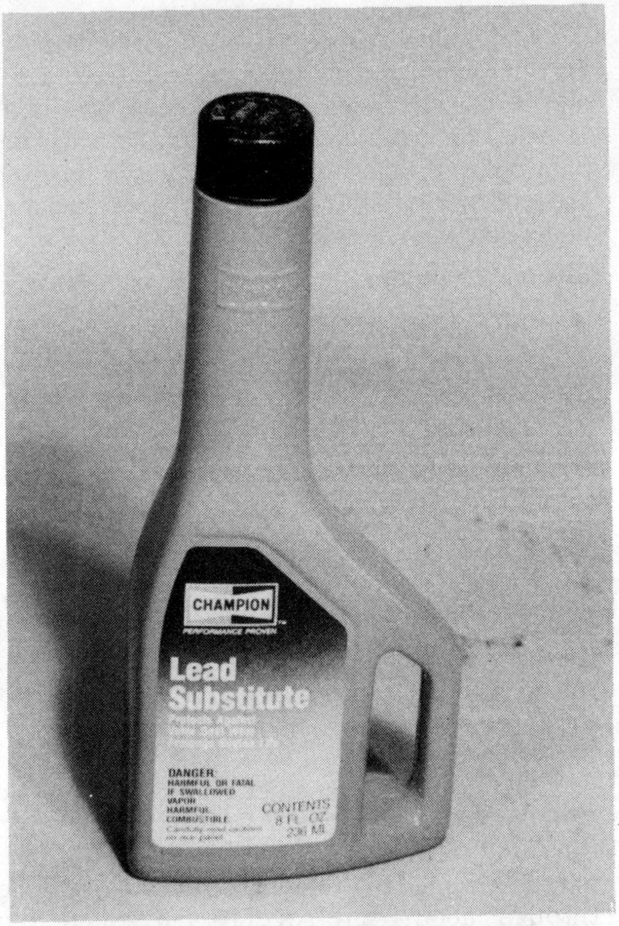

A lead substitute additive -- octane booster -- can help prevent detonation when unleaded gasoline is used in a personal watercraft.

Removing Fuel from the System

For many years there has been the widespread belief that simply shutting off the fuel at the tank and then running the engine until it stops is the proper procedure before storing the engine for any length of time. Right? **WRONG.**

It is **NOT** possible to remove all of the fuel in the carburetor by operating the engine until it stops. Some fuel is trapped in passages. The **ONLY** guaranteed method of removing **ALL** of the fuel is to take the time to remove the carburetor, and drain the fuel.

If the engine is operated with the fuel supply shut off until it stops the fuel and oil mixture inside the engine is removed, leaving bearings, pistons, rings, and other parts with little protective lubricant, during long periods of storage.

Proper procedure involves: Shutting off the fuel supply at the tank; disconnecting the fuel line at the tank; operating the engine until it begins to run **ROUGH;** then stopping the engine, which will leave some fuel/oil mixture inside; and finally removing and draining the carburetor. By disconnecting the fuel supply, all **SMALL** passages are cleared of fuel even though some fuel is left in the carburetor. A light oil should be put in the combustion chamber as instructed in the Owner's Manual.

For short periods of storage, simply running the carburetor dry may help prevent severe gum and varnish from forming in the carburetor. This is especially true during hot weather.

6-5 TROUBLESHOOTING

The following paragraphs provide an orderly sequence of tests to pinpoint problems in the fuel system. It is very rare for the carburetor by itself to cause failure of the engine to start.

FUEL PROBLEMS

Many times fuel system troubles are caused by a plugged fuel filter, a defective fuel pump, or by a leak in the line from the fuel tank to the fuel pump. A defective choke may also cause problems. **WOULD YOU BELIEVE,** a majority of starting troubles which are traced to the fuel system are the result of an empty fuel tank, clogged fuel filter, or aged "sour" fuel.

Fuel Filter and Sediment Bowl Service

Some models are equipped with a fuel filter and sediment bowl mounted inside the engine compartment.

WORDS OF CAUTION

Be sure to loosen the fuel filler cap on the fuel tank **BEFORE** removing the sediment bowl to relieve any pressure built up inside the fuel system.

Have a shop cloth handy to catch any spilled fuel. Unthread the ring nut and pull down the sediment bowl with the fuel filter inside. A large **O**- ring seals the bowl to the filter cap and ring nut. The function of this **O**-ring is to prevent fuel and air leaks. If the **O**-ring is defective, the fuel pump will draw air into the system and the engine will either not start, faulter, or die. Sometimes, this problem is not apparent with a full fuel tank, because the fuel is gravity fed to the carburetor. When the fuel level drops in the tank, the pump takes over, but its effectiveness is impaired by the air lead at the sediment bowl **O**-ring.

STA-BIL gasoline Stabilizer and Conditioner -- or equal substitute — may be used to prevent fuel from "souring" for up to twelve full months.

Clean out the sediment bowl and rinse the filter screen in solvent every 25 hours of engine operation.

"Sour" Fuel

Under average conditions (temperate climates), fuel will begin to breakdown in about four months. A gummy substance forms in the bottom of the fuel tank and in other areas. The filter screen between the tank and the carburetor and small passages in the carburetor will become clogged. The gasoline will begin to give off an odor similar to rotten eggs. Such a condition can cause the owner much frustration, time in cleaning components, and the expense of replacement or overhaul parts for the carburetor.

Even with the high price of fuel, removing gasoline that has been standing unused over a long period of time is still the easiest and least expensive preventative maintenance possible. In most cases, this old gas can be used without harmful effects in an automobile using regular gasoline.

A gasoline preservative additive, such as Stabil Fuel Conditioner and Stabilizer for 2 and 4-Cycle Engines, will keep the fuel "fresh" for up to twelve months. If this particular product is not available in your area, other similar additives are produced under various trade names.

Choke Problems

When the engine is hot, the fuel system can cause starting problems. After a hot engine is shut down, the temperature inside the fuel bowl may rise to 200°F and cause the fuel to actually boil. All carburetors are vented to allow this pressure to escape to the atmosphere. However, some of the fuel may percolate over the high-speed nozzle.

If the choke should stick in the open position, the engine will be hard to start. If the choke should stick in the closed position, the engine will flood making it very difficult to start.

In order for this raw fuel to vaporize enough to burn, considerable air must be added to lean out the mixture. Therefore, the only remedy is to remove the spark plugs; ground the leads; crank the engine through about ten revolutions; clean the plugs; install the plugs again; and start the engine.

Rough Engine Idle

If an engine does not idle smoothly, the most reasonable approach to the problem is to perform a tune-up to eliminate such areas as: fouled or faulty spark plugs.

Other problems that can prevent an engine from running smoothly include: An air leak in the intake manifold; uneven compression between the cylinders; and sticky or broken reeds.

Of course any problem in the carburetor affecting the air/fuel mixture will also prevent the engine from operating smoothly at idle speed. These problems usually include: a weak float arm return spring; leaking needle valve and seat; defective choke; and improper adjustments for idle mixture or idle speed.

Excessive Fuel Consumption

Excessive fuel consumption can be the result of any one of two conditions, or both may apply.

1- Inefficient engine operation.

2- Poor boating habits of the operator.

If the fuel consumption suddenly increases over what could be considered normal, then the cause can probably be attributed to the engine or watercraft and not the operator.

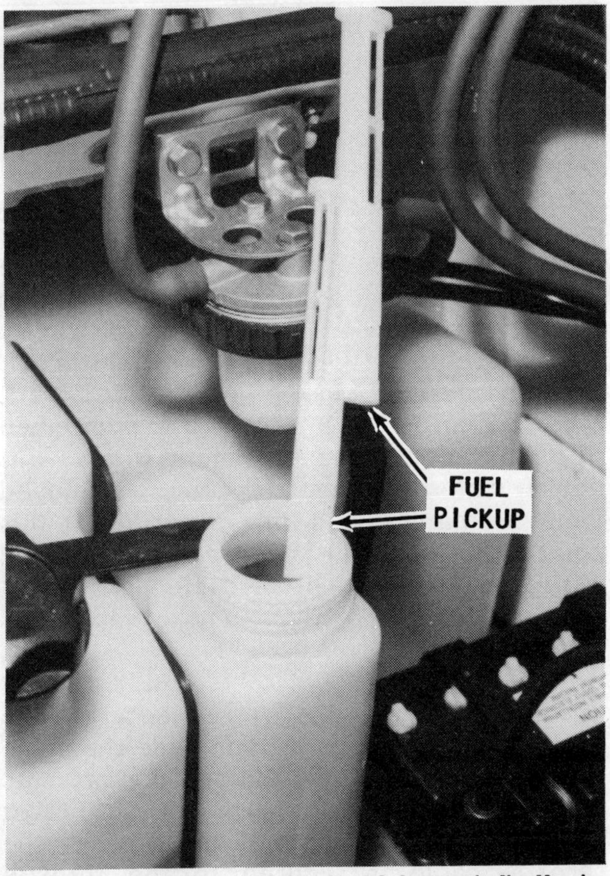

The fuel pick-up tubes should be periodically inspected and cleaned.

Check the fuel system for possible leaks. A leak between the tank and the pump many times will not appear when the engine is operating, because the suction created by the pump drawing fuel will not allow the fuel to leak. Once the engine is turned off and the suction no longer exists, fuel may begin to leak.

If a minor tune-up has been performed and the condition still exists, then the problem most likely is in the carburetor and an overhaul is in order.

Engine Surge

If the engine operates as if the load on the craft is being constantly increased and decreased, even though an attempt is being made to hold a constant engine speed, the problem can most likely be attributed to the fuel pump, or a restriction in the fuel line between the tank and the carburetor.

6-6 CARBURETOR MODELS

Two diaphragm type carburetors, one with an integral fuel pump, and one with a remote fuel pump, are used by all Kawasaki personal watercraft.

Mikuni Carburetor

The single cylinder 300 series engine, the twin cylinder 440, and 550 series engines, and the 650 series (1991 and on), not including the JetMate, are equipped with the following Mikuni carburetors:

BN34 bore -- 28mm venturi
BN38 bore -- 34mm venturi
BN44 bore -- 40mm venturi

All these Mikuni carburetors are almost identical except for the diameter of the venturi and those installed on the 550 series engines have a fourth fuel fitting for the rpm limiter function of the carburetor. The following procedures are valid for all the listed Mikuni carburetors.

Complete detailed procedures for removal, disassembly, cleaning and inspecting, assembling including bench adjustments, installation, and operating adjustments for the Mikuni carburetor are outlined in Section 6-7.

The Mikuni carburetor, with an integral fuel pump, is used on the Kawasaki 300, 440, 550 Series and on the 650 Series after 1990.

The Keihin carburetor, with a remote fuel pump is used on the 650 Series prior to 1991 and on all Jet-Mates.

Keihin Carburetor

The twin cylinder 650 series engine -- prior to 1991, including the JetMate, are equipped with a Keihin carburetor -- the CDK34 with a 28mm venturi.

Complete detailed procedures for removal, disassembly, cleaning and inspecting, assembling including bench adjustments, installation, and operating adjustments for the Keihin carburetor installed on all 650 series units prior to 1991, including the JetMate, are presented in Section 6-8.

Complete detailed procedures for removal, disassembly, cleaning and inspecting, assembling and installation of the remote fuel pump installed on all 650 series units prior to 1991, including the JetMate, are presented in Section 6-9.

6-7 SERVICE MIKUNI CARBURETOR

This section provides complete detailed procedures for removal, disassembly, cleaning and inspecting, assembling including bench adjustments, installation, and operating adjustments for the Mikuni carburetor installed on the single cylinder 300 series engine, the twin cylinder 440, and 550 series engines, and the twin cylinder 650 (1991 and on).

REMOVAL AND DISASSEMBLING

Good shop practice dictates a carburetor repair kit be purchased and new parts be installed any time the carburetor is disassembled.

Make an attempt to keep the work area organized and to cover parts after they have been cleaned. This practice will prevent foreign matter from entering passageways or adhering to critical parts.

Throttle and Choke Linkage Disconnect
First these words:

Observe the method used in securing the throttle or choke cable end to the shaft. If the cable is threaded through a hole in a barrel, and then the end anchored in place with a small screw through the barrel, the cable must be marked at the point it enters the barrel, to ensure installation in exactly the same position. Obtain some "whiteout", or some masking tape, or equivalent. Mark the point on the cable at which it is inserted into the barrel. Make a mark which will not be disturbed when the cable is removed from the barrel. Marking the cable at this point will ensure the cable can be installed with the same amount of cable tension and with the same throttle or choke valve settings.

1- Loosen the small screw securing the choke cable to the barrel and slide the cable end free of the barrel. Loosen the two nuts securing the cable to the mounting bracket and pull the cable free of the slot. If the throttle cable is secured to the throttle shaft in the same manner, loosen the small screw on the barrel and slide the throttle cable end free. If the throttle cable lies over a curved guide, loosen the two nuts, one on either side of the bracket, and slide the cable free of the bracket. Unhook the cable end from the guide.

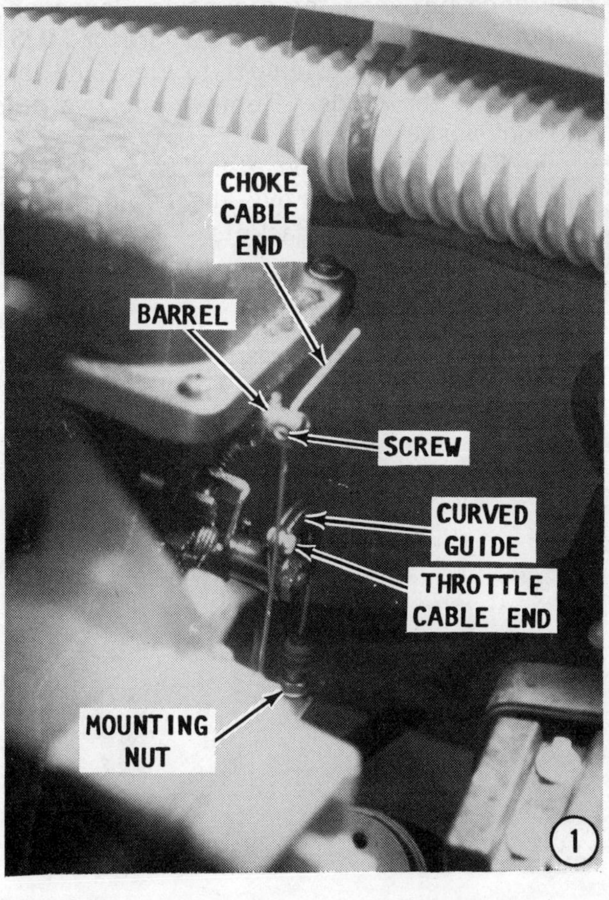

1

Fuel and Vacuum Hose Disconnect

2- If working on a 300, 440 or 650 series: Identify the **THREE** hoses at the carburetor.

The first hose -- attached to the plastic fitting at the outer end of the fuel pump --is the fuel supply hose. The other end of this hose is attached to an in-line fuel filter mounted on the engine compartment wall.

The second hose -- attached to the lower part of the pump -- is a vacuum hose. See illustration **2A.** The other end of this hose is attached to the crankcase.

The third hose -- attached to the carburetor body -- is an excess fuel return hose. See illustration **2B.** This hose leads back to the fuel tank.

Mark the three hoses and their respective fittings on the carburetor and fuel pump **BEFORE** they are disconnected.

Take time to plug the disconnected ends of all three hoses with a golf tee, pencil end, or similar item to prevent contaminants from entering the fuel system.

2A

If working on a 550 series: Identify the **FOUR** hoses at the carburetor, as identified in the accompanying line drawing.

The first hose -- attached to the plastic fitting at the outer end of the fuel pump --is the fuel supply hose. The other end of this hose is attached to an in-line fuel filter mounted on the engine compartment wall.

The second hose -- attached to the lower part of the pump -- is a vacuum hose. The other end of this hose is attached to the crankcase.

The third hose -- attached to the left of the fuel pump is the excess fuel return hose.

2B

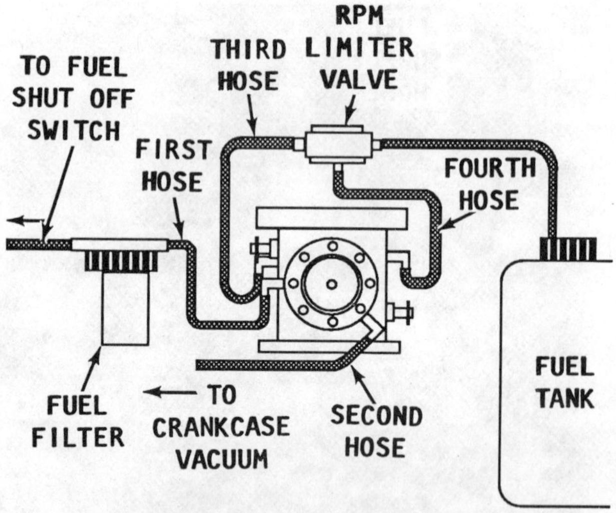

Functional line drawing to graphically identify the first hose, second hose, third hose, and fourth hose, described in the text.

The other end of this hose is connected to the rpm limiter. The fuel is then returned to the fuel tank. Disconnect this hose at the fitting on the carburetor.

The fourth hose -- attached to the fitting on the right of the fuel pump is the actual rpm limiter hose. The other end of this hose is attached to the fuel limiter valve at the center fitting.

Mark the **FOUR** hoses and their respective fittings on the carburetor and on fuel pump **BEFORE** they are disconnected.

Take time to plug the disconnected ends of all four hoses with a golf tee, pencil end, or similar item to prevent contaminants from entering the fuel system.

3- Remove the four Phillips head screws securing the flame arrestor cover to the carburetor. Lift the arrestor screen free of the lower cover.

WORDS OF WISDOM

DO NOT attempt to remove the three Phillips head screws securing the lower flame arrestor cover to the carburetor at this time - while the carburetor body is still attached to the intake manifold. If a screw or washer is accidentally dropped into the lower funnel shaped cover, the screw/washer will almost certainly find its way down through the carburetor, the intake manifold and into the crankcase. If allowed to remain there, the screw/washer will reek havoc on pistons, rings and cylinder walls! If a screw or washer is lost in this manner, the engine **MUST** be completely disassembled to locate the part to avoid expensive damage. The moral of this story is -- leave the lower cover in place until the carburetor is on the workbench!

4- Back off the two carburetor mounting nuts. Lift the carburetor slightly while the nuts are loosened. There is very little clearance between the mounting flange and carburetor. Therefore the carburetor **MUST** be raised to enable the nuts to clear the last threads on the studs. Remove the two nuts and lockwashers.

5- Lift the carburetor free of the mounting studs. Remove the cable mounting bracket and two gaskets -- one on each side of the bracket.

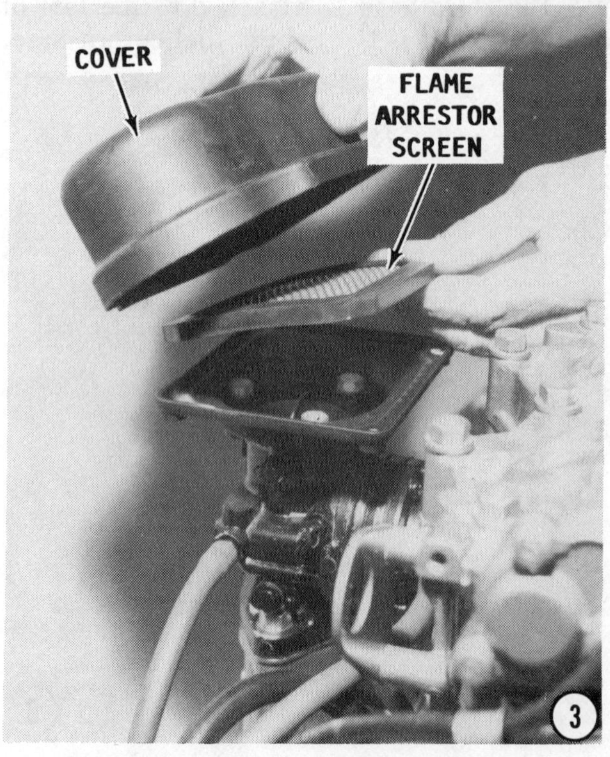

Special Words on Insulators

Some carburetors may be equipped with an insulator -- spacer plate -- between the intake manifold and the carburetor. If the carburetor being serviced is equipped with an insulator and a gasket beneath it, take time to make a note of which item is on top to ensure the gasket will be installed first with the insulator on top. If the insulator and gasket is not installed on top of the intake manifold, the cable mounting bracket will not align properly.

6- Remove the bolts and lockwashers securing the lower flame arrestor cover to the carburetor.

SPECIAL FUEL PUMP WORDS

This design fuel pump consists of many thin, delicate component layers. Some of these components may stick together and appear to be one gasket instead of several different items. If the fuel pump is to be returned to service without replacing any of the fragile gaskets and check valves made of plastic film, then **DO NOT ATTEMPT** to separate any components stuck together. Such action will surely rip or stretch delicate pieces.

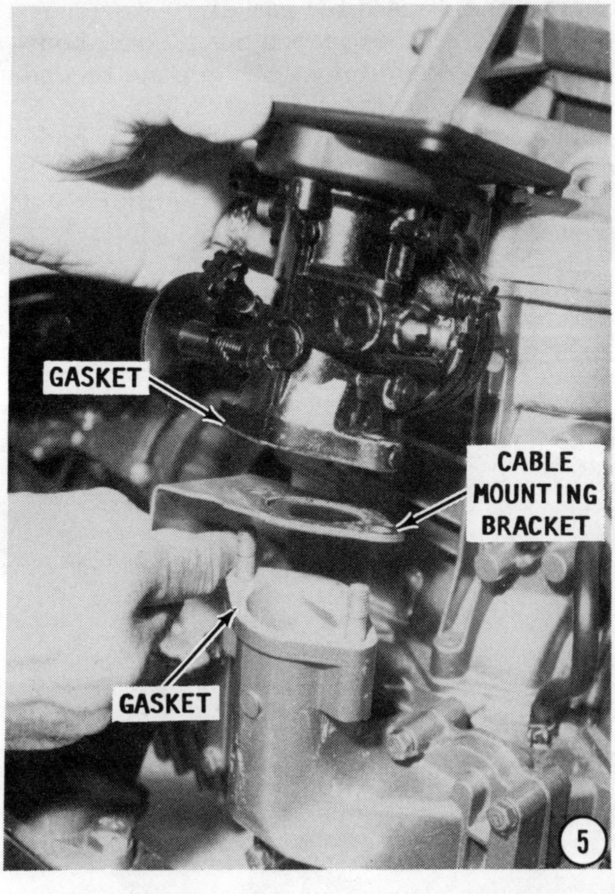

5

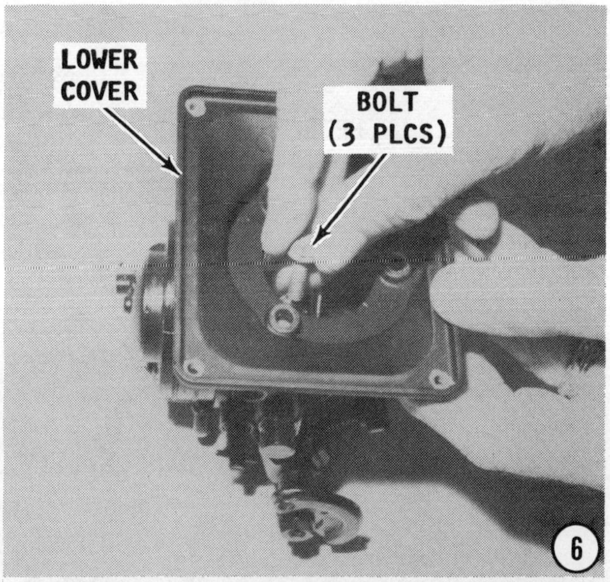

6

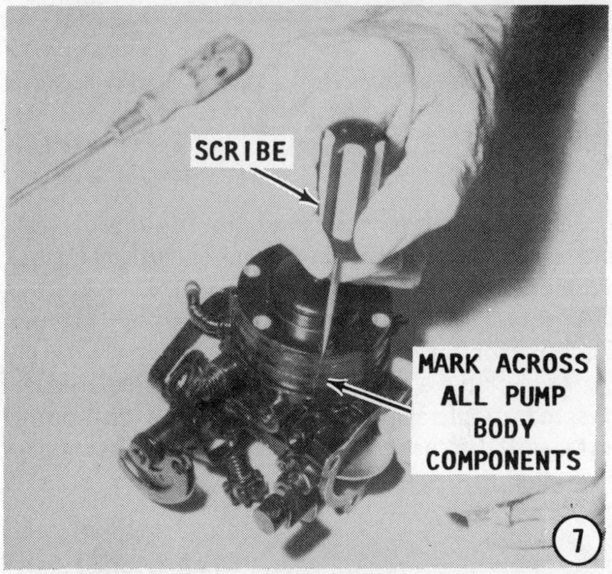

7

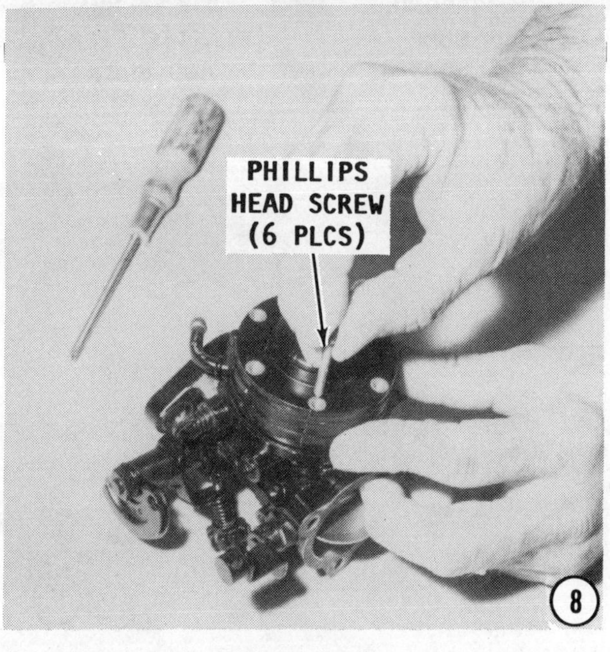

8

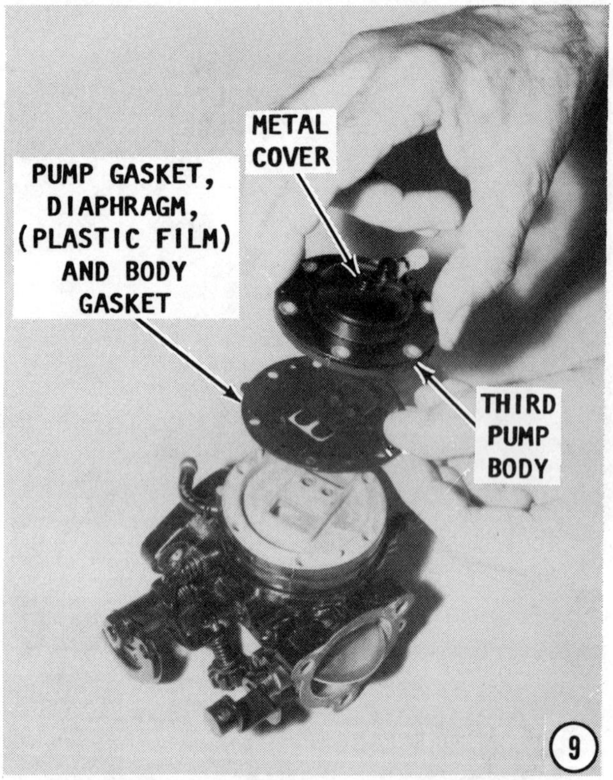

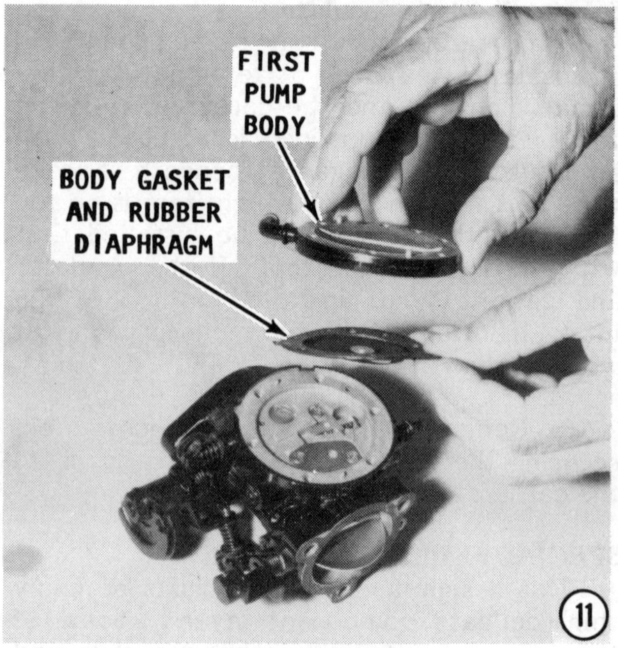

If the fuel pump is to be replaced, lay the old parts out in the order of removal, as an aid to assembling the new parts.

Carburetor Disassembling

7- Scribe a mark across the entire thickness of the fuel pump to ensure all components will be assembled in the correct position during installation.

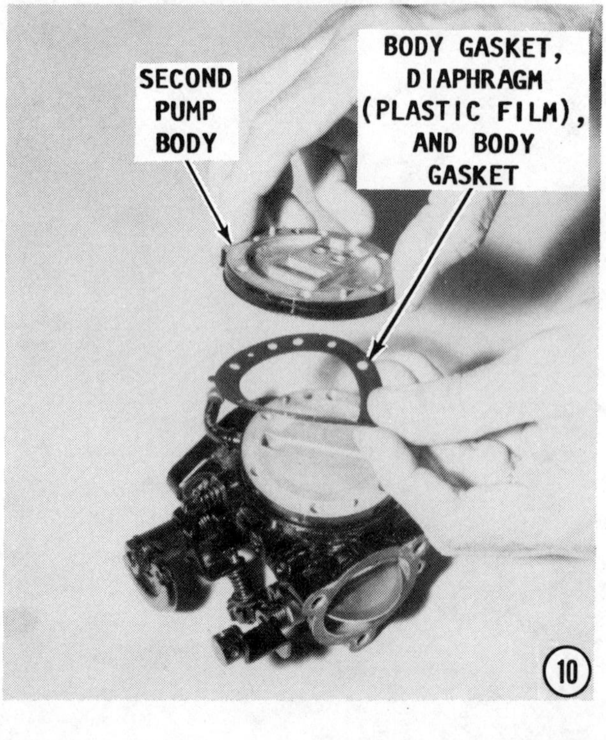

8- Remove the six Phillips head screws and lockwashers securing the third (outer) pump body to the carburetor.

9- Lift off the following items: the third pump body, pump gasket, check valve (plastic film), and body gasket.

Remove the Phillips head screw securing the metal cover to the third pump body. Remove and discard the gasket.

10- Lift off the second pump body, body gasket, diaphragm (plastic film), and second body gasket from the carburetor.

11- Remove the first pump body, body gasket, and rubber diaphragm.

12- Remove the two small Phillips head screws from the top cover, as shown.

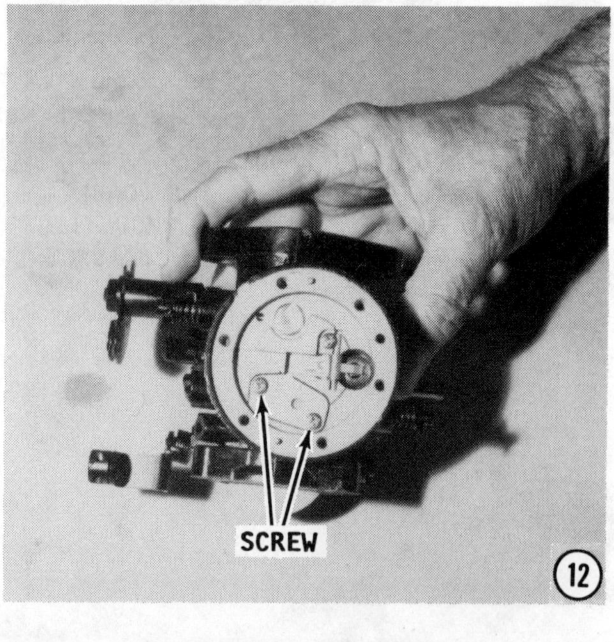

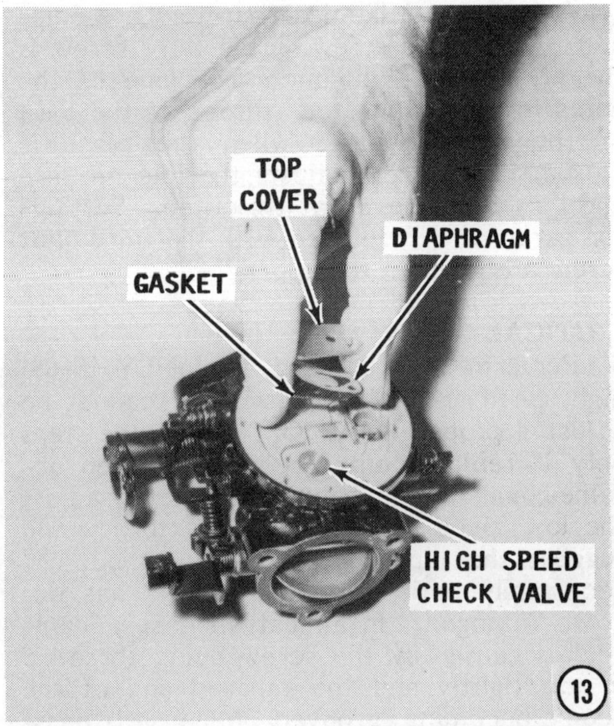

13- Lift off the top cover, diaphragm and gasket from the high speed check valve. The high speed check valve is pressed into the fuel chamber and **CANNOT** be removed.

14- Obtain a stubby Phillips head screwdriver and a hammer. Now, tap lightly on the end of the screwdriver while attempting to remove the hinge screw. **DO NOT** attempt to remove this screw "cold" --without tapping the screwdriver -- because the head will almost certainly be stripped!

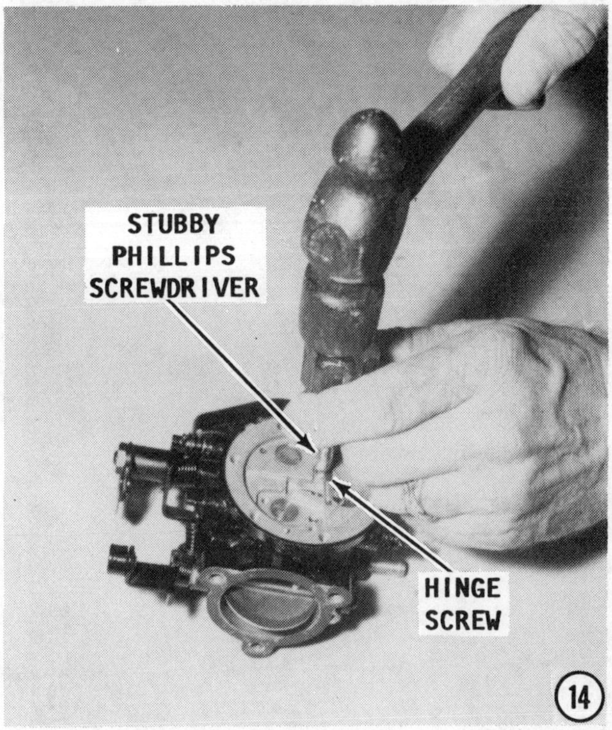

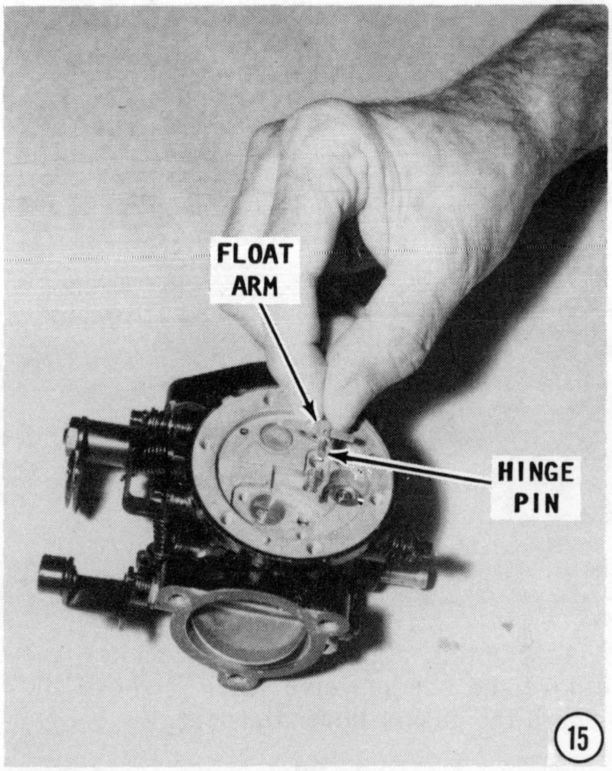

15- Lift out the hinge pin with float arm attached from the groove in the carburetor surface.

16- Lift out the small spring.

17- Using a pair of needlenose pliers, remove the needle valve from the seat.

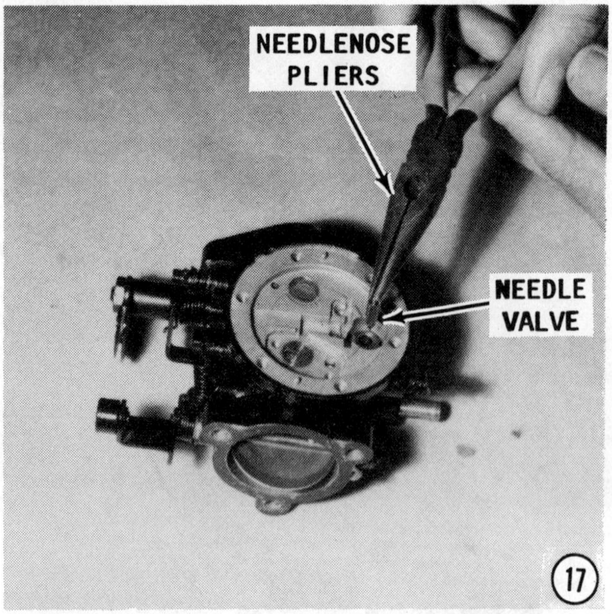

18- Obtain a narrow walled socket and remove the needle valve seat. Remove and discard the gasket under the seat.

SPECIAL WORDS
ADJUSTING SCREWS

The low and high speed mixture screws must **NOT** be removed unless damage to the tapered end is suspected and the screws are to be replaced. Under normal service these two screws should remain undisturbed.

Each Mikuni carburetor is individually adjusted at the factory using a flow meter. After the adjustment is completed, a limiter cap is installed over the head of the screw to prevent accidental tampering.

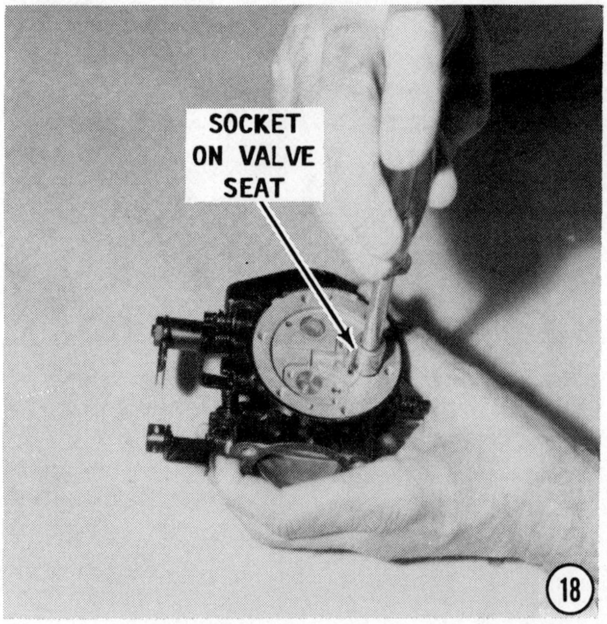

A very few Mikuni carburetors are euipped with a third screw. This third screw is the pilot screw which controls one of the holes in the carburetor throat at the base of the throttle plate when the plate is closed. This screw adjusts the idle mixture and may be removed at this time. Settings for carburetors equipped with the third pilot screw are given in the appendix.

CRITICAL WORDS

Perform Steps 19 and 20, **ONLY** if the high speed or low speed mixture cannot be adjusted properly. **ALSO,** perform the steps only if replacement screws have been obtained and are on hand. Failure to adjust the low speed or high speed mixture would suggest the tapered end of one -- but not necessarily both -- of the screws has become damaged. In almost all cases, "damage" is caused by the screw being threaded in too tightly and the tapered end of the needle becoming "grooved", because it passed through the orifice. Once the needle is "grooved", proper mixture cannot be obtained.

19- Back out the high speed mixture screw from the carburetor body.

20- Back out the low speed mixture screw.

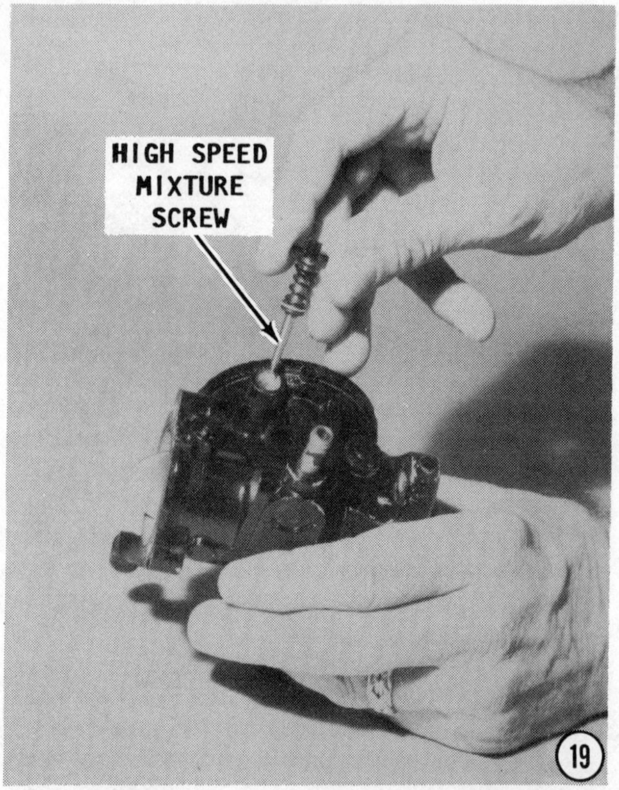

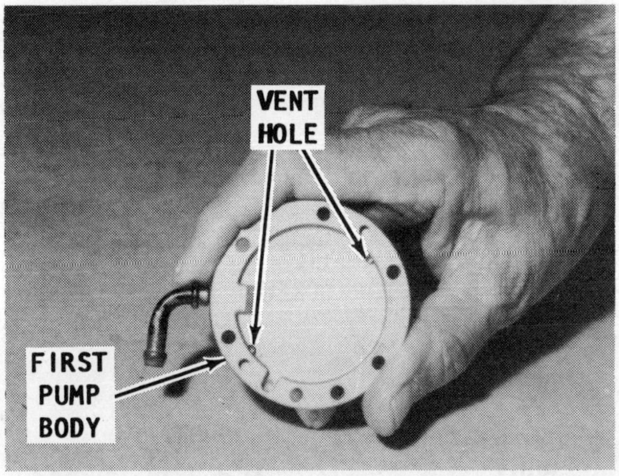

*The vent holes in the first pump body are vented to the atmosphere. These holes **MUST** be kept clean and clear to **ENSURE** efficient carburetor operation.*

CLEANING AND INSPECTING

NEVER dip rubber parts, plastic parts, or diaphragms in carburetor cleaner. These parts should be cleaned **ONLY** in solvent, and then blown dry with compressed air.

Place all metal parts in a screen-type tray and dip them in carburetor cleaner until they appear completely clean, then blow them dry with compressed air.

Blow out all passages in the castings with compressed air. Check all parts and passages to be sure they are not clogged or contain any deposits. **NEVER** use a piece of wire or any type of pointed instrument to clean drilled passages or calibrated holes in a carburetor.

Move the throttle shaft back and forth to check for wear. If the shaft appears to be too loose, replace the complete throttle body because individual replacement parts are **NOT** available.

CRITICAL AREA

Inspect the condition of both the choke and throttle shafts. Most shaft designs incorporate a cutout portion to enable the butterfly valve to lay flat against the shaft. Two small holes are drilled in this cutout portion to accommodate the small screws securing the valve to the shaft. The weakest point of the shaft is at the screw holes. Small cracks form at the holes due to metal fatigue. These cracks elongate the threaded holes and the small screws shake loose. Once loose, they are sucked into the carburetor, then the intake manifold, and into the crankcase -- eventually finding their way into the combustion chamber. One of these small screws can then cause very expensive damage to pistons, rings and cylinder walls. Therefore, check the cutout in the shaft **CLOSELY** for any evidence of a small crack.

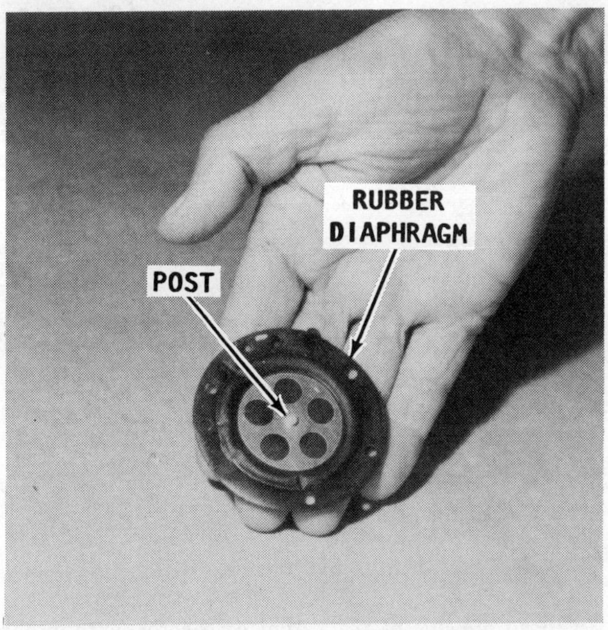

*A damaged diaphragm -- with even the tiniest hole --**MUST** be replaced to obtain full engine performance.*

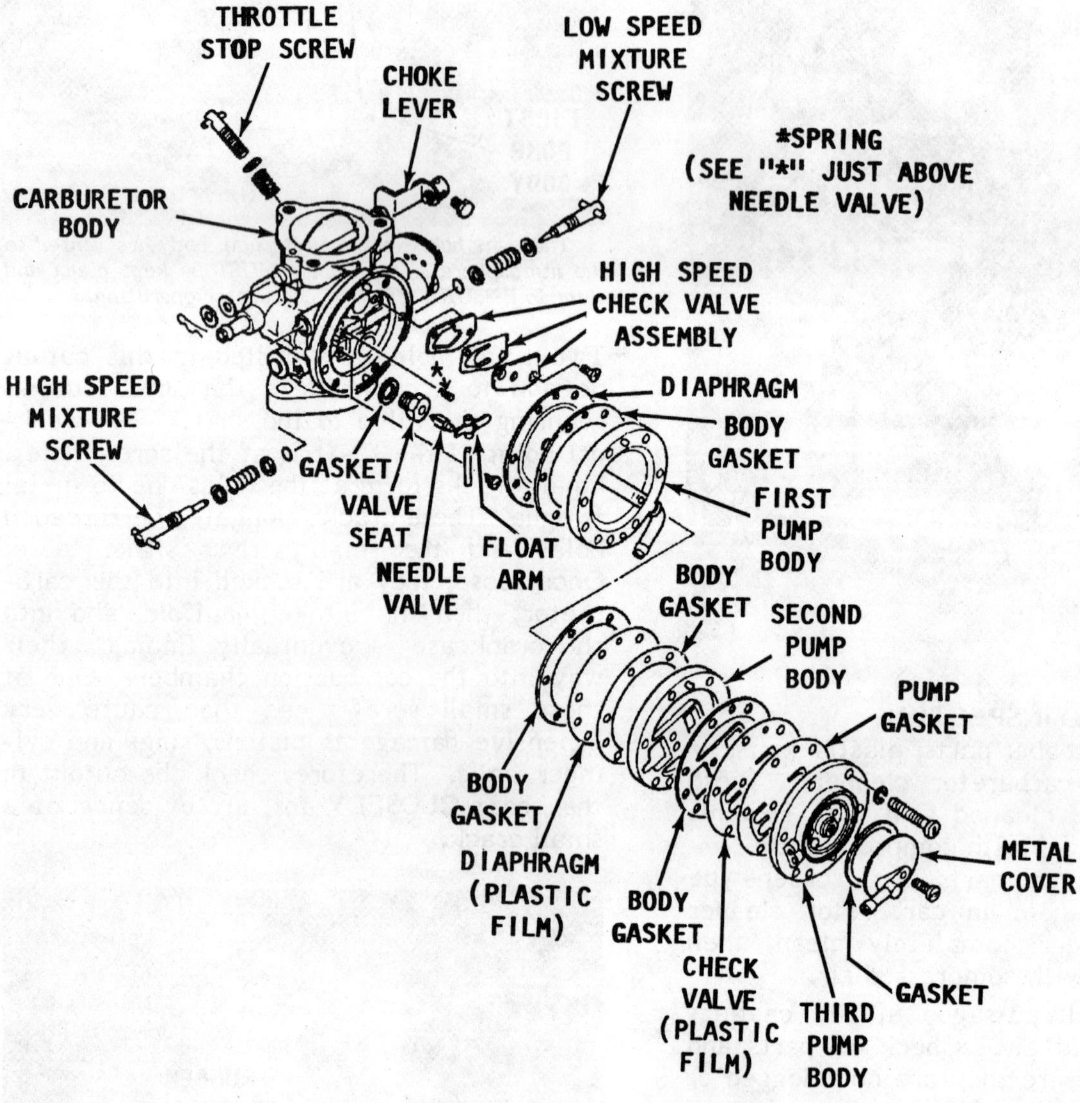

Exploded drawing of the Mikuni carburetor, with integral fuel pump, used on 300 Series, 440 Series, 550 Series, and the 650 Series after 1990. Major parts are identified.

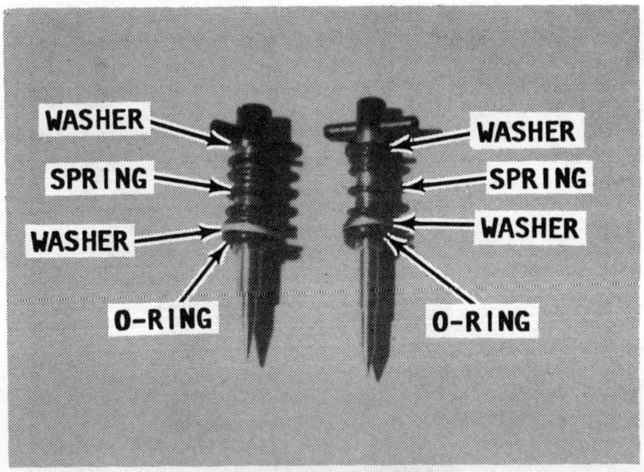

Check to be sure the low speed adjustment screw and the high speed adjustment screw are properly assembled, prior to installation.

Inspect the main body, airhorn, and venturi cluster gasket surfaces for cracks and burrs which might cause a leak. Check to be sure the float arm return spring has not been stretched. Check the float arm needle contacting surface and replace the diaphragm if this surface has a groove worn in it.

As previously mentioned, most of the parts which should be replaced during a carburetor overhaul are included in overhaul kits available from your local marine dealer. One of these kits will contain a matched fuel inlet needle and seat. This combination should be replaced each time the carburetor is disassembled as a precaution against leakage.

Make a thorough inspection of the fuel chamber and fuel pump diaphragms for the tiniest pin hole. If one is discovered, the hole will only get bigger. Therefore, the diaphragms must be replaced in order to obtain full performance from the engine.

Inspect the two vent holes in the first pump body. These holes **MUST** be clean. If these two holes are not perfectly clear, the diaphragm in the fuel pump will not function properly -- full fuel volume will not be pumped.

ASSEMBLING

Special Words on Mixture Screws

Observe the low and high speed mixture screws. They appear to be very similar, but they are **NOT** interchangeable. New screws should be identified on or in the package. To identify the correct screw, place them side by side as shown in the accompanying illustration. The low speed screw is shorter than the high speed screw. Both screws are threaded into the carburetor at the base of the fuel pump. The low speed screw is installed on the linkage side -- the high speed screw on the opposite side of the carburetor, close to the fuel return fitting.

Both screws are adjusted later at the completion of the assembly procedures. Therefore, thread each screw in just a few turns, at this time.

1- Install the low speed mixture screw just a few turns on the linkage side of the carburetor, with the washer, spring, second washer and O-ring on the shaft as indicated in the accompanying illustration.

2- Install the high speed mixture screw just a few turns on the same side as the fuel return fitting, with the supporting parts assembled on the shaft in the same order as the low speed screw.

SPECIAL WORDS
PILOT SCREW — IF EQUIPPED

Install the pilot screw into the carburetor body. Lightly seat the screw and then back it out the specified number of turns from this lightly seated position. Refer to the Appendix for model specifications.

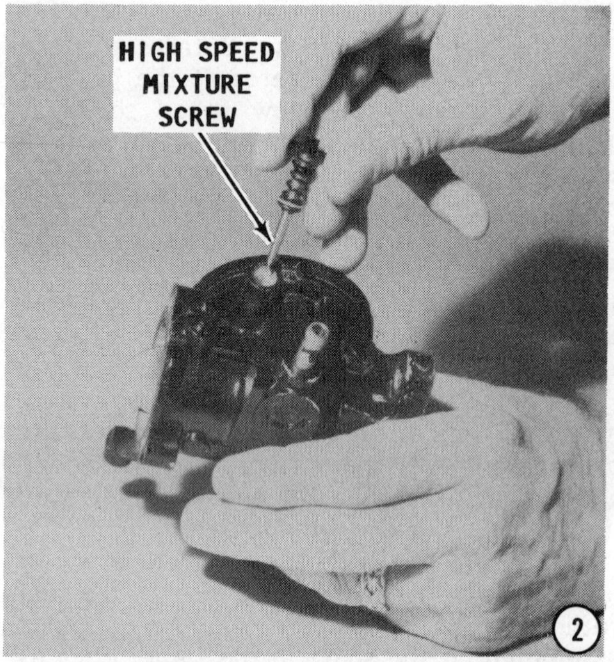

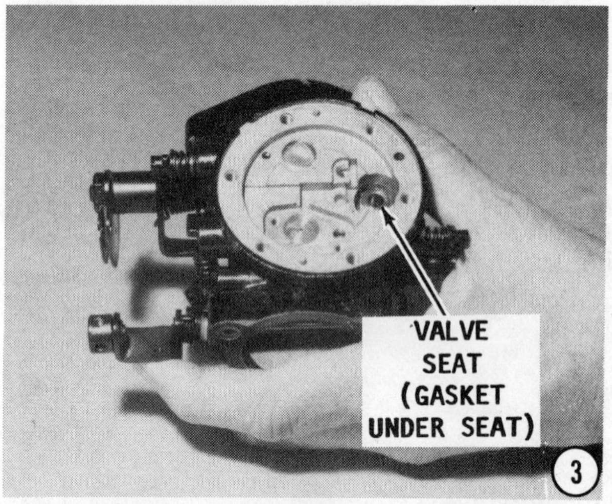

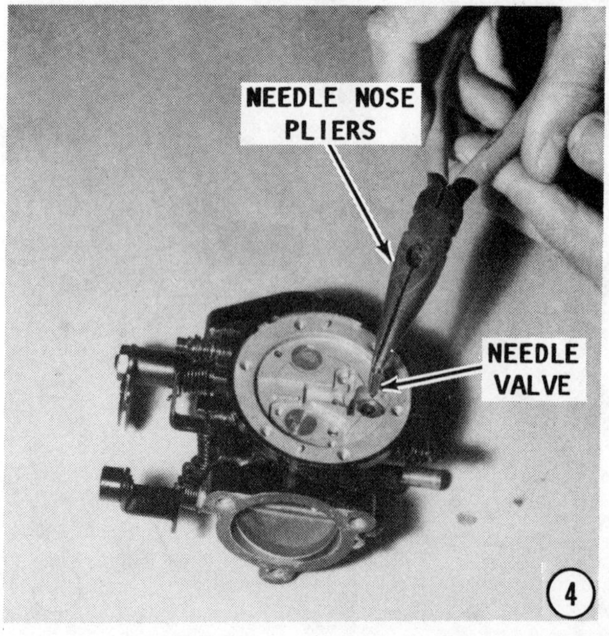

3- Place a new gasket over the needle valve seat and install the seat and gasket into the fuel chamber. Tighten the seat to a torque value of 2.9 ft lb (4Nm).

4- Lower the needle valve into the seat.

5- Slide the spring into the small hole in the fuel chamber.

6- Place the float arm at one end of the hinge pin and lower both into position over the installed spring. One end of the float arm **MUST** rest squarely on top of the needle valve, and the other end must lightly compress the spring.

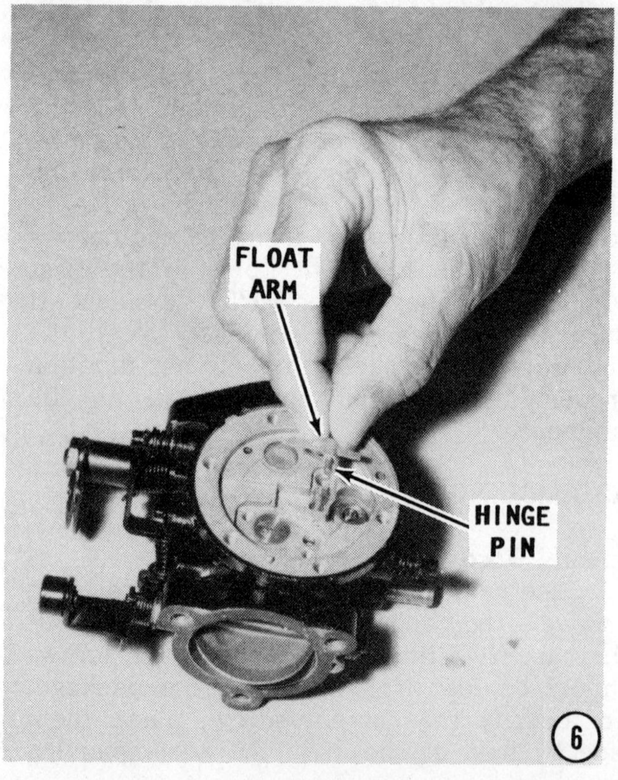

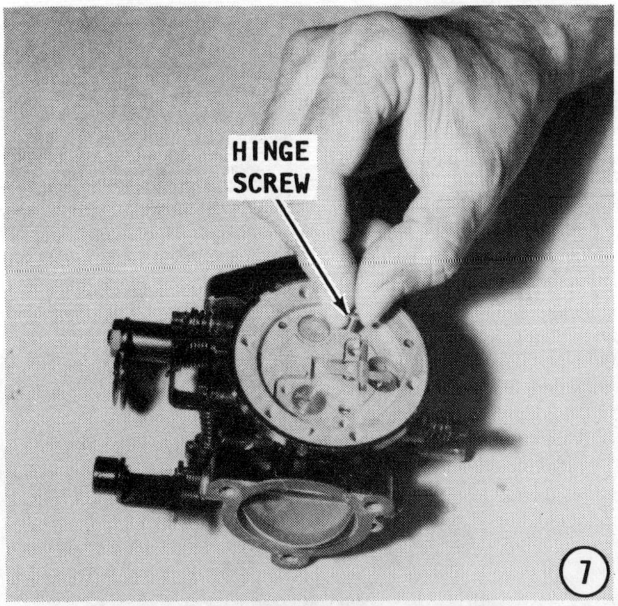

7

7- Install the hinge pin retaining screw. Tighten the screw securely.

8- Position the top cover gasket, diaphragm, and top cover over the high speed check valve.

9- Install and tighten the two small Phillips head screws securing the top cover to the fuel chamber.

Float Arm Adjustment

10- The top of the float arm must be parallel to the fuel chamber surface. If not, replace the return spring. Do not attempt to bend the float arm or to change the characteristics of the old return spring.

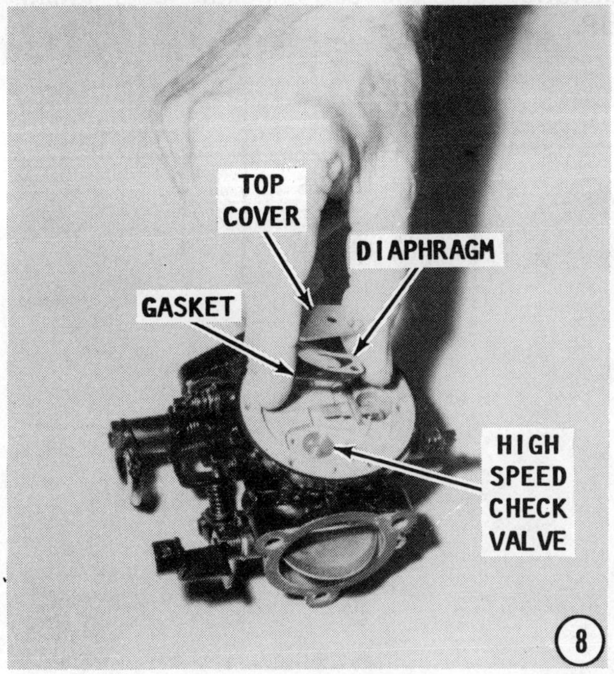

8

9

Fuel Pump Assembling

SPECIAL WORDS

Take care to align all **EIGHT** layers of diaphragms, gaskets and check valves (made of plastic film), with their tabs indexing in the cutout in the fuel chamber base. This cutout is offset from two aligning pins. A component is correctly installed when the tab is aligned with the cutout and the two holes are indexed over the two pins.

The first pump body is equipped with a vacuum fitting which takes the place of the indexing tab. The second and third pump bodies are equipped with an indexing tab. The first and second pump bodies are also equipped with a set of aligning pins. There are eleven fuel pump components to a complete assembly. All **ELEVEN** items must be carefully stacked and held together until the securing screws may be passed through the "stack" and threaded into the fuel pump body.

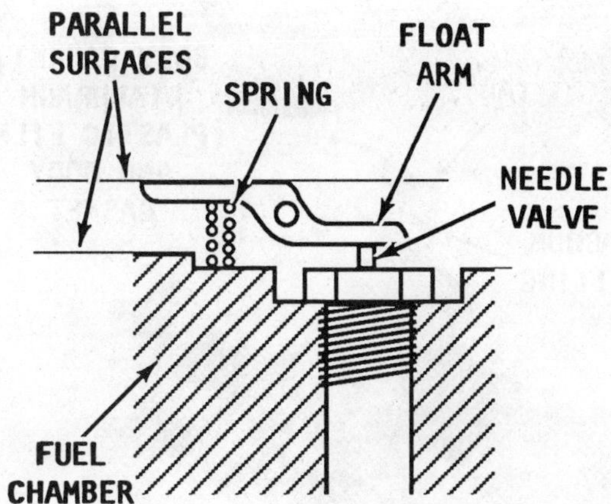

10

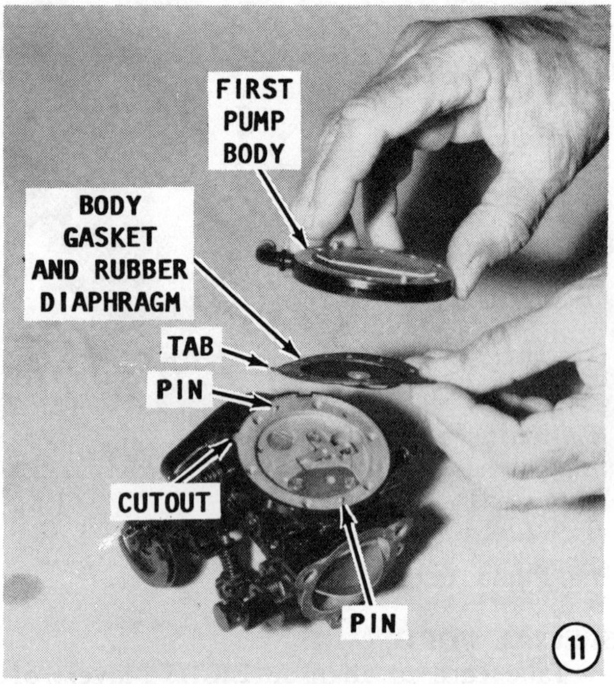

If a gasket, diaphragm or check valve is askew, the slightest pressure from the end of the securing screw will tear the component and affect performance of the fuel pump.

11- Position the rubber diaphragm and gasket over the two indexing pins with the tabs aligned with the cutout on the fuel chamber base. Next, position the first pump body over the gasket with the vacuum fitting aligned with the cutout on the fuel chamber base and the two indexing pins facing **UP.**

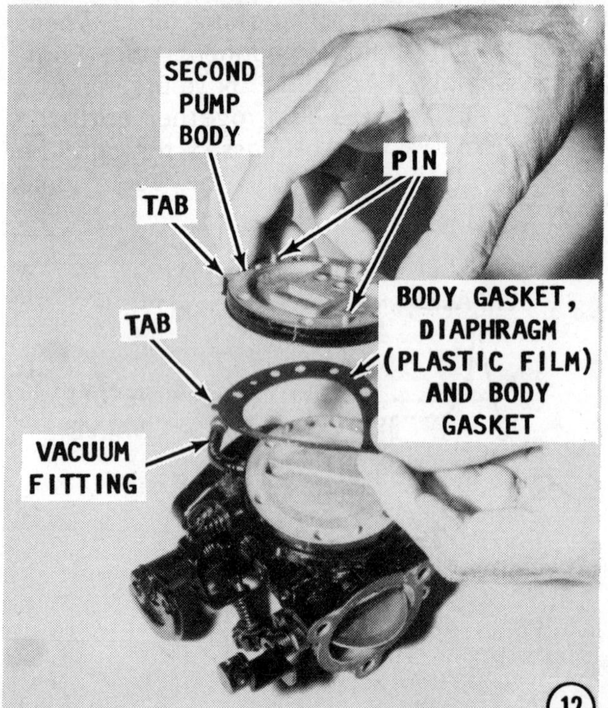

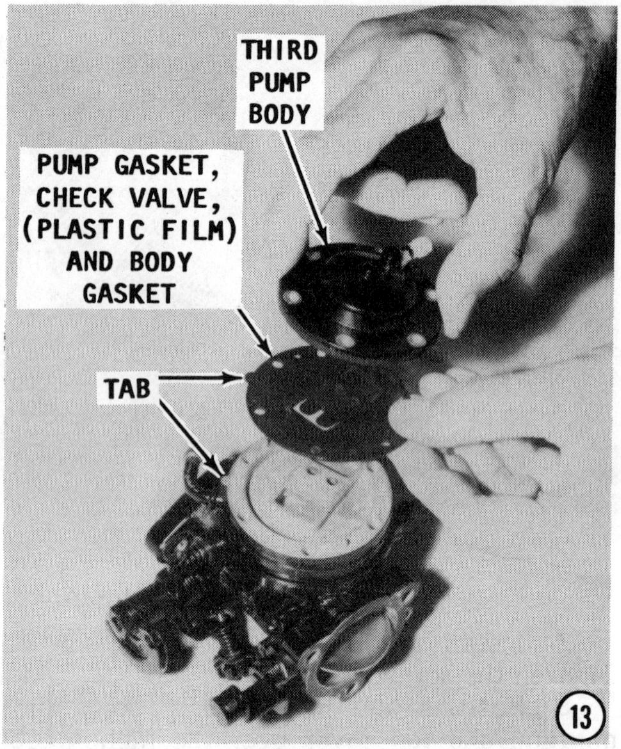

12- Place the body gasket, diaphragm (plastic film), and second body gasket over the first pump body with the tabs aligned with the vacuum fitting on the pump body and the holes in the gaskets and diaphragm sliding down over the two indexing pins. Now, place the second pump body over the gasket, with the tab on the pump body aligned with the tabs on the parts already installed and the two indexing pins facing **UP.**

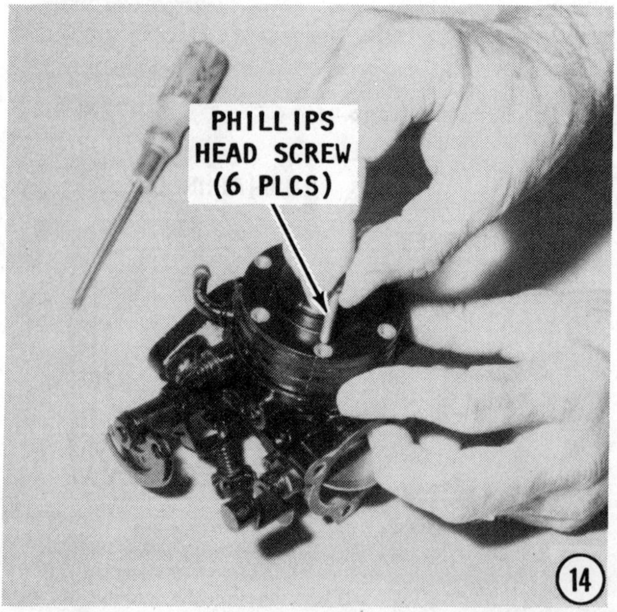

13- Install another body gasket, followed by the check valve (plastic film), the final body gasket and the third fuel pump body over the two indexing pins on the second pump body. Check to be sure all tabs are aligned.

14- Visually inspect the screw holes to verify all components are squarely in place and the path for all six screws are clear. Install the six Phillips head screws starting with two directly opposite each other. Continue to hold the stack of components tightly together until all six screws are started into the fuel chamber body. After all six screws have been started, tighten them securely -- working alternately and evenly.

15- Install the flame arrestor base onto the carburetor. Secure the base with the bolts and washers. Tighten these bolts to a torque value of 87 in lb (10 Nm). If any of the bolts should happen to vibrate loose, they will be sucked into the carburetor, the intake manifold, the crankcase and eventually find their way into the combustion chamber. Once in the combustion chamber, they will cause very extensive and expensive damage to pistons, rings, and cylinder walls.

Special Words on Insulators

Some carburetors may be equipped with an insulator -- spacer plate -- between the intake manifold and the carburetor. If an insulator was removed during the disassembling procedures, ensure the insulator is installed at this time with a gasket beneath it before installing the carburetor.

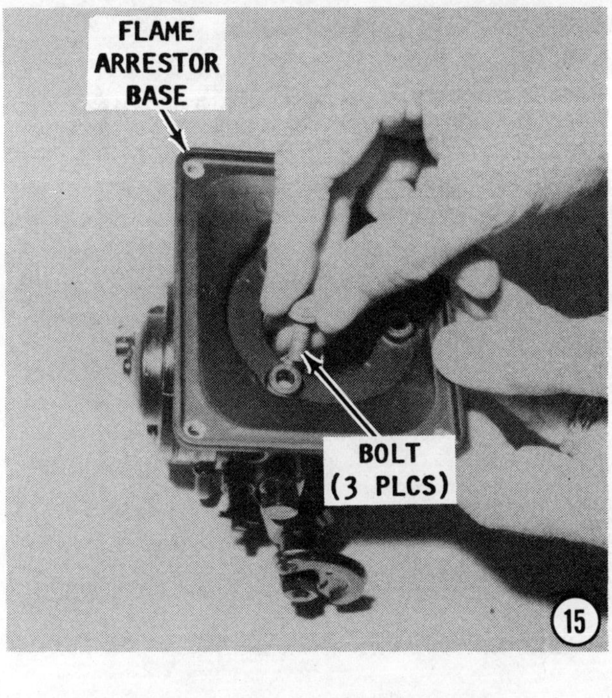

Carburetor Installation

16- Place a new carburetor mounting gasket over the intake manifold studs. Slide the cable mounting bracket over the studs with the cable supports on the outside, followed by another new mounting gasket. Place the carburetor over the mounting studs.

Place a lockwasher over each of the two carburetor mounting studs. Raise the carburetor slightly on the studs until enough clearance exists to start the two mounting nuts. Tighten the nuts securely.

Some models have an upper bracket to help prevent carburetor vibration. If so equipped, install this upper bracket, with a gasket under it. Secure the bracket in place with the two head nuts and tighten the nuts to a torque value of 16 ft lb (22Nm).

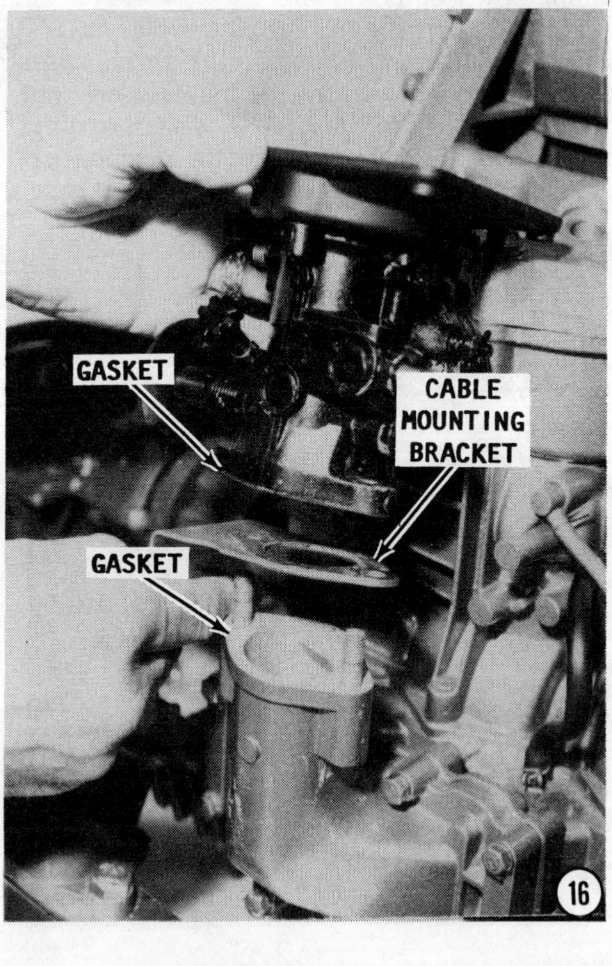

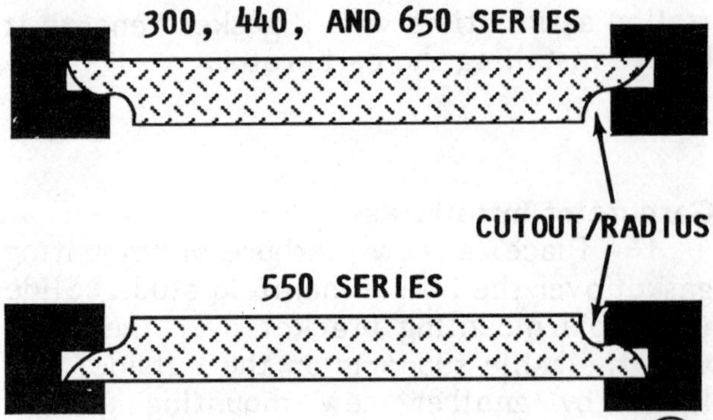

17

17- Install the flame arrestor base gasket.

If working on a 300, 440 or 650 series: Position the flame arrestor screen over the lower cover with the cutout/radius facing downward, as shown. Install the top cover and secure the two halves together with the attaching bolts.

If working on a 550 series: Install the flame arrestor assembly. Tighten the bolts to a torque value of 52 in lb (6Nm). If the arrestor screen was removed for service, install the screen with the cutout/radius facing upward, as shown.

18- If working on a 300, 440, or 650 series carburetor: Connect the three hoses to the carburetor. If the hoses were not marked as directed in Step 2, disassembling, proceed as follows: Locate the disconnected vacuum hose leading from the crankcase. Connect this hose to the vacuum fitting on the first fuel pump body -- the fitting mid-way in the fuel pump. Locate the disconnected hose leading from the in-line fuel filter mounted on the engine compartment wall. Connect this fuel line to the fuel fitting on the third fuel pump body -- the outermost fitting on the pump. See illustration **18A.** Locate the third and last disconnected hose leading from the fuel tank. Connect this hose to the fuel fitting on the carburetor body. See illustration **18B.**

If working on a 550 series carburetor: Connect the four hoses to the carburetor. If the hoses were not marked as directed in Step 2, disassembling, proceed as follows: Locate the disconnected vacuum hose leading from the crankcase. Connect this hose to the vacuum fitting on the first fuel pump body -- the fitting mid-way in the fuel pump. Locate the disconnected hose leading from in-line fuel filter mounted on the engine compartment wall. Connect this fuel line to the fuel fitting on the third fuel pump body -- the outermost fitting on the pump. See illustration **18A.** Locate the third disconnected hose leading from the side fitting on the rpm limiter. This hose is the excess fuel return hose and leads back to the fuel tank -- the fuel being routed first through the rpm limiter. Attach this hose to the fitting to the left of the fuel pump.

18 A

18 B

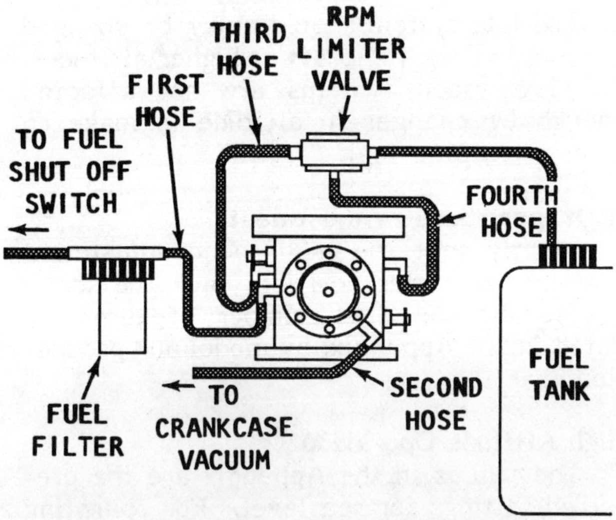

Functional line drawing to graphically identify the first hose, second hose, third hose, and fourth hose, described in the text.

The fourth hose is the rpm limiter hose attached to the carburetor just right of the fuel pump. The other end of this hose is also attached to the fuel limiter valve at the center fitting.

19- Slide both the throttle and choke cables into their respective slots in the cable mounting bracket. Tighten the two nuts on each cable to secure the cables to the bracket. If the throttle cable must be

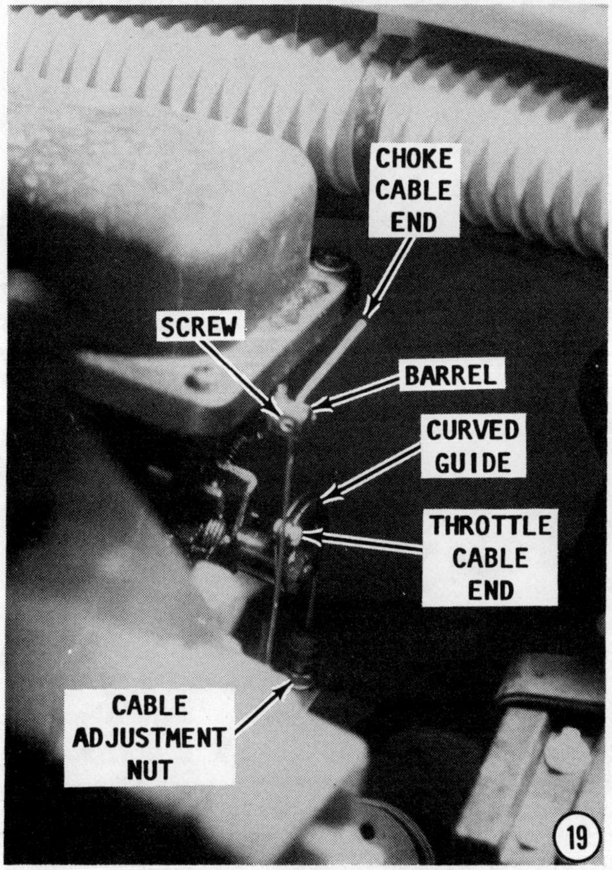

fitted into a curved guide, hook the cable end into the guide before sliding the cable into the slot in the mounting bracket. If either the throttle or choke cable end was secured in a small barrel pivoted on the linkage, pass the cable end through the barrel until the mark, made in Step 1 of disassembling, is up against the barrel. Tighten the small screw on the barrel to secure the cable in place. If no mark was made, insert the cable into the barrel until just a small amount of slack remains in the cable, and then temporarily tighten the screw to secure the cable within the barrel. Throttle and choke cable adjustments begin on Page 6-27.

Carburetor Priming

The carburetor must be primed before starting the engine to save the battery from excessive cranking. Pull off the fuel return hose at the carburetor. Place a shop towel under the fitting. Blow through the hose until fuel starts to drip from the fuel return fitting on the carburetor. The carburetor is now full of fuel and the engine is ready for cranking after the hose has been reconnected.

Mixture Screws with Limiter Caps

Since about 1989, the manufacturer has installed small black plastic limiter caps which snap over both the low speed and high speed adjustment screws. The purpose of these caps is to prevent any unauthorized tampering with the factory setting. Each cap has a tab and the tab fits into a slot in the carburetor body at the base of the mixture screw. Therefore, the optimum position of each screw is with the tab down and indexed into the slot.

However, if the mixture screws are removed during service procedures, the low and high speed mixtures must be adjusted and both caps installed over the screws with the tabs on the caps indexed into slots in the carburetor body.

Low Speed Screw Adjustment

Move the craft to a test tank, to a body of water, or connect a flush attachment and hose to the engine. For maximum performance, the idle rpm should be adjusted under actual operating conditions. Connect a tachometer to the engine.

Lightly seat the low speed adjustment screw. From this position, back the screw out 3/4 of a turn as a preliminary adjustment.

Start the engine and allow it to warm to operating temperature.

CAUTION

Water must circulate through the jet pump -- to and from the engine, anytime the engine is operating. Circulating water will prevent overheating -- which could cause damage to moving engine parts and possible engine seizure.

NEVER, AGAIN NEVER, operate the engine at high speed with a flush device attached. An engine operating at high speed with such a device attached, would **RUNAWAY** from lack of a load on the impeller shaft, causing extensive damage.

With the engine running, slowly rotate the throttle stop screw until the engine idles at 1,700 rpm for the 300 series, 1,800 rpm for the 440 series and 1,500 rpm for the 550 and 650 series. Next, rotate the low speed screw in or out 1/8th of a turn -- at a time -- until the cylinder/s fire evenly and engine rpm increases. Readjust the throttle stop screw to reduce the rpm to idle speed.

If the engine idle is set too low, the self-circling radius increases and the craft may lose its self-circling feature -- described in Chapter 1.

If the engine idle is set too high, the craft may circle too fast for a displaced rider to climb aboard the craft.

Close view to clearly indicate location of the throttle stop screw and the low speed adjustment screw on the Mikuni carburetor.

The idle system should never be changed when operating the craft at higher altitudes. The low speed systems are not affected enough by changes in altitude to make an adjustment practical.

High Speed Screw Adjustment

Lightly seat the high speed adjustment screw. From this position, back the screw out to the specified number of turns, as listed in the Appendix, by model and production year.

High Altitude Operation

The values in the Appendix are the preferred settings for sea level. For operating at higher elevations, the required leaner mixture is obtained by rotating the screw the specified number of turns **CLOCKWISE** as follows:

3,300 ft (1,000 m) -- 1/8
6,600 ft (2,000 m) -- 1/4
9,900 ft (3,000 m) -- 3/8

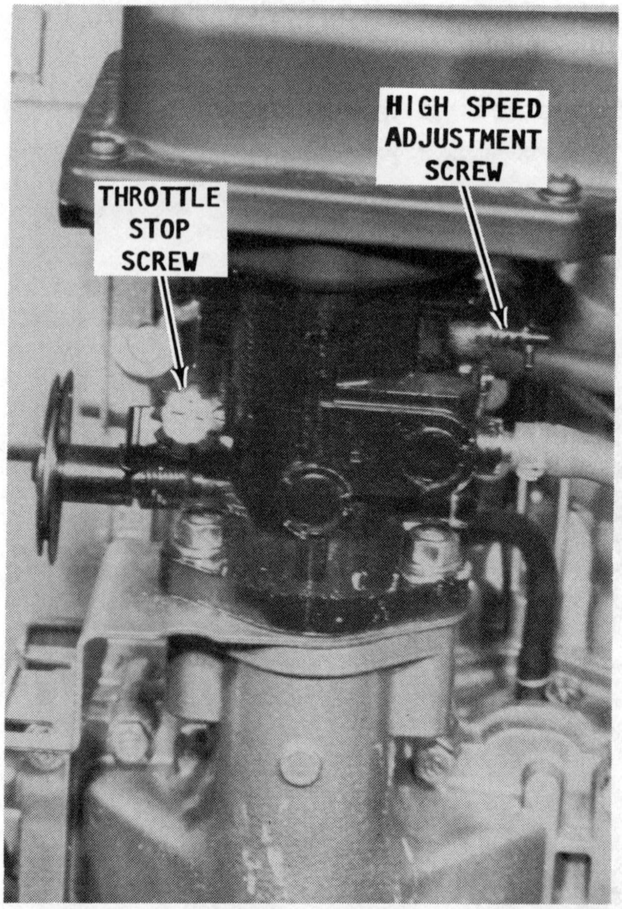

View showing location of the high speed adjustment screw on the Mikuni carburetor. The throttle stop screw is also shown from a different angle than in the illustration to the left.

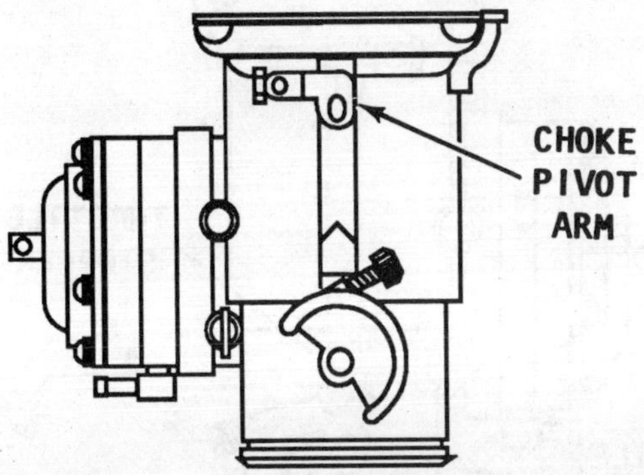

OPEN

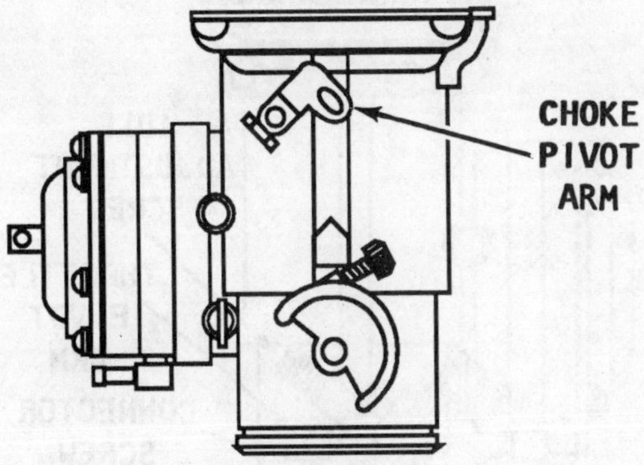

CLOSED

Line drawings to indicate the choke cable adjustment on the 300 Series, with the choke open (top), and when closed (bottom). The 650 Series carburetor is almost identical.

Choke Cable Adjustment
All Models

Push the choke knob inward. Observe the choke pivot arm on the side of the carburetor. The arm should be in the position shown in the accompanying illustration, with minimum slack in the choke cable. If the choke needs adjustment: Loosen the small screw securing the choke cable end to the barrel on the choke lever. Push the choke knob inward until it rests against the control panel. Set the choke pivot arm in the fully open position, and then tighten the screw securing the cable to the barrel.

Choke Cable -- Excessive Slack

If there is too much slack in the cable, loosen the top nut around the cable at the cable support bracket and tighten the bottom nut until only a small amount of slack remains.

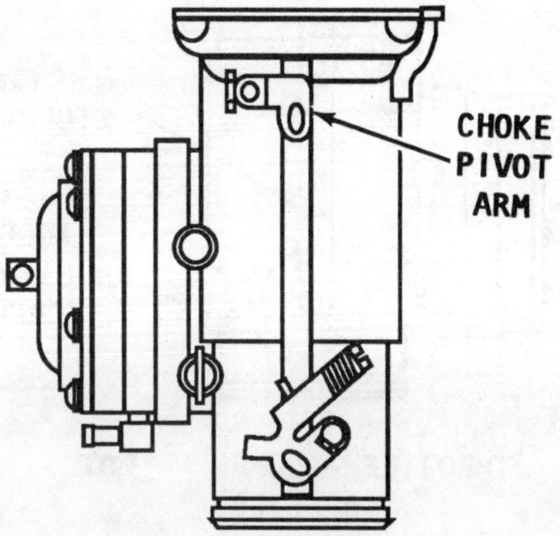

OPEN

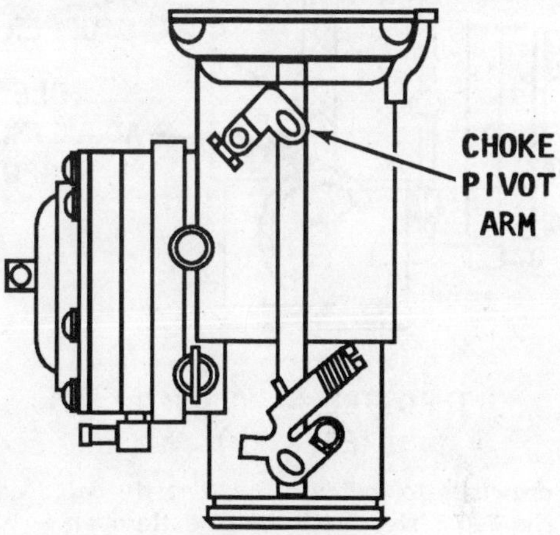

CLOSED

Line drawings to indicate the choke cable adjustment on the 440 Series and the 550 Series, with the choke open (top), and closed (bottom).

If the cable is tight -- with no slack -- loosen the bottom nut around the cable at the cable support bracket and tighten the top nut until just a small amount of slack remains in the cable.

If the cable has stretched, or has become kinked or frayed, or tends to stick, the entire cable should be replaced.

Throttle Cable Adjustment
300, 440 Series -- All Models
650 Series — 1991 and On

The idle speed must be adjusted **BEFORE** adjusting the throttle lever free "play".

With the throttle lever released and the throttle closed, the idle adjustment screw should rest against the stop on the carburetor with a slight amount of slack in the

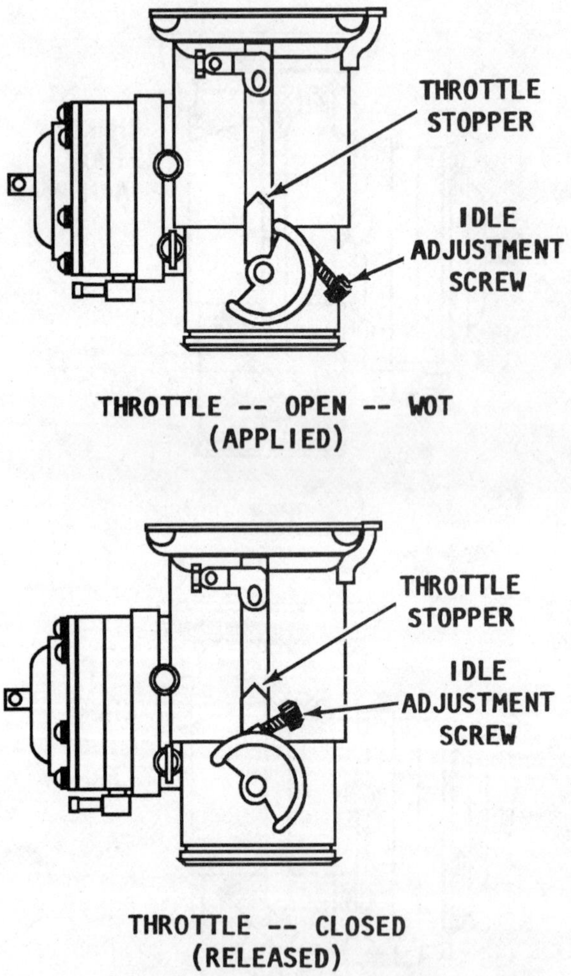

Line drawings to indicate the throttle cable adjustment on the 300 Series, with the throttle open — WOT -- (top), and closed (bottom). The 650 Series since 1991 is almost identical.

cable, as shown in the accompanying illustration.

With the throttle lever in the WOT position, the arm on the curved guide should rest against the stop on the carburetor.

If the throttle cable needs adjustment: Loosen and rotate the two locknuts around the throttle cable at the cable support bracket until the idle adjustment screw rests against the throttle stopper with the throttle released, and with a minimum amount of slack in the cable. Tighten the two locknuts to hold this new adjustment.

Throttle Cable Adjustment 550 Series — All Models

The idle speed must be adjusted **BEFORE** adjusting the throttle lever free "play".

With the throttle lever released and the throttle closed, the throttle linkage on the carburetor should rest against the idle ad-

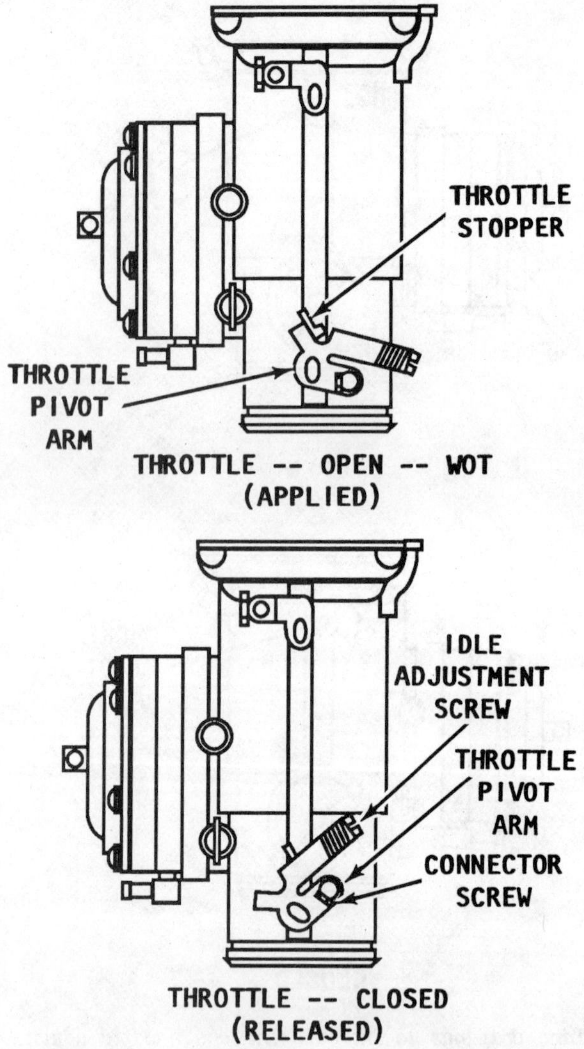

Line drawings to indicate the throttle cable adjustment on the 440 Series and the 550 Series, with the throttle open -- WOT -- (top), and closed (bottom).

justment screw, with a slight amount of slack in the cable, as shown in the accompanying illustration.

With the throttle lever in the WOT position, the throttle pivot arm should be down against the throttle stopper.

If the throttle cable needs adjustment: Loosen the set screw on the throttle pivot arm connector. Set the throttle lever to the WOT position and at the same time push the pivot arm down against the throttle stopper. Tighten the set screw on the throttle pivot arm to hold this new adjustment.

6-8 SERVICE KEIHIN CARBURETOR

This section provides complete detailed procedures for removal, disassembly, cleaning and inspecting, assembling including bench adjustments, installation, and operating adjustments for the Keihin carburetor.

The Keihin carburetor with remote fuel pump used on the 650 Series prior to 1991 and on all JetMates.

The Keihin carburetor is identified by the manufacturer as the CDK34 with a 28mm venturi and is installed on the twin cylinder 650 series engine, prior to 1991. This carburetor is also used on all JetMate craft since 1989.

Complete detailed procedures to service the remote fuel pump installed as an associated unit with the Keihin carburetor will be found in the next section -- 6-9, Remote Fuel Pump Service.

REMOVAL AND DISASSEMBLING

Good shop practice dictates a carburetor repair kit be purchased and new parts be installed any time the carburetor is disassembled.

The remote fuel pump used with all Keihin carburetors on the 650 Series prior to 1991 and on all JetMates.

Make an attempt to keep the work area organized and to cover parts after they have been cleaned. This practice will prevent foreign matter from entering passageways or adhering to critical parts.

1- Disconnect the fuel supply and return hoses from the carburetor. Take time to plug the disconnected ends of both hoses with a golf tee, pencil end, or similar item to prevent contaminants from entering the fuel system.

2- Loosen the two locknuts on the choke cable -- one on each side of the cable support bracket.

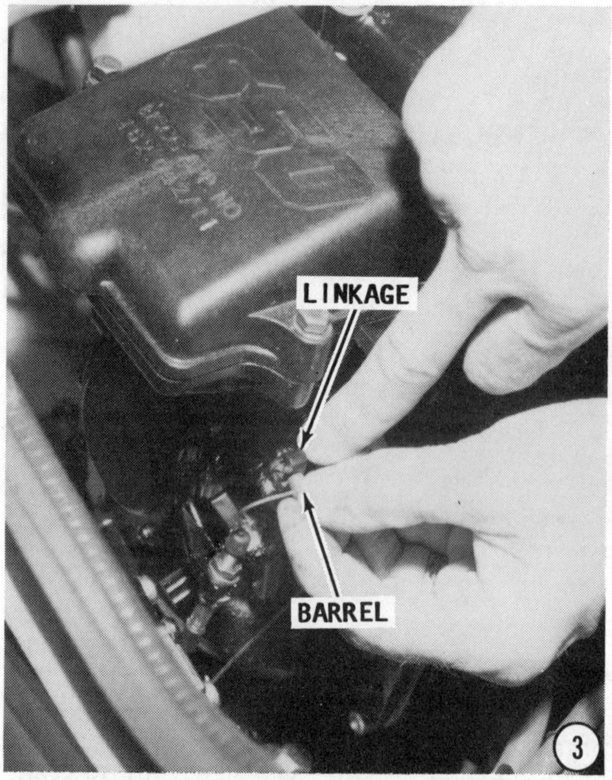

3- Depress the choke linkage and unhook the barrel from the hole in the linkage.

Repeat these last two steps to disconnect the throttle cable.

4- Remove the bolts securing the flame arrestor cover. Lift off the cover and the arrestor screen.

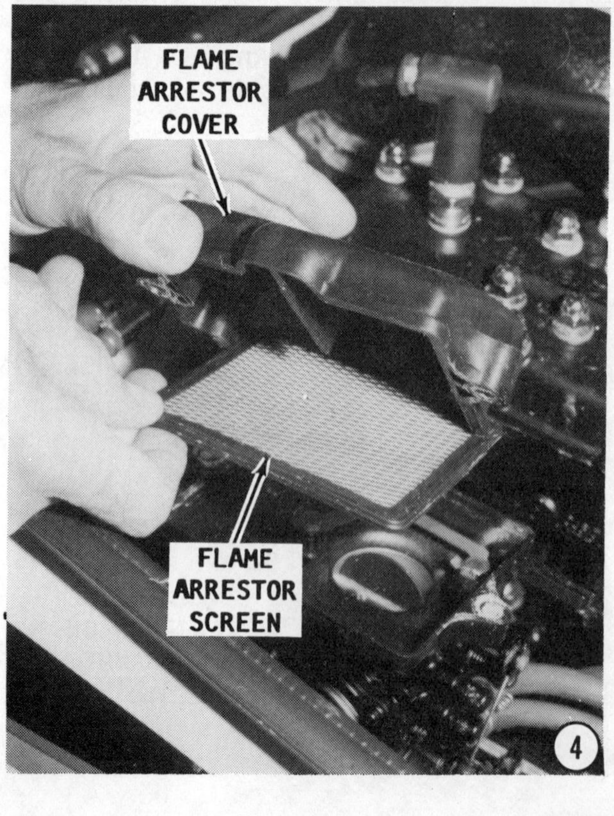

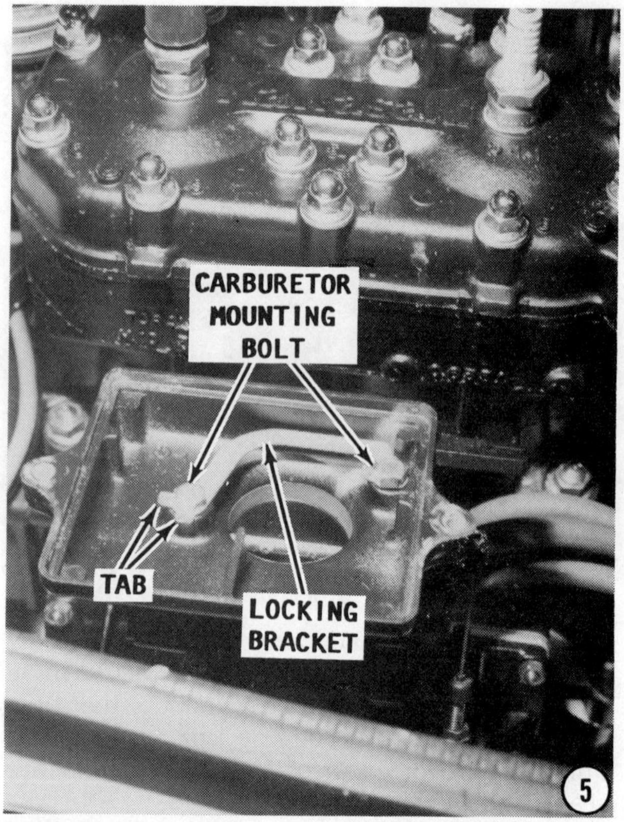

5- Bend down the two tabs against the sides of the two carburetor mounting bolts. Loosen the two bolts.

6- Lift off the lower air intake cover with the locking bracket and bolts still in place, and then remove the carburetor.

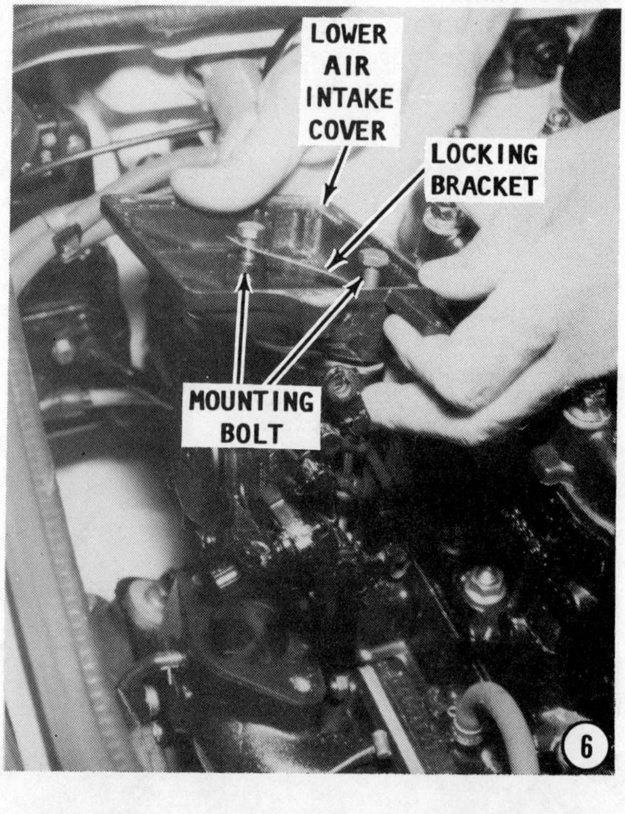

7- Remove the carburetor base gasket.

Carburetor Disassembling

8- Remove the four Phillips head screws securing the diaphragm cover to the carburetor and lift off the diaphragm cover.

Remove the diaphragm from the fuel chamber.

9- Remove the screw securing the diaphragm housing to the carburetor.

10- Lift off the diaphragm housing.

11- Remove and discard the sealing ring.

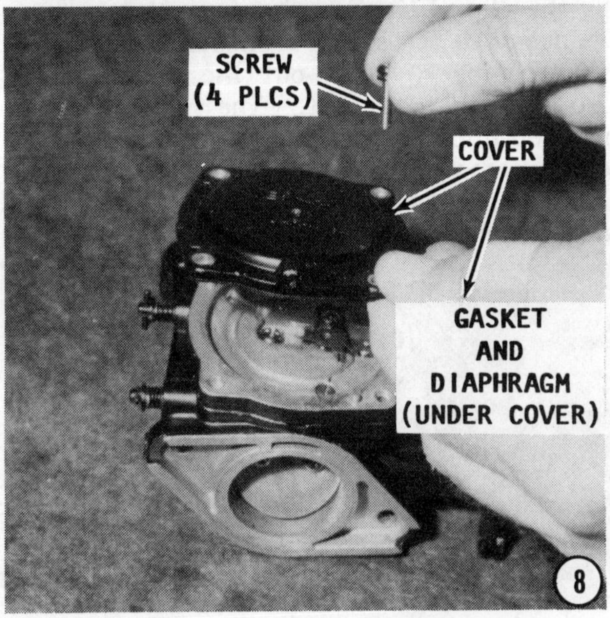

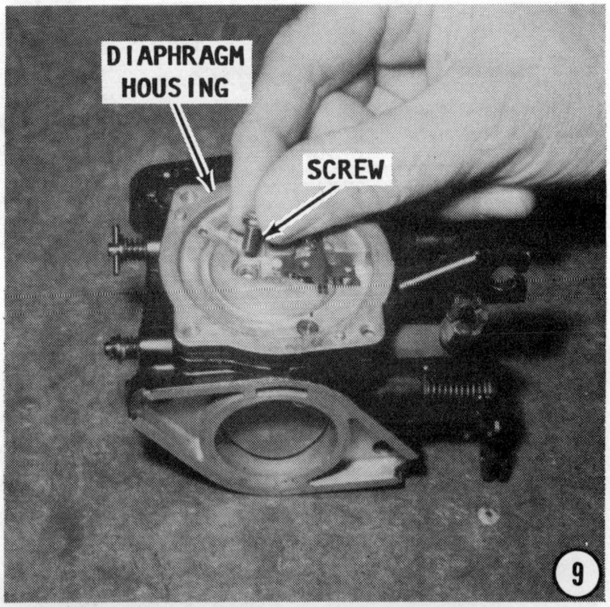

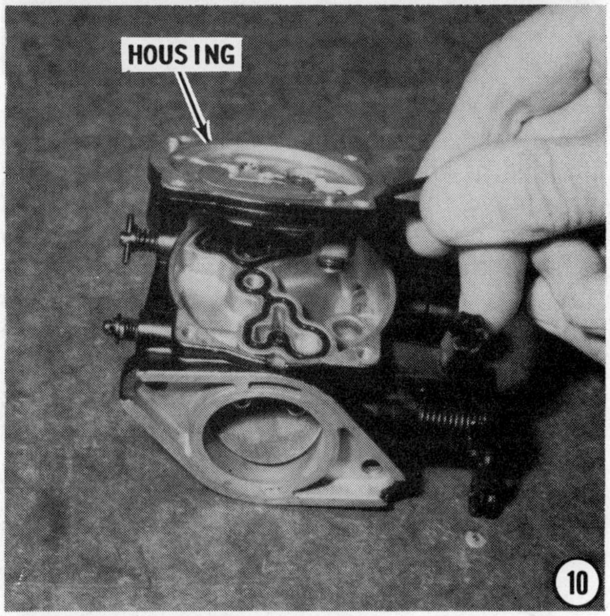

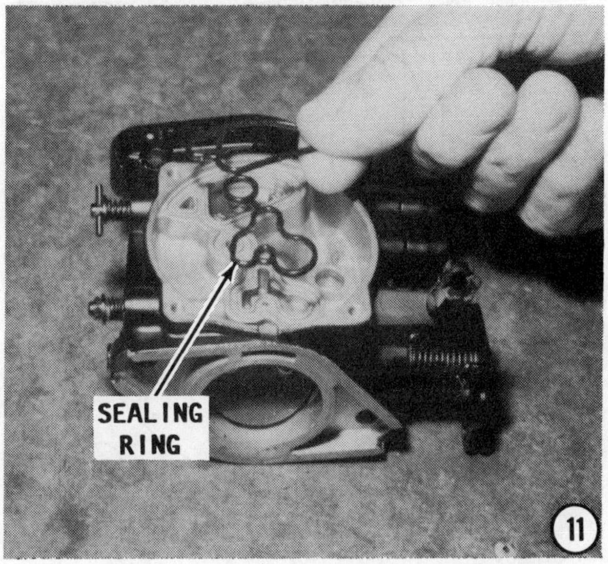

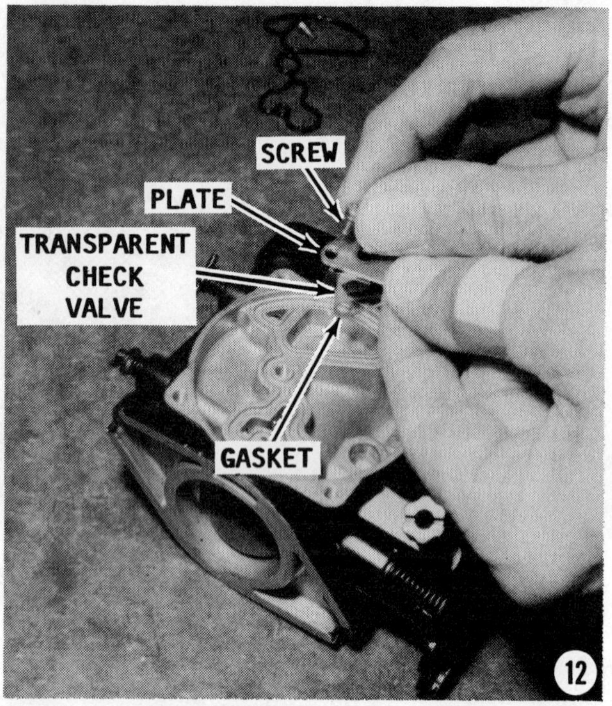

12- Obtain a stubby Phillips head screwdriver and a hammer. Now, tap lightly on the end of the screwdriver while attempting to remove the two screws securing the high speed check valve plate. Do not attempt to remove these screws "cold" -- without tapping the screwdriver -- because the heads will almost certainly be stripped! Remove the plate.

CAREFULLY lift off the transparent check valve. **TAKE CARE** not to lose this item. This valve controls the flow of fuel in one direction and prevents the engine from "faultering" when the throttle is suddenly

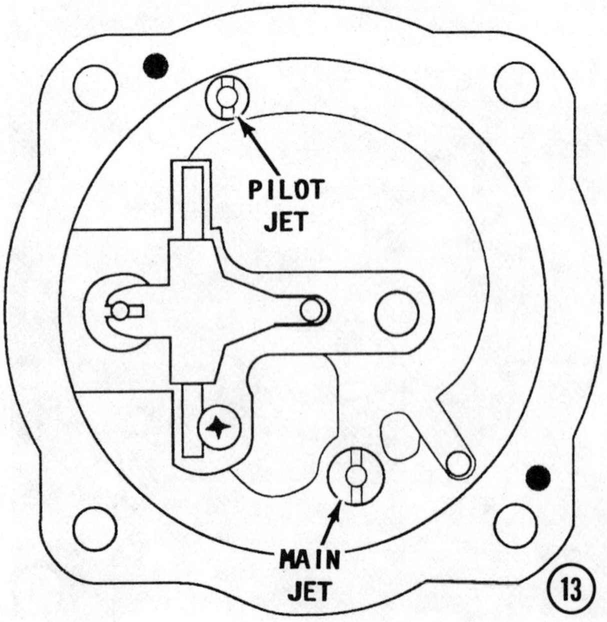

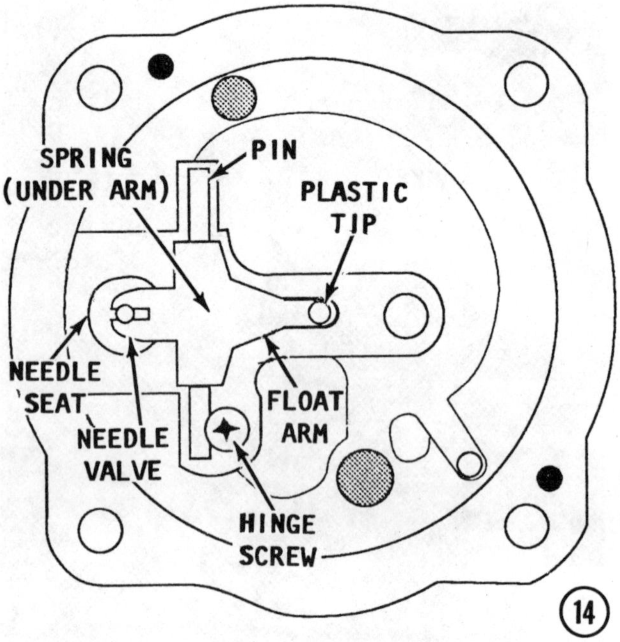

changed from wide open, to idle, and then back to wide open again. Remove the gasket under the check valve.

13- Use the correct size screwdriver to remove the main jet -- the larger of the two jets -- from the fuel chamber. Again, use the correct size screwdriver and remove the pilot jet from the chamber.

14- Use the stubby Phillips head screwdriver and a hammer from Step 12 and lightly tap the end of the screwdriver while attempting to remove the hinge screw. Do not attempt to remove this screw "cold" -- without tapping the screwdriver -- because the head will almost certainly be stripped!

Lift out the float arm and float pin. The needle valve will remain hanging on the pin by the clip. Lift the valve from the seat.

Remove the spring from its recess in the fuel chamber. Lift out the needle valve from the needle seat. The needle seat is pressed into the carburetor and **CANNOT** be easily removed.

15- Turn the housing over and remove and discard the O-ring around the seat.

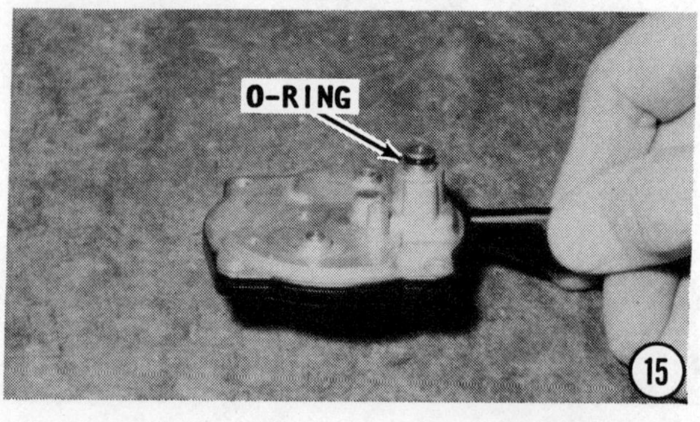

SPECIAL WORDS ADJUSTING SCREWS

The low and high speed mixture screws must **NOT** be removed unless damage to the tapered end is suspected and the screws are to be replaced. Under normal service these two screws should remain undisturbed.

Each Keihin carburetor is individually adjusted at the factory using a flow meter. After the adjustment is completed, a limiter cap is installed over the head of the screw to prevent accidental tampering.

Under normal service these two screws should remain undisturbed.

CRITICAL WORDS

Perform Steps 16 and 17, **ONLY** if the high speed or low speed mixture cannot be adjusted properly. **ALSO**, perform the steps only if replacement screws have been obtained and are on hand. Failure to adjust the low speed or high speed mixture would suggest the tapered end of one -- but not necessarily both -- of the screws has become damaged. In almost all cases, "damage" is caused by the screw being threaded in too tightly and the tapered end of the needle becoming "grooved", because it passed through the orifice. Once the needle is "grooved", proper mixture cannot be obtained.

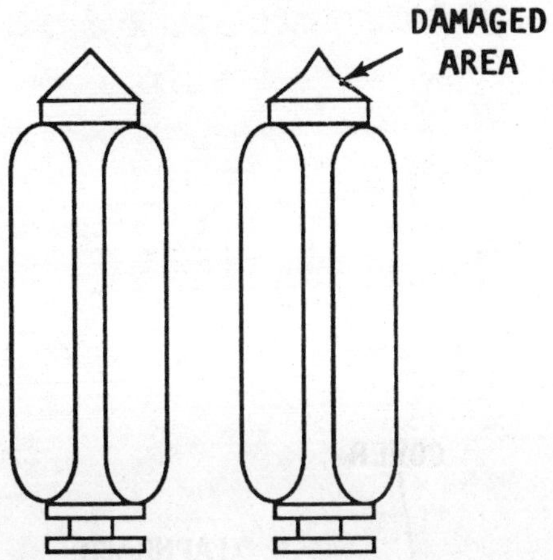

Line drawing of a new valve needle (left), with a smooth tapered surface compared with a damaged needle (right) -- unfit for further service.

16- Pull off the limiter cap on the low speed mixture adjusting screw -- the screw next to the embossed **"L"**. Back out the low speed mixture screw.

17- Pull off the limiter cap on the high speed mixture adjusting screw -- the screw next to the embossed **"H"**. Back out the high speed mixture adjusting screw.

CLEANING AND INSPECTING

NEVER dip rubber parts, plastic parts, or diaphragms in carburetor cleaner. These parts should be cleaned **ONLY** in solvent, and then blown dry with compressed air.

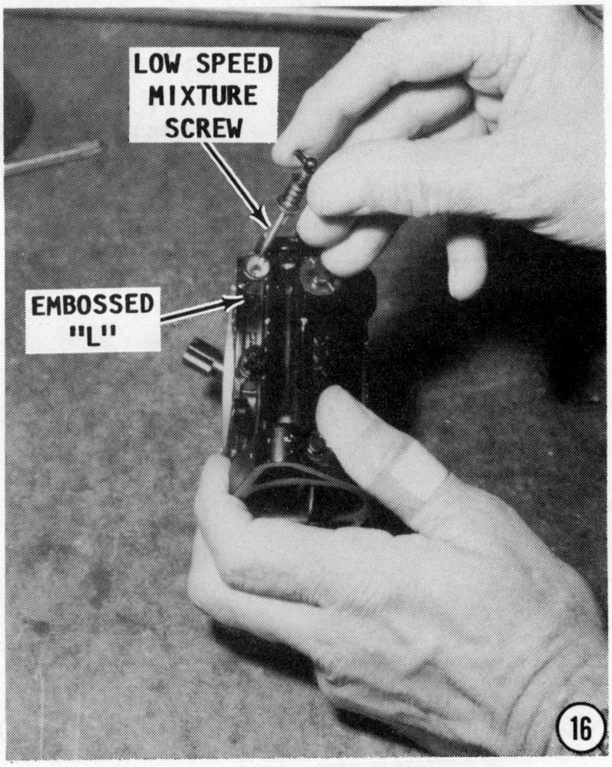

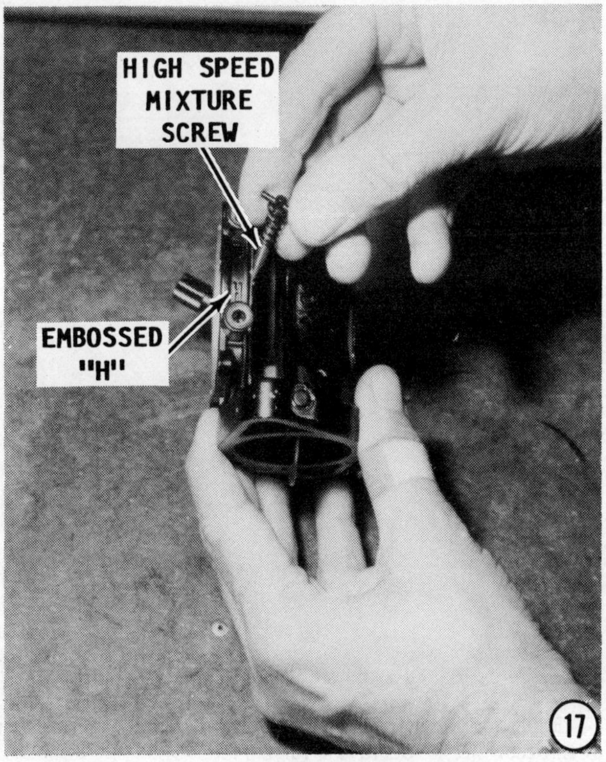

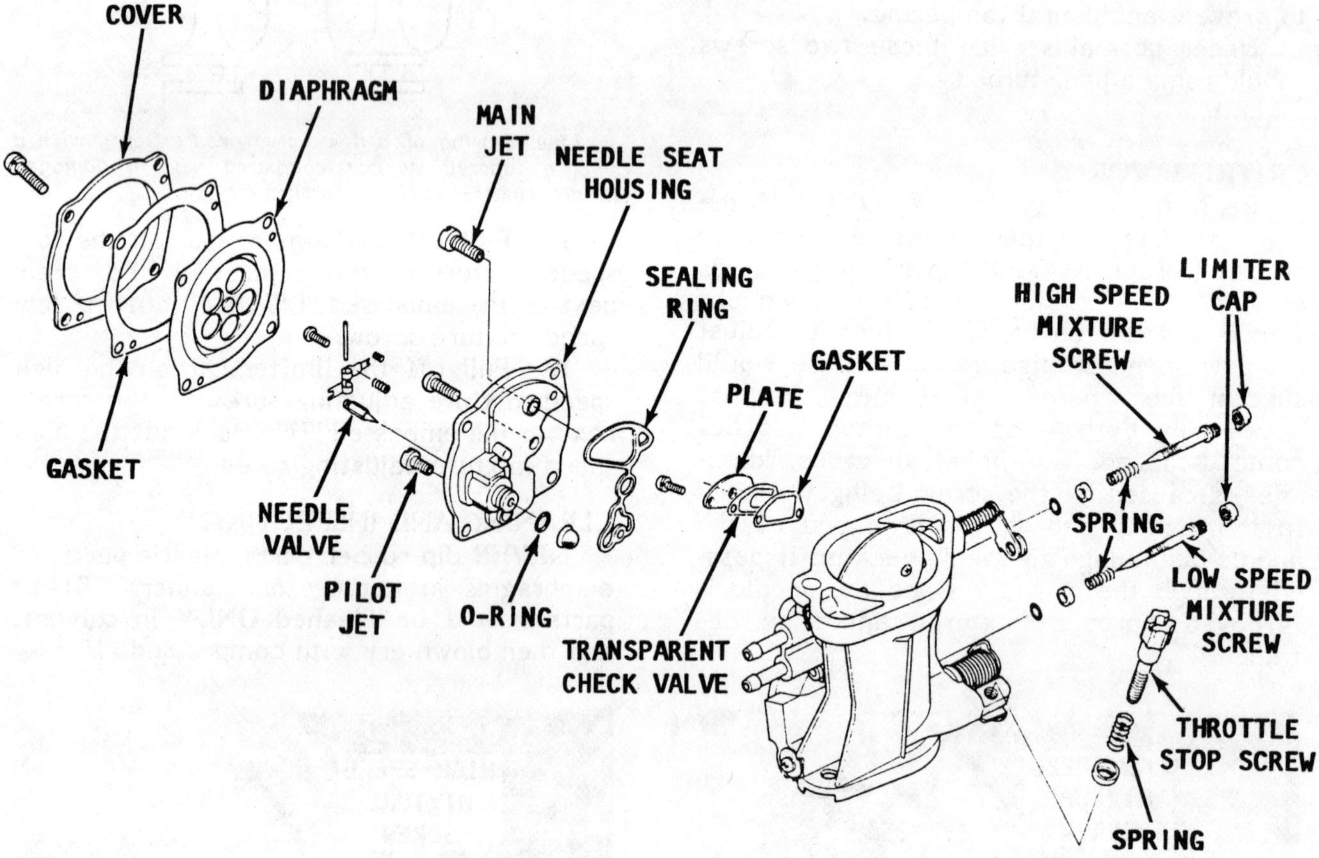

Exploded drawing of the Keihin CDK34 diaphragm carburetor used on the 650 Series prior to 1991 and on all JetMates. Major parts are identified.

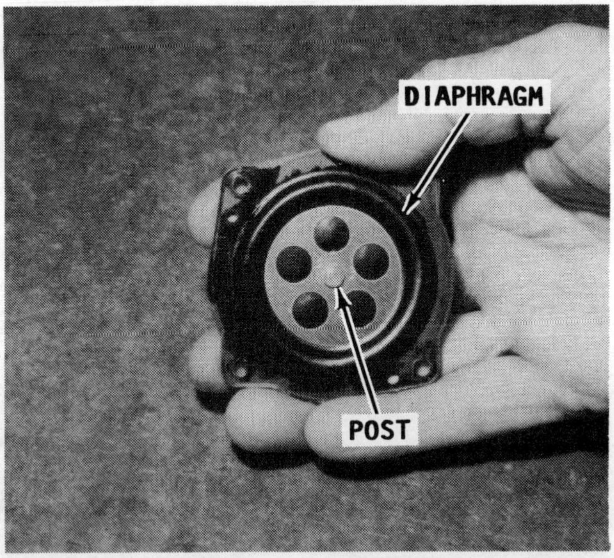

*A damaged diaphragm -- with even the tiniest hole --**MUST** be replaced to obtain full engine performance.*

Place all metal parts in a screen-type tray and dip them in carburetor cleaner until they appear completely clean, then blow them dry with compressed air.

Blow out all passages in the castings with compressed air. Check all parts and passages to be sure they are not clogged or contain any deposits. **NEVER** use a piece of wire or any type of pointed instrument to clean drilled passages or calibrated holes in a carburetor.

Move the throttle shaft back and forth to check for wear. If the shaft appears to be too loose, replace the complete throttle body because individual replacement parts are **NOT** available.

CRITICAL AREA

Inspect the condition of both the choke and throttle shafts. Most shaft designs incorporate a cutout portion to enable the butterfly valve to lay flat against the shaft. Two small holes are drilled in this cutout portion to accommodate the small screws securing the valve to the shaft.

The weakest point of the shaft is at the screw holes. Small cracks form at the holes due to metal fatigue. These cracks elongate the threaded holes and the small screws shake loose. Once loose, they are sucked into the carburetor, then the intake manifold, through the reed valves, and into the crankcase -- eventually finding their way into the combustion chamber. One of these small screws can then cause very extensive and expensive damage to reed valves, pistons, rings, and cylinder walls. Therefore,

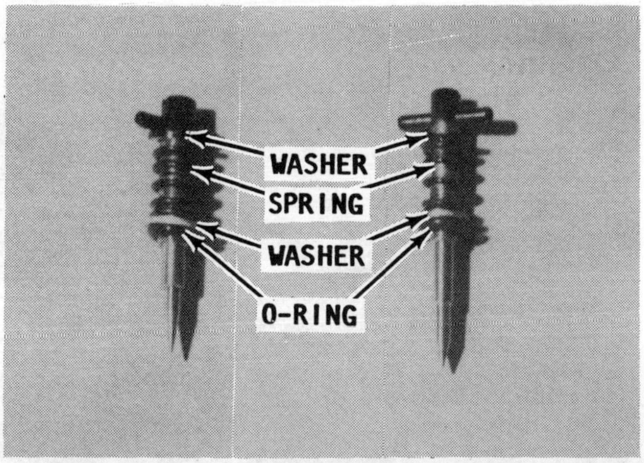

Check to be sure the low speed adjustment screw and the high speed adjustment screw are properly assembled, prior to installation.

check the cutout in the shaft **CLOSELY** for any evidence of a small crack.

Inspect the main body, airhorn, and venturi cluster gasket surfaces for cracks and burrs which might cause a leak. Check to be sure the float arm return spring has not been stretched. Check the float arm needle contacting surface and replace the diaphragm if this surface has a groove worn in it.

As previously mentioned, most of the parts which should be replaced during a carburetor overhaul are included in overhaul kits available from your local marine dealer. One of these kits will contain a fuel inlet needle and seat. The needle should be replaced each time the carburetor is disassembled as a precaution against leakage.

Make a thorough inspection of the fuel chamber and fuel diaphragms for the tiniest pin hole. If one is discovered, the hole will only get bigger. Therefore, the diaphragm must be replaced in order to obtain full performance from the engine.

ASSEMBLING

Special Words on Mixture Screws

Observe the low and high speed mixture screws. They appear to be very similar, but they are **NOT** interchangeable. To identify the correct screw, place them side by side as shown in the accompanying illustration. The low speed screw is shorter than the high speed screw. Both screws are threaded into the carburetor at the base of the fuel pump. The low speed screw is installed next to the embossed letter **"L"**, and the high speed screw is installed next to the embossed letter **"H"**.

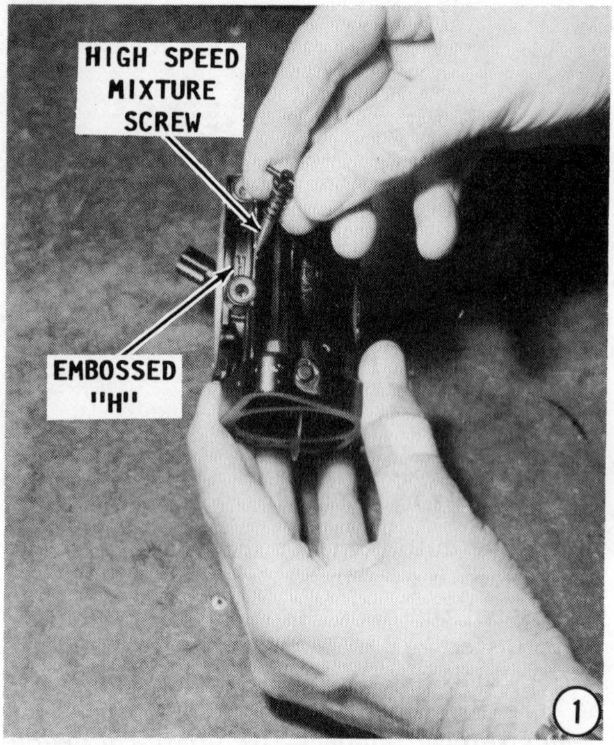

Both screws are adjusted later at the completion of the assembly procedures. Therefore, at this time, thread each screw in just a few turns. Check to make sure both screws are equipped with a washer, a spring, another washer, and an O-ring.

1- Install the high speed screw into the opening embossed with the letter **"H"**.
2- Install the low speed screw into the opening embossed with the letter **"L"**.

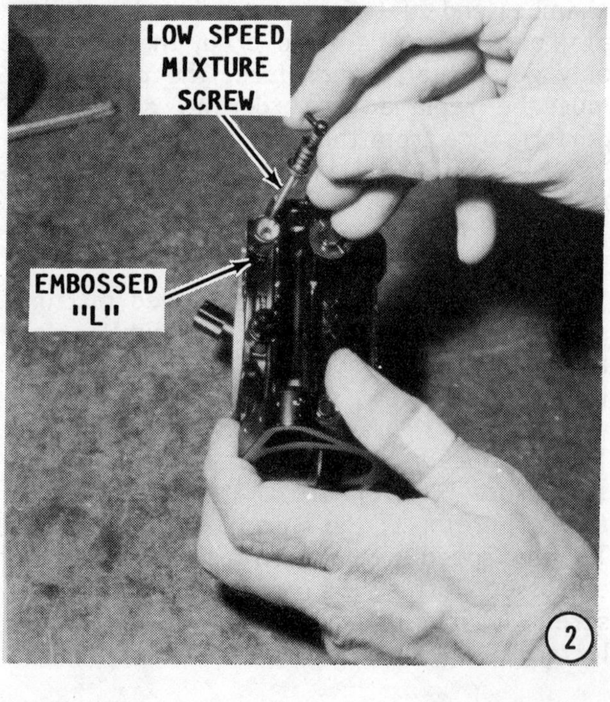

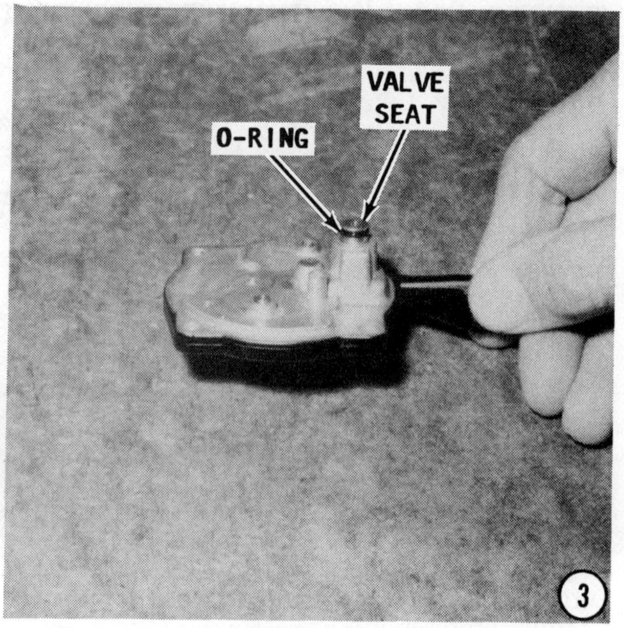

3- Install a new O-ring around the valve seat on the underside of the float chamber housing.

4- Install the small spring into the hole in the center of the float chamber. Slide the hinge pin into the float arm and hook the needle valve into the slot in the arm.

Lower the needle valve into the needle seat and index the dimple in the float arm over the spring. Position the hinge pin into place in the elongated slot in the float chamber and secure it with the small Phillips head screw. Tighten this screw very securely.

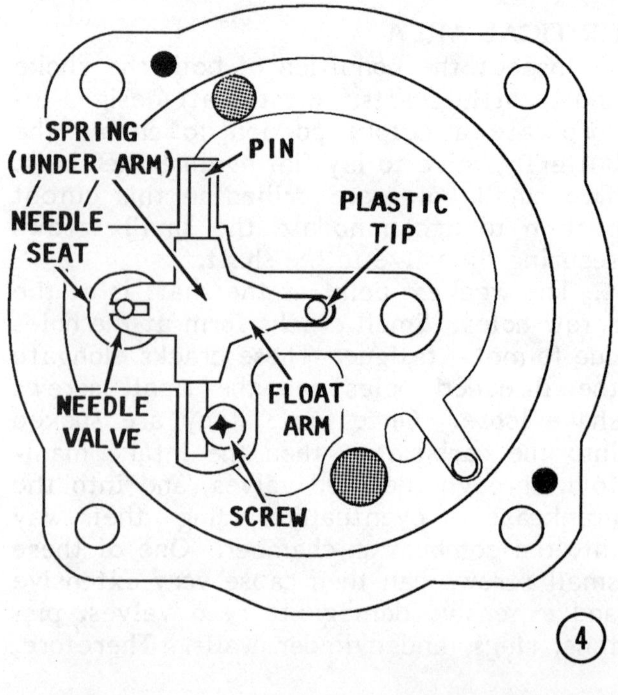

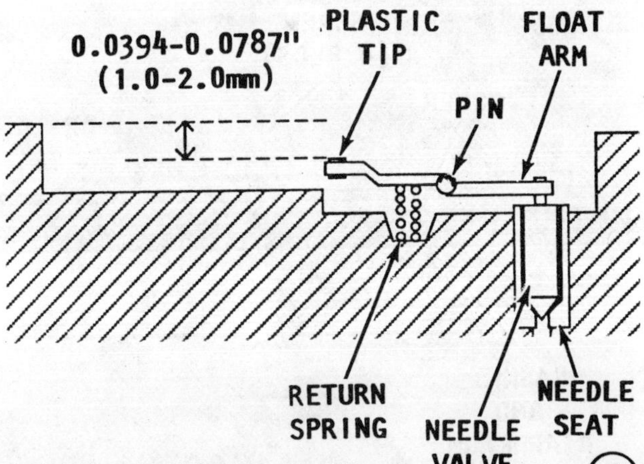

5

Float Arm Adjustment

5- Three conditions must be satisfied in order to achieve correct float arm adjustment.

a) The top of the float arm must be parallel to the fuel chamber surface.

b) The distance between the top surface of the float arm and the diaphragm contact surface of the carburetor must be between 0.0394-0.0787" (1.0-2.0mm).

c) The float arm must make contact with the needle valve, without applying any pressure to the valve.

If any of these conditions are not satisfied, then replace the float arm and/or return spring. **DO NOT** attempt to bend the float arm or to change the characteristics of the old return spring.

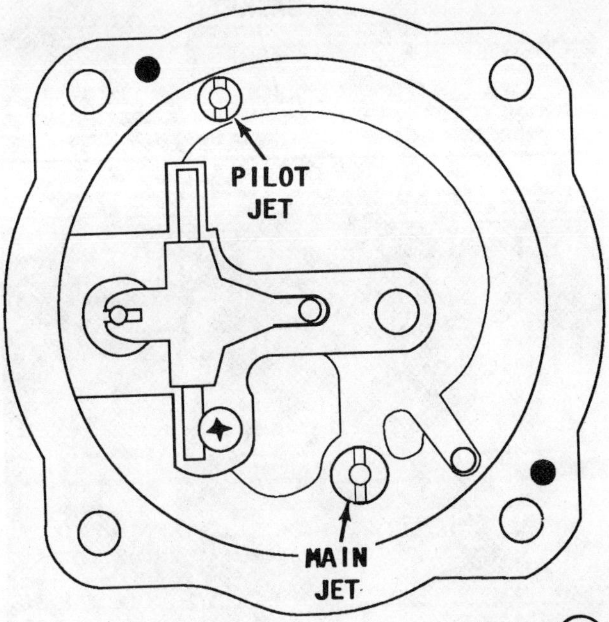

6

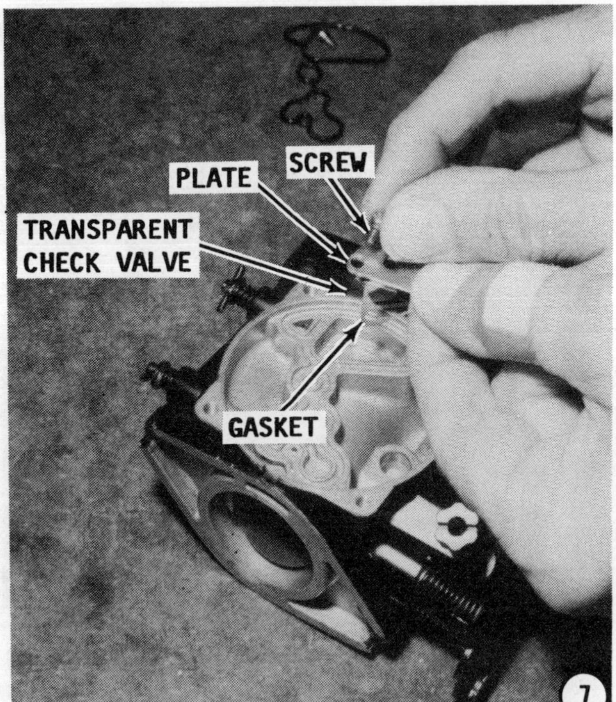

7

Depress the rounded end of the float arm. The needle valve should lift smoothly and return inside the seat when the arm is released.

6- Install the pilot jet -- the smaller of the two -- into the jet opening in the fuel chamber. Tighten the jet securely using the proper size screwdriver.

Install the main jet into the jet opening in the fuel chamber. Again, using the proper size screwdriver, tighten the jet securely.

7- Position the high speed check valve gasket into the recess in the fuel chamber, followed by the transparent check valve. Install the plate over these two items and secure it in place with two Phillips head

8

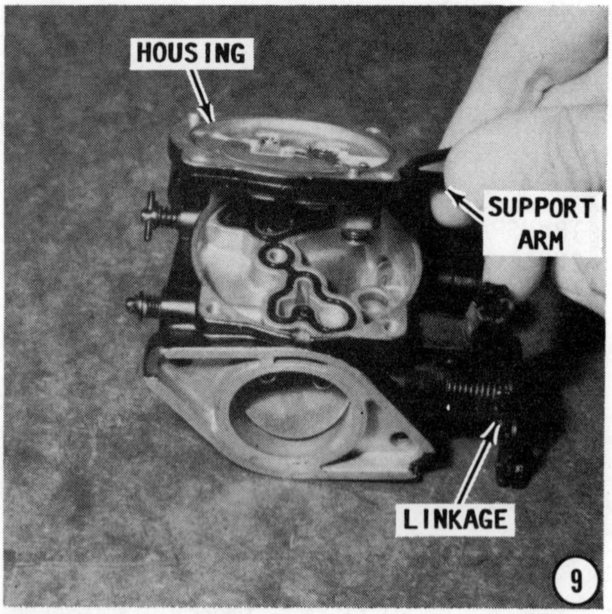

screws. Tighten the screws to a torque value of 1.5 ft lb (2Nm).

8- Position a **NEW** sealing ring over the fuel chamber.

9- Lower the housing over the fuel chamber, with the choke cable support arm on the same side as the linkage.

10- Install the Phillips head screw securing the housing to the chamber. Tighten the screw securely.

11- Place the diaphragm over the fuel chamber with the float post on the underside facing **DOWN** to rest against the float arm. Position the diaphragm cover over the diaphragm. Secure the cover to the carburetor with the four Phillips head screws.

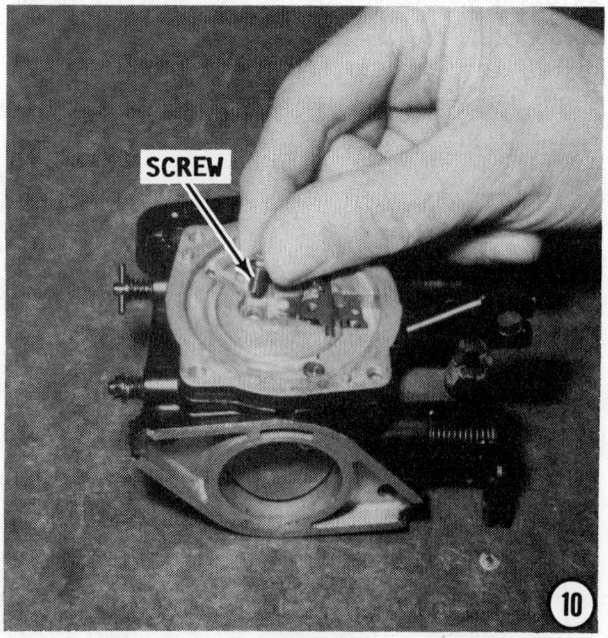

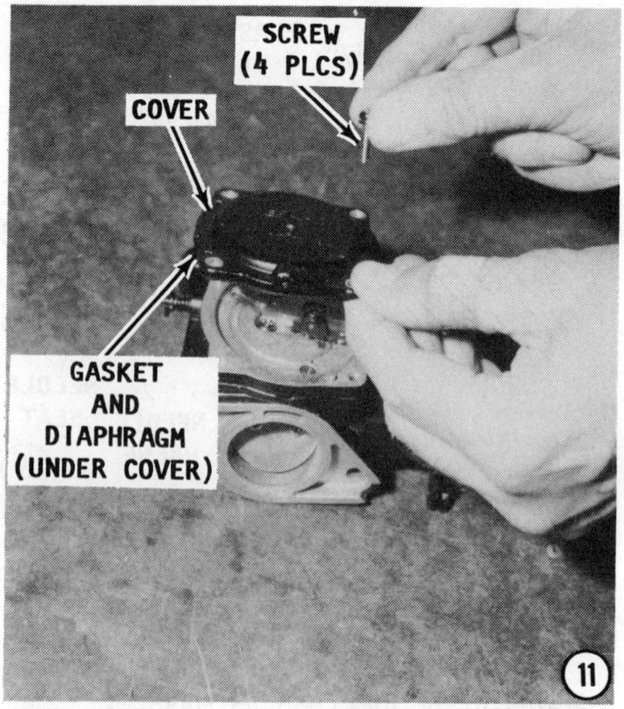

Tighten the screws alternately and evenly to a torque value of 3 ft lb (4Nm).

Carburetor Installation

12- Place a new carburetor mounting gasket over the intake manifold.

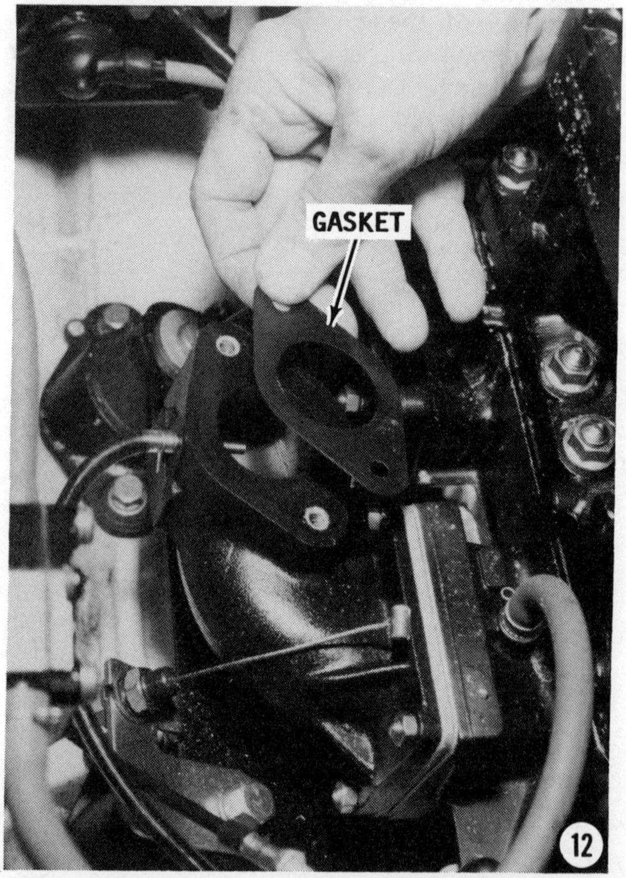

13- Place the carburetor over the intake manifold and gasket, aligning all bolt holes. Position the assembled lower air intake cover over the carburetor and guide the two bolts through the carburetor and gasket as the cover is lowered into place.

14- Tighten the two carburetor mounting bolts. After the bolts have been installed and tightened securely, bend the tab on

the locking bracket up against the side of the bolts to prevent the bolts from backing out. If the tab does not align with the side of a bolt, loosen the bolt very slightly until a tab does align.

15- Position the flame arrestor screen over the lower cover with the cutout/radius edge facing **UPWARD.** Install the top cover and secure the two halves together with the two bolts.

Choke and Throttle Cable Installation

16- Slide the choke cable into the cable support arm on the carburetor, with a lock-

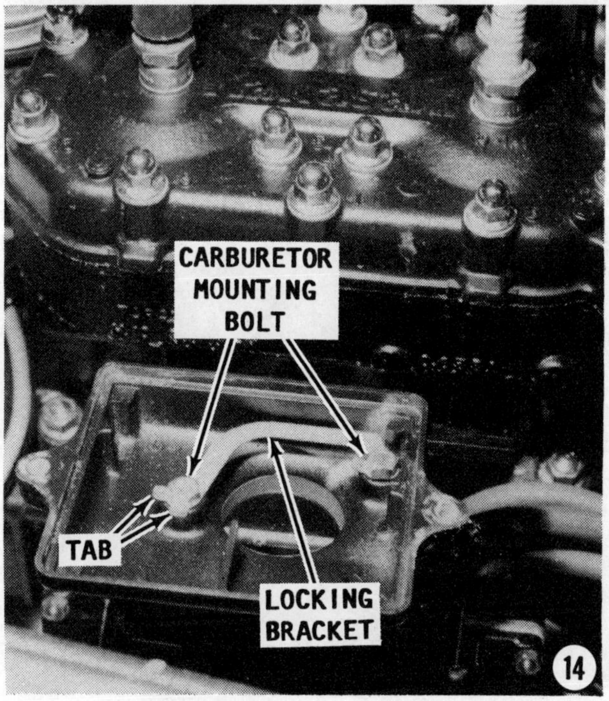

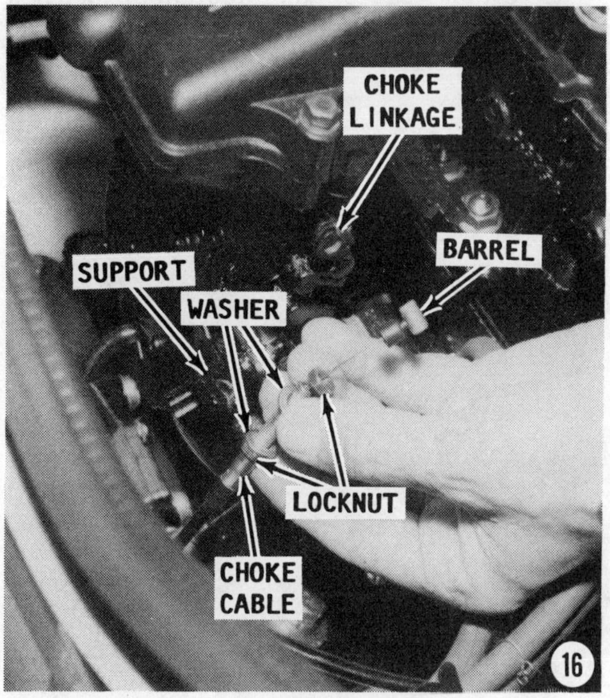

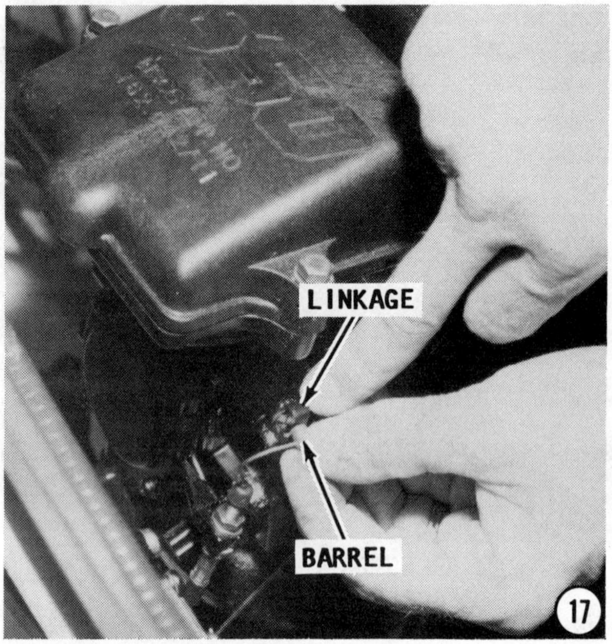

nut and washer on each side of the support. Temporarily, hand tighten the two locknuts against each other.

17- Depress the choke linkage on the carburetor. Push the barrel at the end of the choke cable into the slotted hole in the choke linkage.

18- Adjust the position of the two locknuts to give an acceptable amount of slack in the cable, and then tighten the locknuts against each other to hold the cable in place on the support bracket.

Repeat these last three steps for the throttle cable.

If either cable has stretched, or has become kinked or frayed, or tends to stick, the entire cable should be replaced.

19- Connect the fuel supply hose to the lower of the two fittings on the carburetor.

Carburetor Priming

The carburetor must be primed before starting the engine to save the battery from excessive cranking.

20- Place a shop towel under the fuel return hose fitting. Blow through the hose until fuel starts to drip from the fuel return fitting on the carburetor. The carburetor is now full of fuel and the engine is ready for cranking after the hose has been reconnected.

Mixture Screws with Limiter Caps

Since about 1989, the manufacturer has installed small black plastic limiter caps which snap over both the low speed and high speed adjustment screws. The purpose of these caps is to prevent any unauthorized tampering with the factory setting.

Each carburetor is individually adjusted at the factory using a flow meter. After the adjustment is completed, a limiter cap is installed over the head of the screw to prevent accidental tampering.

Each cap has a tab and the tab fits into a slot in the carburetor body at the base of the mixture screw. Therefore, the optimum position of each screw is with the tab down and indexed into the slot.

However, if the mixture screws are removed during service procedures, the low and high speed mixtures must be adjusted and both caps installed over the screws with the tabs on the caps indexed into slots in the carburetor body.

Idle Adjustment Screw

Identify the idle adjustment screw located on the throttle linkage just below the high speed and low speed mixture screws.

Move the craft to a test tank, to a body of water, or connect a flush attachment and hose to the engine. For maximum performance, the idle rpm should be adjusted under actual operating conditions. Connect a tachometer to the engine.

Start the engine and allow it to warm to operating temperature.

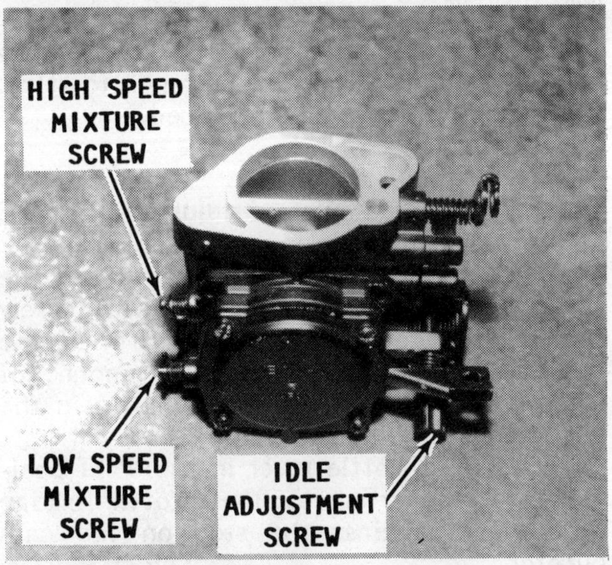

Location of the high speed adjustment screw and the low speed adjustment screw on a Keihin carburetor used on the 650 Series prior to 1991 and on all JetMates.

CAUTION

Water must circulate through the jet pump — to and from the engine, anytime the engine is operating. Circulating water will prevent overheating — which could cause damage to moving engine parts and possible engine seizure.

NEVER, AGAIN NEVER, operate the engine at high speed with a flush device attached. An engine operating at high speed with such a device attached, would **RUNAWAY** from lack of a load on the impeller shaft, causing extensive damage.

With the engine running, slowly rotate the idle adjustment screw until the engine idles at 1,150 to 1,350 rpm.

If the engine idle is set too low, the self-circling radius -- described in Chapter 1 -- increases and the craft may lose its self-circling feature.

If the engine idle is set too high, the craft may circle too fast for a displaced rider to climb aboard the craft.

Low and High Speed Mixture Adjustment

Each carburetor is individually adjusted at the factory using a flow meter and should not need any adjustment. Hovever, if replacement screws were installed the manufacturer suggests a starting point from which fine tuning on the water may be necessary.

Lightly seat both mixture screws. From this position back out the low speed screw 1-1/8 turns. Back out the high speed screw 5/8 of a turn.

High Altitude Operation

The idle system should never be changed when operating the craft at higher altitudes -- the low speed systems are not affected enough by changes in altitude to make an adjustment practical.

The idle adjustment just made is the optimum setting for sea level. When operating at higher elevations, the required leaner mixture is obtained by rotating the high speed screw the specified number of turns **CLOCKWISE** according to the following list.

3,300 ft (1,000 m) -- 1/8
6,600 ft (2,000 m) -- 1/4
9,900 ft (3,000 m) -- 3/8

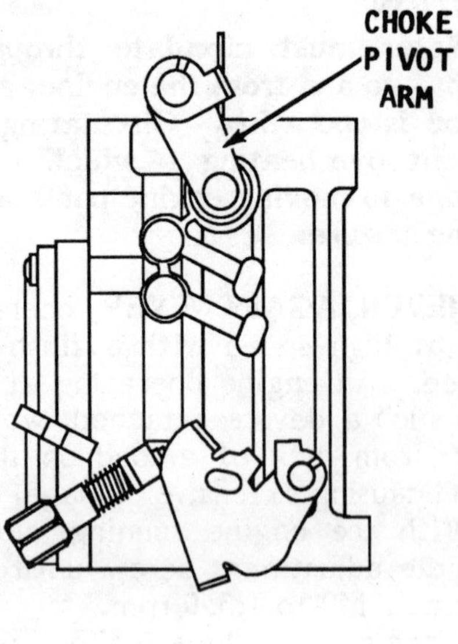

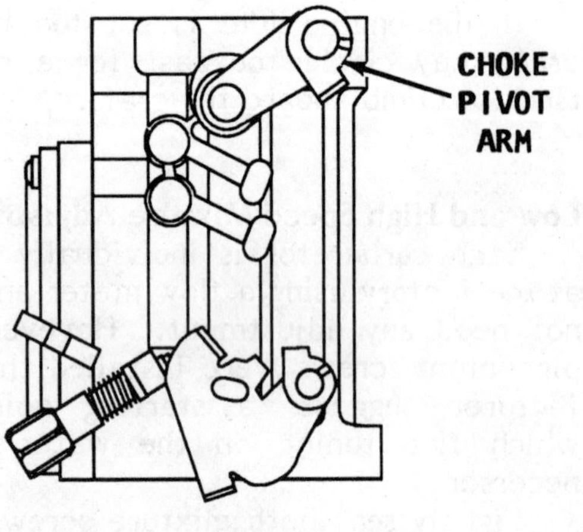

Line drawings to indicate the choke cable adjustment on the 650 Series prior to 1991 except the JetMate, with the choke open (top), and when closed (bottom).

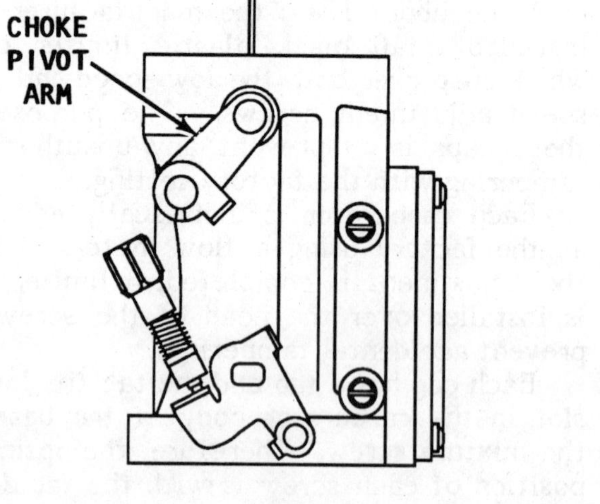

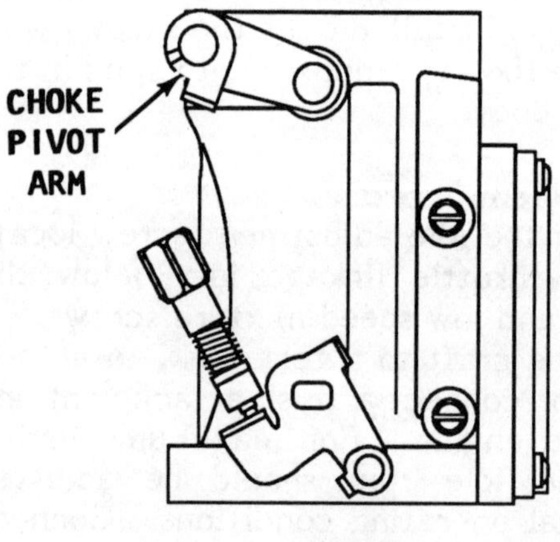

Line drawings to indicate the choke cable adjustment on the JetMate, with the choke open (top), and closed (bottom).

Choke Cable Adjustment

Rotate the choke knob to the full counterclockwise position. Observe the choke pivot arm on the side of the carburetor. The arm should be in the position shown in the accompanying illustration, with a minimum of slack in the choke cable, with the choke butterfly valve fully open.

If the choke needs adjustment: Rotate the choke knob to the full clockwise position. Loosen the two locknuts, one on each side of the cable support bracket. Set the choke pivot to the fully closed position, as shown. Tighten the two locknuts on the support bracket to hold this position, allowing a small amount of slack in the cable.

If the choke cable has stretched, or has become kinked or frayed, or tends to stick, the entire cable must be replaced.

Throttle Cable Adjustment

The idle speed must be adjusted **BEFORE** adjusting the throttle lever free "play".

With the throttle lever released and the throttle closed, the idle adjustment screw should rest against the stopper arm of the throttle linkage, with a slight amount of slack in the throttle cable, as shown in the accompanying illustration.

With the throttle lever in the WOT position, the stopper arm of the throttle linkage should rest against the stop on the carburetor.

If the throttle cable needs adjustment: Loosen and rotate the two locknuts around

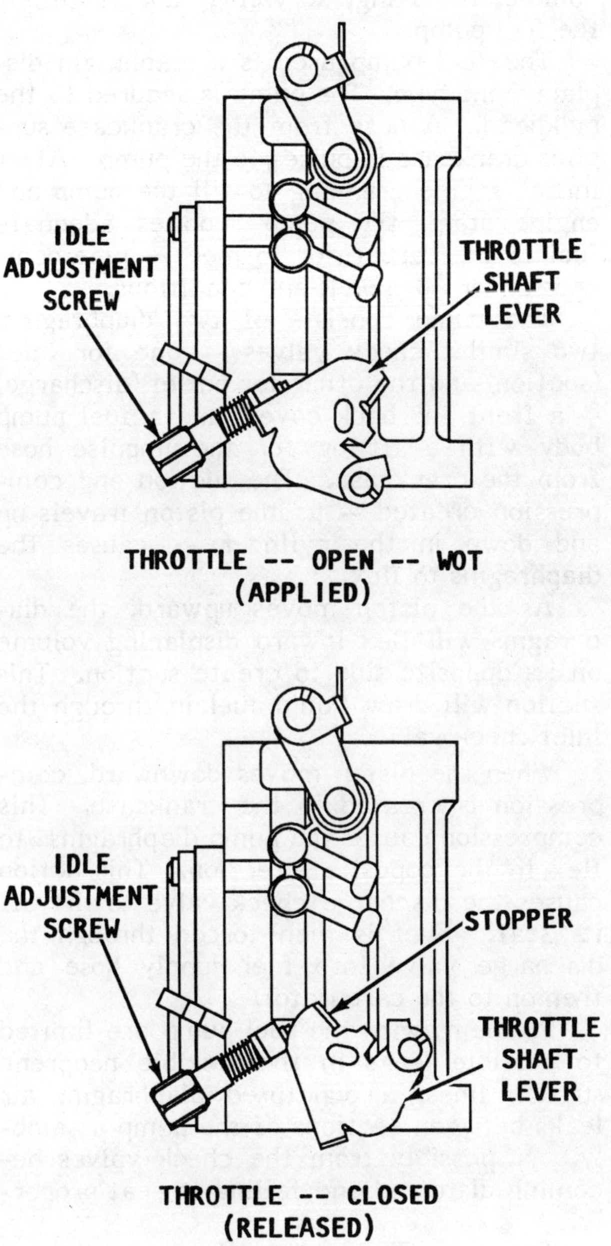

Line drawings to indicate the Keihin carburetor throttle cable positions on the 650 Series prior to 1991, except the JetMate, with the throttle open -- WOT -- (top), and closed (bottom). The JetMate adjustment if made at the cable support bracket as shown in the center illustration — next column.

the throttle cable at the cable support bracket until, with a minimum amount of slack in the cable, the idle adjustment screw rests against the throttle stopper with the throttle released. Tighten the two locknuts to hold this new adjustment.

If the throttle cable has stretched, or has become kinked or frayed, or tends to stick, the entire cable must be replaced.

Start the engine and check the completed work.

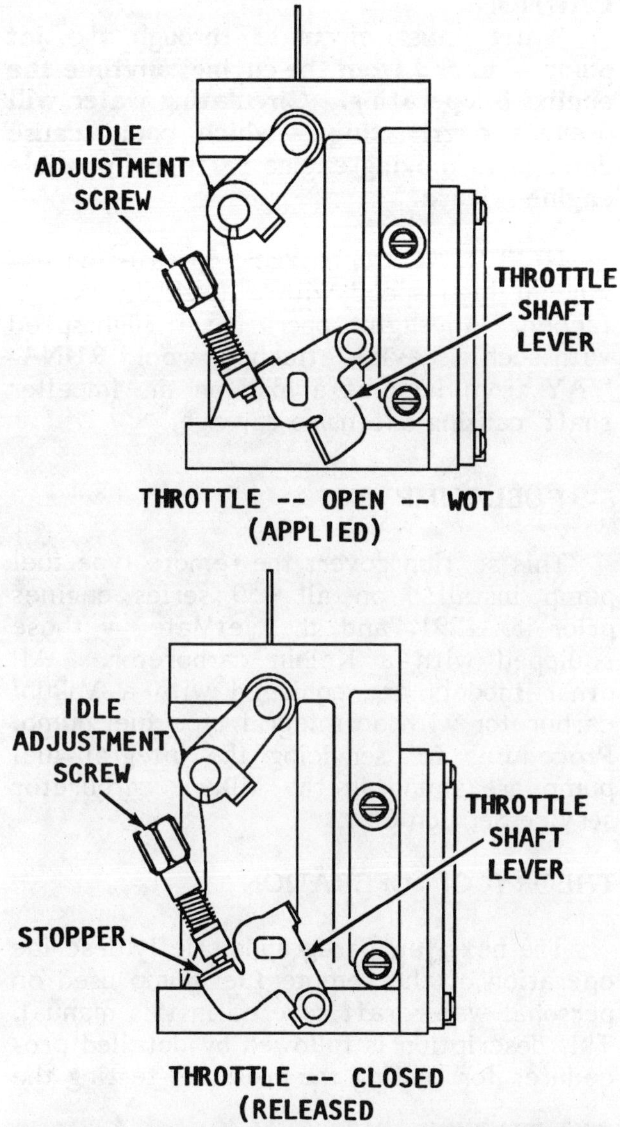

Keihin carburetor throttle cable positions on the JetMate, with the throttle open — WOT -- (top), and closed (bottom). Next illustration indicates adjustment.

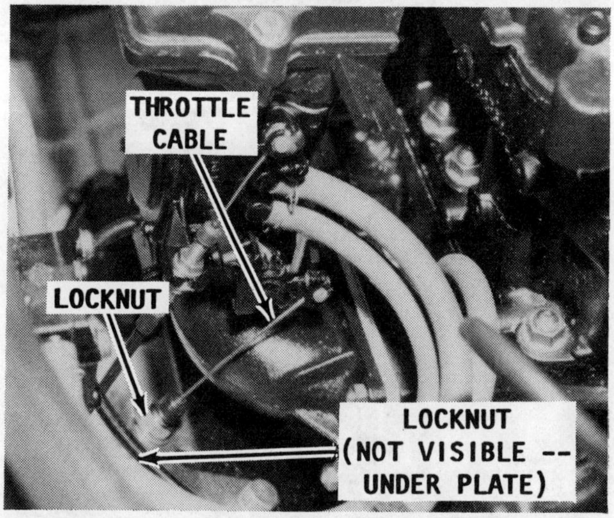

Keihin carburetor throttle cable adjustment, as explained in the text and illustrated in the line drawings above.

CAUTION

Water must circulate through the jet pump -- to and from the engine, anytime the engine is operating. Circulating water will prevent overheating — which could cause damage to moving engine parts and possible engine seizure.

NEVER, AGAIN NEVER, operate the engine at high speed with a flush device attached. An engine operating at high speed with such a device attached, would **RUNAWAY** from lack of a load on the impeller shaft, causing extensive damage.

6-9 FUEL PUMP

This section covers the remote type fuel pump installed on all 650 series engines prior to 1991, and the JetMate -- those equipped with a Keihin carburetor. All other models are equipped with a Mikuni carburetor with an integral type fuel pump. Procedures for servicing the integral fuel pump are found in the Mikuni carburetor service Section 6-7.

THEORY OF OPERATION

The next few paragraphs briefly describe operation of the remote fuel pump used on personal watercraft covered in this manual. This description is followed by detailed procedures for testing the pressure, testing the volume, removing, servicing, and installing the fuel pump.

Location of the remote fuel pump on the 650 Series prior to 1991, including the JetMate.

The fuel pump used is a diaphragm displacement type. The pump is secured to the bulkhead. A hose from the crankcase supplies crankcase impulses to the pump. After initial engine cranking to fill the pump and engine start, the pump supplies adequate fuel to the carburetor to meet engine demands under all speeds and conditions.

The pump consists of two diaphragms, two similar check valves -- one for inlet (suction) and the other for outlet (discharge) -- a front and back cover, and a fuel pump body with a fitting for the impulse hose from the crankcase. The suction and compression created -- as the piston travels up and down in the cylinder -- causes the diaphragms to flex.

As the piston moves upward, the diaphragms will flex inward displacing volume on its opposite side to create suction. This suction will draw liquid fuel in through the inlet check valve.

When the piston moves downward, compression is created in the crankcase. This compression causes the pump diaphragms to flex in the opposite direction. This action causes the discharge check valve to lift off its seat. Fuel is then forced through the discharge valve into fuel supply hose and then on to the carburetor.

Problems with the fuel pump are limited to possible leaks in the flexible neoprene suction lines; a punctured diaphragm; air leaks between sections of the pump assembly, or possibly from the check valves becoming distorted and failing to seat properly.

The pump is activated by one cylinder. If this cylinder indicates a wet fouled condition, as evidenced by a wet fouled spark plug, be sure to check the fuel pump diaphragms for possible puncture or leakage.

Even the tiniest pin hole in a diaphragm will cause loss of engine performance.

PUMP PRESSURE CHECK

Lack of an adequate fuel supply will cause the engine to operate lean, lose rpm, or cause piston scoring.

Check the fuel strainer on the end of the pickup in the fuel tank to be sure it is not too small or is not clogged. Be sure to check the in-line filter at the carburetor. If

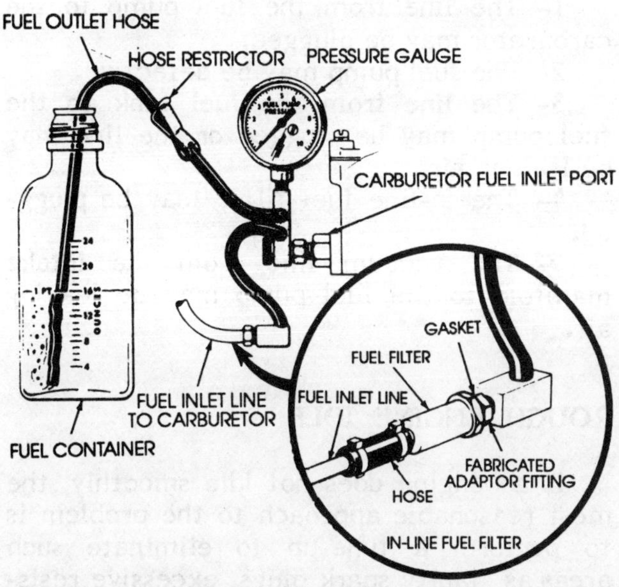

Test setup to check fuel pump pressure, as described in the text.

this screen becomes clogged, adequate fuel cannot pass through into the carburetor to meet engine demands.

TO TEST

Install a fuel pressure gauge in the fuel line between the fuel pump and the carburetor.

Start the engine and allow it to warm to operating temperature and then check the fuel pressure.

The fuel filter should be kept as clean as possible to ***ENSURE*** *an adequate fuel supply to the carburetor under all engine demands.*

CAUTION

Water must circulate through the jet pump -- to and from the engine, anytime the engine is operating. Circulating water will prevent overheating -- which could cause damage to moving engine parts and possible engine seizure.

NEVER, AGAIN NEVER, operate the engine at high speed with a flush device attached. An engine operating at high speed with such a device attached, would **RUNAWAY** from lack of a load on the impeller shaft, causing extensive damage.

Operate the enigne at full throttle and check the pressure reading. The gauge should indicate at least 2 psi.

PUMP VOLUME CHECK

First, These Words

The following test applies only to personal watercraft equipped with a remote type fuel pump separate from the carburetor.

CAUTION

Gasoline will be flowing in the engine area during this test. Therefore, guard against fire by grounding the high-tension leads to prevent possible sparking, which could easily ignite fuel or fuel vapors.

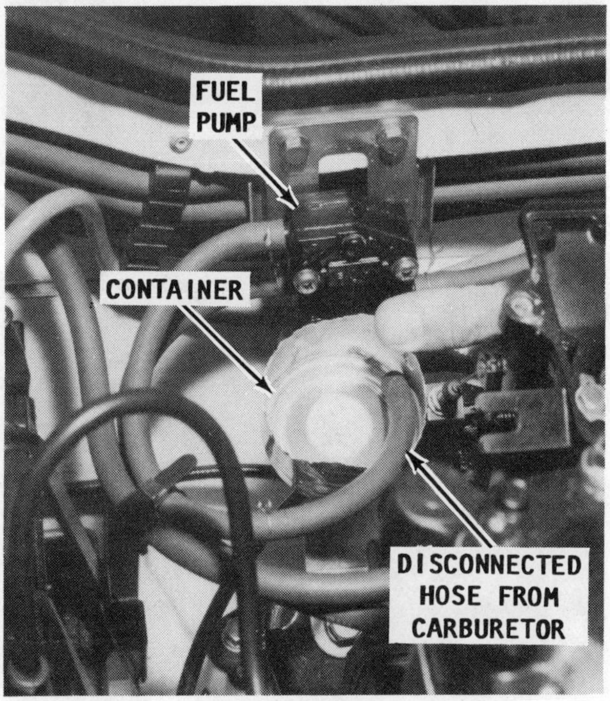

Test setup to check fuel volume, as outlined in the text.

A good safety procedure is to ground both spark plug leads. Disconnect the fuel line at the carburetor. Place a suitable container over the end of the fuel line to catch discharged fuel. Crank the engine and observe the flow of fuel discharged into the container.

If no fuel is discharged from the hose, check the crankcase vacuum hose connections and the fuel hose for leaks. Another possible problem may be a break or obstruction in the fuel line.

If the fuel pump is operating properly, a healthy stream of fuel should pulse from the fuel hose.

Continue cranking the engine and catching fuel for about 15 pulses to determine if the amount of fuel decreases with each pulse or maintains a constant amount. A decrease in the discharge indicates a restriction in the line. If the fuel line is plugged, the fuel stream may stop.

If there is fuel in the fuel tank but no fuel is discharged while the engine is being cranked, the problem may be in one of five areas:

If air is allowed to enter the vacuum line between the intake manifold and the fuel pump, the pump will fail to operate properly.

1- The line from the fuel pump to the carburetor may be plugged.

2- The fuel pump may be defective.

3- The line from the fuel tank to the fuel pump may be plugged or the line may be leaking air.

4- The in-line fuel filter may be plugged.

5- The vacuum line from the intake manifold to the fuel pump may be leaking air.

ROUGH ENGINE IDLE

If an engine does not idle smoothly, the most reasonable approach to the problem is to perform a tune-up to eliminate such areas as faulty spark plugs, excessive resistance in high tension leads, or timing out of adjustment.

Other problems contributing to rough engine idle may include:

An air leak in the intake manifold.
Uneven compression between cylinders.
Sticky or broken reeds, if so equipped.

SERVICING THE FUEL PUMP

The fuel pump should **NOT** be disassembled unless internal damage is suspected or the pump is leaking fuel. A fuel pump rebuilding kit must be on hand to ensure faulty parts will be replaced. Most fuel pump kits include new diaphragms, gaskets and check valves.

Disassembling and assembling should be performed on a clean work surface. Make every effort to prevent foreign material from entering the fuel pump or adhering to the diaphragms.

REMOVAL AND DISASSEMBLY

The pump consists of fragile diaphragms which are easily damaged. The pump should not be disassembled unless replacement diaphragms and gaskets are on hand.

Disconnect the vacuum hose, the inlet hose and the outlet hose from the fuel pump. Take time to plug the disconnected ends of all three hoses to prevent contaminants from entering the crankcase and fuel

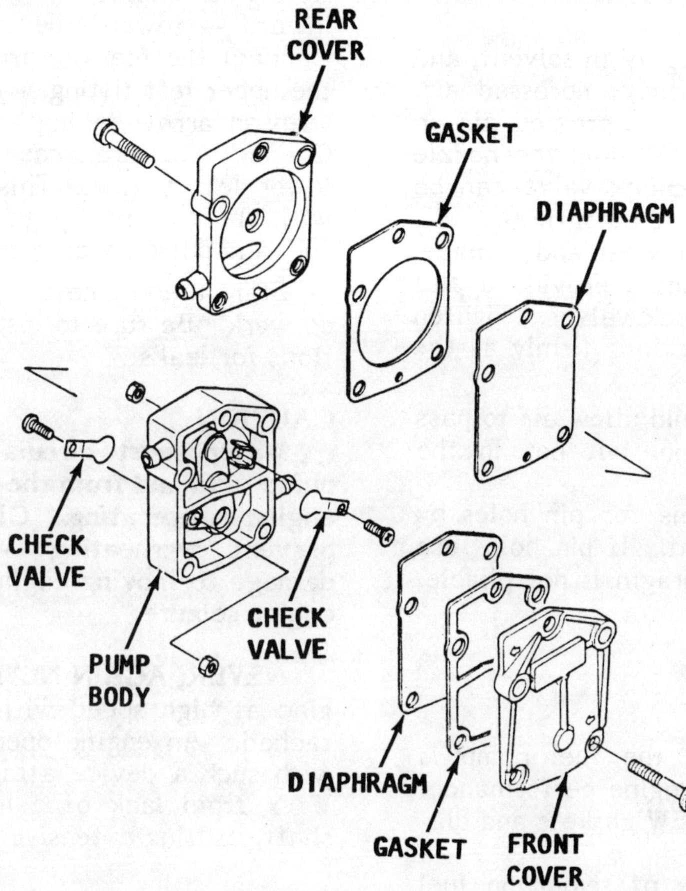

Exploded drawing of the remote fuel pump used on all 650 Series prior to 1991 and all JetMate watercraft, with major parts identified.

system. Observe the five screws on the fuel pump cover. Two of the screws are silver and the remaining three are black. Only the two silver screws are used to mount the pump to the bracket, the black screws hold the internal components of the the pump together. Therefore, remove **ONLY** the two silver screws.

Remove the pump and move it to a suitable clean work surface. Refer to the accompanying exploded drawing of the fuel pump as an aid to disassembling and assembling. Remove the three black screws securing the pump parts together.

Now, **CAREFULLY** separate the parts and keep them in **ORDER** as an assist in assembling. As parts are removed, take time to observe and remember how each check valve faces, because it must be installed in exactly the same manner, or the pump will not function.

*Head on view of the fuel pump clearly showing the **TWO** mounting screws and the three screws to be left in place -- as instructed in the text.*

CLEANING AND INSPECTING

Wash all parts thoroughly in solvent, and then blow them dry with compressed air. **USE CARE** when using compressed air on the check valves. **DO NOT** hold the nozzle too close because the check valve can be damaged from an excessive blast of air.

Inspect each part for wear and damage. Verify that the valve seats provide a flat contact area for the check valves. Tighten both check valve connections firmly as the valves are replaced.

The check valves should allow air to pass through in one direction, but not in the opposite direction.

Check the diaphragms for pin holes by holding it up to the light. If pin holes are detected or if the diaphragm is not pliable, it **MUST** be replaced.

ASSEMBLING

Proper operation of the fuel pump is essential for maximum engine performance. Therefore, always use **NEW** gaskets and diaphragms.

NEVER use any type of sealer on fuel pump gaskets.

TAKE CARE not to damage the very fragile flat surface of the check valve. Secure each check valve in place with a Phillips head screws. Tighten the screws securely.

Install one of the diaphragms and the gasket with the large circle cut out, over the rear pump body and then install the rear cover -- with the single crankcase pulse fitting.

Hold these parts together and turn the pump over. Install the second diaphragm, followed by the partitioned gasket and outer pump cover onto the front side of the pump body.

Check to be sure the holes for the screws are all aligned through the covers, diaphragms, and gaskets. If the diaphragms are not properly aligned, a tear would surely develop when the screws are installed.

Install the three black Phillips head screws through the various parts and tighten the screws securely.

Secure the pump to the bracket with the two silver Phillips head screws.

Connect the fuel line from the filter to the fitting on the lower right side. This fitting is embossed with an arrow pointing inward -- toward the center of the pump. Connect the fuel line from the carburetor to the upper left fitting -- the fitting embossed with an arrow facing away from the pump. Connect the crankcase pulse hose to the lower left fitting. This fitting is embossed with the "infinity" sign (∞). Secure the fuel and pulse hoses with wire type clamps.

Start the engine and check the completed work. Be sure to inspect all fuel connections for leaks.

CAUTION

Water must circulate through the jet pump -- to and from the engine, anytime the engine is operating. Circulating water will prevent overheating — which could cause damage to moving engine parts and possible engine seizure.

NEVER, AGAIN NEVER, operate the engine at high speed with a flush device attached. An engine operating at high speed with such a device attached, would **RUNAWAY** from lack of a load on the impeller shaft, causing extensive damage.

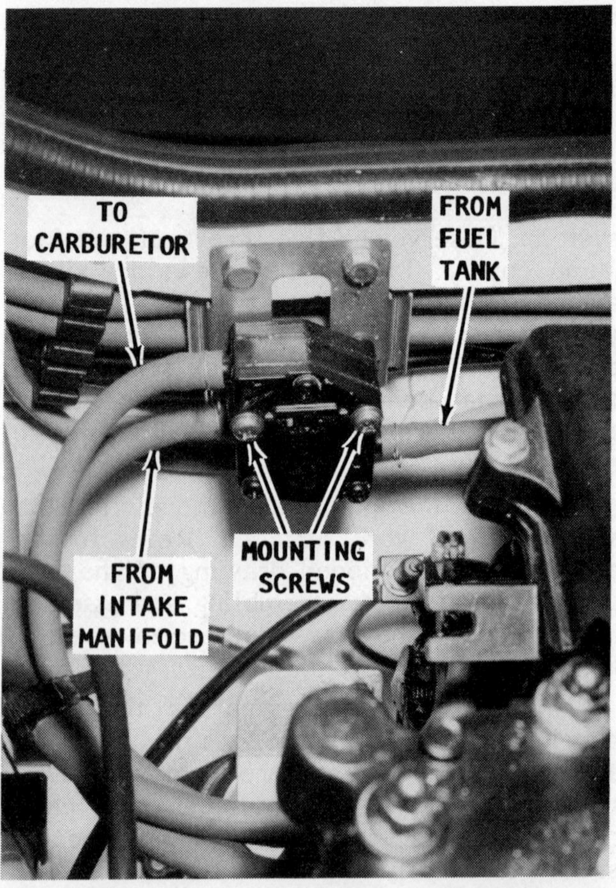

Hose identification for the remote fuel pump.

6-10 OIL INJECTION

Oil injection is standard equipment on the 300 and 650 series engines. The 440 and 550 use a pre-mix -- oil mixed with the fuel in the tank.

In recent years a more sophisticated method of mixing oil with the fuel for internal lubrication has become more popular with the manufacturers of personal watercraft. This oil injection system replaces the age old method of manually mixing oil with the fuel for lubrication of internal moving parts in the engine.

Kawasaki engineers designed and developed their oil injection system as an independent unit -- separate from the fuel system until the lubricant enters the intake manifold. This system injects oil directly into the intake manifold between the carburetor and the reed valve. Such an arrangement provides immediate oil delivery response to changes in rpm and load demands. Because the oil bypasses the carburetor, gummy deposits in the carburetor passages are avoided.

Oil Mixture

Fill the oil tank to the top line embossed on the tank. Use only two cycle N.M.M.A. engine oil (or an equivalent oil with a BIA certified rating TC-W). This oil is suitable for use through a temperature range of 14°F to 140°F (-10°C to 60°C).

The rate of oil injection is controlled by engine rpm. The oil pump is directly coupled to the top of the crankshaft. After the initial break-in period, oil is injected at a rate giving approximately a 40:1 mixture at low and idle speed. The mixture increases as the throttle is advanced to WOT (wide open throttle).

In order to obtain extra lubrication during the first 10 hours of break-in, the manufacturer recommends oil to be added directly into the fuel tank to give a mixture of 25:1.

Professional mechanics, who own oil injected units, constantly add an extra ounce of oil per each gallon of fuel after the break-in period. The addition of this extra oil may prevent engine seizure in the event of an oil pump failure or clog in an oil delivery line.

Mechanics are quite cost conscious. They realize all too well how much cheaper replacing fouled spark plugs can be when weighed against an engine overhaul to replace expensive parts damaged from lack of adequate lubrication.

Use only outboard marine oil in the mixture, never automotive oils. Four stroke automotive engine oil is not formulated to burn completely, only to lubricate. Therefore, automotive oil, if used in a two stroke engine, will leave undesirable residue.

SYSTEM COMPONENTS

Principle components of the oil injection system include:

Main oil tank.
Oil gauge — some models.
Two one-way check valves.
A breather valve.
An oil injection pump.
A network of hoses.

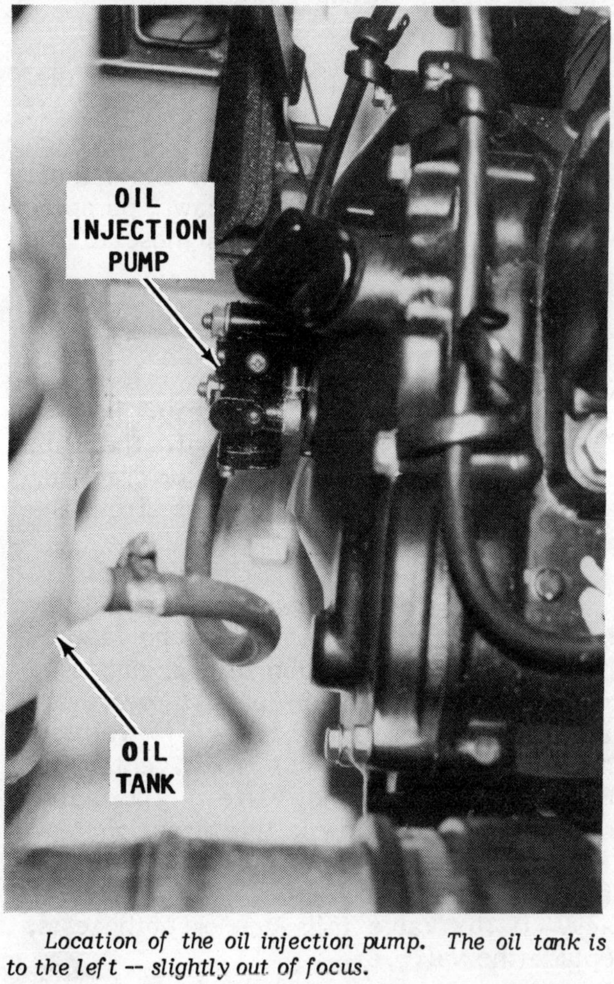

Location of the oil injection pump. The oil tank is to the left -- slightly out of focus.

Oil Tank

The oil tank is mounted inside the engine compartment, with the filler easily accessible when the engine cover is removed. The capacity of this tank is as follows:

JS300A1	2.2L
JS300B1	1.7L
JS300B2	1.7L
JS300B3	1.7L
JS650A1	2.8L
JS650A2	2.8L
JetMate	5.5L

Oil supply to the oil injection pump is gravity fed from the oil tank. Therefore, a one-way check valve is mounted on top of the tank. The other end of the valve is open to the atmosphere. The check valve allows air to enter the tank replacing oil used, but will not allow oil to flow out of the tank -- making the tank "spill-proof".

Oil Injection Pump

The oil injection pump is a positive displacement type unit driven by a short shaft indexed into the crankshaft. With each revolution of the crankshaft the short shaft moves a plunger cam up and down, pumping oil through a single transparent hose to the intake manifold.

System Inspection

Remove the oil tank filler cap. Inspect the condition of the seal beneath the cap, and replace the seal if it is worn or cracked.

Disconnect the ventilation hose from the fitting on the oil tank.

To test operation of the check valve:

First, pull off the two hoses and blow through the fitting centered on one face of the valve, in the direction of the embossed arrow. The air should pass through the valve **ONLY** in this direction. Now, blow through the fitting which is off-center, on the other face of the valve. The valve should prevent air flow in this direction.

If the valve passes both tests, install the valve with the embossed arrows toward the tank. If the valve fails one or both tests, replace the valve.

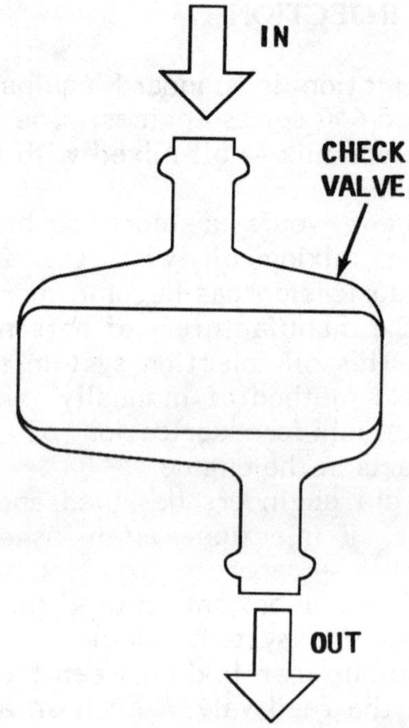

Very simple line drawing to depict the one-way check in the oil tank. This valve allows air to pass into the tank to replace oil used, but will prevent oil from leaving the tank, even if the craft is rotated 360°.

If the arrows are missing, the fitting in the center of the valve face is the air "in" fitting -- air should pass through in only one direction. The off-center fitting should have the hose connected to the oil tank.

A faulty check valve will starve the engine of lubrication and quickly cause extensive and expensive damage.

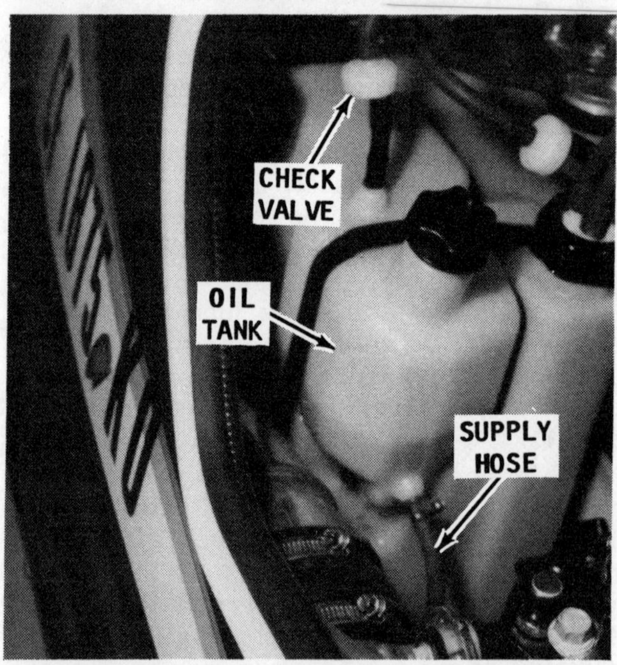

Arrangement of the oil tank, one-way check valve to allow air to enter the tank, and the supply hose to the oil injection pump.

Oil Pump Output Test

Prepare about a ten minute supply of 40:1 pre-mix using a good grade of two-stroke oil. "Jury-rig" a setup to provide this mixture to the fuel pump during this purging procedure. Check the level of oil in the oil tank and replenish as necessary.

Obtain a transparent measuring cylinder graduated in mL. If such a container is not available, use a measuring cup graduated in ounces. A conversion table on the first page of the Appendix may be used to convert ounces to milli-litres.

Start the engine and allow it to warm to operating temperature.

CAUTION

Water must circulate through the jet pump -- to and from the engine, anytime the engine is operating. Circulating water will prevent overheating — which could cause damage to moving engine parts and possible engine seizure.

NEVER, AGAIN NEVER, operate the engine at high speed with a flush device attached. An engine operating at high speed with such a device attached, would **RUNAWAY** from lack of a load on the impeller shaft, causing extensive damage.

SPECIAL WORDS

Because the fuel and oil premix is being fed through the carburetor, in addition to oil from the oil injection system also being fed directly into the cylinders, the engine will be operating on a heavy mixture of oil.

Using a pair of needlenose pliers, squeeze the clamp on the transparent delivery hose. Push the clamp up the hose. Gently pull the hose from the fitting on the intake manifold and immediately hold it over the container.

Observe the flow of oil for four full minutes to verify the pump is functioning properly. With the engine operating at 3000 rpm for 2 minutes, the pump will deliver 3.3-4.1mL of oil for a 300 series engine or 5.9-7.2mL of oil for a 650 series engine.

MORE SPECIAL WORDS

The oil/fuel mixture prepared earlier will provide adequate lubrication while the system is being tested.

A steady slow pulsing flow with no air bubbles may be expected. Reconnect the line. Shut down the engine and remove the flushing device.

Replenish the oil tank with two-stroke oil.

Troubleshooting

Unfortunately, most problems with the oil injection system are only discovered after it is too late. If either of the two hoses becomes clogged, kinked, or restricted, the oil delivery to the cylinders will be reduced.

Insufficient oil delivery to the engine will cause the oil level to drop slower than normal; the engine will overheat due to inadequate lubrication; and moving internal parts will wear more quickly or be severely damaged.

Excessive oil delivery to the cylinders will cause the oil level to drop at a much faster rate than normal. The engine will smoke -- especially at idle speed -- and the spark plugs will become fouled, causing the engine to misfire.

Oil is injected through a nozzle into the intake manifold under the carburetor. Here, the carburetor has been removed to show the oil injection nozzle.

First Checks
Oil Delivery
If excessive or reduced oil delivery occurs, as described above, one or more of the following areas may require immediate attention:

Defective oil pump.
Clogged oil tank check valve.
Kinked oil hose.
Blocked oil hose.
Air in the system.
Leaking oil hose.
Improper idle fuel adjustment.
Poor quality of fuel.
Poor quality of two-stroke oil.

BAD NEWS
No overhaul kits or spare parts are available for the oil pump used on these units. Therefore, a defective oil pump must be replaced.

Systematically check each oil line and connection. A free flow of oil through an unrestricted line is **CRITICAL** for adequate engine lubrication.

CAUTION
Any time the oil tank hose is disconnected, the oil injection pump **MUST** be purged ("bled") of any trapped air. Failure to "bleed" the system could lead to engine seizure due to lack of adequate lubrication.

Instructions for "bleeding" -- purging -- air from the system follows.

Purging Air From Oil Injection System
The following procedures are to be performed any time the oil injection system has been opened (other than to add oil to the tank) and air has entered the system.

Prepare about a ten minute supply of 40:1 premix using two-stroke oil. "Jury-rig" a setup to provide this mixture to the fuel pump during this purging procedure. The oil/fuel mixture prepared will provide adequate lubrication while the system is being purged. Check the level of oil in the oil tank and replenish as necessary.

Start the engine and allow it to warm to operating temperature.

CAUTION
Water must circulate through the jet pump -- to and from the engine, anytime the engine is operating. Circulating water will prevent overheating -- which could cause damage to moving engine parts and possible engine seizure.

NEVER, AGAIN NEVER, operate the engine at high speed with a flush device attached. An engine operating at high speed with such a device attached, would **RUNAWAY** from lack of a load on the impeller shaft, causing extensive damage.

SPECIAL WORDS
Because the fuel and oil premix is being fed through the carburetor, in addition to oil from the oil injection system also being fed directly into the manifold, the engine will be operating on a heavy mixture of oil.

Obtain a small container and have a shop towel handy to catch any oil drips. Loosen the bleed screw on the oil pump a few turns and wait until a slow pulsing flow of oil emerges without any air bubbles. Observe the flow of oil for four full minutes to verify the pump is functioning properly and all air has been expelled from the system. Tighten the screw securely.

Replenish the oil tank with two-stroke oil.

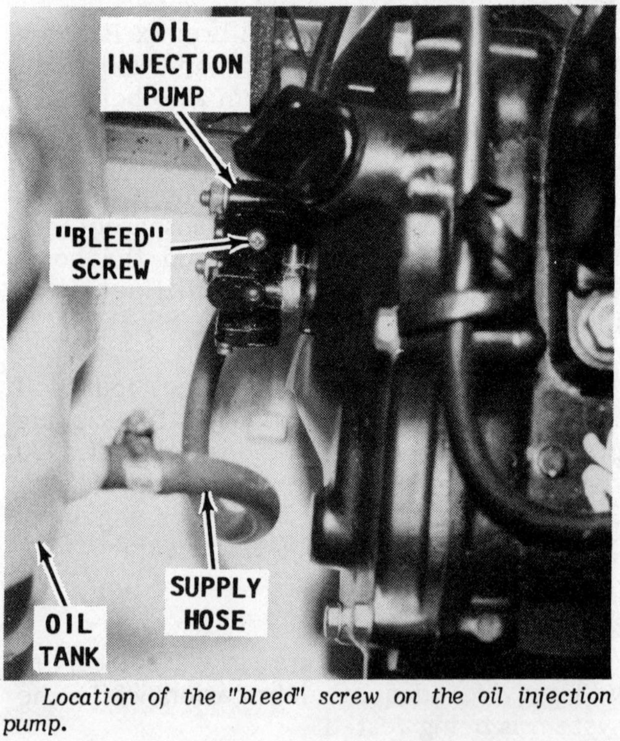

Location of the "bleed" screw on the oil injection pump.

7
IGNITION

7-1 INTRODUCTION AND CHAPTER COVERAGE

The less a marine engine is operated, the more care it needs. Allowing a marine engine to remain idle will do more harm than if it is used regularly. To maintain the engine in top shape and always ready for efficient operation at any time, the engine should be operated every 3 to 4 weeks throughout the year.

The carburetion and ignition principles of two-cycle engine operation **MUST** be understood in order to perform a proper tune-up on a marine engine.

If you have any doubts concerning your understanding of two-cycle engine operation, it would be best to study the operation theory section in the first portion of Chapter 8, before tackling any work on the ignition system.

A Capacitor Discharge Ignition (CDI) is used on all engine models covered in this manual.

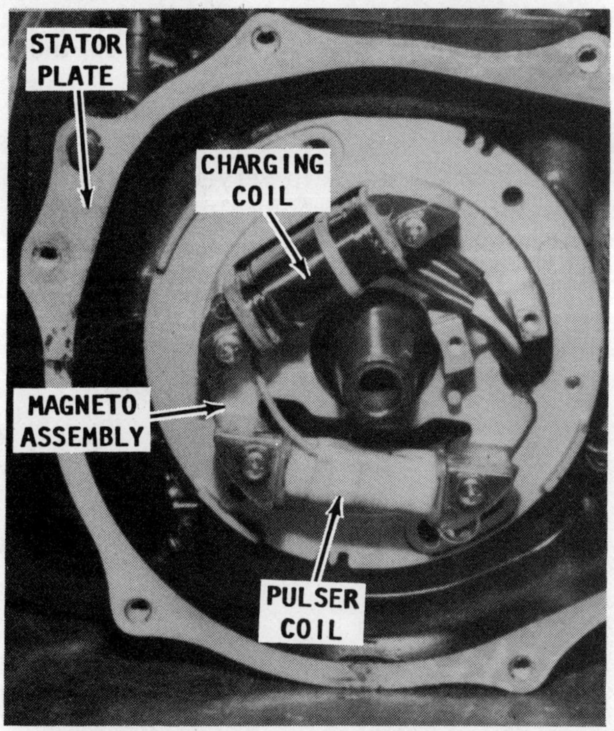

The magneto assembly, consisting of the pulser coil and charging coil, is mounted on the stator plate.

7-2 SPARK PLUG EVALUATION

Removal

Remove the spark plug wires by pulling and twisting on only the molded cap. **NEVER** pull on the wire or the connection inside the cap may become separated or the boot damaged. Remove the spark plugs. **TAKE CARE** not to tilt the socket as you remove the plug or the insulator may be cracked.

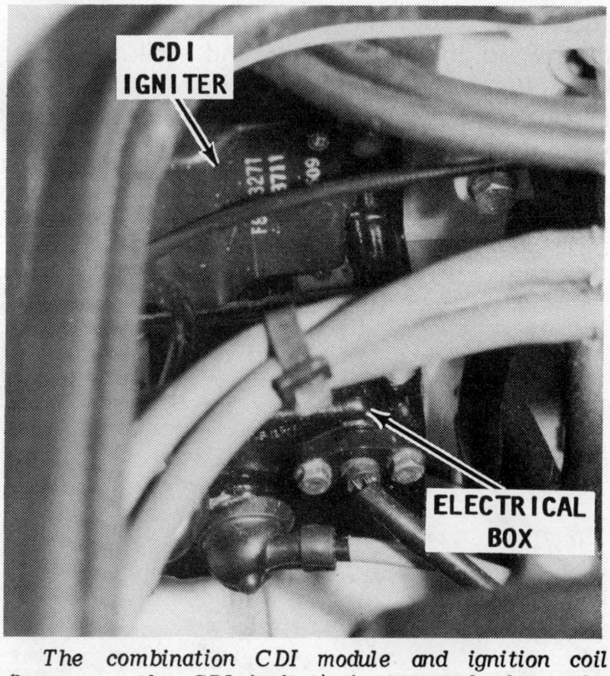

The combination CDI module and ignition coil (known as the CDI igniter), is mounted above the electrical box in the engine compartment, on this 650 Series engine.

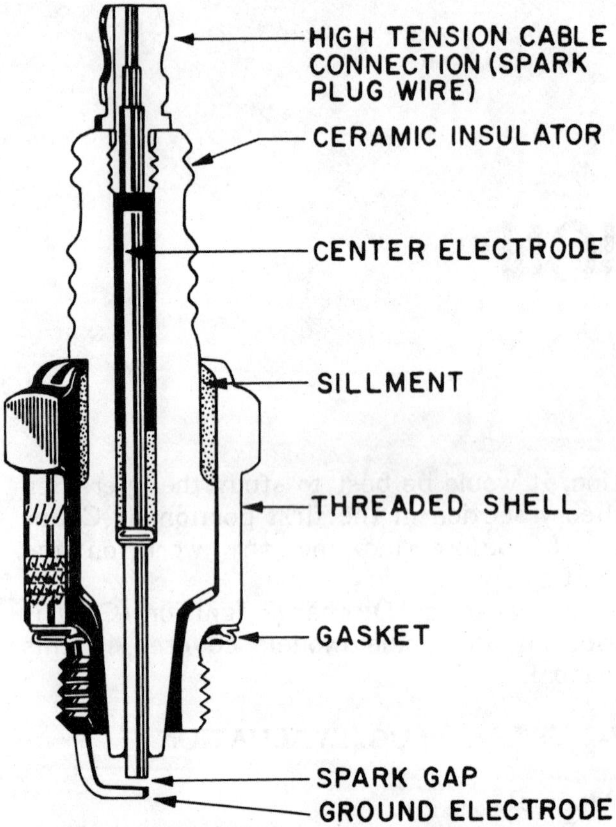

Cross-section drawing of a typical spark plug, with major parts identified.

Examine

Carefully examine both plugs to determine the firing conditions in the cylinder. If the side electrode is bent down onto the center electrode, the piston is traveling too far upward in the cylinder and striking the spark plug. Such damage indicates the piston pin or the rod bearing is worn excessively.

In most cases, an engine overhaul is required to correct the condition. To verify the cause of the problem, rotate the flywheel by hand. As the piston moves to the full up position, push on the piston crown with a screwdriver inserted through the spark plug hole, and at the same time rock the flywheel back-and-forth. If any play in the piston is detected, the engine must be rebuilt.

Correct Color

A proper firing plug should be dry and powdery. Hard deposits inside the shell indicate too much oil is being mixed with the fuel. The most important evidence is the light gray to tan color of the porcelain, which is an indication this plug has been running at the correct temperature. This means the plug is one with the correct heat range and also that the air/fuel mixture is correct.

Rich Mixture

A black, sooty condition on both spark plug shells and the porcelain is caused by an excessively rich air/fuel mixture, both at low and high speeds. The rich mixture lowers the combustion temperature so the spark plug does not run hot enough to burn off the deposits.

Deposits formed only on the shell is an indication the low-speed air/fuel mixture is too rich. At high speeds with the correct mixture, the temperature in the combustion chamber is high enough to burn off the deposits on the insulator.

Too Cool

A dark insulator, with very few deposits, indicates the plug is running too cool. This condition can be caused by low compression or by using a spark plug of an incorrect heat

This spark plug is foul from operating with an overrich air/fuel mixture, possibly caused by an improper carburetor adjustment.

This spark plug has been operating too-cold, because it is rated with a too-low heat range for the engine.

range. If this condition shows on only one plug it is most usually caused by low compression in that cylinder. If both plugs have this appearance, then it is probably due to the plugs having a too-low heat range.

Fouled

A fouled spark plug may be caused by the wet oily deposits on the insulator shorting the high-tension current to ground inside the shell. The condition may also be caused by ignition problems which prevent a high-tension pulse from being delivered to the spark plug.

Carbon Deposits

Heavy carbon-like deposits are an indication of excessive oil in the fuel. This condition may be the result of worn piston rings or excessive ring end gap.

Overheating

A dead white or gray insulator, which is generally blistered, is an indication of overheating and pre-ignition. The electrode gap wear rate will be more than normal and in the case of pre-ignition, will actually cause the electrodes to melt. Overheating and pre-ignition are usually caused by overadvanced timing, detonation from using too-low an octane rating fuel, an excessively lean air/fuel mixture, or problems in the cooling system.

Electrode Wear

Electrode wear results in a wide gap and if the electrode becomes carbonized it will form a high-resistance path for the spark to jump across. Such a condition will cause the engine to misfire during acceleration. If both plugs are in this condition, it can cause an increase in fuel consumption and very poor performance at high-speed operation. The solution is to replace the spark plugs with a rating in the proper heat range and gapped to specification.

Red rust-colored deposits on the entire firing end of a spark plug can be caused by water in the cylinder combustion chamber. This can be the first evidence of water entering the cylinders through the exhaust manifold because of an accumulation of scale. This condition **MUST** be corrected at the first opportunity. Refer to Chapter 8, Engine.

*Damaged spark plugs. Notice the broken electrode on the left plug. The missing part **MUST** be found and removed before returning the engine to service. If overlooked, the part could cause serious damage to expensive internal parts.*

7-3 CDI (CAPACITOR DISCHARGE IGNITION) AND CHARGING SYSTEM

A Capacitor Discharge Ignition (CDI) system is used on all Kawasaki Personal Watercraft covered in this manual. This CDI system incorporates an automatic electronic spark advance, which should not require adjustment **UNLESS** the magneto assembly becomes misaligned or an ignition component is replaced.

About half of the ignition circuit is housed and well protected behind the flywheel. The remainder of the system components are mounted inside the engine compartment.

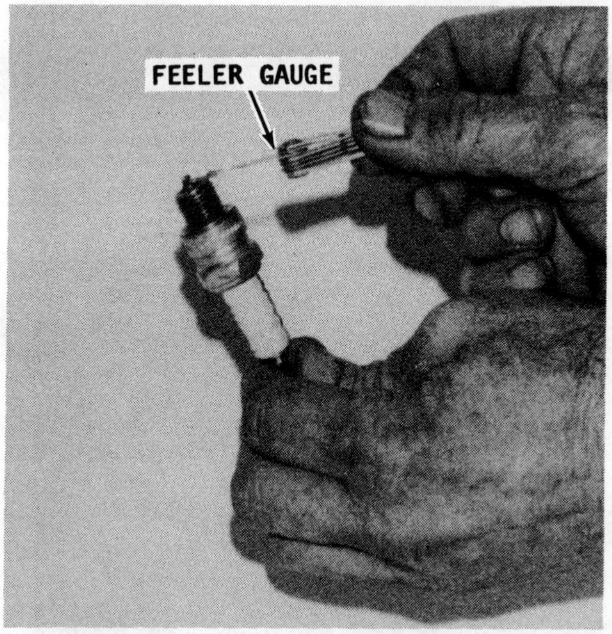

The spark plug gap should always be checked with a wire-type feeler gauge before installing new or used spark plugs.

SPECIAL WORDS

The charging coils are mounted next to the pulser coils behind the flywheel. Therefore, the charging system is presented in this chapter and in this section because once the flywheel is removed the components for both system are exposed.

DESCRIPTION AND OPERATION IGNITION CIRCUIT

The following paragraphs describe -- in layman's language -- the CDI system and how it functions. Every effort has been made to keep "technical" terms to a minimum.

Basically, two major components comprise the CDI system -- namely an exciter coil and a "black box" CDI igniter. The exciter coil is housed behind the flywheel. The CDI igniter is mounted in the engine compartment next to the electrical box.

The exciter coil is sometimes called a "trigger coil". Some series units have a pulser coil in the ignition circuit. The exciter coil and the trigger coil provide a source of AC current and are used to sequentially time electrical impulses.

As mentioned, the CDI igniter is a sealed "black box" type unit and is a combination CDI module and ignition coil.

The ignition module circuitry converts AC current to DC current. The module stores this DC current and releases it to the ignition coil automatically at the correct time.

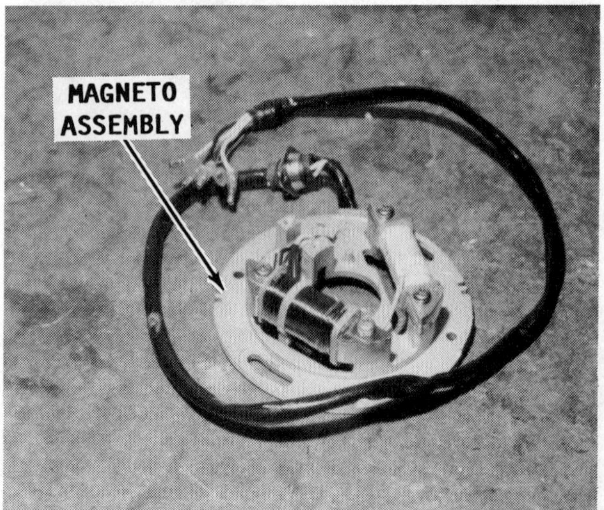

Ignition and charging system components installed on the magneto assembly.

A series of diodes, an SCR gate, and a capacitor comprise the module circuitry. A diode is a solid state device designed to allow current flow in one direction but preventing flow in the opposite direction. An SCR gate is a solid state electronic switching device which permits voltage to flow only after it has been triggered by another source.

The ignition coil circuitry boosts the DC voltage instantly to approximately 20,000 volts. This high voltage is passed on to the spark plugs through high tension leads.

Operation

Consider the engine running and the flywheel rotating. At the point in time when the ignition timing marks align, an alternating coil is induced in the exciter coil. This voltage charges a capacitor in the CDI igniter circuit.

As the flywheel continues to rotate, another voltage is induced in the pulser coil, if so equipped. If the system does not have a pulser, the same voltage is induced in the exciter coil. This voltage flows through a diode and on to the SCR gate. The SCR gate signals the capacitor to discharge the voltage to the primary coils inside the CDI igniter. The voltage in the primary windings of the ignition coil induces a high voltage in the secondary windings and causes a spark to jump to ground across the spark plug electrodes.

In this manner, a spark at the spark plug may be accurately timed by the marks on the flywheel relative to the magnets in the flywheel to provide as many as 100 sparks per second for an engine operating at about 6,000 rpm.

The CDI igniter is a combination of CDI module and ignition coil.

Timing Advance

All Kawasaki Personal Watercraft covered in this manual are equipped with an automatic advance type CDI system with no moving or sliding parts. Therefore, adjustment is not required. Instead, ignition advance timing is accomplished electronically and signal output increases proportionately to engine speed increase.

The timing is correctly set at the factory. As mentioned earlier, the timing should never require adjustment **UNLESS** the magneto assembly is misaligned or an ignition component has been replaced.

Detailed ignition timing procedure is presented in Section 7-7 of this chapter, beginning on Page 7-37.

Charging System

The charging system consists of the flywheel magnets, a charging coil, a rectifier/-regulator, and the battery. As the flywheel magnets rotate, voltage is induced in the charging coil mounted on the magneto assembly under the flywheel. The alternating current passes through a series of four diodes inside the rectifier/regulator and is transformed to DC current.

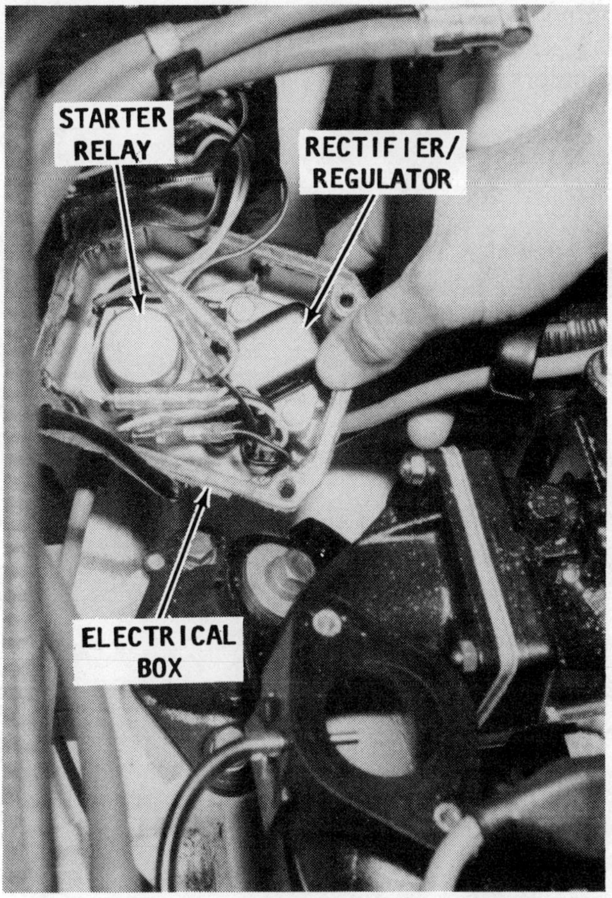

The starter relay and the rectifier/regulator are housed and protected inside the electrical box.

The rectifier/regulator allows a maximum of 14 volts to pass on to the battery. Any excess voltage is passed to a SCR gate and back through the charging coil. In this manner, overcharging the battery is avoided. The life of components in the charging circuit is greatly extended due to the absence of power surges.

The rectifier/regulator is a sealed "black box" type unit. The regulator circuit contains four diodes. If any one of the diodes becomes defective, the complete rectifier/-regulator must be replaced.

TROUBLESHOOTING IGNITION CDI SYSTEM

Always attempt to proceed with the troubleshooting in an orderly manner. The "shot in the dark" approach will only result in wasted time, incorrect diagnosis, replacement of unnecessary parts, and frustration.

Begin the ignition system troubleshooting with the spark plug and continue through the system until the source of trouble is located.

Spark Plugs

1- Check the plug wires to be sure they are properly connected. Check the entire

length of the wires from the plugs to the magneto assembly on the stator plate. If the wires are to be removed from the spark plug, **ALWAYS** use a pulling and twisting motion as a precaution against damaging the connection.

Attempt to remove the spark plug by hand. This is a rough test to determine if the plug is tightened properly. The attempt to loosen the plug by hand should fail. The plug should be tight and require the proper socket size tool. Remove the spark plug and evaluate its condition as described in Section 7-2.

2- Use a spark tester and check for spark. If a spark tester is not available, hold the plug wire about 1/4" (6.4mm) from the engine. Crank the engine through a few revolutions using the cranking motor and check for spark. A strong spark over a wide gap must be observed when testing in this manner, because under compression a strong spark is necessary in order to ignite the air/fuel mixture in the cylinder. This means it is possible to think a strong spark is present, when in reality the spark will be too weak when the plug is installed. If there is no spark, or if the spark is weak, the trouble is most likely under the flywheel in the stator assembly.

Compression

3- Before spending too much time and money attempting to trace a problem to the ignition system, a compression check of the both cylinders should be made. If a cylinder does not have adequate compression, troubleshooting and attempted service of the ignition or fuel system will fail to give the desired results of satisfactory engine performance.

Remove the spark plug wire by pulling and twisting **ONLY** on the molded cap. **NEVER** pull on the wire because the connection inside the cap may be separated or the boot may be damaged. Remove the spark plug. Insert a compression gauge into the cylinder spark plug hole. Ground the spark plug leads to prevent damage to the ignition coil. If a lead is not grounded, the coil will attempt to match the demand created by the spark trying to jump from the electrode to the nearest ground.

Crank a cold engine through several revolutions with the electric cranking motor. Note the compression reading.

An acceptable pressure reading for an engine covered in this manual is 142 psi (981 kPa) or more. The readings between cylinders should not vary by more than 10%.

TROUBLESHOOTING CHARGING SYSTEM

As mentioned earlier, the charging coil is mounted on the magneto assembly located under and protected by the flywheel. The charging coil seldom causes problems in the charging circuit. Most problems can be traced to the rectifier/regulator, the battery, or to the connecting wiring.

The following three conditions will cause damage to the rectifier/regulator.

a- Battery cables connected in reverse.

b- Operating the engine with one or both battery cables disconnected.

c- A loose or broken connection in the charging circuit.

The voltage generated by the charging coil is intended to charge the battery. Anytime the circuit is incomplete, the voltage has actually no where to go. As a result, the voltage will actually reverse itself.

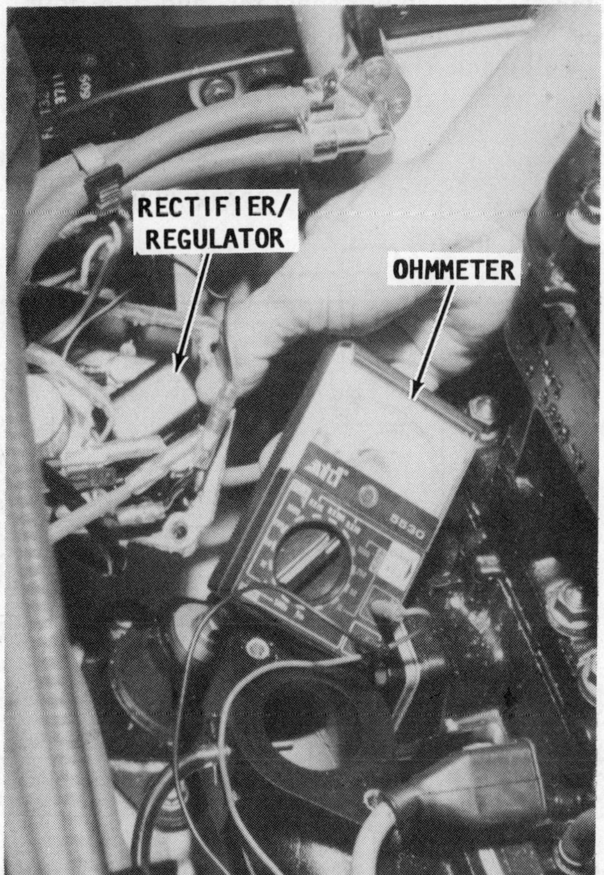

After making a visual inspection, the rectifier/regulator in the charging circuit are tested.

Such reversed voltage will damage the internal structure of a diode in the circuit. The diode will then have a scorched or burned appearance.

Perform a visual inspection of the following areas before making any resistance tests on charging circuit components.

a- Check to be sure the battery cables are connected to the correct terminals.

b- Check condition of the battery and recharge, if necessary.

c- Inspect the wiring and the connections between the charging coil, the rectifier/regulator, and the battery. Replace any damaged or deteriorated cable insulation. Payparticular attention for loose, corroded, or disconnected terminals orconnections.

7-4 TESTING IGNITION COMPONENTS

General Information

Due to the complex nature of the circuitry and high energy output pulses of microsecond duration, conventional testing devices such as a Volt/Ohm/Ammeter will not measure electrical output with the degree of accuracy required. A Kawasaki Ignition Tester, as used by the local Kawasaki dealer is an electrical device capable of measuring the peak energy output of capacitor discharge ignition units, as well as the exciter and pulser coils. This instrument was designed to troubleshoot the CDI ignition system. Unfortunately this tester may not be easily accessible. Therefore, this chapter includes only those tests which may be performed with instruments normally and easily available.

WORDS ON TESTERS

Resistance tests performed with a Volt/-Ohmmeter are not as accurate or as conclusive as tests performed using a Kawasaki tester

To replace a component, the flywheel must be "pulled" -- see Section 7-6, this chapter.

WORDS FROM EXPERIENCE

During the tests, if a reading is slightly different from the specifications, but the engine still operates, then there is no real need to replace the affected component, until it actually fails. Bear in mind, in **MOST** cases electrical components are not returnable, once the item leaves the store. Therefore, make every attempt to avoid the **BUY** and **TRY** method of troubleshooting. Such a practice will lead only to wasted time, unneeded cash outlay, and frustration.

Intermittent or Multiple Problems

Many ignition problems occur only during engine operation -- when the component is subject to vibration, the engine is operating under a load, and/or there is a marked increase in temperature -- from the engine operating.

SPECIAL WORDS ON IGNITION AND CHARGING CIRCUIT TESTING

Most of the components in the ignition and charging circuits have leads with quick-disconnect fittings or harness plugs located inside the electrical box. Therefore, testing begins inside the electrical box. Some components in these circuits are housed inside the box -- others are housed under the flywheel. If a component inside the electrical box is found to be defective, it may be replaced immediately -- usually through attaching hardware. If a component under the flywheel -- exciter, pulser, or charge coil -- is found to be defective, the flywheel must be "pulled" in order to remove and replace the item. See Chapter 7, Section 7-6, beginning on Page 7-27 for instuctions to "pull" the flywheel.

__NEVER__ attempt to verify the charging circuit by operating the engine with the battery disconnected. Such action will damage the rectifier/regulator.

CRITICAL WORDS

Never attempt to verify the charging circuit by operating the engine with the battery disconnected. Such action would force current -- normally directed to charge the battery -- back through the rectifier and damage the diodes in the rectifier.

Electrical Box Removal
300, 440, and 550 Series

1- Disconnect the battery ground cable from the battery. Remove the spark plug lead/s from the spark plug/s. Remove the two bolts securing the connector cover to the electrical box. Pull off the cover. Disconnect the 4-pin connector and the four free leads at their quick-disconnect fittings.

Pull back the two boots over the large cables at the cranking motor relay. Discon-

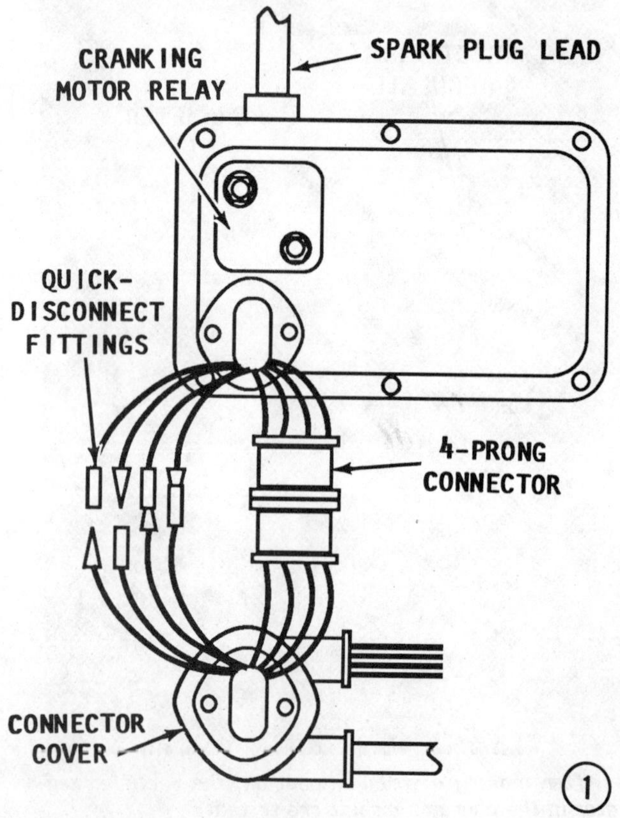

1

nect both large cables from the relay terminals. Remove the six bolts securing the electrical box to the engine compartment bulkhead. Remove the two bolts securing the two halves of the box together, and then open the box.

Electric Box Removal
650 Series Only

2- Disconnect the battery ground cable from the battery. Remove the spark plug leads from the spark plugs. Pull back the large boot over both large cables at the cranking motor relay. Disconnect both cables from the relay terminals. Remove the four bolts securing the cover to the electrical box on the engine compartment bulkhead. Note there are two different bolt sizes.

Exciter Coil Resistance Test
300 Series Only

Identify the magneto assembly harness with a Red, Black, and two light Green leads. Disconnect the two halves of the 4-prong plug. Select the Rx 100 ohm scale on an ohmmeter. Make contact with the Red meter lead to the Red male prong in the plug from the magneto assembly. Make contact with the Black meter lead to the Black male prong in the plug from the magneto assembly. The meter should register 168-252 ohms.

If the reading is less than expected, a short probably exists in the coil and the coil must be replaced.

If the meter indicates an infinite resistance, an open circuit exists in the coil and the coil must be replaced.

If the coil passes this resistance test, but the problem is still determined to be in the magneto assembly, the cause is most likely a loss of magnetism in the flywheel magnets. The flywheel must be replaced.

Exciter and Pulser Coils Resistance Tests
440 and 550 Series Only

Testing resistance of the exciter coil: identify the magneto assembly harness with a Red, Black, Grey, and two light Green leads. Disconnect the two halves of the 6-prong plug -- one empty connector. Select the Rx 100 ohm scale on an ohmmeter.

Make contact with the Red meter lead to the Red male prong in the plug from the magneto assembly. Make contact with the Black meter lead to the Black male prong in the plug from the magneto assembly. The meter should register 100-180 ohms.

Testing resistance of the pulser coil: select the Rx 10 ohm scale on the ohmmeter. Keep the Red meter lead on the Red male prong in the plug from the magneto assembly. Make contact with the Black meter lead to the Grey male prong in the plug from the magneto assembly. The meter should register 14.4-21.6 ohms.

Results for either coil: if the reading is less than expected, a short probably exists in the coil and it must be replaced. If the meter indicates an infinite resistance, an open circuit probably exists in the coil and it must be replaced. If the coil passes the resistance test, but the problem is still determined to be in the magneto assembly, the cause is most likely a loss of magnetism in the flywheel magnets. The flywheel must be replaced.

Exciter Coil Resistance Test
650 Series Only

Identify the magneto assembly harness. The manufacturer used different colored leads for the exciter coil. Look for these color combinations in the harness lead: Black/Red and Black; Black/Yellow and Black/Red; or Purple and Black/Yellow.

Select the Rx 100 ohm scale. Make contact with either meter lead to either exciter coil lead. The meter should register 250-380 ohms for the first two color combinations or 1,000-1600 ohms for the third color combination.

If the reading is less than expected, a short probably exists in the coil and the coil must be replaced.

If the meter indicates an infinite resistance, an open circuit exists in the coil and the coil must be replaced.

If the coil passes this resistance test, but the problem is still determined to be in the magneto assembly, the cause is most likely a loss of magnetism in the flywheel magnets. The flywheel must be replaced.

CDI Igniter Resistance Test
300 Series Only

The ignition coil portion of the CDI igniter can be checked for a shorted or broken winding using an ohmmeter meter, as follows. Select the Rx 1K (1,000) ohm scale on the meter. Make contact with the Red meter lead to the spark plug terminal inside the boot of the high tension lead. Make contact with the Black meter lead to a good ground on the engine. The meter should register 5,040-7,560 ohms. If the reading is not as expected, the CDI igniter must be replaced.

To gain access to the magneto assembly to replace a coil, the flywheel must first be "pulled".

CDI Igniter Removal
300 Series Only

To remove the CDI igniter, open the electrical box, as described in ignition and charge circuit testing, beginning on Page 7-8. Disconnect the Red and Black/White leads at their quick-disconnect fittings. Remove the Black cap from the White threaded fitting on the side of the electrical box. Pull out the small Black grommet from inside the threaded fittings. The high tension lead is an integral part of the CDI igniter. Pull off the spark plug boot and remove the terminal from the end of the high tension lead. Remove the two attaching screws and lift the igniter free of the electrical box, and then thread the high tension lead through the White threaded grommet cap in the side of the box.

CDI Igniter Installation
300 Series Only

Coat the threads of all securing bolts and screws with a non-permanent locking substance, just prior to installation.

Pass the high tension lead through the White threaded fitting on the side of the electrical box. Position the igniter over the two mounting holes and secure it in place with the two Phillips head screws. Apply a coating of multi-purpose water resistant lubricant to the small Black grommet, and then install the grommet followed by the Black grommet cap onto the end of the high tension lead. Push the grommet into the White fitting and install the Black grommet cap over the grommet. Connect the Red and Black/White leads, matching color-to-color, between the CDI igniter and the magneto harness 4-prong connector and the controller. Install the spark plug terminal onto the end of the high tension lead, followed by the spark plug boot. Install the electrical box. Tighten the securing bolts to a torque value of 69 in lb (7.8Nm).

CDI Igniter Resistance Test
440 Series and 550 Series

The ignition coil portion of the CDI igniter can be checked for a shorted or broken winding using an ohmmeter. Select the Rx 1K (1,000) ohm scale on the meter. Make contact with either meter lead to the spark plug terminal inside the boot of one spark plug high tension lead. Make contact with the other meter lead to the other spark plug terminal in the other high tension lead. The meter should register 5,040-7,560 ohms. If the reading is not as expected, replace the CDI igniter.

CDI Igniter Removal
440 Series and 550 Series

To remove the CDI igniter: First, open the electrical box, as outlined in Ignition and Charge Circuit Testing beginning on Page 7-8.

Next, disconnect the Red, Black, Grey, and Black/White leads at their quick-disconnect fittings. For each high tension lead --back off the Black cap from the White threaded fitting on the side of the electrical box. Pull out the small Black grommet from inside the threaded fitting. The high tension lead is an integral part of the CDI igniter.

Now, pull off the spark plug boot and remove the terminal from the end of the high tension lead. Remove the two attaching screws and lift the igniter up and out of the electrical box, and at the same time feed the high tension leads through the White threaded grommet caps in the side of the box.

CDI Igniter Installation
440 Series and 550 Series

Apply a coating of non-permanent locking substance to the threads of all securing bolts and screws, just prior to installation.

Begin by passing the high tension leads through the White threaded fittings on the side of the electrical box. Next, position the igniter over the two mounting holes and secure it in place with the two Phillips head screws.

For each high tension lead: Apply a coating of multi-purpose water resistant lubricant to the small Black grommet, and then install it onto the end of the high tension lead followed by the Black grommet cap.

Connect the Red, Black, Grey, and the Black/White leads, matching color-to-color, between the CDI igniter and the magneto harness 6-prong connector and the stop switch relay. Install the spark plug terminals onto the ends of the high tension leads, followed by the spark plug boot.

Install the electrical box Tighten the bolts to a torque value of 52 in lb (6Nm).

CDI Igniter Resistance Test
650 Series -- Except JetMate

The CDI igniter on the 650 Series -- except JetMate -- is located inside the electrical box and can be statically tested using an ohmmeter. To open the electrical box -- see the procedures beginning on Page 7-8. The manufacturer used different CDI igniters on various models of the 650 Series. Therefore, different resistance values are presented for CDI igniters with either three or four leads. A CDI igniter with four leads also has two different color combinations.

The CDI igniter may remain installed inside the electrical box while resistance tests are performed. Disconnect all three or four igniter leads at their quick-disconnect fittings. If the igniter is equipped with a short ground lead, disconnect the lead from its ground connection and take adequate precautions to prevent the lead from coming into contact with anything during the test.

Select the Rx 1K (1,000) ohm scale on the meter. Two separate accompanying resistance tables are presented. One table for a CDI igniter with three leads and the other table for the CDI igniter with four

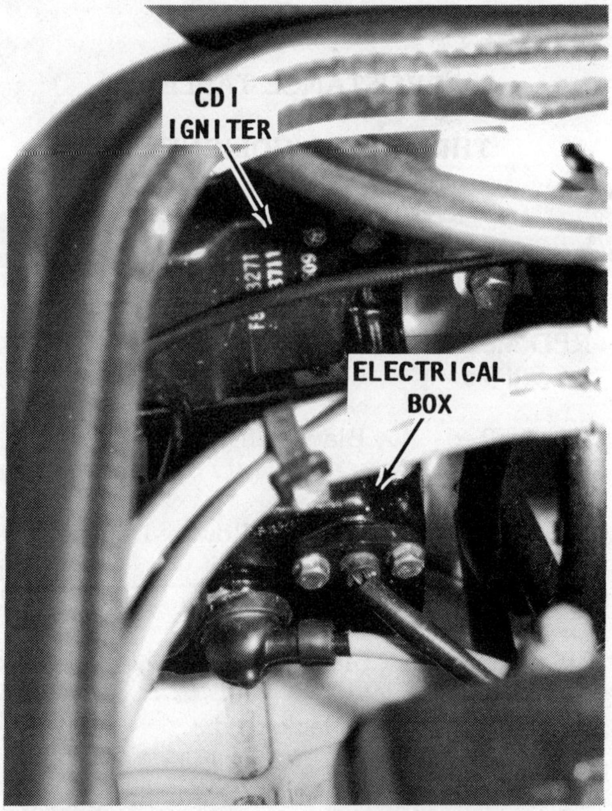

The CDI igniter is mounted above the electrical box on a 650 Series engine.

A 650 Series igniter ready for bench testing.

leads. This "four lead" table has two sets of readings for some meter connections. The first reading is valid for the **FIRST** time the test is performed. If it is necessary to repeat the test -- wait for at least 30 seconds. The small amount of current used in making the first test --supplied from the internal battery of the ohmmeter -- is sufficient to charge the internal capacitor within the CDI igniter. The **SECOND** reading is now valid.

If the meter fails to register the expected readings -- as listed in the tables -- the igniter must be replaced.

RESISTANCE TABLE

THREE LEAD CDI IGNITER

Red/Black, Black/White, and Black/Yellow Leads

RED Meter Lead To	BLACK Meter Lead To	Meter Reading
Black/Red	Black/White	No Continuity
Black/Red	Black/Yellow	1,000-5,000
Black/White	Black/Red	1,000-5,000
Black/White	Black/Yellow	5,000-17,000
Black/Yellow	Black/Red	Above 500,000
Black/Yellow	Black/White	No Continuity

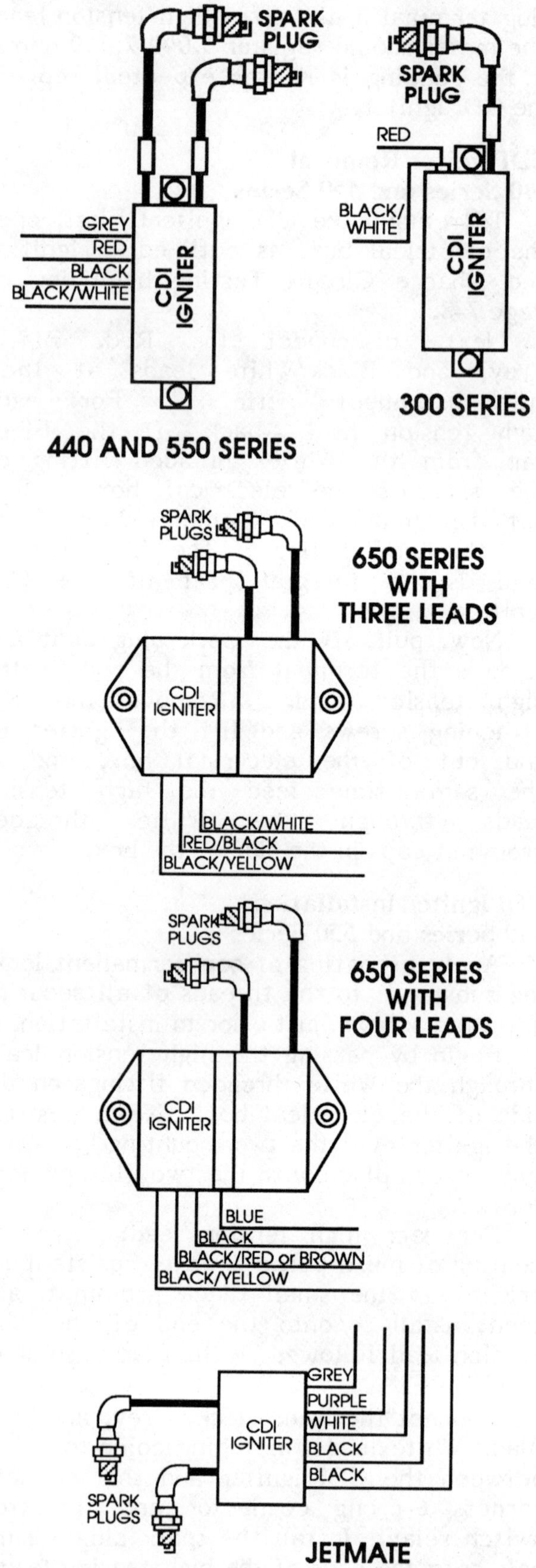

CDI Igniter and lead identification for the series engines indicated.

RESISTANCE TABLE

FOUR LEAD CDI IGNITER

Black/Red or Brown, Black/Yellow, Black, and Blue Leads

RED Meter	BLACK Meter	Meter Reading
Black/Red or Brown	Black/Yellow	3,000-6,000
Black/Red or Brown	Black	1st Test 15,000-25,000 2nd Test 50,000-100,000
Black/Red or Brown	Blue	3,000-6,000
Black/Yellow	Black/Red or Brown	1st Test 50,000 2nd Test No Continuity
Black/Yellow	Black	1st Test 15,000-25,000 2nd Test 50,000-100,000
Black/Yellow	Blue	0
Black	Black/Red or Brown	No Continuity
Black	Black/Yellow	No Continuity
Black	Blue	No Continuity
Blue	Black/Red or Brown	1st Test 50,000 2nd Test No Continuity
Blue	Black/Yellow	0
Blue	Black	1st Test 15,000-25,000 2nd Test 50,000-100,000

RESISTANCE TABLE
CDI IGNITER
JETMATE ONLY

RED Meter	BLACK Meter	Meter Reading
Purple	Green	1st Test 500,000 2nd Test No Continuity
Purple	Eyelet Black	3,000-4000
Purple	White	1st Test 40,000-150,000 2nd test No Continuity
Purple	Female Black	3,000-4,000
Green	Purple	1st Test 500,000 2nd Test No Continuity
Green	Eyelet Black	20,000-30,000
Green	White	1st Test 14,000-22,000 2nd Test 50,000-150,000
Green	Female Black	20,000-30,000
Eyelet Black	Purple	1st Test 500,000 2nd Test No Continuity
Eyelet Black	Green	1st Test 500,000 2nd Test No Continuity
Eyelet Black	White	1st Test 14,000-22,000 2nd Test 50,000-150,000
Eyelet Black	Female Black	0
White	Purple	No Continuity
White	Green	No Continuity
White	Eyelet Black	No Continuity
White	Female Black	No Continuity
Female Black	Purple	1st Test 500,000 2nd Test No Continuity
Female Black	Green	1st Test 500,000 2nd Test No Continuity
Female Black	Eyelet Black	0
Female Black	White	1st Test 14,000-22,000 2nd Test 50,000-150,000
High Tension Lead	High Tension Lead	2,100-3,100

CDI Igniter Removal
650 Series — Except JetMate

Begin by opening the electrical box as outlined in Ignition and Charge Circuit Testing beginning on Page 7-8. Disconnect the three or four igniter leads at their quick-disconnect fittings. Remove the two Phillips head screws securing the igniter to the electrical box.

Disconnect the two high tension leads from the spark plugs and feed the grommet, with the small leads attached, through the hole in the electrical box.

CDI Igniter Installation
650 Series -- Except JetMate

Apply a coating of non-permanent locking substance to the threads of all securing bolts and screws, just prior to installation.

Insert the grommet, with the CDI leads attached, into the recess in the side of the electrical box. Secure the igniter to the side of the electrical box with the two Phillips head screws. If the unit being serviced has a ground lead, make the connection.

Connect the three or four igniter leads -- matching color-to-color. If difficulty is encountered with the lead connections. refer to the wiring diagrams in the Appendix for the unit being serviced. Connect the two high tension leads to the spark plugs.

Install the electrical box. Tighten the attaching bolts to a torque value of 69 in lb (7.8Nm).

CDI Igniter Resistance Tests
JetMate Only

The CDI igniter on a JetMate is located inside the electrical box and can be statically tested using an ohmmeter. Open the electrical box according to the procedures beginning on Page 7-8.

GOOD WORDS

It is not necessary to remove the CDI igniter while performing this resistance test. Disconnect all four igniter leads at their quick-disconnect fittings. The igniter is equipped with a short ground lead. Disconnect this lead from the ground connection on one of the mounting screws. Take extra precautions to ensure this ground lead does not come in contact with anything during the test.

Select the Rx 1K (1,000) ohm scale on an ohmmeter. The accompanying "four lead" table has two sets of readings for some meter connections. The first reading is valid for the **FIRST** time the test is performed. If it is necessary to repeat the test -- wait for at least 30 seconds. The small amount of current used in making the first test --supplied from the internal battery of the ohmmeter -- is sufficient to charge the internal capacitor within the CDI igniter. The **SECOND** reading is now valid.

If the meter fails to register the expected readings -- as listed in the table --the igniter must be replaced.

CDI Removal
JetMate Only

Remove the four bolts securing the cover of the electrical box, and then open the box. Disconnect the four leads at their quick-disconnect fittings. Snip the tie strap from each spark plug cap and **CAREFULLY** pull off the caps from the ends of the high tension leads. Remove the spark plug terminals from the lead ends. Remove the two bolts on the outside of the electrical box securing the grommet plate to the box. Work the two high tension leads free of the two grommets in the grommet plate.

Remove the two bolts securing the CDI igniter to the electrical box. One of these bolts also has the igniter grounding lead attached. Lift out the igniter, and at the same time, free the leads through the two holes in the side of the electrical box.

CDI Installation
JetMate Only

Apply a coating of non-permanent locking substance to the threads of all securing bolts and screws, just prior to installation.

Position the igniter inside the electrical box. Feed the two high tension leads through the holes in the side of the box. Secure the igniter with the two bolts -- one bolt also secures the Black ground lead. Tighten the bolts to a torque value of 69 in lb (7.8Nm).

Feed the high tension leads through the two grommets in the grommet plate until the plate makes contact with the box. Install and tighten the two attaching bolts to a torque value of 69 in lb (7.8Nm). Connect the four igniter leads, matching color-to-color. If necessary, refer to the wiring diagram in the Appendix.

Insert the spark plug terminals into the ends of the high tension leads. Push the spark plug caps over the terminals and secure them in place with new tie straps. Close the electrical box. Tighten the four cover bolts alternately and evenly to a torque value of 69 in lb (7.9Nm).

7-5 CHARGING CIRCUIT

DESCRIPTION AND OPERATION

The charging circuit consists of the following items:

- Permanent magnets -- attached to inside perimeter of the flywheel.
- Charging coil -- installed on the magneto assembly.
- A rectifier/regulator -- located either inside or attached to the electrical box.
- An external battery.
- Necessary wiring -- to connect the components together.

The **NEGATIVE** side of the rectifier/regulator is always grounded. The **POSITIVE** side is always connected to the battery. The negative terminal on the battery must be connected to a good ground on the engine.

The alternating current generated in the charging coil windings passes to the rectifier/regulator. The rectifier portion changes the alternating current (AC) to direct current (DC) to charge the 12-volt battery. A second function of the rectifier is to smooth the pulses of current to form a steady DC current flow.

The regulator portion of the rectifier/regulator limits the current produced and sends any excess current back through the charging coil windings.

Electronic components in the charging circuit, including the battery, will last much longer if subjected to a low steady voltage. A sudden voltage surge could overheat and damage charging circuit components.

The magneto assembly is located behind and protected by the flywheel. Therefore, the magneto assembly, including the charging coil, seldom causes problems in the charging circuit. When problems do occur, they can usually be traced to the rectifier/regulator or to the battery.

If either the charging coil or the rectifier/regulator fails troubleshooting tests, the defective unit cannot be repaired, it **MUST** be replaced.

TESTING CHARGING COIL OUTPUT

Charging Coil Output Test
300 Series Only

1- A voltmeter capable of registering 50 volts AC is required for this test. Select the Vx10 volts scale on the meter.

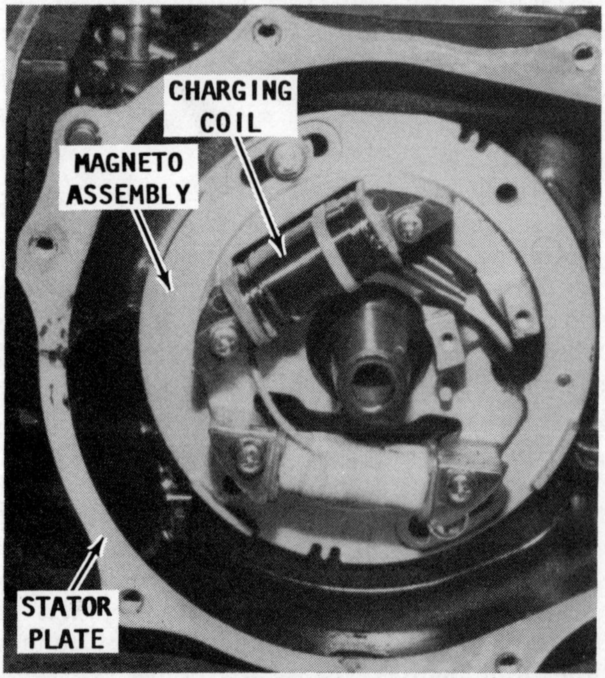

The charging coil is mounted on the magneto assembly. The magneto assembly is secured to the stator plate.

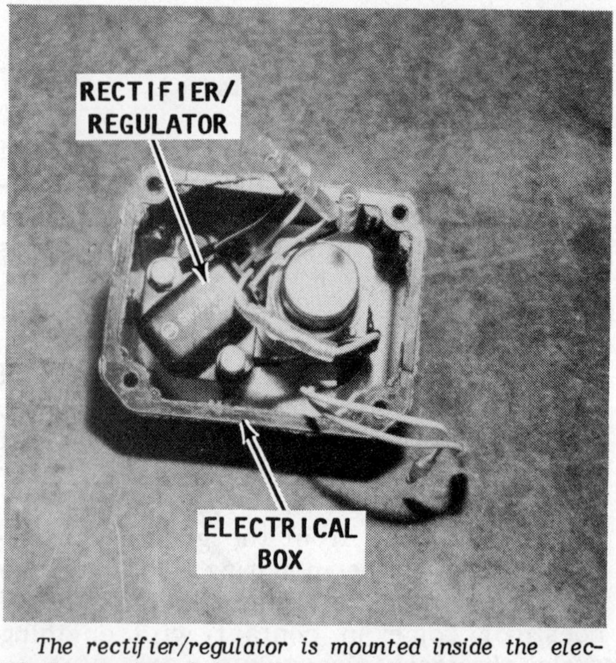

The rectifier/regulator is mounted inside the electrical box, where it is well protected.

Remove the connector cover at the electrical box. Disconnect the 4-prong magneto harness connector, as described in Ignition and Charge Circuit Testing beginning on Page 7-8.

Using a pair of needle nose pliers, reach into the magneto end of the connector half and push out the male prongs at the end of the two light Green leads to free the leads from the connector. Temporarily connect the two halves of the connector without the two light Green leads.

PLAN AHEAD

Each of the disconnected leads will be tested with a voltmeter connected across one light Green lead and a suitable ground connection, and then the other light Green lead and a suitable ground connection with the engine running. Therefore, position the leads to make the test quickly and efficiently and decide on the ground connection to be used. Have a pencil and paper at hand to record both readings.

SAFETY WORDS

The manufacturer warns this test **MUST NOT** be performed with the craft in a body of water due to the possibility of electric shock to the individual performing the test. In light of this warning, this test **MUST** be performed in a test tank with the person performing the test on dry land.

Therefore, move the craft to a test tank or secure it in a body of water next to a float, as in a regular boat "slip". Secure the craft fore and aft to prevent movement during the following tests.

CAUTION

Water must circulate through the jet pump -- to and from the engine, anytime the engine is operating. Circulating water will prevent overheating -- which could cause damage to moving engine parts and possible engine seizure.

NEVER allow the engine to operate without cooling water for more than 15 seconds.

Now, with the engine operating, advance the throttle to reach about 6,000 rpm. Make contact with the Red voltmeter lead to one of the disconnected light Green leads. Make contact with the Black voltmeter lead to a suitable ground. Record the reading. Keep the Black meter lead to the engine ground and move the Red meter lead to the other disconnected light Green lead. Again, record the meter reading.

Shut down the engine. The two voltmeter readings should be 41.4-50.6 and 45.9-56.1 volts AC, but not necessarily in that order.

If the meter readings are as expected, the problem in the charging system is probably in the rectifier/regulator. Proceed directly to Step 3.

If the meter readings are below the values given, the charging coil output voltage is too low. Proceed to Step 2 to check the charging coil under the flywheel.

Install the two light Green leads into their original locations inside the harness connector. Pull the male prongs at the end of the leads into position with a pair of needle nose pliers.

Charging Coil Resistance Test
300 Series Only

2- Select the Rx1 ohm scale on an ohmmeter. Make contact with the Red meter lead to either light Green male prongs in the plug from the magneto assembly. Make contact with the Black meter lead to the Black male prong in the same plug. Repeat the test for the other light Green lead.

The meter should register 1.12-1.68 ohms and 1.28-1.92 ohms for the two tests, but not necessarily in that order.

Make contact with either meter lead across the two light Green male prongs. The meter should register 2.4-3.6 ohms.

If the meter readings are not as expected, but the charging coil must be replaced. See Section 7-6 to "pull" the flywheel.

If the meter readings are as given, but the charging coil output test (Step 1), gives a low voltage, the problem is in the charging system, most likely loss of magnetism in the permanent magnets in the flywheel. In such a case, the flywheel must be replaced.

Rectifier/Regulator Resistance Test
300 Series Only

The rectifier/regulator is located inside the electrical box. See Step 1, on Page 7-8 to open the electrical box.

The manufacturer recommends three different test for the rectifier/regulator. The first test is a resistance test. The rectifier/regulator may remain installed inside the electrical box.

The second test is a rectifier circuit test. The rectifier/regulator must be removed from the electrical box for this test to be valid.

The third and most conclusive test is to replace the rectifier/regulator with a known good unit. The manufacturer admits -- a unit may pass the first two tests, but nevertheless be defective.

RESISTANCE TABLE

THREE LEAD CDI IGNITER

300 SERIES ONLY

RED Meter Lead to	BLACK Meter Lead to	Meter Reading
Orange	Black	No Continuity
Orange	First Light Green	Continuity
Orange	Second Light Green	Continuity
Black	Orange	No Continuity
Black	First Light Green	Continuity
Black	Second Light Green	Continuity
First Light	Orange	No Continuity
First Light Green	Black	No Continuity
First Light Green	Second Light Green	No Continuity
Second Light Green	Orange	No Continuity
Second Light Green	Black	No Continuity
Second Light Green	First Light Green	No Continuity

3- As mentioned earlier, the rectifier/-regulator may remain installed inside the electrical box for the following resistance test.

Disconnect all four leads -- Orange, Black, and the two light Green leads at their quick-disconnect fittings. Remove the short black lead from its ground connection. Take extra precautions to ensure this ground lead does not make contact with anything during the test.

Select the Rx100 ohm scale on the ohmmeter. Specific resistance values are not given for the results of these tests. The meter will either indicate continuity or no continuity.

If the rectifier/regulator fails any of the resistance tests, it must be replaced. If the rectifier/regulator passes all the resistance tests, proceed to the following tests.

Rectifier Circuit Test
300 Series Only

4- Obtain two 12-volt bulbs rated between 3 and 6 watts; two 12-volt batteries; and five test leads.

SPECIAL WORDS

DO NOT attempt to use an ammeter or multimeter in place of the bulbs. The purpose of the bulb is to limit the flow of current through the rectifier/regulator.

Perform each test quickly and in the order presented. **DO NOT** allow the battery or batteries to be connected in the circuit for more than just a few seconds at a time.

Remove the rectifier/regulator from the electrical box.

a- Using one of the bulbs, three leads, and a single battery -- connect the test light and battery to the rectifier/regulator between the orange and first Green lead, as shown.
If the bulb lights -- the rectifier/regulator is defective and must be replaced.
If the bulb fails to light -- proceed to the next test.

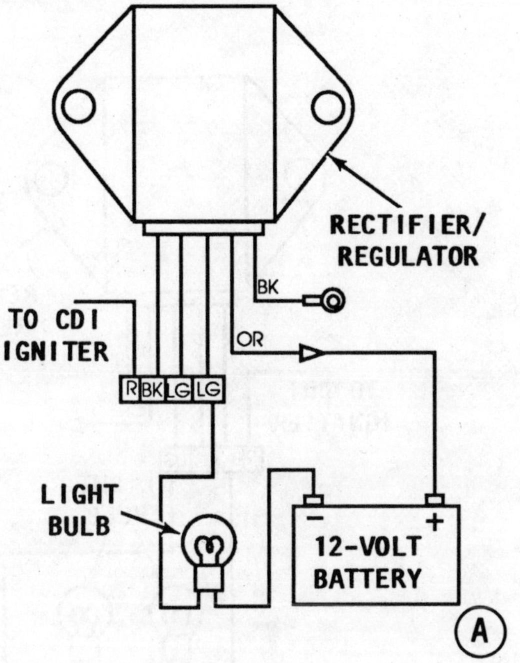

b- Again, using one bulb, three leads and a single battery -- connect the test light and battery between the Orange and second Green lead, as shown.
If the bulb lights -- the rectifier/regulator is defective and must be replaced.
If the bulb fails to light -- proceed to the next test.

c- Still using one bulb, three leads, and a single battery -- connect the test light and battery to the rectifier/regulator between the Orange and first Green lead, as shown. These are the

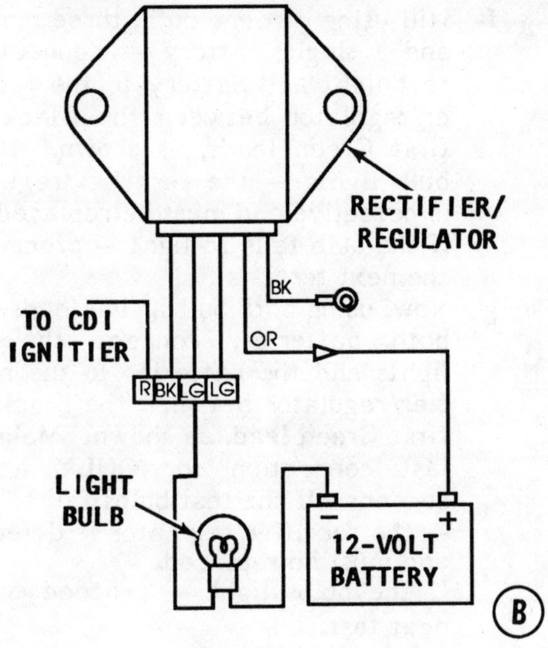

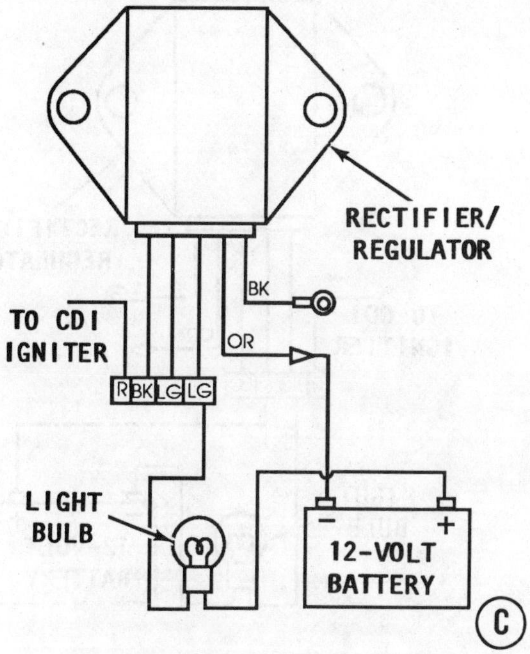

same connections as in Test "a" above, except the battery connections are reversed.
If the bulb lights -- the rectifier/regulator is defective and must be replaced.
If the bulb fails to light -- proceed to the next test.

d- Still using one bulb, three leads, and a single battery -- connect the test light and battery to the rectifier/regulator between the Orange and second Green lead, as shown. These are the same connections as in Test "b" above, except the battery connections are reversed.

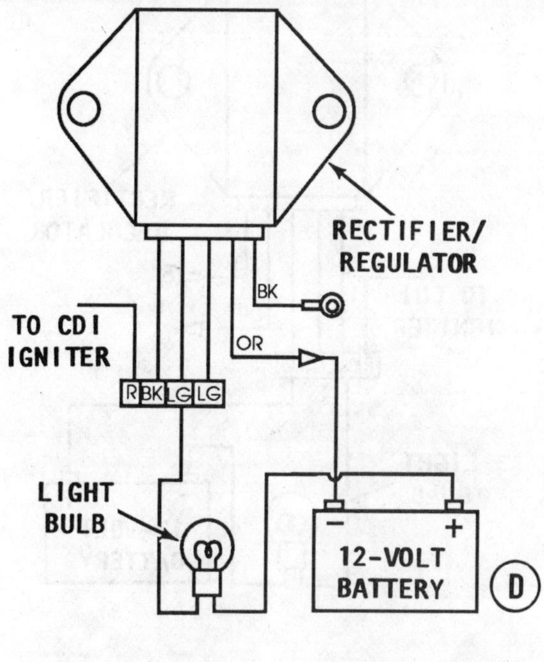

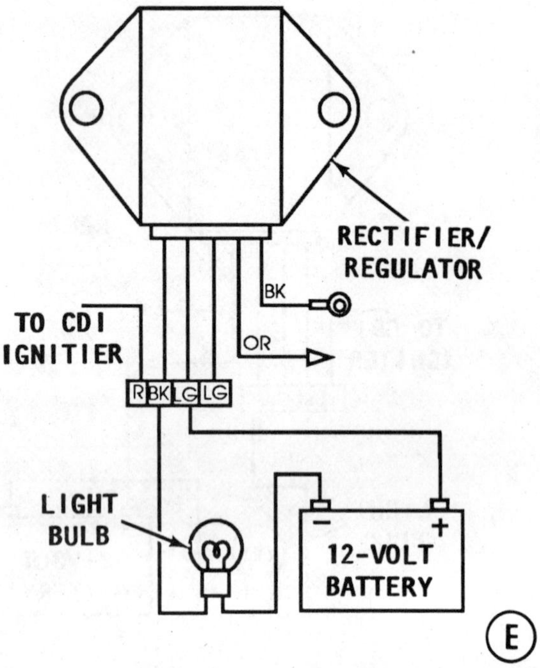

If the bulb lights -- the rectifier/regulator is defective and must be replaced.
If the bulb fails to light -- proceed to the next test.

e- Still using just one bulb, three leads, and a single battery -- connect the test light and battery to the rectifier/regulator between the Black and second Green leads, as shown. If the bulb lights -- the rectifier/regulator is defective and must be replaced.
If the bulb fails to light -- proceed to the next test.

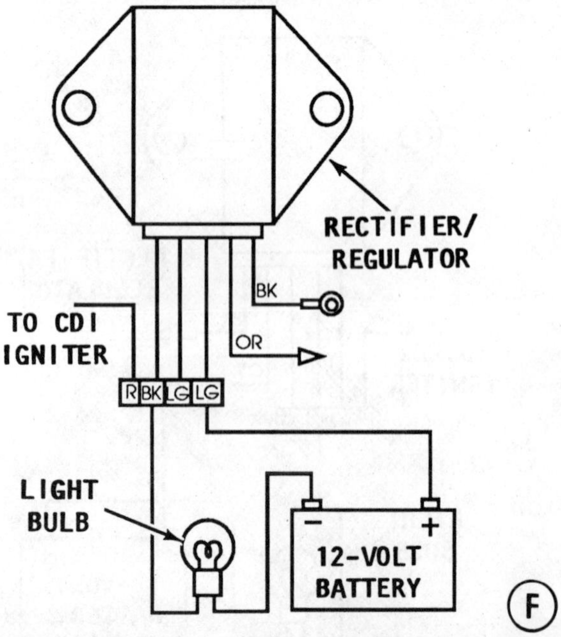

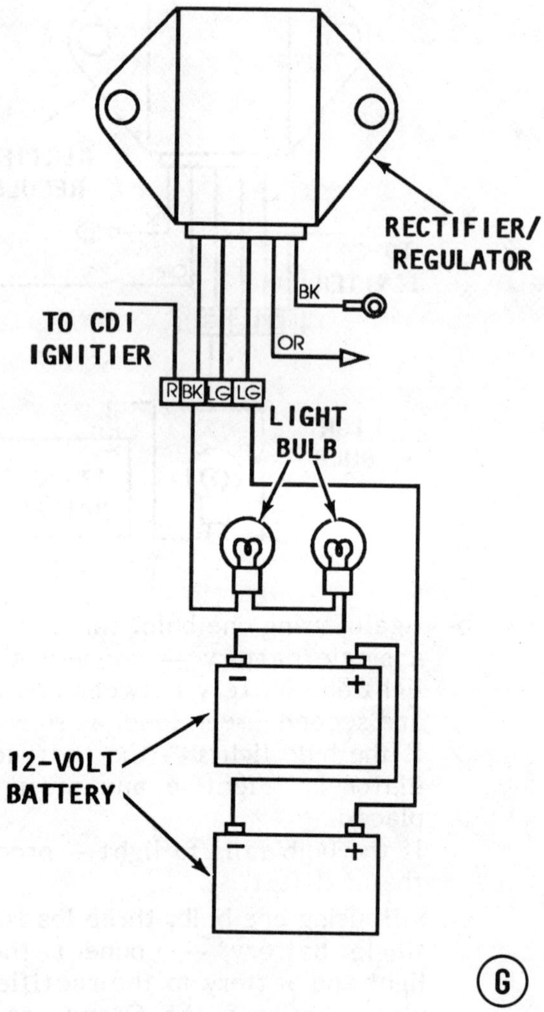

f- Still using just one bulb, three leads, and a single battery -- connect the test light and battery to the rectifier/regulator between the Black and first Green leads, as shown. If the bulb lights -- the rectifier/regulator is defective and must be replaced.
If the bulb fails to light -- proceed to the next test.

g- Now, using both bulbs, five leads, and both batteries, connect the test lights and the batteries to the rectifier/regulator between the Black and first Green lead, as shown. Make the last connection for **ONLY** a few seconds. If the test bulbs fail to light -- the rectifier/regulator is defective and must be replaced.
If the bulbs light -- proceed to the next test.

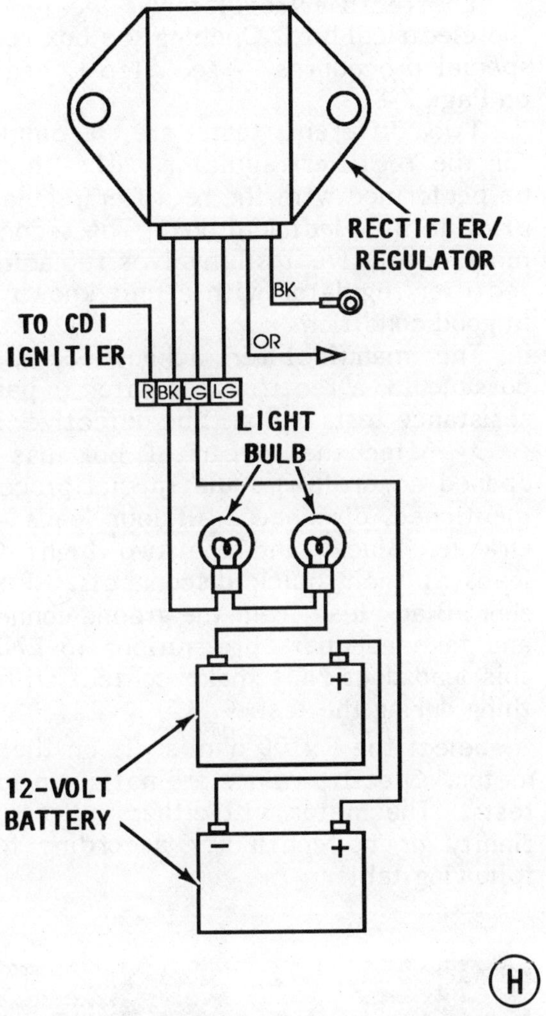

h- Again, using both bulbs, five leads, and both 12-volt batteries, connect the test lights and the batteries to the rectifier/regulator between the Black and second Green leads, as shown. Make the last connection for **ONLY** a few seconds. If bulbs light -- the rectifier/regulator is defective and must be replaced. If the bulbs fail to light -- proceed to the next test.

SPECIAL WORDS

As stated earlier, a defective rectifier/-regulator may pass the resistance tests -- Step 3, and the rectifier circuit tests in Step 4. If the charging system still does not function properly after checking all components in the circuit, including the battery, replace the rectifier with a known good unit.

TESTING 440 SERIES AND 550 SERIES COMPONENTS

Charging Coil Output Test 440 Series & 550 Series

1- A voltmeter capable of registering 50 volts AC is required for this test. Remove the connector cover of the electrical box. Disconnect the 6-prong magneto harness connector, as described in Ignition and Charge Circuit Testing on Page 7-8. This connector has only five leads -- one position is empty. Using a pair of needle nose pliers, reach into the magneto end of the connector half and push out the male prongs at the end of the two Light Green leads to free the leads from the connector.

Select the Vx10 volts AC scale on the voltmeter.

Temporarily connect the two halves of the connector without the two Light Green leads.

PLAN AHEAD

Each of the disconnected leads will be tested with a voltmeter connected across one Light Green lead and a suitable ground connection -- with the engine running. Therefore, position the leads to make the test quickly and efficiently. Decide on the ground connection to be used and have a pencil and paper ready to record the readings.

SAFETY WORDS

The manufacturer warns this test **MUST NOT** be performed with the craft in a body of water due to the possibility of electric shock to the individual performing the test. In light of this warning, this test **MUST** be performed in a test tank with the person performing the test on dry land.

Move the craft to a test tank. Start the engine and allow it to reach operating temperature.

CAUTION

Water must circulate through the jet pump -- to and from the engine, anytime the engine is operating. Circulating water will prevent overheating -- which could cause damage to moving engine parts and possible engine seizure.

Advance the throttle to obtain about 6,000 rpm. Make contact with the Red voltmeter lead to one of the disconnected Light Green leads. Make contact with the Black voltmeter lead to a suitable engine ground. Record the meter reading. Keep the Black meter lead in place as the ground and move the Red meter lead to the other disconnected Light Green lead. Again, record the meter reading.

Shut down the engine. The voltmeter reading for both leads must be 12-15 volts AC.

If the meter readings are as expected, the problem in the charging system is probably in the rectifier/regulator. Proceed directly Step 3.

If the meter readings are below the values expected, the charging coil output voltage is too low. Proceed to the next step to check the charging coil under the flywheel.

Install the two Light Green leads into their original locations inside the harness connector. Pull the male prongs at the end of the leads into position with a pair of needle nose pliers.

Charging Coil Resistance Test
440 Series and 550 Series

2- Select the Rx1 ohm scale on an ohmmeter. Make contact with the Red meter lead to either of the two Light Green male prongs in the plug from the magneto assembly. Make contact with the Black meter lead to the Black male prong in the plug from the magneto assembly.

Repeat the test for the other Light Green lead. The meter should register 1.2-1.8 ohms during both tests.

Make contact with either meter lead across the two Light Green male prongs. The meter should indicate 2.4-3.6 ohms.

If the meter readings are not as expected, the charging coil must be replaced. See Section 7-6, beginning on Page 7-27 for proper procedures to "pull" the flywheel.

If the meter readings are as expected, but the charging coil output test (Step 1) indicated low voltage, the problem in the charging system must be loss of magnetism in the permanent magnets in the flywheel. Therefore, the flywheel must be replaced.

Rectifier/Regulator Tests
440 Series and 550 Series

The rectifier/regulator is located inside the electrical box. Opening the box requires special procedures -- see Step 1, beginning on Page 7-8.

Two different tests are recommended for the rectifier/regulator. The first may be performed with the rectifier/regulator in place in the electrical box. The second and most conclusive test involves replacing the rectifier/regulator with a unit known to be in good condition.

The manufacturer acknowledges, it is possible for a rectifier/regulator to pass the resistance tests and still be defective.

3- After the electrical box has been opened according to the special procedures mentioned, disconnect all four leads -- the Orange, Black, and the two Light Green leads at their quick disconnects. Remove short Black lead from the ground connection and take adequate precautions to **ENSURE** this lead does **NOT** make contact with anything during the test.

Select the Rx100 ohm scale on the ohmmeter. Specific values are not given for the tests. The meter will either indicate continuity or no continuity according to the following table.

The rectifier/regulator is housed inside and protected by the electrical box.

RESISTANCE TESTS

Rectifier/Regulator
440 Series and 550 Series

Red Meter Lead To	Black Meter Lead To	Meter Reading
Orange	Black	No Continuity
Orange	First Light Green	Continuity
Orange	Second Light Green	Continuity
Black	Orange	No Continuity
Black	First Light Green	Continuity
Black	Second Light Green	Continuity
First Light Green	Orange	No Continuity
First Light Green	Black	No Continuity
First Light Green	Second Light Green	No Continuity
Second Light Green	Orange	No Continuity
Second Light Green	Black	No Continuity
Second Light Green	First Light Green	No Continuity

SPECIAL WORDS

As stated earlier, a defective rectifier/-regulator may still pass the resistance tests just listed in Step 3. If the charging system still fails to function properly after all components in the system --including the battery -- have been checked, replace the rectifier/regulator with a unit known to be in good condition.

Charging Coil Output Test
650 Series -- Except JetMate

1- Obtain a voltmeter capable of registering 50V AC. Select the Vx10 volt AC scale on the meter.

Open the electrical box according to the special procedures listed in Step 1, beginning on Page 7-8.

Disconnect the 9-prong magneto harness connector. Using a pair of needle nose pliers, reach into the magneto end of the connector half and push out the male prongs at the end of the two Yellow leads to free the leads from the connector. Temporarily connect the two halves of the connector without the two Yellow leads.

PLAN AHEAD

Each of the disconnected leads will be tested first with a voltmeter connected across one Yellow lead and a suitable ground connection and then tested again across the other Yellow lead and a good ground -- with the engine running. Therefore, position the leads to make the test quickly and efficiently. Decide on the gound connection to be used and have a pencil and paper ready to record the readings.

SAFETY WORDS

The manufacturer warns this test **MUST NOT** be performed with the craft in a body of water due to the possibility of electric shock to the individual performing the test. In light of this warnlng, this test **MUST** be performed in a test tank with the person performing the test on dry land.

Move the craft to a test tank. Start the engine and allow it to reach operating temperature.

CAUTION

Water must circulate through the jet pump -- to and from the engine, anytime the engine is operating. Circulating water will prevent overheating — which could cause damage to moving engine parts and possible engine seizure.

Advance the throttle to obtain about 3,000 rpm engine speed. Now, quickly make contact with the Red voltmeter lead to one of the disconnected Yellow leads. Make contact with the Black meter lead to the selected ground on the engine. Record the reading. Keep the Black meter lead in place on the ground and quickly move the Red meter lead to the other disconnected Yellow lead. Again, record the meter reading.

Shut down the engine. The voltmeter reading for both Yellow leads must be 36-40V AC.

If the meter readings are as expected, the problem in the charging system is probably in the rectifier/regulator. Proceed directly to Step 3.

If the meter readings are below the values expected, the charging coil output voltage is too low. Proceed to the next step to check the charging coil under the flywheel.

Install the two Yellow leads into their original locations inside the harness connector. Pull the male prongs at the end of the leads into position with a pair of needle nose pliers.

Charging Coil Resistance Test
650 Series — Except JetMate

2- Select the Rx1 ohm scale on an ohmmeter. Make contact with the Red meter lead to either of the two Yellow male prongs in the plug from the magneto assembly. Make contact with the Black meter lead to the Black male prong in the plug from the magneto assembly. Repeat the test for the other Yellow lead. The meter should register 0.7-1.3 ohms for both tests.

Make contact with either meter lead across the two Yellow male prongs. The meter should register 1.5-2.3 ohms.

If the meter reading is not as expected, the charging coil must be replaced. See Section 7-6, beginning on Page 7-27 for special instructions to "pull" the flywheel.

If the meter readings are as expected, but the charging coil output test (Step 1), gives a low voltage, the problem in the charging system must be loss of magnetism in the permanent magnets in the flywheel. The flywheel must be replaced. See Section 7-6, beginning on Page 7-27 for special instructions to "pull" the flywheel.

Rectifier/Regulator Tests
650 Series -- Except JetMate

The rectifier/regulator is located inside the electrical box. Opening the box requires special procedures -- see Step 1, beginning on Page 7-8.

Two different tests are recommended for the rectifier/regulator. The first may be performed with the rectifier/regulator in place in the electrical box. The second and most conclusive test involves replacing the rectifier/regulator with a unit known to be in good condition.

The manufacturer acknowledges, it is possible for a rectifier/regulator to pass the resistance tests and still be defective.

3- After the electrical box has been opened, disconnect all four leads -- the Orange (JS650SX), or White (JF650), the Black lead, and the two Yellow leads at their quick disconnect. Remove the short Black lead from the ground connection and take adequate precautions to **ENSURE** this lead does not make contact with anything during the test.

Select the Rx1K (1,000) ohm scale on the meter. The meter should register the results indicated for the tests listed in the table on the following page.

SPECIAL WORDS

As stated earlier, a rectifier/regulator may pass the resistance tests in Step 3 and still be defective. If the charging coil still fails to perform properly after all components in the circuit -- including the battery --have been checked, replace the rectifier/regulator with a unit known to be in good condition.

Charging Coil Output Test
JetMate Only

1- Obtain a voltmeter capable of registering 50V AC, and then select the Vx10 volt AC scale.

Open the electrical box according to the special instruction outlined in Step 1, beginning on Page 7-8.

Identify the two Brown leads between the charging coil under the flywheel and the rectifier/regulator. Disconnect these two leads at their quick-disconnect.

PLAN AHEAD

Each of the disconnected leads will be tested first with a voltmeter connected across one Brown lead and a suitable ground connection and then tested again across the other Brown lead and a good ground --with the engine running. Therefore, position the

RESISTANCE TESTS

Rectifier/Regulator
650 Series — Except JetMate

Some units may have Orange (JS650SX)
Others may have White (JF650)

Red Meter Lead To	Black Meter Lead To	Meter Reading
Orange or White	Black	No Continuity
Orange or White	First Yellow	1,000-5,000
Orange or White	Second Yellow	1,000-5,000
Black	Orange or White	No Continuity
Black	First Yellow	2,000-10,000
Black	Second Yellow	2,000-10,000
First Yellow	Orange or White	No Continuity
First Yellow	Black	No Continuity
First Yellow	Second Yellow	No Continuity
Second Yellow	Orange or White	No Continuity
Second Yellow	Black	No Continuity
Second Yellow	First Yellow	No Continuity

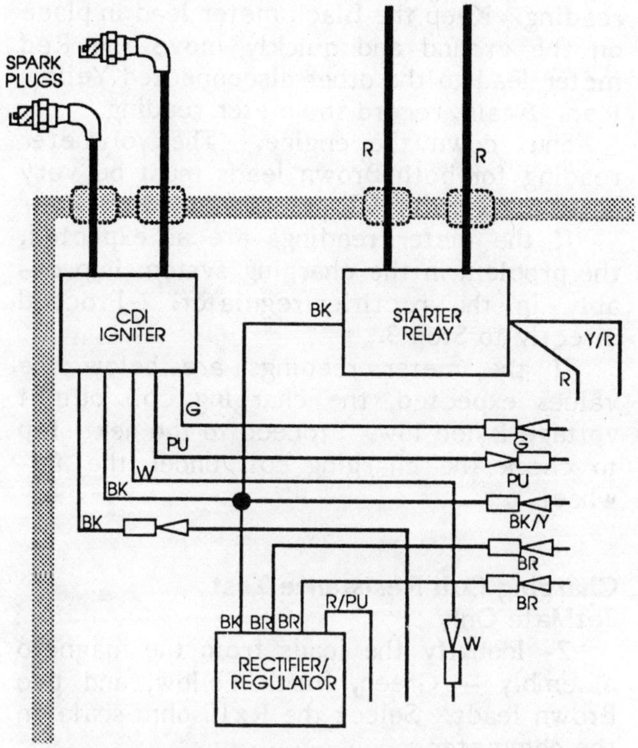

Section of the JetMate wiring diagram, with color identification and routing.

leads to make the test quickly and efficiently. Decide on the gound connection to be used and have a pencil and paper ready to record the readings.

SAFETY WORDS

The manufacturer warns this test **MUST NOT** be performed with the craft in a body of water due to the possibility of electric shock to the individual performing the test. In light of this warning, this test **MUST** be performed in a test tank with the person performing the test on dry land.

Move the craft to a test tank. Start the engine and allow it to reach operating temperature.

CAUTION

Water must circulate through the jet pump -- to and from the engine, anytime the engine is operating. Circulating water will prevent overheating -- which could cause damage to moving engine parts and possible engine seizure.

Advance the throttle to obtain about 3,000 rpm engine speed. Now, quickly make contact with the Red voltmeter lead to one of the disconnected Brown leads. Make contact with the Black meter lead to the selected ground on the engine. Record the reading. Keep the Black meter lead in place on the ground and quickly move the Red meter lead to the other disconnected Yellow lead. Again, record the meter reading.

Shut down the engine. The voltmeter reading for both Brown leads must be very close to 42-volts AC.

If the meter readings are as expected, the problem in the charging system is probably in the rectifier/regulator. Proceed directly to Step 3.

If the meter readings are below the values expected, the charging coil output voltage is too low. Proceed to the next step to check the charging coil under the flywheel.

Charging Coil Resistance Test
JetMate Only

2- Identify the leads from the magneto assembly -- Green, Black/Yellow, and two Brown leads. Select the Rx10 ohm scale on the ohmmeter.

Make contact with either meter lead across the two Brown male prongs. The meter should register 0.3-0.6 ohms.

Make contact with either meter lead across the Green and Black/Yellow leads. The meter should register 13.6-20.5 ohms.

If the meter reading is not as expected, the charging coil must be replaced. See Section 7-6, beginning on Page 7-27 to "pull" the flywheel.

If the meter readings are as expected, but the charging coil output test (Step 1), gave a low voltage, the problem in the charging system must be loss of magnetism in the permanent magnets in the flywheel. The flywheel must be replaced.

Rectifier/Regulator Tests
JetMate Only

3- The rectifier/regulator is located inside the electrical box. Opening the box requires special procedures -- see Step 1, beginning on Page 7-8.

Two different tests are recommended for the rectifier/regulator. The first may be performed with the rectifier/regulator in place in the electrical box. The second and most conclusive test involves replacing the rectifier/regulator with a unit known to be in good condition.

The manufacturer acknowledges, it is possible for a rectifier/regulator to pass the resistance tests and still be defective.

After the electrical box has been opened, disconnect all four leads -- the Red or Purple, Black, and the two Brown leads at their quick-disconnect. Remove the short Black lead from the ground connection and take adequate precautions to **ENSURE** this lead does not make contact with anything during the test.

Select the RxlK (1000) ohm scale on the meter, and then perform the tests outlined in the following table.

RESISTANCE TESTS
Rectifier/Regulator
JetMate Only

Note: Some units may have Red lead, Others may have Purple lead.

Red Meter Lead To	Black Meter Lead To	Meter Reading
Red or Purple	First Brown	1,000-5,000
Red or Purple	Second Brown	1,000-5,000
Red or Purple	Black	2,000-10,000
First Brown	Red or Purple	20,000-100,000
First Brown	Second Brown	30,000-200,000
First Brown	Black	1,000-5,000
Second Brown	Red or Purple	20,000-100,000
Second Brown	First Brown	30,000-200,000
Second Brown	Black	1,000-5,000
Black	Red or Purple	15,000-80,000
Black	First Brown	20,000-100,000
Black	Second Brown	20,000-100,000

SPECIAL WORDS

As stated earlier, a rectifier/regulator may pass the resistance tests in Step 3 and still be defective. If the charging coil still fails to perform properly after all components in the circuit -- including the battery --have been checked, replace the rectifier/regulator with a unit known to be in good condition.

7-6 FLYWHEEL AND MAGNETO

The following short section lists the procedures required to "pull" the flywheel and remove the stator plate in order to service the ignition system. Cleaning and Inspecting procedures in addition to proper assembling and installation steps are also included.

Because so many different series are covered in this manual, it would not be feasible to provide an illustration of each and every unit covered.

Therefore, the accompanying illustrations are of a "typical" unit. The unit being serviced may differ slightly in appearance due to engineering or cosmetic changes but the procedures are valid. If a difference should occur, it will be clearly indicated for the series and possible model year affected.

SPECIAL WORDS ON SPECIAL TOOLS

The following list of special tools are required to properly perform the service work involving "pulling" the flywheel and service of the magneto plate. Some tasks become extremely difficult, if not impossible, without the special tools or some type of equivalent.

A special effort has been made to give some type of equivalent or homemade tool which may be substituted, whenever possible. The cost of a special tool may be offset by money saved if an expensive part is not damaged.

300 Series Tools

Flywheel Holder	P/N 57001-306
Magneto Puller	P/N 57001-259
and center bolt	P/N 57001-1099

440 Series & 550 Series

Flywheel Holder	P/N T57001-276
Flywheel Puller	P/N 57001-259
and center bolt	P/N 57001-1099

650 Series

Flywheel Holder	P/N 57001-1156
Plus three 8mm (hd) x 10mm (long) bolts	
Flywheel Puller	P/N 57001-259
and center bolt	P/N 57001-1099

Alternate Flywheel Holder

If the special flywheel holder listed is not available, a tool may be fabricated from a metal bar and two long bolts. The longer the bar, the more leverage gained. The bolts should be of sufficient length to allow a socket and a torque wrench to be placed over the flywheel nut or bolt and still clear the bar. Drill two holes, large enough for the bolts to pass through, at the end of the bar, with the same distance between as the two holes in the flywheel. Secure the bolts to the bar.

Alternate Flywheel Puller

If the special flywheel puller listed is not available, a similar puller, shown in the accompanying illustration, may be used **PROVIDED** it will pull from the bolt holes in the flywheel and **NOT** from around the per-

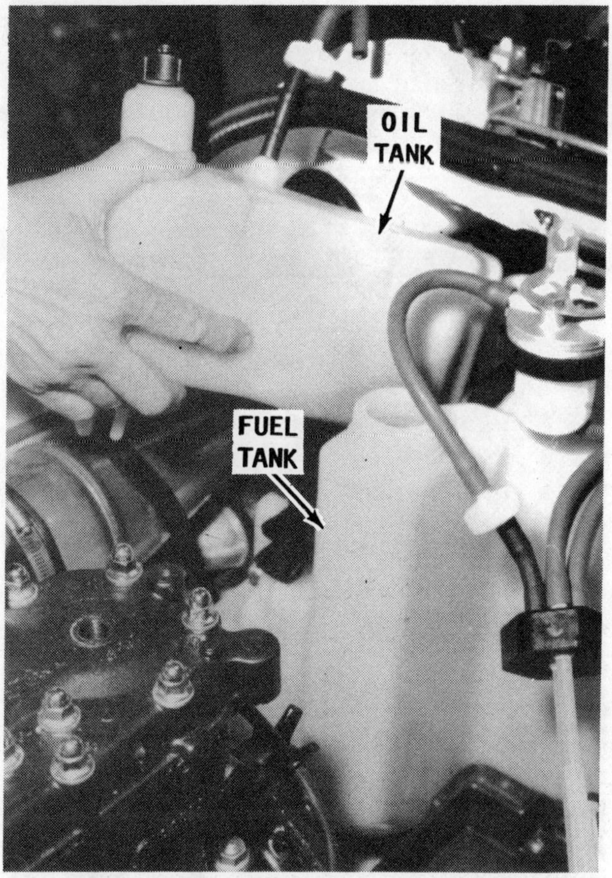

Removing the oil tank and fuel tank on a 650 Series unit to gain access to the flywheel.

imeter of the flywheel. **NEVER** attempt to use a puller which would put a strain on the outside edge of the flywheel.

Preliminary Tasks

If servicing a 300, 400, or 550 Series engine, remove the fuel tank.

If servicing a 300 or 650 Series engine, remove the oil injection pump. See Section 6-10, Oil Injection, beginning on Page 6-49.

If servicing a JetMate, remove the oil tank and bracket, the water box muffler, and the oil pump.

"PULLING" THE FLYWHEEL

1- Remove the spark plugs.

2- Remove the bolts securing the flywheel cover to the stator plate. If servicing a 300 Series engine, remove the two cranking motor mounting bolts.

3- Remove the flywheel cover from the stator plate. If the cover is stuck to the plate, obtain a soft head mallet and tap around the circumference of the cover to jar it free. Remove and discard the sealing ring, only if a replacement is available.

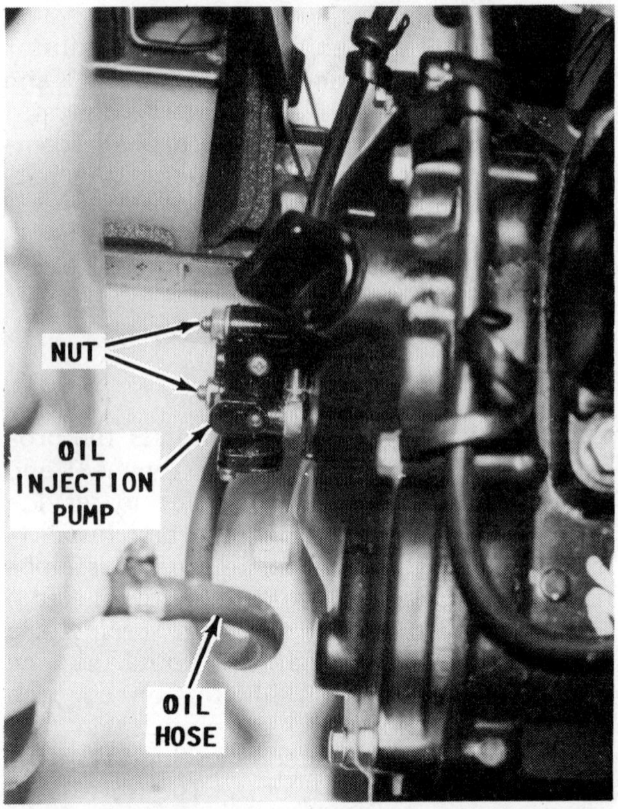

Items to be disconnected or removed on a 300 Series or 650 Series installation before the flywheel cover can be removed.

4- If servicing a 440 or 550 Series engine, bend back the tabs on the lockwasher behind the flywheel nut. Now using flywheel holder tool listed above for the engine being serviced, insert the two indexing pins on the ends of the arms through the holes in the flywheel. Prevent the flywheel from rotating with the special tool. Loosen, and then remove the flywheel nut and washer -- 650 Series; the tabbed lockwasher --440 and 550 Series; flanged bolt -- late 300 Series; or the bolt and washer -- early 300 Series.

5- Install the flywheel puller onto the flywheel, and then using the center bolt, take a strain on the puller with the proper size wrench. Now, continue to tighten the center bolt and at the same time, **SHOCK** the crankshaft with a gentle to moderate tap with a hammer on the end of the center bolt. This shock will assist in "breaking" the flywheel loose from the crankshaft. **DO NOT** "pound" on the center bolt. The force of such action would be transmitted to the seals and bearings along the crankshaft and cause damage. **NEVER** strike the flywheel because this type of shock would surely cause the flywheel magnets to loose their magnetism.

4

3

5

6- Pull the flywheel free of the crankshaft. Remove and save the Woodruff key from the crankshaft keyway.

650 Series Only

Lift out the reduction gear assembly from the splined armature shaft of the cranking motor.

Removing the reduction gear from the armature shaft of the cranking motor on a 650 Series installation.

SPECIAL WORDS ON WOODRUFF KEYS

Inspect the condition of the Woodruff key and its keyway in the crankshaft. If the key shows any signs of shearing, the cause is probably from jumping waves and the effect will be a loss of power. When the craft is clear of the water, the engine RPM approaches the limit set by the RPM limiter in an unloaded condition.

When the craft returns to the water surface, the sudden load on the impeller is transferred along the driveline. Speed of the rotating pump shaft and the crankshaft is suddenly reduced while the flywheel continues to spin rapidly due to its mass. This condition, especially when repeated over a period of time, often shears the Woodruff key and the flywheel becomes repositioned on the crankshaft.

A very small misalignment will result in a change in ignition timing. The craft operator will notice a drastic loss of power. A worse condition would be the flywheel shifting to the degree engine start would not be possible.

SPECIAL WORDS ON COIL IDENTIFICATION AND COIL REPLACEMENT

On some series, a coil on the magneto assembly can be removed and installed with-

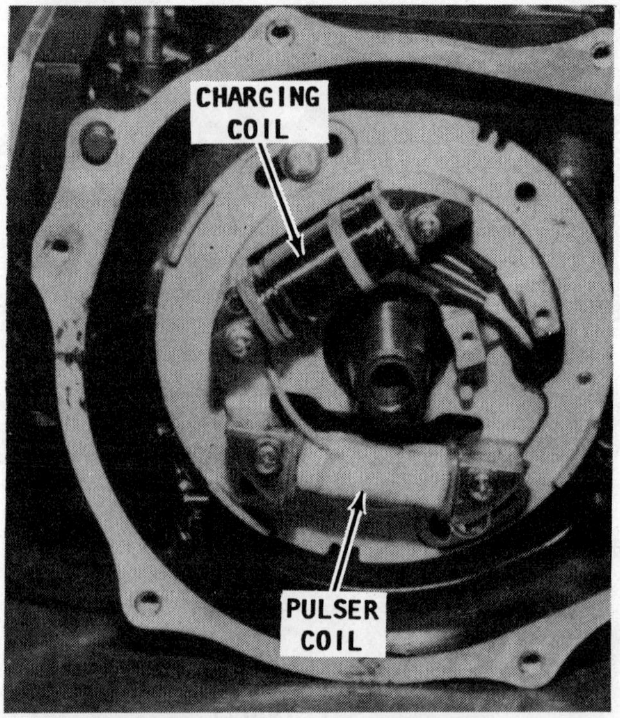

*Arrangement of coils on the stator plate. If a third coil -- an exciter coil -- exists, it would be mounted **UNDER** the pulser coil.*

out removing the assembly from the engine. To remove the coil, simply remove the coil mounting screws and unsolder the connections. To install, solder the connections and secure the coil with the mounting screws.

Kawasaki Series engines have either two or three coils mounted on the magneto assembly. If there are only two coils installed, one coil is the charging coil and will always have the same color leads connected -- possibly two Green leads. The second coil is the exciter coil.

If there are three coils mounted on the magneto assembly, again, one is the charging coil and will always have the same color leads connected --possibly two Green leads. The second and third coils are always stacked, one on top of the other. The top coil is the pulser coil (smallest of the two), and the bottom coil is the exciter coil (larger of the two).

Check with the local Kawasaki dealer to determine if an individual coil is available or if the entire magneto assembly must be replaced. The answer varies from series to series and from model year to model year.

Magneto Assembly Removal

7- The securing screws pass through elongated slots in the magneto assembly. The assembly could be installed as much as 15° off its original position -- drastically affecting the timing characteristics of the engine. Therefore, at the elongated slot in the magneto assembly, make a set of aligned punch marks -- one on the magneto assembly -- the other on the stator plate.

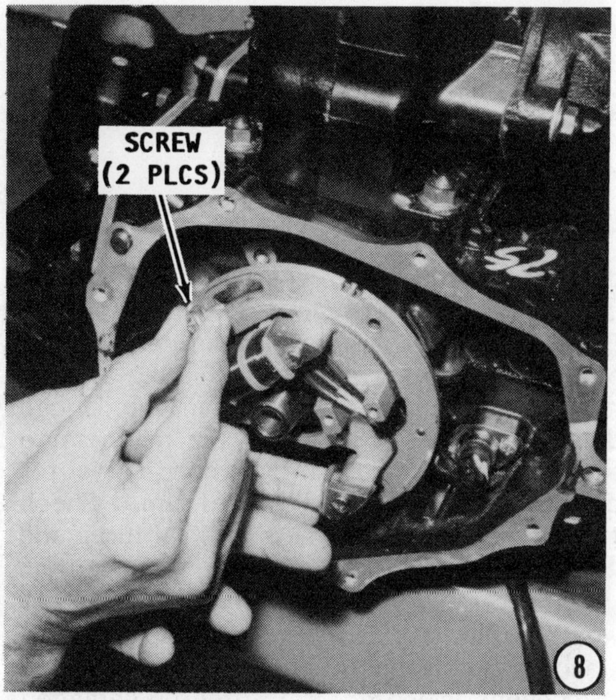

These marks will serve as an aid during installation to **ENSURE** correct positioning of the magneto assembly on the stator plate.

8- Remove the two screws securing the magneto assembly to the stator plate.

9- Remove the two small screws securing the magneto assembly harness to the grommet in the stator plate. Disconnect the other end of the harness either at the connector cover on the electrical box or

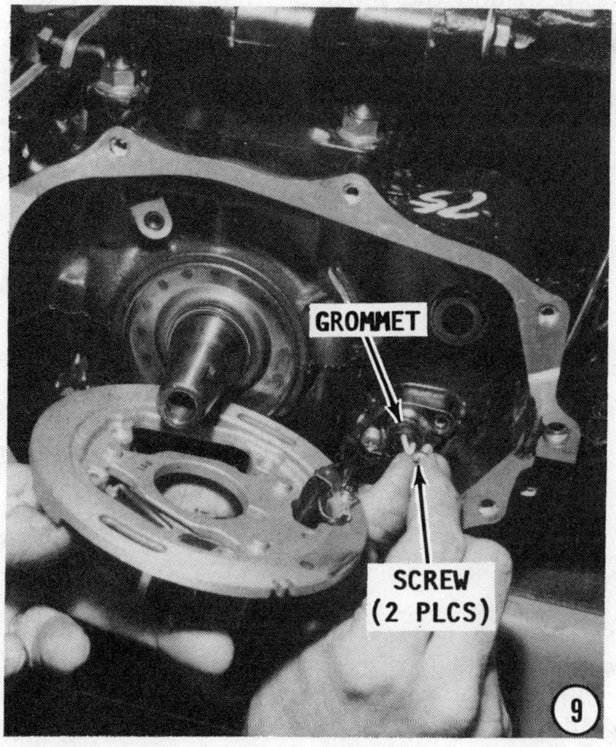

inside the electrical box. Feed the harness and connector end through the opening in the stator plate and remove the magneto assembly from the crankshaft.

CLEANING AND INSPECTING

Inspect the flywheel for cracks or other damage, especially around the inside of the center hub. A damaged flywheel can be **EXTREMELY** dangerous. The forces exerted on the flywheel while it is spinning at high rpm are sufficient to shatter an unbalanced damaged flywheel. Inspect the magnets to see if they are held tightly around the outer circumference of the flywheel rim. Check the surface of the magnets for fractures and grooves.

Grooves around the inner perimeter would indicate a component mounted on the stator plate is not aligned properly or the component is loose. Fractured or damaged magnets cannot be replaced. Both the magnets and flywheel share the same part number, therefore attempting to remove the magnets would be a pointless task. If magnets or the flywheel is defective, the entire flywheel must be replaced.

Check to be sure small metal particles have not become attached to the magnets. Verify each magnet has good magnetism by using a screwdriver or other suitable tool.

Thoroughly clean the inside taper of the flywheel and the taper on the crankshaft to prevent the flywheel from "walking" on the crankshaft during operation.

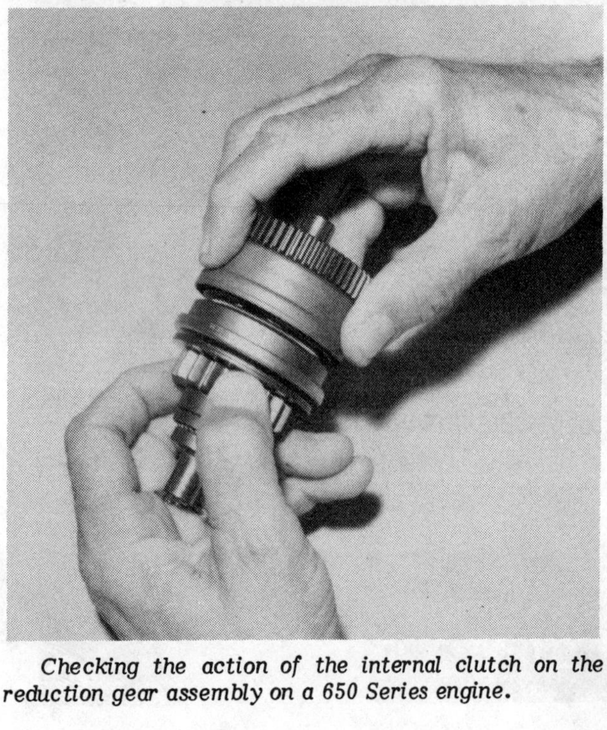

Checking the action of the internal clutch on the reduction gear assembly on a 650 Series engine.

Test the magneto assembly to verify it is not loose. Attempt to lift each component on the plate. There should be little or no evidence of movement.

650 Series Only

Inspect the pinion gear and the inner teeth of the reduction gear assembly. If any signs of damage or excessive wear are discovered, the assembly must be replaced. Check the action of the internal clutch. If the movement of the clutch is not smooth, an internal defect exists in the clutch assembly, and the reduction gear assembly must be replaced.

SPECIAL WORDS ON FLYWHEEL MAGNETS

The outer ends of any magnet are called poles. One end is the north pole and the other end is the south pole. The magnetic field surrounding a magnet is concentrated around these two poles.

Some flywheel magnets are fairly long and curved around the outer perimeter of the underside of the flywheel. Others are short and are mounted around the center hub, depending on the location of the coils mounted on the stator. Magnets are usually installed in pairs with the north pole of one adjacent to the south pole of its neighbor and so on. In this manner continuous magnetic field surrounds the inside of the flywheel.

If a flywheel is accidently dropped, not only could the teeth be damaged, but the impact will weaken the magnetic strength of all the magnets housed in the flywheel.

If one or more of the magnets should break or fracture, two new magnetic poles will be created. A long magnet with two poles will become two short magnets with four poles. The new poles will possess only half the magnetic strength of an original pole. The overall magnetic field will be altered. The new field of the shorter magnets will not extend to cover the area of the flywheel.

Serious consequences apply to a CDI type ignition system in the event of a flywheel magnet fracture. Exciter and pulser coils are energized by the concentrated magnetic field at the magnet poles. If new poles are suddenly created, the coil module will receive conflicting signals from the magnets and may even attempt to fire the cylinder twice in one revolution.

All these reasons require the flywheel to be handled with **CARE.**

If there is evidence of oil on the magneto assembly, the oil seal/s at the forward main bearing inside the crankcase may have failed. Refer to Chapter 8, Engine.

ASSEMBLING AND INSTALLATION

Individual Coil Installation

If individual coils were replaced on the magneto assembly, without removal of the assembly from the plate, solder the new coil leads into place. Coat the soldered connections with Loctite Superflex to insulate the new connections. Install the mounting screws and tighten them securely.

Coil Placement

Kawasaki Series engines have either two or three coils mounted on the magneto assembly. If there are only two coils installed, one coil is the charging coil and will always have the same color leads connected -- possibly two Green leads. The second coil is the exciter coil.

If there are three coils mounted on the magneto assembly, again, one is the charging coil and will always have the same color leads connected --possibly two Green leads. The second and third coils are always stacked, one on top of the other. The top coil is the pulser coil (smallest of the two), and the bottom coil is the exciter coil (larger of the two).

Always check the flywheel carefully to be sure particles of metal have not become stuck to the magnets.

A new magneto assembly -- with coils in place -- ready for installation.

1- Apply a coating of multi-purpose water resistant lubricant to the grommet in place on the stator harness. Feed the harness connector through the opening in the stator plate and position the magneto assembly over the end of the crankshaft. Push the grommet into the opening and

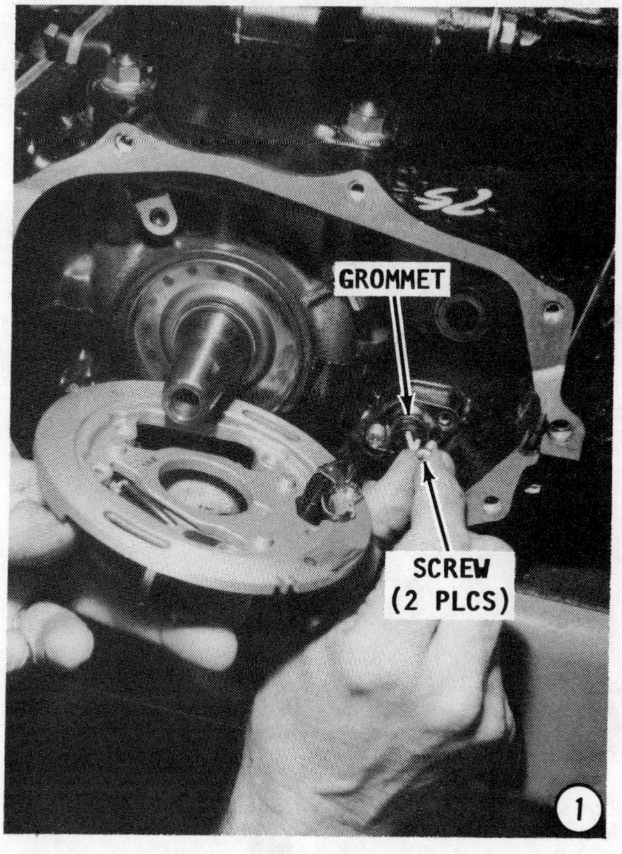

secure it in place with the two attaching screws. Tighten both screws securely.

Connect the other end of the stator harness to the electrical box connector either inside the box or at the connector cover on top of the box. If the connector is not the plug-in type, match the leads color-to-color.

Timing Marks

Identify the two marks made on the magneto assembly and the stator plate prior to removal. These marks serve to correctly position the magneto plate on the stator plate. The securing screws pass through elongated slots in the magneto assembly. Therefore, the assembly could be installed as much as 15° to its original position, drastically affecting the timing characteristics of the engine. If a **NEW** magneto assembly is being installed, match the new assembly to the old assembly and make a new mark as close to the original location as possible. The engine must be timed according to the procedures outlined in Section 7-7, at the end of this chapter.

If **NO** timing marks were made prior to removal of the magneto assembly, as instructed in Step 7 of Disassembling, closely inspect the elongated slots in the magneto assembly and attempt to determine the original positioning of the screws. If no identifiable mark can be found, center the magneto assembly in the slot. This position may or may not be close to the correct position to permit the engine to start. Detailed timing procedure is presented in Section 7-7, at the end of this chapter.

2- Apply a coating of Loctite #572, or equivalent, to the threads of the two magneto securing screws. Align the mark on the magneto assembly with the mark on the stator plate. Install and tighten the two screws securing the assembly to the stator plate **WITHOUT** disturbing the two aligned marks.

Flywheel Installation

FIRST, THESE WORDS

If the engine is to be timed, do not hesitate to install the flywheel using the Woodruff key and tightening the flywheel nut/bolt to the required torque value. The flywheel has cutouts affording access to the two magneto assembly securing screws. Therefore, the engine can be properly timed with the flywheel in place and secured.

650 Series Only

3- Slide the reduction gear over the splined armature shaft of the cranking motor. Apply a coating of multi-purpose water resistant lubricant to the reduction gear.

The timing marks on the magneto assembly ***MUST*** *align with the marks on the stator plate during installation.*

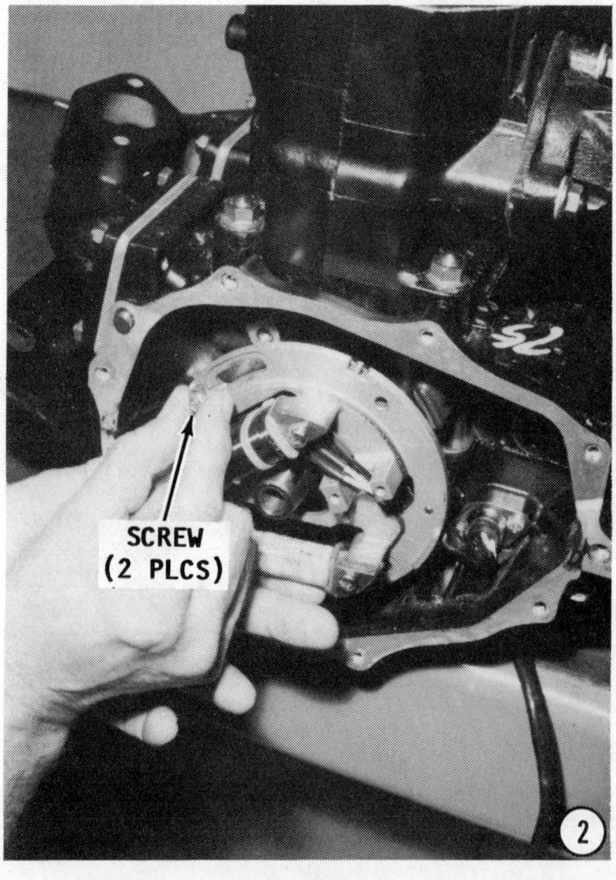

All Series

4- Apply just a "dab" of thick lubricant to the curved surface of the Woodruff key to hold it in place while the flywheel is being installed. Press the Woodruff key into place in the crankshaft recess. Clean any excess lubricant from the shaft to prevent the flywheel from "walking" during engine operation.

Check the flywheel magnets to ensure they are free of any metal particles. Double check the taper in the flywheel hub and the taper in the crankshaft to verify they are clean and contain no oil.

Now, slide the flywheel onto the crankshaft with the keyway in the flywheel aligned with the Woodruff key in place on the crankshaft. Rotate the flywheel **COUNTERCLOCKWISE** and at the same time check to be sure the flywheel does not contact any part of the magneto assembly or any wiring.

5- Thread the flywheel retaining nut/bolt onto the crankshaft.

300 Series -- bolt and washer.
Late 300 Series -- flanged bolt.
440 Series -- tabbed lockwasher and nut.
550 Series -- tabbed lockwasher and nut.
650 Series -- regular washer and nut.

Use the special flywheel holding tool, or a spanner wrench, or the home made tool to prevent the flywheel from rotating while the flywheel nut/bolt is tightened to the following torque value for the series being serviced.

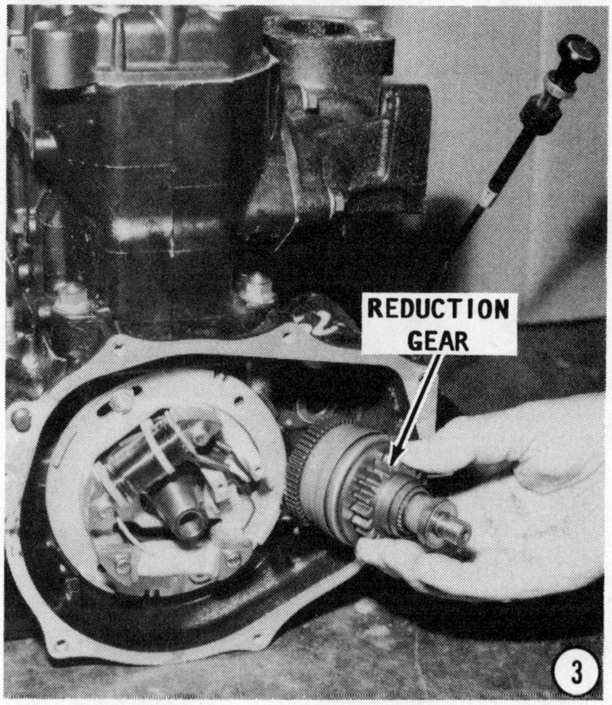

300 Series -- 29 ft lb (39Nm)
440 Series -- 115 ft lb (160Nm)
550 Series -- 115 ft lb (160Nm)
650 Series -- 72 ft lb (98Nm)

If servicing a 440 or 550 Series engine, use a hammer and punch to pry up the tabs on the tabbed lockwasher against the matching faces of the flywheel nut. If no tabs align, back off the nut slightly until the tabs align with a nut face.

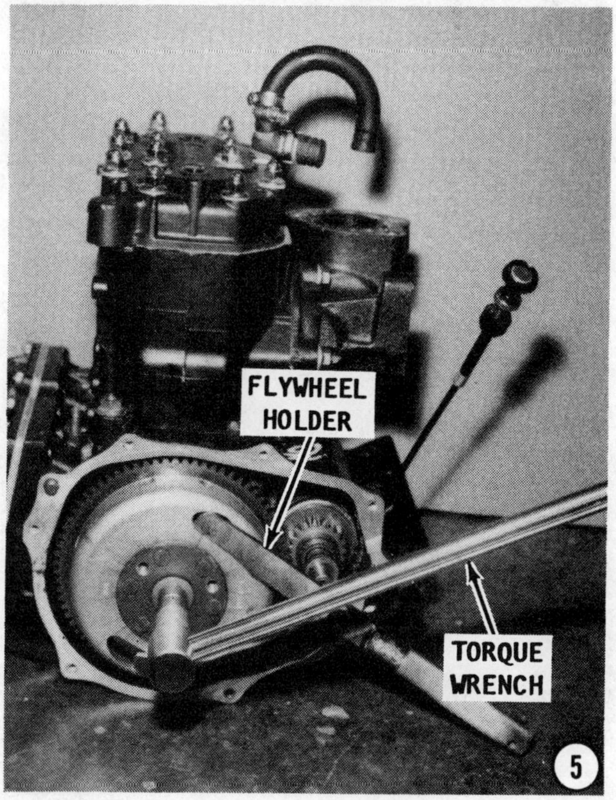

6- Apply a coating of multi-purpose water resistant lubricant to a **NEW** sealing ring. Position the ring over the rim of the stator plate. Install the flywheel cover over the stator plate with the two pins -- one in the cover and the other in the plate indexed into their respective holes.

7- Apply a coating of Loctite Lock N' Seal to the threads of the attaching bolts, and then tighten the bolts in the sequence shown to a torque value of 12 ft lb (16Nm).

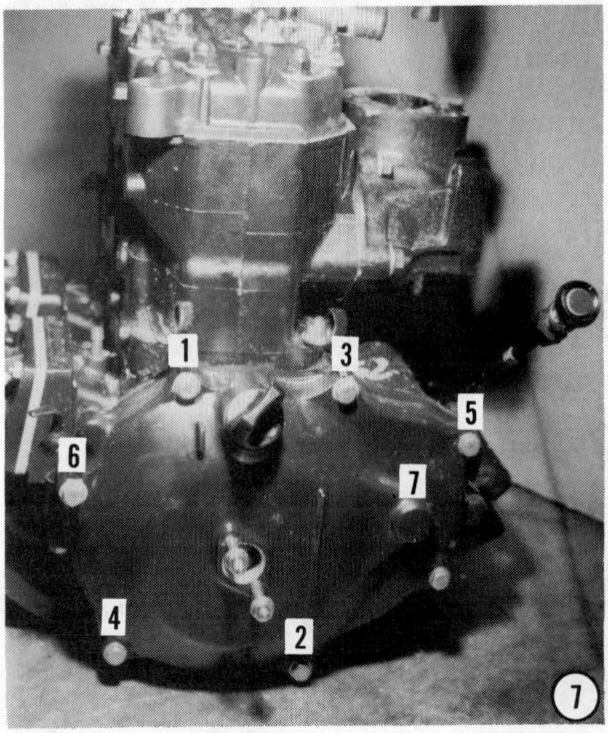

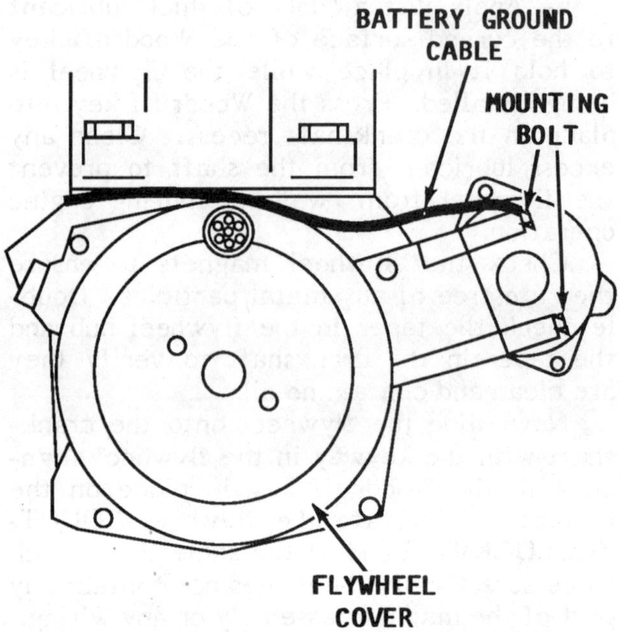

Simple line drawing to depict the flywheel cover installation on a 300 Series engine.

If servicing a 300 Series engine, install the two cranking motor mounting bolts and the battery ground cable onto the flywheel cover, as shown. Tighten the bolts securely.

8- Install and tighten the spark plug/s to a torque value of 14 ft lb (20Nm). Secure the spark plug high tension leads/s to the spark plug/s.

9- If servicing a 300 or 650 Series engine, install the oil injection pump onto the flywheel cover, with the pump indexed with the crankshaft. refer to Section 6-10 -- Oil Injection. The oil injection system **MUST** be "bled" to purge air from the pump and from disconnected hoses. If working on a Jet-Mate, install the oil tank and bracket, and the water box muffler. If servicing a 300, 440, or 550 Series engine, install the fuel tank.

Move the craft to a body of water or connect a garden hose to the engine cooling water supply fitting or flush fitting on the cylinder head.

If the garden hose is used, start the engine and allow the rpm's to stabilize at idle speed **FOR JUST A FEW SECONDS,** and then turn the water on.

Adjust the water flow until a small trickle is discharged from the bypass outlet on the port side of the hull.

When the engine is to be shut down, turn the water off **FIRST** -- raise the aft portion of the hull -- **WHILE THE ENGINE IS OPERATING AT IDLE** -- "rev" the engine just a **COUPLE** times to clear water from the exhaust system -- and then shut it down.

NEVER allow the engine to operate without cooling water for more than 15 seconds.

CAUTION
Water must circulate through the jet pump -- to and from the engine, anytime the engine is operating. Circulating water will prevent overheating — which could cause damage to moving engine parts and possible engine seizure.

NEVER, AGAIN NEVER, operate the engine at high speed with a flush device attached. An engine operating at high speed with such a device attached, would **RUNAWAY** from lack of a load on the impeller shaft, causing extensive damage.

7-7 IGNITION TIMING ADJUSTMENTS

All Kawasaki personal watercraft covered in this manual are equipped with an automatic advance type CDI system. Ignition timing is accomplished electronically and signal output increases proportionately to engine speed increase -- thereby advancing the timing. Such a system has no moving or sliding parts. Therefore, periodic adjustments are not necessary.

The timing is set correctly at the factory and should never need adjustment **UNLESS** the magneto assembly is misaligned or an ignition component -- such as the magneto assembly -- has been replaced.

To adjust the timing on all series, the fuel tank must first be removed. On the 300 Series, the oil injection pump must also be removed.

The oil injection system need not be disturbed on the 650 Series -- including the JetMate -- to adjust the timing.

TEST TANK

The craft must be moved to a test tank or to a body of water where the engine can be run at high speed with restraint lines to prevent the craft from moving.

Fuel Tank Removal
All Models

Begin by loosening the outlet ring nut on the fuel tank to release any vapor pressure built up inside the tank.

If the level of fuel in the tank is above the inlet neck fitting, siphon some of the fuel from the tank to avoid spillage when the tank is removed.

Next, disconnect the oil tank vent hose from the fuel tank vent hose at the tank fitting. Lift out the fuel outlet assembly. Loosen the large hose clamp around the inlet neck and ease the hose from the fitting on the tank. Now, unhook the two rubber straps securing the tank in the hull. Lift the tank up and out of the craft, and then set it down close by. Reconnect the fuel outlet and retainer nut to the tank.

Preliminary Timing Tasks
300 Series Only

First, remove the fuel tank, as outlined in the previous paragraphs. Prepare a 50:1 fuel/oil premix for use in the fuel tank --2.6 oz of Kawasaki Jet Ski Oil or BIA Certified TC-W oil to each gallon of gasoline.

Next, pull the oil pump outlet hose free and plug the end to prevent loss of oil and contaminants from entering the system. Remove the two nuts and washers securing the oil injection pump to the flywheel cover, and then remove the oil injection pump from the cover studs.

Now, remove the two cranking motor mounting bolts on the side of the flywheel cover.

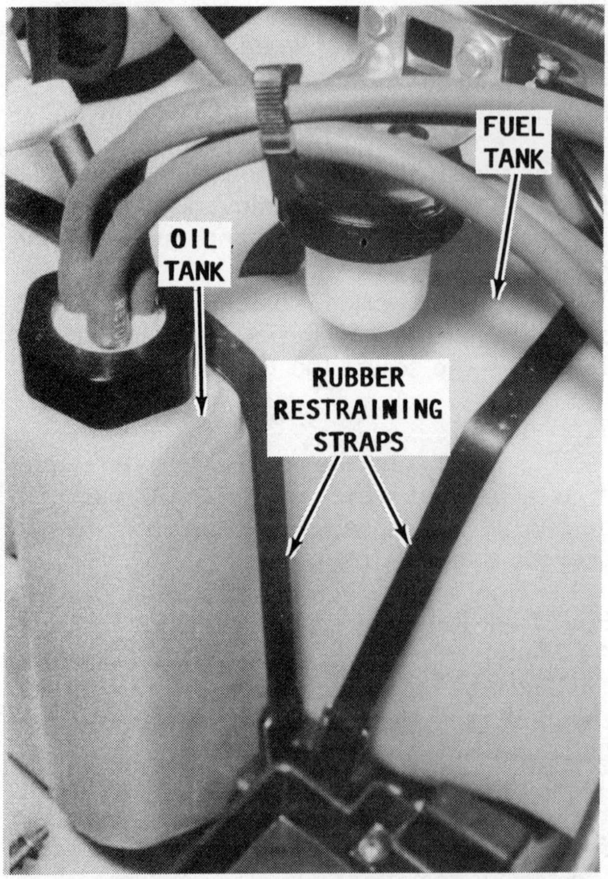

Close view of the fuel tank and oil tank prior to removal to gain access to the flywheel cover, flywheel, and ignition components behind the flywheel.

Preliminary Timing Tasks
440 Series and 550 Series

First, remove the fuel tank, as outlined in the paragraphs above under "Fuel Tank Removal".

Next, remove the flywheel cover. Remove and save the sealing ring.

Preliminary Timing Tasks
650 Series Only

First, remove the fuel tank, as outlined in the previous paragraphs above under "Fuel Tank Removal". Remove the magneto breather plug on the flywheel cover. There is no need to disturb the oil injection system on the 650 Series for this timing procedure.

Determining TDC (Top Dead Center)
All Series

Remove the spark plug -- 300 Series, or the forward spark plug -- all other Series. Install a dial indicator into the spark plug opening and determine TDC of the piston.

If a dial indicator is not available, use a pencil, straw, or equivalent, resting on top of the piston dome. Determine the highest point to which the inserted object rises when the crankshaft is rotated.

The timing marks on a 650 Series engine may be viewed through the magneto breather plug hole, as described in the text.

Timing Marks
All Series

Locate the timing mark -- may be a line, a dot, a letter, or even a projection embossed onto the flywheel. If necessary apply whiteout or other white substance to the mark to make it more visible.

With the piston at TDC -- as explained in the previous paragraphs -- make an aligning mark on a stationary surface near the timing mark using whiteout. This stationary mark may be on the rim of the stator plate on a 300, 440, or 550 Series, or the mark can be on the edge of the flywheel cover on the 650 Series engine. These two marks will be compared during the timing procedure. Therefore, make them as close together as practical -- extend the embossed mark, if necessary.

Connect a timing light to the battery and hook the pickup around the forward spark plug high tension lead. Check to be sure the direction of the embossed arrow on the timing light pickup is **TOWARD** the spark plug.

Connect a tachometer to the engine.

SPECIAL WORDS ON TACHOMETERS

The manufacturer recommends the use of a Kawasaki Electronic Tachometer. Instructions for connecting the tachometer leads are provided with the meter.

However, on a two-stroke engine with CDI type ignition, the tachometer leads are usually connected across the charging coil leads. As a general rule, the two leads of the same color in the magneto plate harness lead at the electrical box, are the leads from the charging coil.

In a test tank or with the craft in a body of water with restraint lines to prevent the craft from moving -- start the engine and allow it to warm to operating temperature.

CAUTION

Water must circulate through the jet pump -- to and from the engine, anytime the engine is operating. Circulating water will prevent overheating -- which could cause damage to moving engine parts and possible engine seizure.

NEVER, AGAIN NEVER, operate the engine at high speed with a flush device attached. An engine operating at high speed with such a device attached, would **RUNAWAY** from lack of a load on the impeller shaft, causing extensive damage.

DYNAMIC TIMING CHECK

After the engine has reached normal operating temperature -- with the timing light and the tachometer connected --advance the throttle lever until the engine has reached 6000 rpm. Aim the timing light at the two timing marks. If the mark on the flywheel aligns with the stationary mark, the timing is set correctly. However, if the mark on the flywheel is not aligned with the stationary mark -- a timing adjustment is required. Shut down the engine and perform the following procedures outlined under "Adjusting Ignition Timing".

Adjusting Ignition Timing

If servicing a 650 Series unit, remove the flywheel cover with the oil injection pump still attached.

On **ALL** series engines -- observe the two cutouts in the forward face of the flywheel. Rotate the flywheel until the two magneto assembly securing screws align with the cutouts.

Loosen, but **DO NOT REMOVE** these two screws. If this is the first work performed

Aim the timing light at the two timing marks. If the marks made at the factory cannot be found, the text explains, in detail, how to establish new marks.

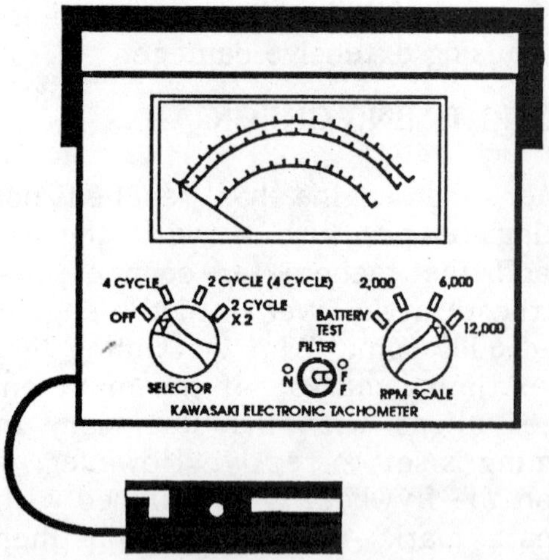

Maximum engine performance can only be obtained through proper tuning using a tachometer.

since the unit left the factory, these screws are almost impossible to loosen "cold". While attempting to rotate the screwdriver **COUNTERCLOCKWISE,** tap the end of the screwdriver with a hammer to jar the screw loose.

Once the screws have been loosened slightly, very carefully -- working through the flywheel cutouts -- rotate the magneto assembly either clockwise or counterclockwise, as necessary to correct the timing. When the two marks align, and this process is "trial and error", tighten the two securing screws very securely. If these screws should happen to vibrate loose, they could jam between the rotating flywheel magnets and the coils on the magneto assembly -- causing very serious and expensive damage.

If the series being serviced does not have cutouts in the face of the flywheel allowing access to the magneto assembly screws, the flywheel must be removed. See Section 7-6, beginning on Page 7-27, "Flywheel Removal".

WORDS FROM EXPERIENCE

The two most important items in establishing and maintaining perfect timing on an engine are the magneto assembly securing screws and the crankshaft Woodruff key.

Magneto Assembly Securing Screws

The magneto assembly is mounted on top of the stator plate and has two elongated slots for the securing screws to pass. If the magneto assembly is allowed to rotate just a small amount because the securing screws have worked loose, the ignition timing may be off as much as 15°. The symptoms of this misalignment are loss of power and difficult starting.

Crankshaft Woodruff Key

The same symptoms apply to a sheared Woodruff key allowing the flywheel -- actually the flywheel magnets -- to become out of position on the crankshaft.

During wave jumping or using a wet ramp, when the craft clears the water, the engine RPM approaches the limit set by the RPM limiter in an unloaded condition.

*The two magneto assembly securing screws **MUST** be tight to prevent the magneto plate from rotating. Any movement of the plate will disturb the ignition timing.*

A sheared Woodruff key will allow the flywheel to lose its original position on the crankshaft. Any change will drastically affect ignition timing.

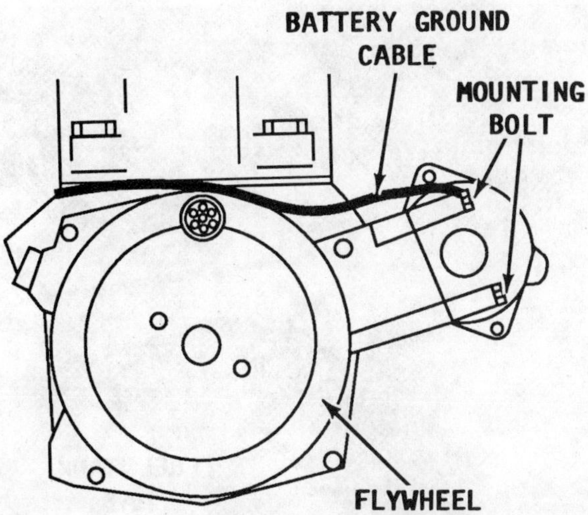

Simple line drawing to depict the flywheel cover installation on a 300 Series engine.

Once the craft returns to the water surface, the sudden load on the impeller is transferred along the driveline. Speed of rotating pump shaft and the crankshaft is suddenly reduced while the flywheel continues to spin rapidly due to its mass. This condition, especially when repeated over a period of time, often shears the Woodruff key and the flywheel becomes repositioned on the crankshaft.

A very small misalignment will result in a change in ignition timing. The craft operator will notice a drastic loss of power. A worse condition would be the flywheel shifting to the degree engine start would not be possible.

Closing Tasks
300 Series

If the flywheel was removed to adjust the position of the magneto assembly, install the flywheel and the flywheel cover according to the procedures outlined in Section 7-6. Install the two cranking motor mounting bolts into the side of the flywheel cover. Tighten both bolts securely.

Place the oil injection pump over the two studs on the flywheel cover and secure it in place with the two washers and nuts. Tighten the nuts securely. Connect the oil pump outlet hose to the intake manifold. Secure the hose with a hose clamp.

If there is sufficient fuel/oil mixture in the fuel tank to run the engine for a few more minutes, do not drain the mixture. The mixture will lubricate the engine while air trapped in the oil injection system passes through the transparent hose into the manifold.

It is not necessary to purge the system of air if the pump outlet hose was the only component in the system disconnected when the pump was set aside.

Disconnect the fuel outlet and retainer nut from the fuel tank. Lift the tank into the craft and set it in position on the two tank mats. Hook the two rubber restraining straps around the tank securing the tank to the hull. Install the inlet neck onto the tank fitting, and then secure it in place with a large hose clamp. Install the fuel outlet assembly. Make a final inspection of all fuel fittings and check valves. On the safe side, to purge the fuel system -- see Section 6-10, beginning on Page 6-49.

Closing Tasks
440 Series and 550 Series

If the flywheel was removed to adjust the position of the magneto assembly, install the flywheel and the flywheel cover -- see Section 7-6.

Disconnect the fuel outlet and retainer nut from the fuel tank. Lift the tank into the craft and set it in position on the two tank mats. Hook the two rubber restraining straps around the tank securing the tank to the hull. Install the inlet neck onto the tank fitting, and then secure it in place with a large hose clamp. Install the fuel outlet assembly. Make a final inspection of all fuel fittings and check valves.

Closing Tasks
650 Series

If the flywheel was removed to adjust the position of the magneto assembly, install the flywheel, the flywheel cover, and the oil pump -- see Section 7-6. Install the magneto to breather plug into the flywheel cover.

Disconnect the fuel outlet and retainer nut from the fuel tank. Lift the tank into the craft and set it in position on the two tank mats. Hook the two rubber restraining straps around the tank securing the tank to the hull. Install the inlet neck onto the tank fitting, and then secure it in place with a large hose clamp. Install the fuel outlet assembly. Make a final inspection of all fuel fittings and check valves. On the safe side, to purge the fuel system -- see Section 6-10, beginning on Page 6-49.

Installing the flywheel cover after the engine has been properly timed and any necessary adjustments have been completed.

*Tank mats in place, ready for installation of the fuel tank and the oil tank on a 650 Series engine. Both tanks **MUST** be well secured with restraining straps.*

8
ENGINE

8-1 INTRODUCTION AND CHAPTER ORGANIZATION

The following Kawasaki Series are covered in this chapter from early model years thru 1991.

JS300	300 series	294cc
300SX	300 series	294cc
JS440	440 series	436cc
JS550	550 series	530cc
550SX	550 series	530cc
JS650	650 series	635cc
JF650	650 series	635cc
JB650	650 series	635cc
650SX	650 series	635cc
JetMate	650 series	635cc
X-2	650 series	635cc
TS	650 series	635cc
SC	650 series	635cc

This chapter is divided into five main working sections as follows:

8-2 Description and operation -- Two-Cycle Engine
8-3 Removal and Disassembling -- All Engines
8-4 Assembling and Installation -- All Engines
8-5 Cleaning and inspecting -- All Engines
8-6 Sealants, Adhesives, Lubricants, and Fuel Stabilizers

The carburetion and ignition principles of two-cycle engine operation **MUST** be understood in order to perform proper service work on the marine engines covered in this manual.

Repair Procedures

Service and repair procedures will vary slightly between individual models, but the basic instructions are almost identical. Any variations between models are clearly indicated. Special tools may be called out in certain instances. These tools may be purchased from the local dealer. In certain instances, an alternate tool may be obtained or fabricated.

Torque Values

All torque values must be met when they are specified. Torque values for various parts of each engine are given in the text.

A torque wrench is essential to correctly assemble the engine. **NEVER** attempt to assemble an engine without a torque wrench. Attaching bolts **MUST** be tightened to the required torque value in two progressive sequences, following the specified tightening order. On the first sequence, tighten the bolts to 1/2 the torque value. On the second sequence, tighten the bolts to the full torque value.

Other Engine Components

Service procedures for the carburetor, fuel pumps, cranking motor and other components are given in separate Chapters. See the Table of Contents.

Reed Valve Service
300 Series and 650 Series

The reeds on engines listed in the heading are contained in an externally mounted reed valve block. Therefore, the engine need not be disassembled in order to replace a broken reed.

440 Series and 550 Series

Due to the unique design of these engines, reeds are not necessary.

Cleanliness

Make a determined effort to keep parts and the work area as clean as possible.

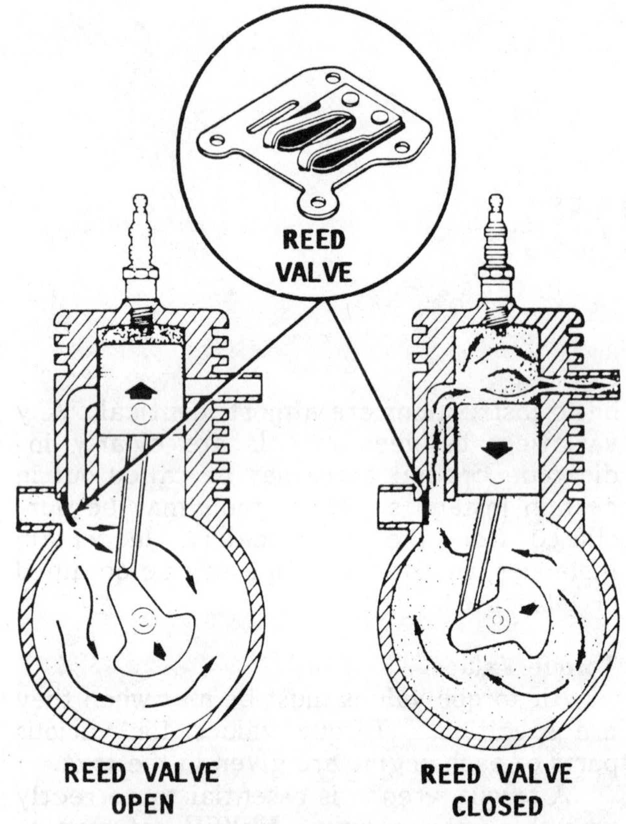

Reed valves are used to control the flow of air/fuel into the crankcase and eventually into the cylinder. As the piston moves upward in the cylinder, the resulting suction in the crankcase overcomes the spring tension of the reed. The reed is pulled free from its seat and the air/fuel mixture is drawn into the crankcase.

Parts **MUST** be cleaned and thoroughly inspected before they are assembled, installed, or adjusted. Use proper lubricants, or their equivalent, whenever they are recommended.

8-2 TWO-CYLCE ENGINE DESCRIPTION AND OPERATION

Intake/Exhaust

Two-cycle engines utilize an arrangement of port openings to admit fuel to the combustion chamber and to purge the exhaust gases after burning has been completed. The ports are located in a precise pattern in order for them to be opened and closed at an exact moment by the piston as it moves up and down in the cylinder. The exhaust port is located slightly higher than the fuel intake port. This arrangement opens the exhaust port first as the piston starts downward and therefore, the exhaust phase begins a fraction of a second before the intake phase.

Lubrication

A two-cycle engine is lubricated by mixing oil with the fuel. Therefore, various parts are lubricated as the fuel mixture passes through the crankcase and the cylinder. On the 300 Series and the 650 Series engines, an oil injection system delivers oil directly into the intake manifold. On the 440 Series and 550 Series engines, oil is mixed with the gasoline in the fuel tank in the age old manner.

Physical Laws

The two-cycle engine is able to function because of two very simple physical laws.

One: Gases will flow from an area of high pressure to an area of lower pressure. A tire blowout is an example of this principle. The high-pressure air escapes rapidly if the tube is punctured.

Two: If a gas is compressed into a smaller area, the pressure increases, and if a gas expands into a larger area, the pressure is decreased.

If these two laws are kept in mind, the operation of the two-cycle engine will be more easily understood.

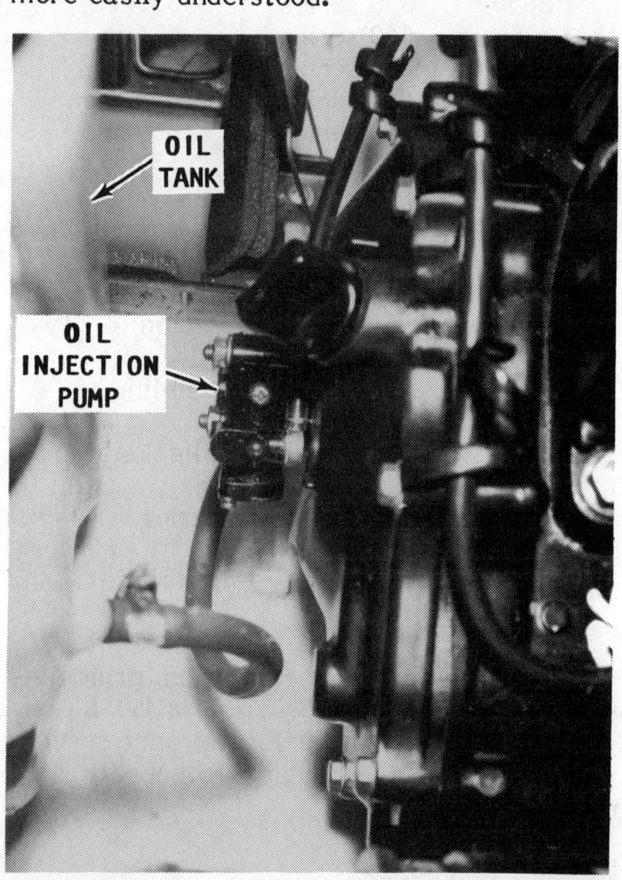

The 300 Series and 650 Series engines are equipped with an oil injection system. The 440 Series and 550 Series engines still use an oil and gasoline premix.

ACTUAL OPERATION TWO-CYCLE ENGINE

Beginning with the piston approaching top dead center on the compression stroke: The intake and exhaust ports are closed by the piston; the reed valve is open; the spark plug fires; the compressed air/fuel mixture is ignited; and the power stroke begins. The reed valve was open because as the piston moved upward, the crankcase volume increased, which reduced the crankcase pressure to less than the outside atmosphere.

As the piston moves downward on the power stroke, the combustion chamber is filled with burning gases. As the exhaust port is uncovered, the gases, which are under great pressure, escape rapidly through the exhaust ports. The piston continues its downward movement. Pressure within the crankcase increases, closing the reed valves against their seats. The crankcase then becomes a sealed chamber. The air/fuel mixture is compressed ready for delivery to the combustion chamber. As the piston continues to move downward, the intake port is uncovered. A fresh air/fuel mixture rushes through the intake port into the combustion chamber striking the top of the piston where it is deflected along the cylinder wall. The reed valve remains closed until the piston moves upward again.

When the piston begins to move upward on the compression stroke, the reed valve opens because the crankcase volume has been increased, reducing crankcase pressure to less than the outside atmosphere. The intake and exhaust ports are closed and the fresh fuel charge is compressed inside the combustion chamber.

Pressure in the crankcase decreases as the piston moves upward and a fresh charge of air flows through the carburetor picking up fuel. As the piston approaches top dead center, the spark plug ignites the air/fuel mixture, the power stroke begins and one full cycle has been completed.

TIMING -- TWO-STROKE ENGINES

The exact time of spark plug firing depends on engine speed. At low speed the spark is retarded, fires later than when the piston is at or beyond top dead center. Engine timing is built into the unit at the factory. At high speed, the spark is advanced, fires earlier than when the piston is at top dead center.

Summary

More than one phase of the cycle occurs simultaneously during operation of a two-cycle engine. On the downward stroke, power occurs above the piston while the ports are closed. When the ports open, exhaust begins and intake follows. Below the piston, fresh air/fuel mixture is compressed in the crankcase.

On the upward stroke, exhaust and intake continue as long as the ports are open. Compression begins when the ports are closed and continues until the spark plug ignites the air/fuel mixture. Below the piston, a fresh air/fuel mixture is drawn into the crankcase ready to be compressed during the next cycle.

8-3 SERVICE — ALL ENGINES

ADVICE

Before commencing any work on the engine an understanding of two-cycle engine operation will be most helpful. Therefore, it would be well worth the time to study the principles of two-cycle engines, as outlined briefly in the previous section -- 8-2. A Polaroid, or equivalent instant-type camera is an extremely useful item, providing the means of accurately recording the arrangement of parts and wire connections **BEFORE** the disassembly work begins. Such a record is invaluable during assembling.

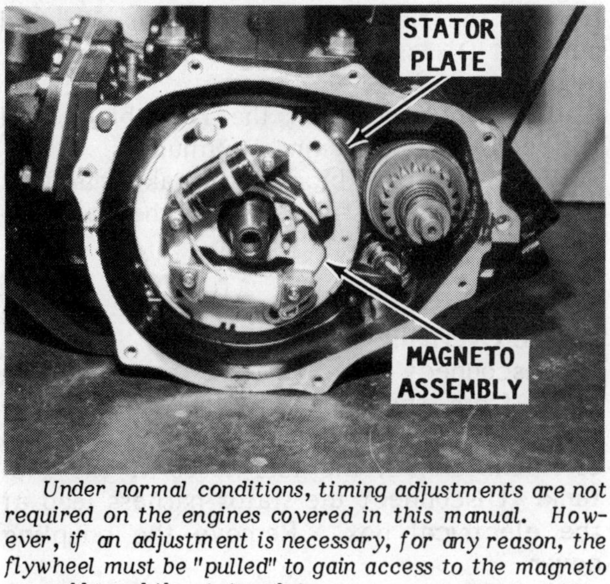

Under normal conditions, timing adjustments are not required on the engines covered in this manual. However, if an adjustment is necessary, for any reason, the flywheel must be "pulled" to gain access to the magneto assembly and the stator plate.

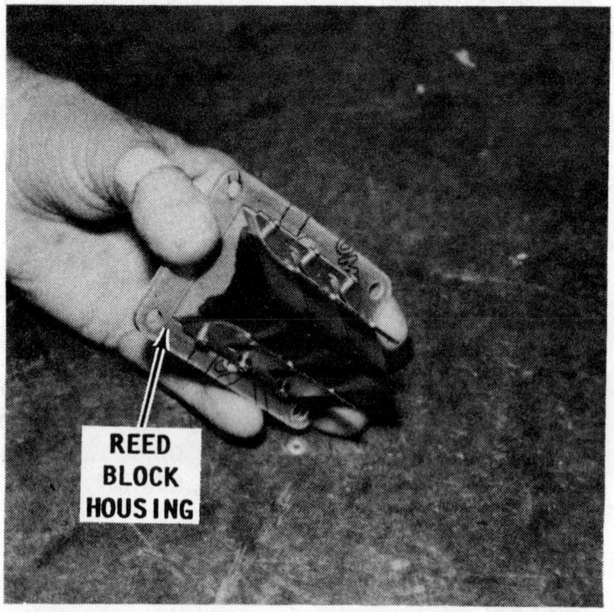

The reed block on the 300 Series and 650 Series engines can be serviced while the engine remains in place.

IMPORTANT WORDS

The following series of illustrations were taken of a 650 Series engine. The 440 Series and 550 Series engines differ slightly in appearance, but have the same basic components. The 300 Series engines are basically the same, except they have only a single cylinder. Where differences occur, such as bolt tightening patterns, the differences will be clearly identified.

REED BLOCK WORDS
300 AND 650 SERIES ONLY

If the only work to be performed on the engine is servicing the reed block assembly, remove the flame arrestor, carburetor, and perform Disassembling Steps 9, 10, and 11, beginning on Page 8-16.

Clearance between the reed block and the reed valve is 0.2mm maximum.

After the reed block and intake manifold assemblies are installed, install the carburetor and the flame arrestor.

Preliminary Tasks
Engine Overhaul

Disconnect the negative and positive leads from the battery. Disconnect the throttle and choke cables from the carburetor. Disconnect the fuel hoses at the fuel tank. Disconnect the stator harness lead at the electrical box. Remove the coupling cover.

Remove the carburetor.

NOW, THESE WORDS

As mentioned earlier, the following procedures and supporting illustrations were developed while servicing a 650 Series engine. Any differences between the 650 Series and other 2-cylinder series engines **OR** a single cylinder engine, except for the number of cylinders, the differences are clearly identified.

REMOVAL

1- Loosen, but do not remove the two large hose clamps at the forward end of the expansion chamber.

2- Release the pliable metal clamps securing the electrical leads to the engine block.

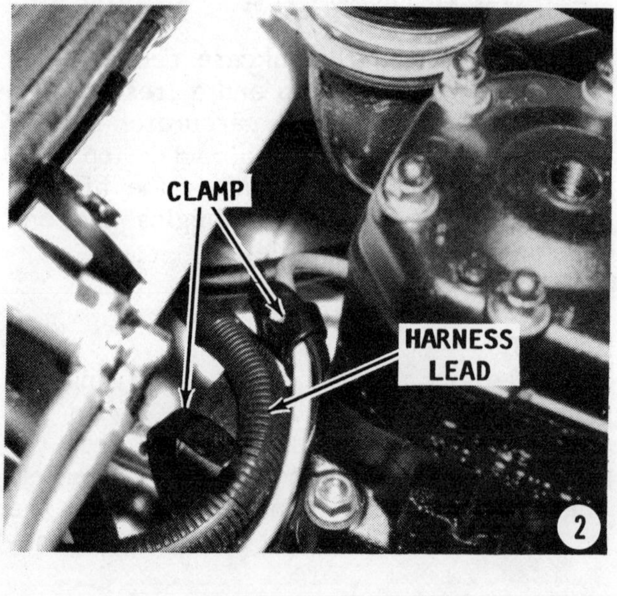

3- Disconnect the pulse hose from the fitting on the intake manifold.

300 Series and 650 Series
With Oil Injection

4- Pinch the oil supply hose, and then disconnect the hose from the oil injection pump. Insert a screw into the disconnected hose end to prevent oil loss and a mess in the bilge.

Disconnect the oil delivery hose from the fitting on the intake manifold at the base of the carburetor.

5- Remove the two nuts and washers securing the oil injection pump to the flywheel cover. Pull the pump free of the studs.

All Series Engines

6- Release the rubber "bungee" type straps securing the fuel tank to the tank tray. Remove the fuel tank cap, and set it to one side.

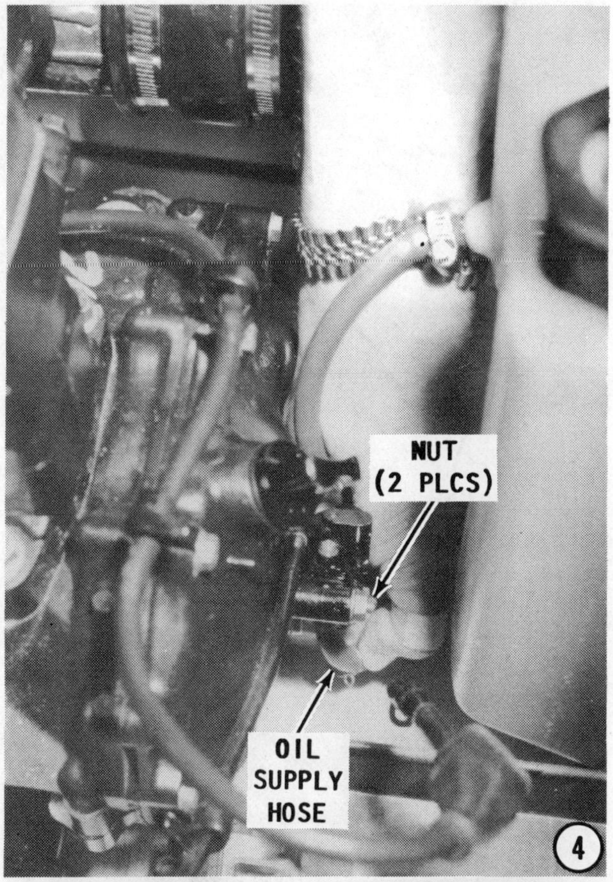

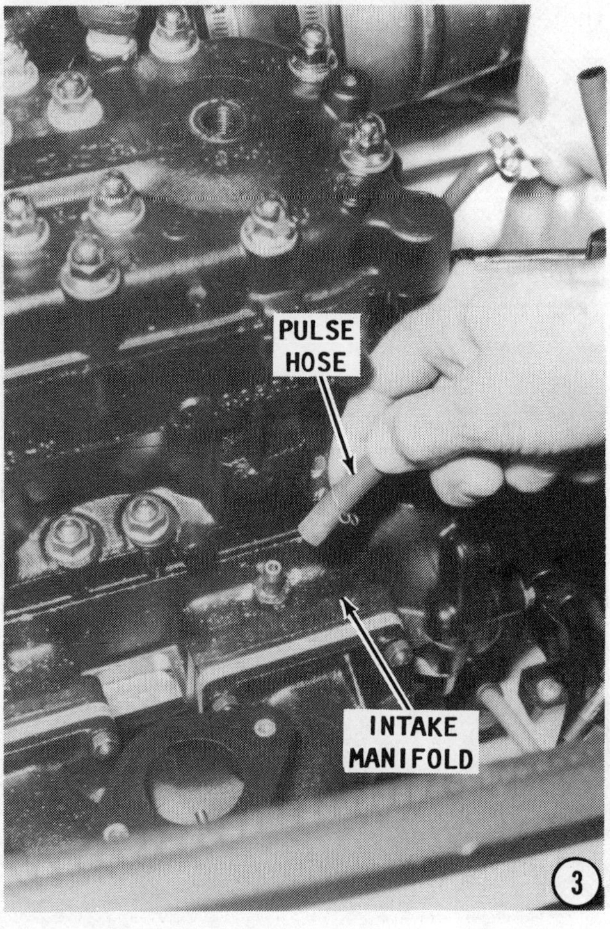

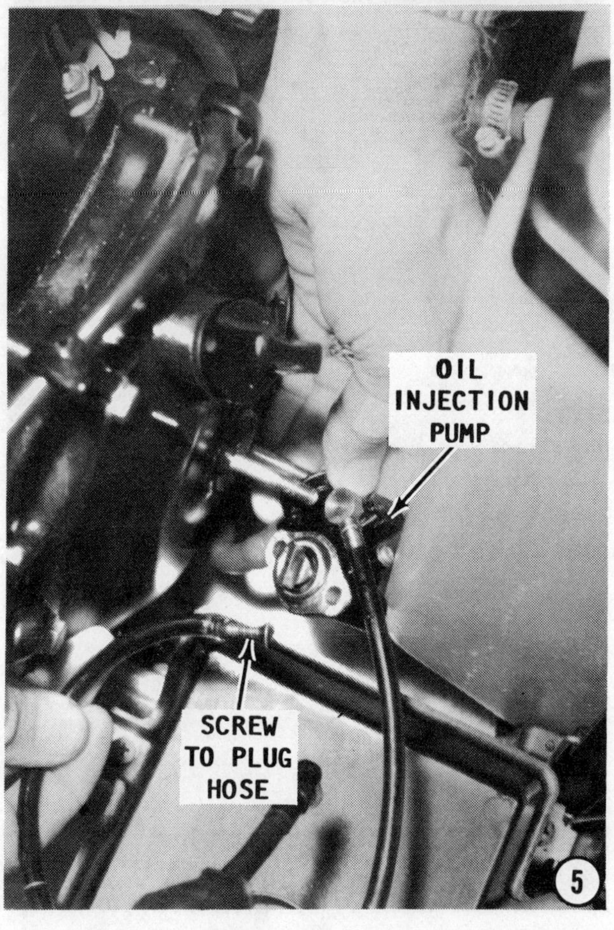

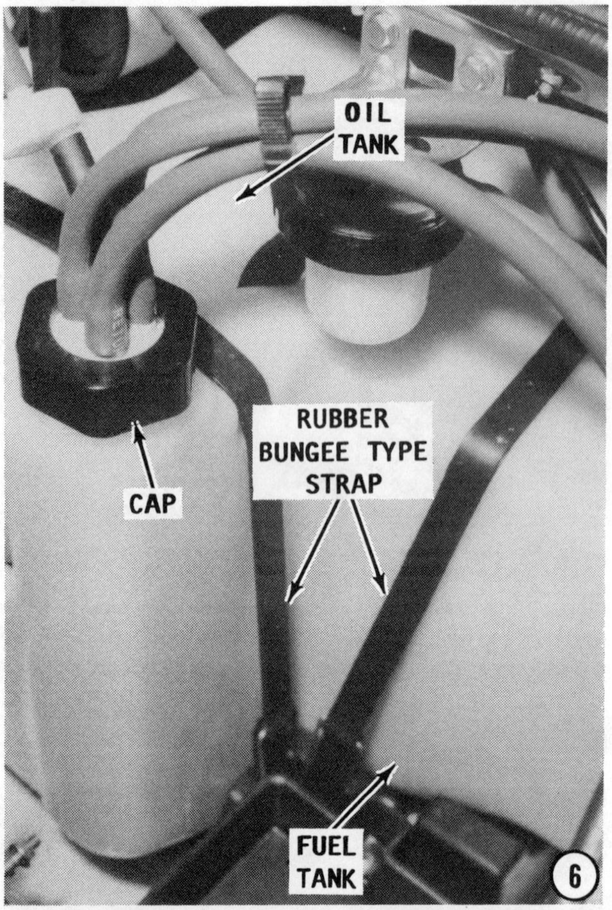

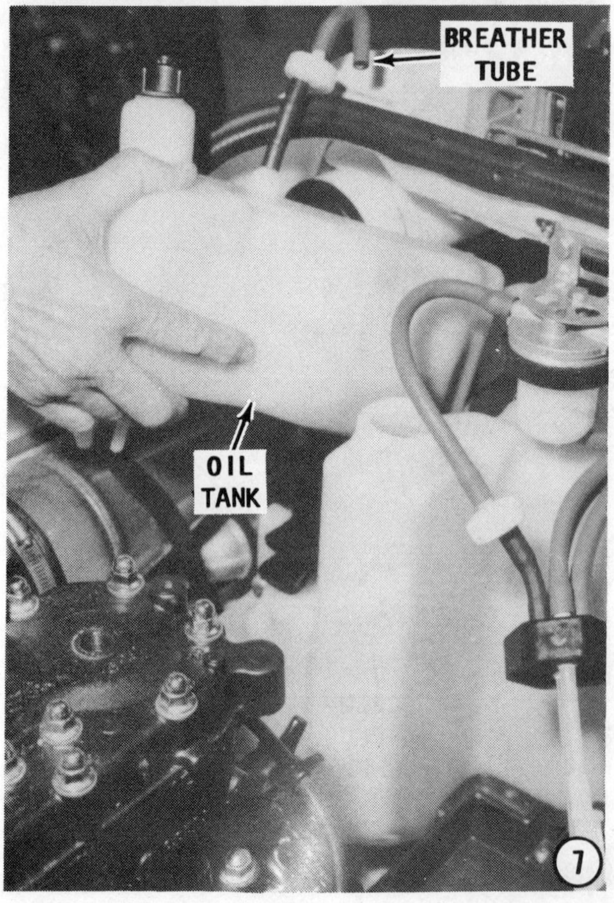

300 Series and 650 Series

7- Lift the oil tank, with the breather tube attached up and out of the engine compartment.

All Series Engines

8- Reach behind the fuel tank and loosen both the filler tube clamps around the large filler hose. Work the hose free of the tank fitting. Lift the fuel tank out of the engine compartment.

SPECIAL WORDS

The exhaust system is now exposed. Therefore, any service work required to this system, including the water box muffler and exhaust pipe may now be performed.

9- Now, reach under the exhaust manifold and disconnect the cooling water supply hose from the fitting on the manifold.

10- Disconnect the bypass hose from the fitting on the exhaust elbow.

11- Remove the large single bolt securing the expansion chamber to the exhaust manifold. Remove the bolts and washers securing the exhaust elbow to the manifold.

Disconnect the water valve cable from the exhaust elbow flange.

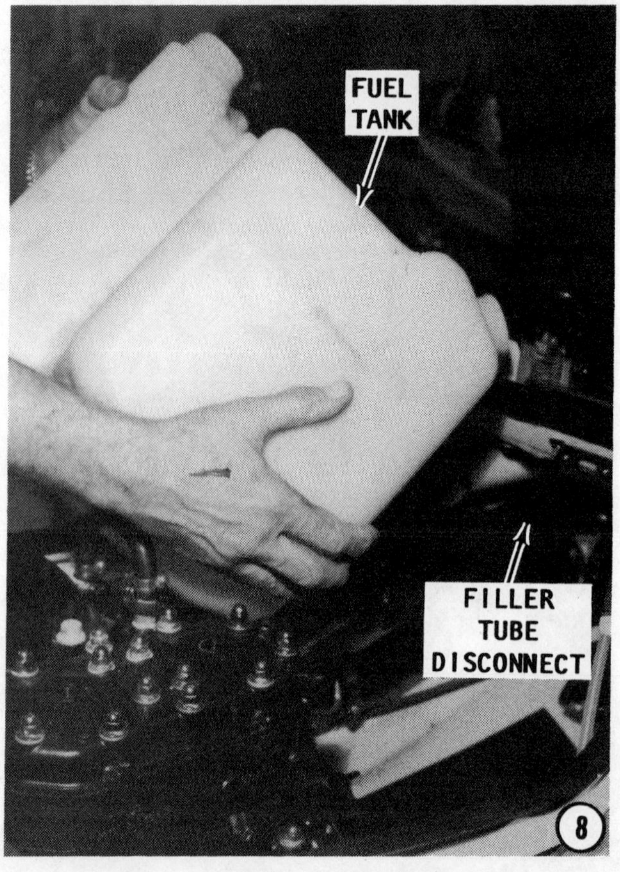

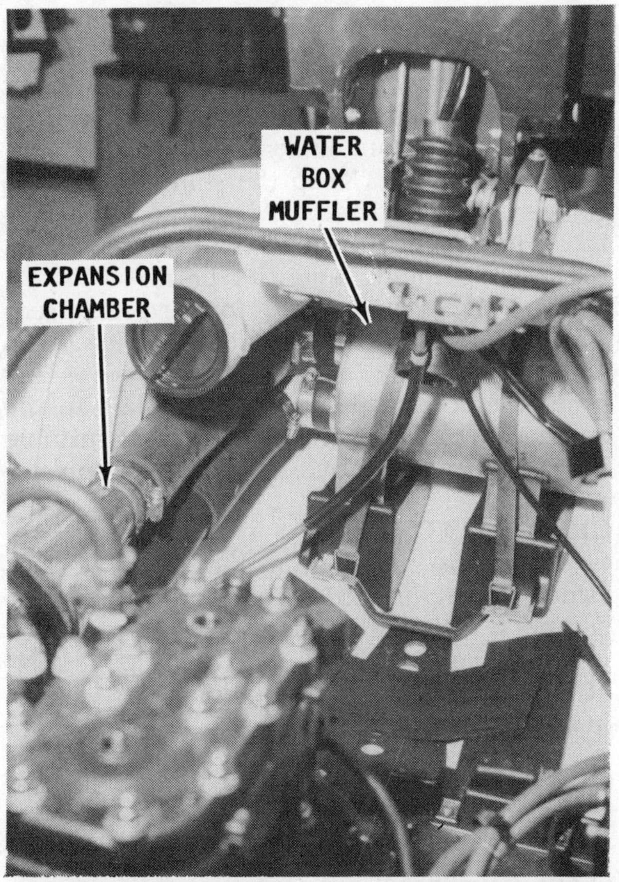

The fuel tank and the oil tank must be removed to gain access for service work on the exhaust system.

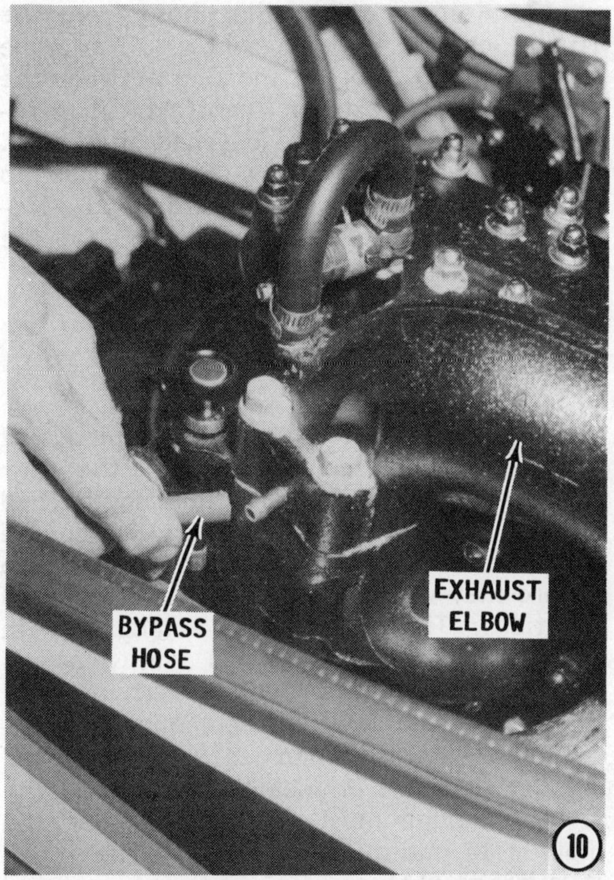

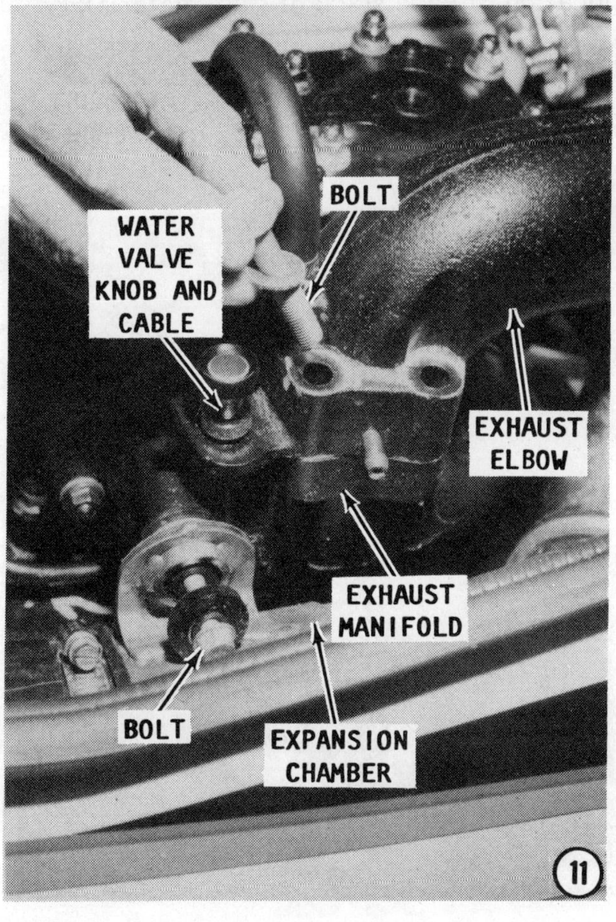

12- Remove the cooling hose from the fitting on the exhaust elbow. Work the exhaust elbow to break the seal between the elbow and the exhaust manifold. The restraining clamps were loosened in Step 1. Move the exhaust elbow to one side to provide clearance for engine removal. Remove the gasket from the surface of the manifold.

VERY IMPORTANT SHIM PACK WORDS

Make a note of the number of pieces of shim material used at each of the four engine mounts -- between the engine bed bracket and the rubber dampers. Keep the four stacks of shim material separate to ensure they will be installed in their original locations. The engine **MUST** be aligned with the impeller shaft to avoid vibration and coupler wear. The amount of shim material at each of the four engine mounts, necessary for accurate alignment, is determined at the factory or dealer/installer. The **SAME** amount must be placed under the mounts to maintain alignment. Therefore, it is **ESSENTIAL** to identify the amount and location of each stack of shim material.

13- Loosen the four engine mounting bolts. Two adults should have no difficulty in lifting this engine. Have an assistant lift the engine slightly, and then slide the shim material out from under each of the four engine brackets. Keep the shim material in groups and identified. After the shim material has been removed for all four mounts, remove the engine mounting bolts.

14- Move the engine forward to disengage the rubber coupler from the jet pump coupler. Lift the engine and watch for any hose or electrical leads still attached to the engine. Transport the engine to a suitable work surface. Remove the rubber coupler.

CRITICAL WORDS

Again, **SAVE** the shim material used between the engine bed bracket and the rubber dampers. Keep the shim material together in groups and identify each group to **ENSURE** proper engine alignment during installation.

Engine Preparation

The following instructions pickup the work after the preliminary tasks listed above have been completed and the engine is on a suitable work surface.

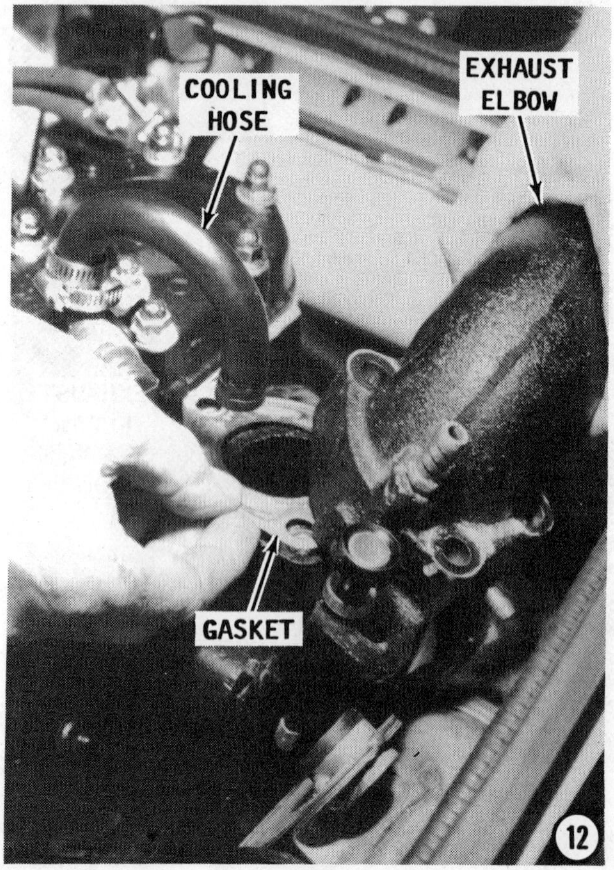

FIRST, THESE WORDS

Procedural steps are given to remove and disassemble virtually all items of the engine. However, as the work moves along, if certain items, i.e. bearings, bushings, seals, etc. are found to be fit for further service, simply skip the disassembly steps involved. Proceed with the required tasks to disassemble the necessary components.

COMPLETE and **DETAILED** procedures to clean and service all parts of the engine are outlined in Section 8-5, beginning on Page 8-46.

DISASSEMBLING

SPECIAL WORDS ON SPECIAL TOOLS

The following list of special tools are required in order to perform some of the necessary work on the engine. Some tasks become extremely difficult, if not impossible, without the special tools or some type of equivalent. Manufacturers' part numbers are given. In specific steps of the procedures -- whenever possible -- a special effort has been made to give some type of equivalent or homemade tool which may be substituted. However, the cost of a special tool may be offset by money saved if an expensive part is not damaged.

300 Series

Piston Pin Puller	P/N 57001-910
Magneto Puller	P/N 57001-259
Flywheel Holder	P/N 57001-306
Coupler Holder	P/N 57001-1230
Crankcase Splitting Tool	P/N 57001-1098
Bearing Driver Set	P/N 57001-1129
Crankshaft Installing Jig	P/N 57001-1174
Rotor Puller	P/N 57001-1216

440 Series and 550 Series

Piston Pin Puller	P/N 57001-910
Coupler Holder	P/N T57001-276

650 Series

Piston Pin Puller	P/N 57001-910
Magneto Puller	P/N 57001-259
Flywheel Holder	P/N 57001-1156
Coupler Holder	P/N 57001-1230

SPECIAL WORDS ON REAR MAIN OIL SEALS

If the rear main oil seals require replacement, the metal coupler must first be removed. This metal coupler is removed while the special tool is in place to remove the flywheel nut.

If the rear main oil seal does not require service, the metal coupler can remain in place. Therefore, skip the instructions under "Special Words" following Step 3, and continue with Step 4.

If the rear main oil seals are to be replaced, the metal coupler must be removed while the flywheel is still in place. There is virtually no way on this green, green earth to remove the coupler, without the flywheel installed.

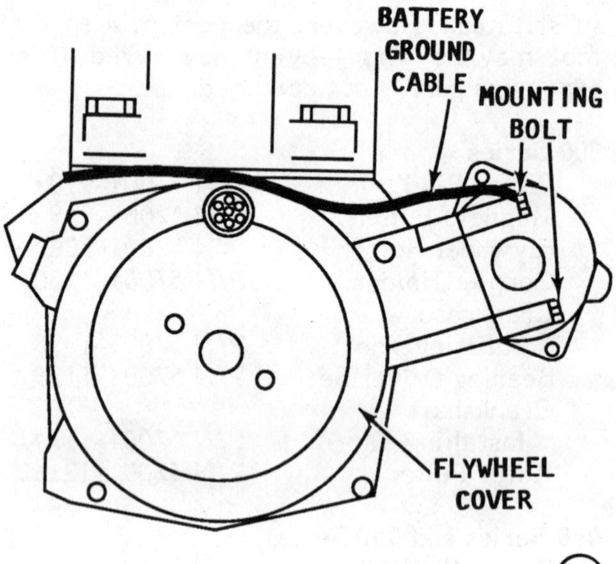

1

If the rear main oil seals require replacement and the metal coupler must be removed, perform the special instructions under "Special Words" following Step 3, and then continue with the work as outlined in Step 4.

Once the flywheel is removed, it is impossible to remove the metal coupler because the crankshaft cannot be prevented from rotating.

Cranking Motor Removal
300 Series

1- Remove the two cranking motor mounting bolts from the flywheel cover. One of these bolts also secures the battery ground cable to the engine. Remove the flywheel cover securing bolts and lift the

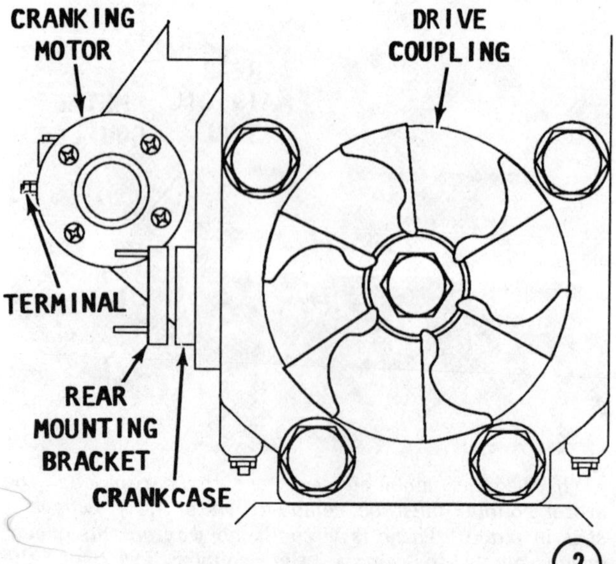

2

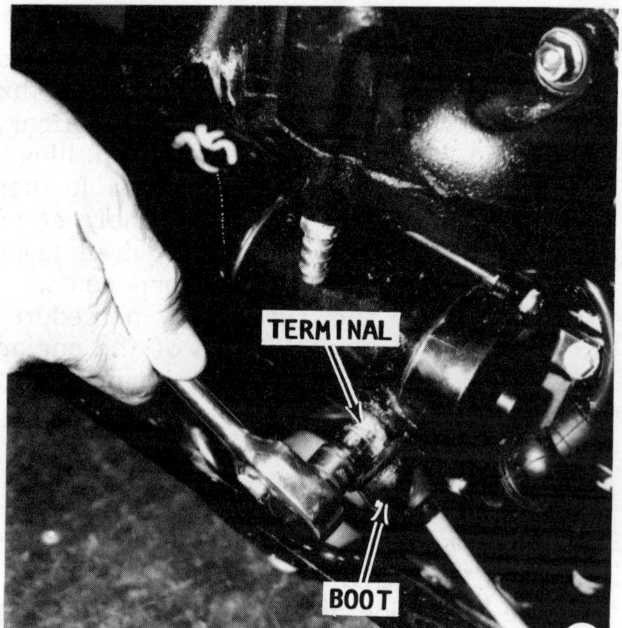

3

cover free. **SAVE** the flywheel sealing ring, if a replacement is not at hand. Pull the cranking out of the rear gear housing -- the armature shaft indexes with an idle gear at the rear of the engine.

Cranking Motor Removal
440 Series and 550 Series

2- Remove the two rear cranking motor mounting bolts. Remove and **SAVE** any shim material installed between the rear mounting bracket and the crankcase. Remove the two front cranking motor mounting bolts. Slide the cranking motor aft --toward the

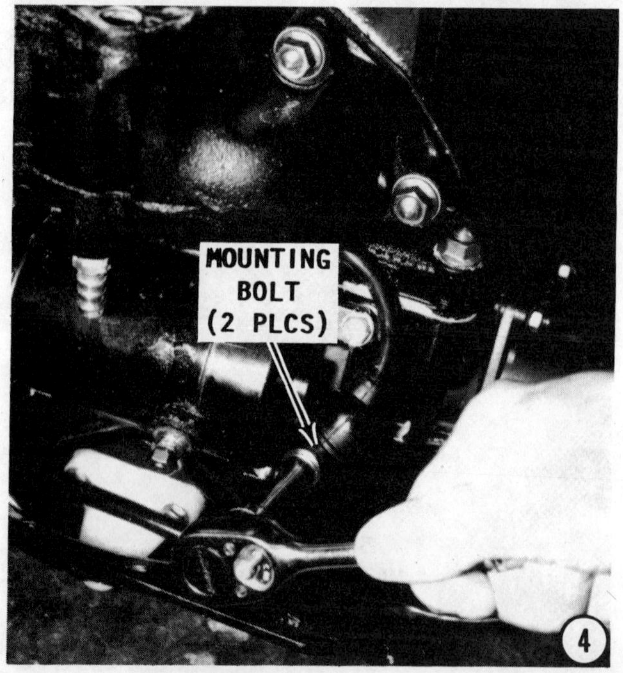

4

rear of the engine. Disconnect the large Red cable from the terminal on the motor.

Cranking Motor Removal
650 Series

3- Push back the protective boot from the terminal on the cranking motor and remove the large Red cable.

4- Remove the two mounting bolts securing the cranking motor to the crankcase. On some 650 Series engines, the lower mounting bolt also secures the battery ground cable to the crankcase.

5- Move the cranking motor aft and the armature shaft will slide free of the reduction gear inside the flywheel cover.

"PULLING" THE FLYWHEEL

1- Remove the spark plugs.

NOTE

If the flywheel cover has not been removed, perform Step 2 and Step 3. If the cover has been removed, proceed directly to "Special Words", just before Step 4.

2- Remove the bolts securing the flywheel cover to the stator plate.

3- Remove the flywheel cover from the stator plate. If the cover is stuck to the plate, obtain a soft head mallet and tap around the circumference of the cover to

jar it free. Remove and discard the sealing ring, only if a replacement is available.

SPECIAL WORDS

If the two rear main oil seals are to be removed, install the flywheel holder to the flywheel and insert the coupler holder into the metal coupler. Prevent the flywheel from rotating and unthread the metal coupler from the rear of the crankshaft.

4- If servicing a 440 or 550 Series engine, bend back the tabs on the lockwasher behind the flywheel nut. Now using flywheel holder tool listed above for the engine being serviced, insert the two indexing pins on the ends of the arms through the holes in the flywheel. Prevent the flywheel from rotating with the special tool. Loosen, and then remove the flywheel nut and washer -- 650 Series; the tabbed lockwasher -- 440 and 550 Series; flanged bolt -- late 300 Series; or the bolt and washer -- early 300 Series.

5- Install the flywheel puller onto the flywheel, and then using the center bolt, take a strain on the puller with the proper size wrench. Now, continue to tighten the center bolt and at the same time, **SHOCK** the crankshaft with a gentle to moderate tap with a hammer on the end of the center bolt. This shock will assist in "breaking" the flywheel loose from the crankshaft.

DO NOT "pound" on the center bolt. The force of such action would be transmitted to the seals and bearings along the crankshaft and cause damage. **NEVER** strike the flywheel because this type of shock would surely cause the flywheel magnets to loose their magnetism.

6- Pull the flywheel free of the crankshaft. Remove and save the Woodruff key from the crankshaft keyway.

SPECIAL WORDS
WOODRUFF KEYS

Inspect the condition of the Woodruff key and its keyway in the crankshaft. If the key shows any signs of shearing, the cause is probably from jumping waves and the effect will be a loss of power. When the craft is clear of the water, the engine RPM approaches the limit set by the RPM limiter in an unloaded condition.

When the craft returns to the water surface, the sudden load on the impeller is transferred along the driveline. Speed of the rotating pump shaft and the crankshaft is suddenly reduced while the flywheel

continues to spin rapidly due to its mass. This condition, especially when repeated over a period of time, often shears the Woodruff key and the flywheel becomes repositioned on the crankshaft.

A very small misalignment will result in a change in ignition timing. The craft operator will notice a drastic loss of power. A worse condition would be the flywheel shifting to the degree engine start would not be possible.

650 Series Only

Lift out the reduction gear assembly.

Magneto Assembly Removal

7- The securing screws pass through elongated slots in the magneto assembly. The assembly could be installed as much as 15° off its original position -- drastically affecting the timing characteristics of the engine. Therefore, at the elongated slot in the magneto assembly, make a set of aligned punch marks -- one on the magneto assembly -- the other on the stator plate. These marks will serve as an aid during installation to **ENSURE** correct positioning of the magneto assembly on the stator plate.

8- Remove the two screws securing the magneto assembly to the stator plate.

Removing the reduction gear from the armature shaft of the cranking motor on a 650 Series engine.

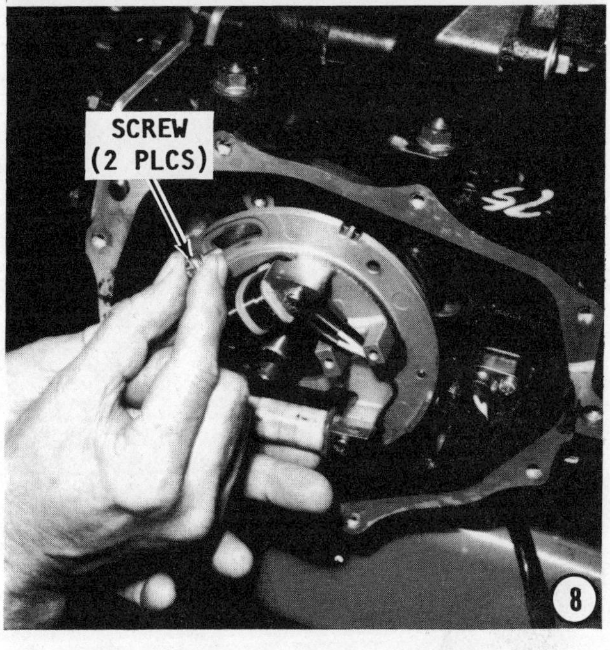

9- Remove the two small screws securing the magneto assembly harness to the grommet in the stator plate. Feed the harness and connector end through the opening in the stator plate and remove the magneto assembly from the crankshaft.

BLOCK DISASSEMBLING

Water Valve
650 Series only

1- Remove the two bolts securing the water drain valve to the block. Pull the valve free of the block.

2- Separate the control valve housing from the reed valve cover. Remove the two

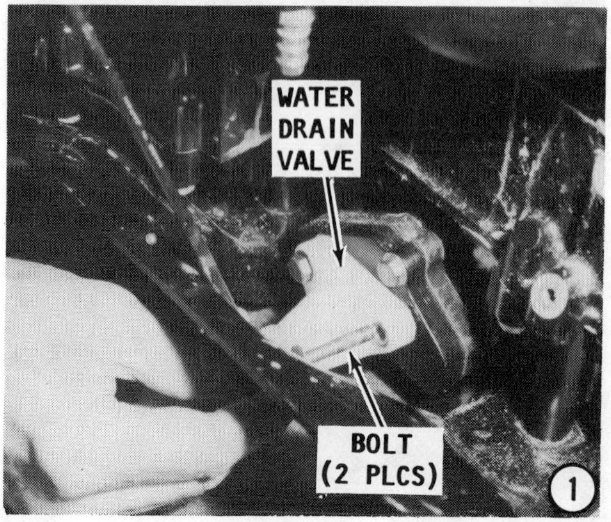

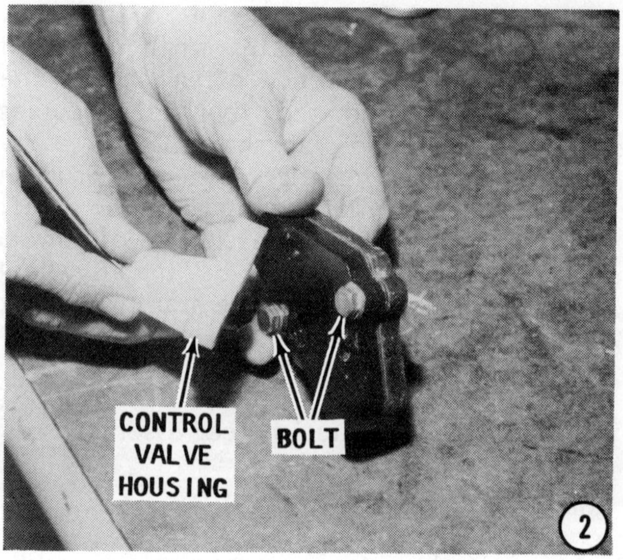

bolts securing the cover to the reed valve assembly.

3- Separate the cover from the reed valve assembly.

Cylinder Head Removal
All Series Engines

4- Remove the nuts and washers securing the cylinder head to the block. Pry the head loose from the block. **TAKE CARE** not to damage the smooth machined mating surfaces of the block or the head. If the head is stuck to the block, it may be necessary to tap the head with a soft head mallet to "jar" the head loose.

5- After the head has been "jarred" loose, keep the head as level as possible and lift it **STRAIGHT** up and off the studs.

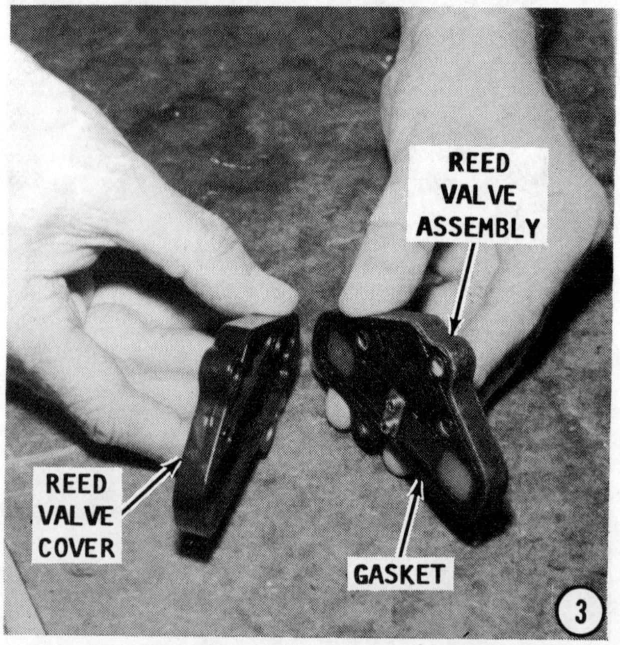

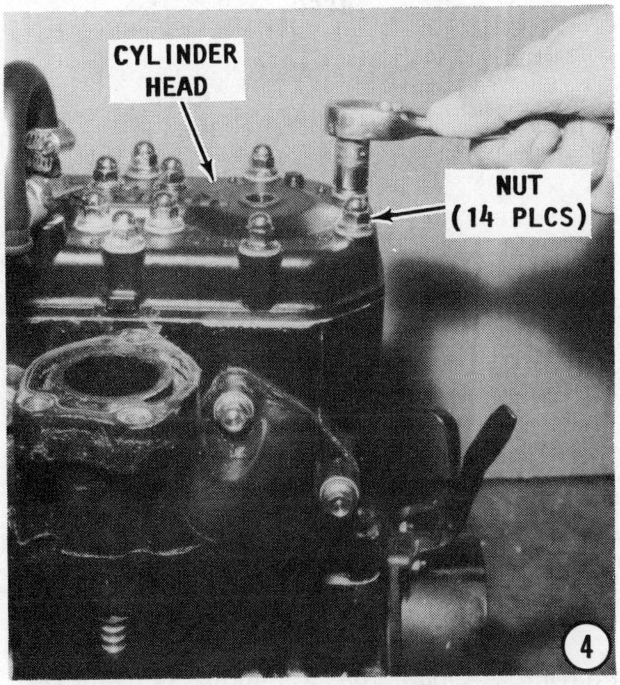

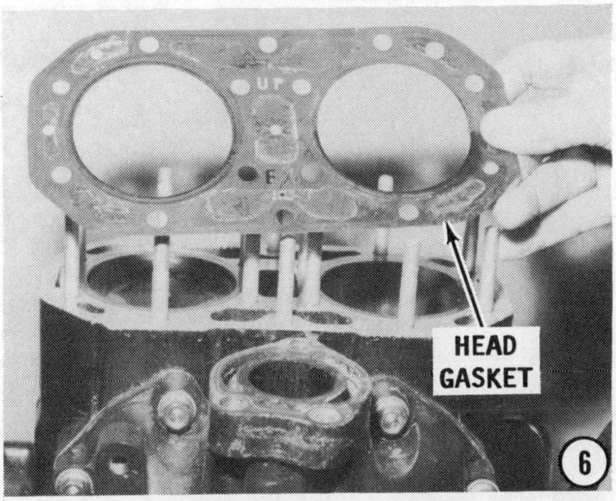

6- Remove and discard the head gasket. The head gasket can **NEVER** be used a second time.

Exhaust Manifold Removal

7- Remove the four nuts/bolts and washers securing the exhaust manifold to the block. "Jar" the manifold loose from the block.

8- Slide the exhaust manifold free of the studs. Remove all traces of gasket material from the two mating surfaces.

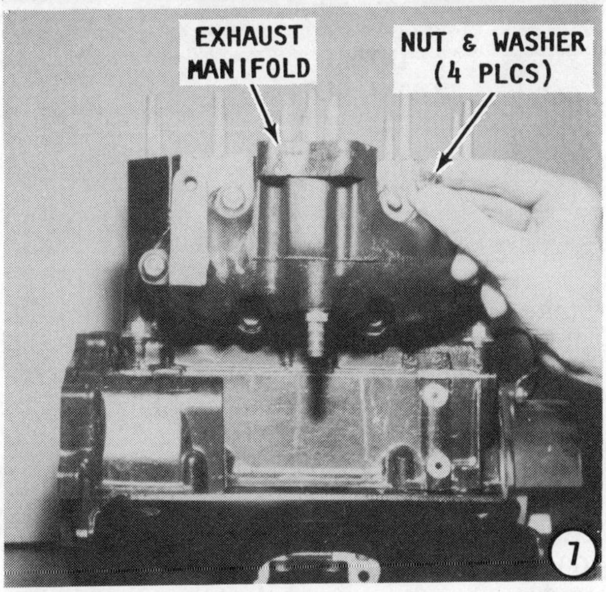

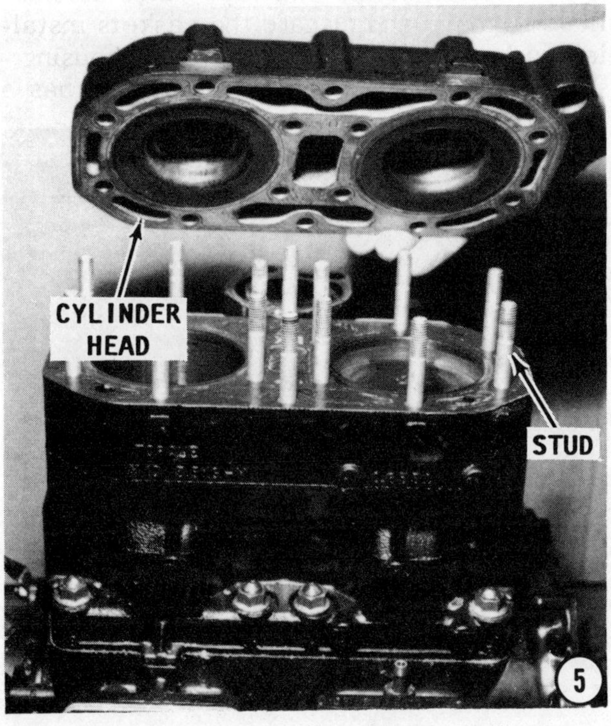

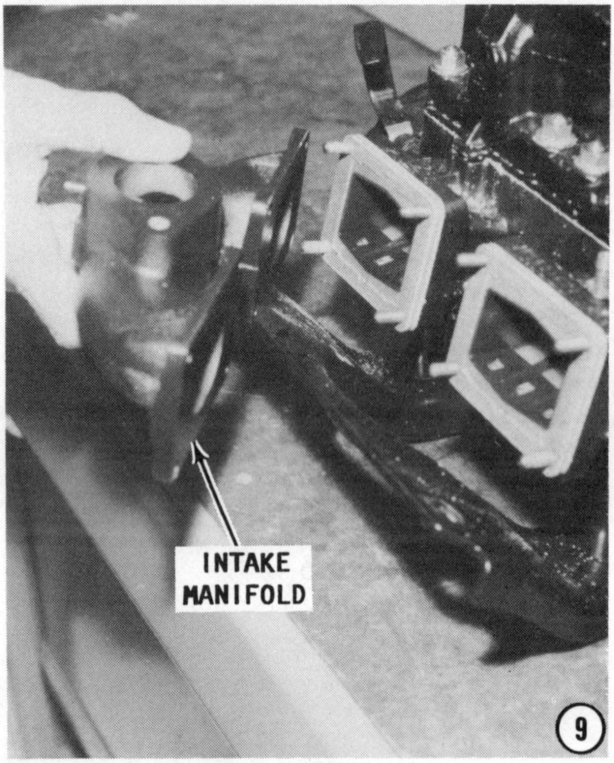

Intake Manifold Removal
All Series, Except 300 Series

9- Remove the nuts and washers securing the intake manifold to the block. Lift the manifold free of the block.

CRITICAL WORDS
REED BLOCK ASSEMBLIES
300 AND 650 SERIES ONLY

Take great care in the next step when handling reed valve assemblies. The reed valve housing is coated with a special rubber substance. **TAKE CARE** not to damage this

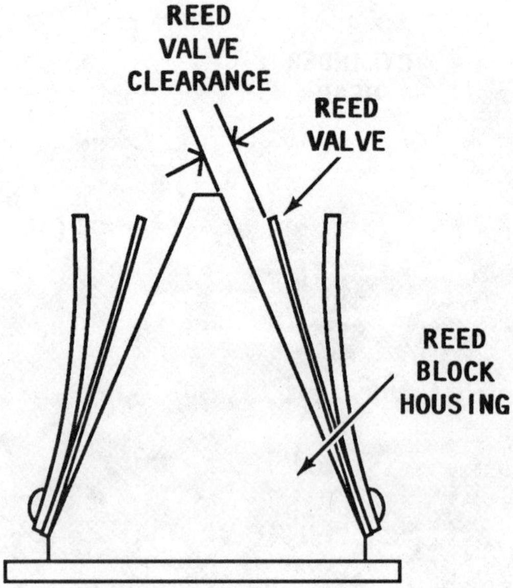

The maximum reed valve clearance for the 300 Series and the 650 Series is 0.007" (0.2mm).

coating. Once the assembly is removed, keep it away from sunlight, moisture, dust, and dirt. Sunlight can deteriorate valve seat rubber seals. Moisture can easily rust stoppers overnight. Dust and dirt -- especially sand or other gritty material -- can break reed petals if caught between stoppers and reed petals.

Make special arrangements to store the reed valves to keep them isolated from the elements while further work is being performed on the engine.

10- Lift out the two reed valve assemblies. Remove and discard the gaskets installed on both sides of the reed block housing -- four gaskets in all. The 300 Series has a

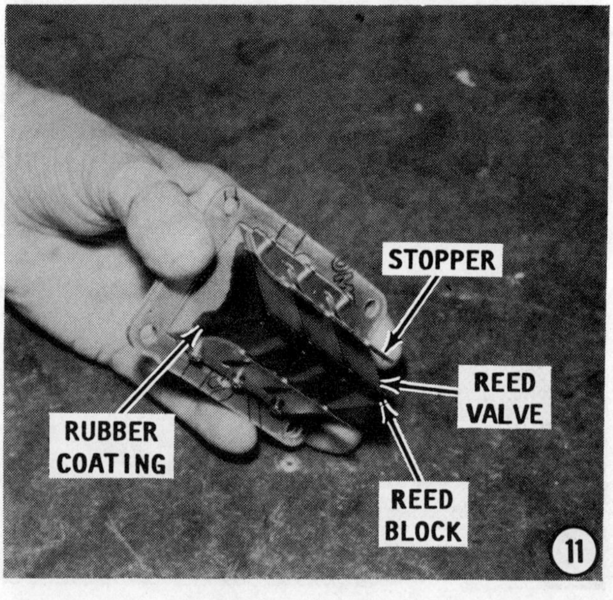

single reed valve assembly, with two gaskets -- one on each side of the reed block.

11- Measure the reed valve warpage with a feeler gauge. If the distance between the reed block housing and the reed valve is more than 0.2mm, the reed valve assembly **MUST** be replaced.

READ AND BELIEVE

The reeds must never be turned over in an attempt to correct a problem. Such action would cause the reed to flex in the opposite direction and the reed would break in a very short time.

Block Disassembling

12- If servicing a 650 Series engine, remove the nuts securing the cylinder block to the upper crankcase. For all other series engines, the block is secured between the head and the upper crankcase with the head nuts.

13- Keep the cylinder block as level as possible and lift it **STRAIGHT** up and clear of the studs. Two locating pins are indexed on one side of the 650 Series block and upper crankcase. These pins may remain with the cylinder block as it is lifted or the pins may remain with the lower crankcase. **TAKE CARE** not to lose these pins.

14- Remove and discard the base gasket.

15- Scribe a mark on the inside of the piston skirt to identify the No. 1 and No. 2 piston **BEFORE** removal from the connecting rod as described in the next step. Naturally, this mark is not necessary, if working on a single cylinder engine.

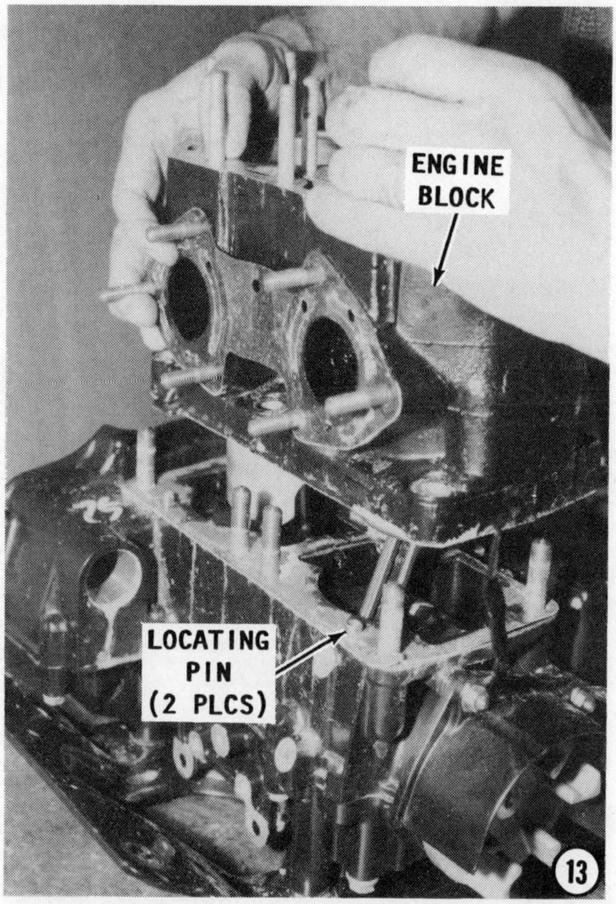

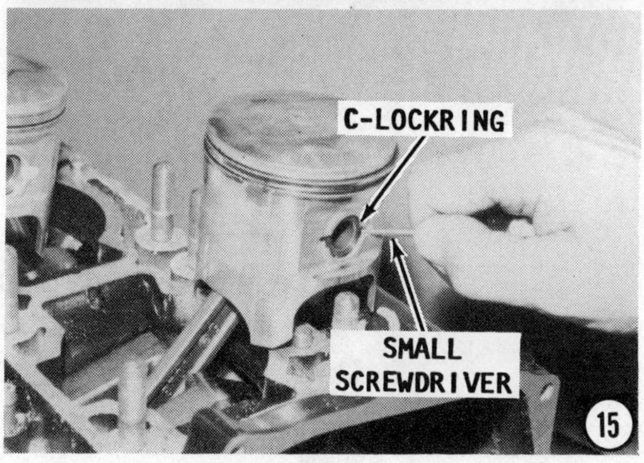

*C-lockrings are under tremendous tension during removal and installation, presenting a potential hazard. As an eye protection measure, safety glasses or a shield should **ALWAYS** be worn during work with such snap rings. Warn others in the area such work is in progress.*

SAFETY WORDS

The piston pin C-lockrings are made of spring steel and may pop out of the groove with considerable force. Therefore, warn other people in the area and **WEAR** eye protection glasses while removing the piston pin lockrings.

Remove the C-lockring from both ends of the piston pin using a screwdriver inserted into the small groove next to the piston

pin opening. Discard the C-lockrings. These rings stretch during removal and must **NEVER** be used a second time.

16- Push the piston pin out free of the piston. If the pin is tight and refuses to budge, use the piston pin puller assembly, listed at the beginning of this chapter.

17- Remove the caged needle bearings from the small end of the connecting rod.

ADVICE
ALL SERIES ENGINES

New needle bearings should be installed in the connecting rods, even though they may appear to be in serviceable condition. New bearings will ensure lasting service after the overhaul work is completed. If it is necessary to install the used bearings, keep them separate and identified to **ENSURE** they will be installed onto the same connecting rod from which they were removed.

CRITICAL WORDS

The rod is an integral part of the crankshaft. The two are manufactured together and **CANNOT** be separated.

Slowly rotate the bearing races. If rough spots are felt, a new crankshaft must be purchased. The bearings are installed at the factory and are not replaceable.

GOOD WORDS

Good shop practice dictates to replace the rings during an engine overhaul. However, if the rings are to be used again, expand them **ONLY** enough to clear the piston and the grooves, because used rings are brittle and break very easily.

18- Gently spread the top piston ring enough to pry it out and up over the top of the piston. No special tool is required to remove the piston rings. Remove the lower ring in a similar manner. These rings are **EXTREMELY** brittle and have to be handled with care if they are intended for further service.

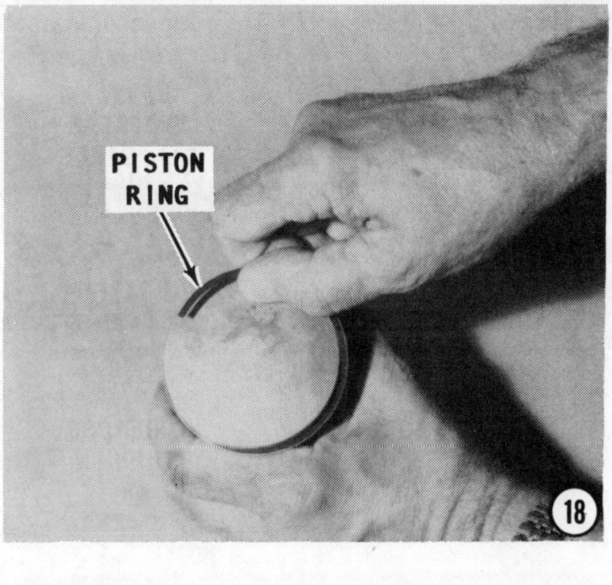

2-CYLINDER ENGINES

The following four steps apply only to 2-cylinder engines. Peform Steps 19 thru 22, and then proceed to the Cleaning and Inspecting work in Section 8-5, beginning on Page 8-46.

SINGLE-CYLINDER ENGINES

Skip Steps 19 thru 22 and proceed directly to the disassembling instructions beginning on the next page.

Engine Bed Bracket Removal

19- Turn the engine upside down on the work surface. Remove the four bolts securing the engine bed bracket to the lower crankcase.

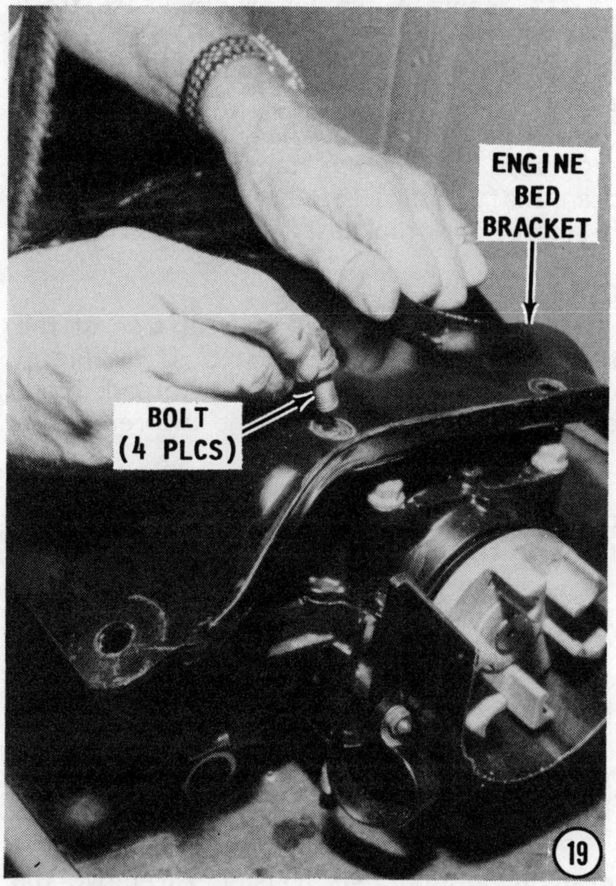

Crankcase Separation

20- Remove the seven 6mm bolts and eight 8mm bolts securing the two halves of the crankcase together. Two prying tabs are cast into the crankcase to assist in separating the two halves. **ALWAYS** pry **ONLY** at these tab locations. Prying elsewhere could very likely damage an aluminum sealing surface.

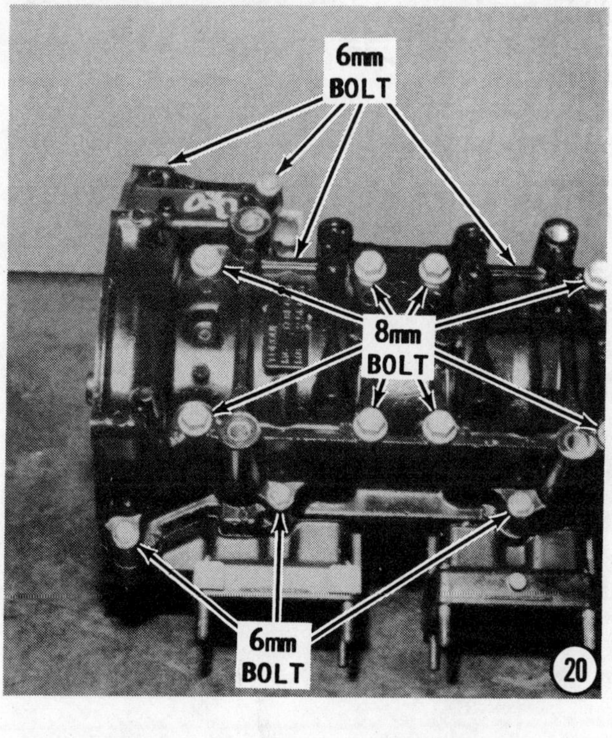

21- Separate the lower crankcase from the upper crankcase. **TAKE CARE** not to lose the two locating pins indexed diagonally across from each other. These pins may remain in either half when the crankcase is separated. Tap the tapered end of the crankshaft with a soft head mallet to "jar" it free of the crankcase.

Crankshaft Disassembling

22- Pull the forward oil seal free of the tapered end of the crankshaft. If the metal coupler was removed from the threaded end of the crankshaft, remove the two rear main oil seals.

Now proceed to Section 8-5, beginning on Page 8-46 for complete detailed instructions on cleaning and inspecting the engine.

DISASSEMBLING SINGLE-CYLINDER ENGINES

Exploded Drawing

Refer to the exploded drawing on Page 8-23, to identify components and as an aid in disassembling.

Position a suitable container under the cranking motor gear housing. Remove the drain bolt and drain the oil from the housing.

Remove the bolts securing the housing cover to the block. Remove and discard the sealing ring, if a replacement is at hand.

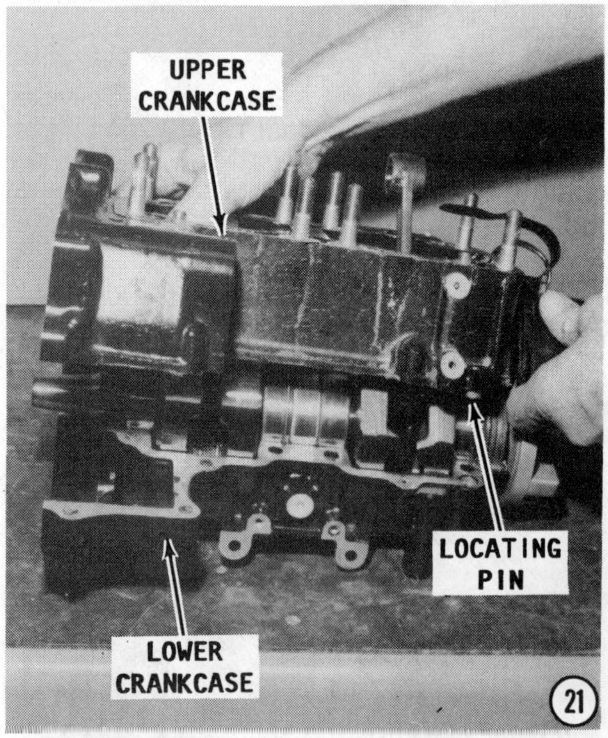

Pull the collar free of the crankshaft. Remove the O-ring.

Attach the flywheel puller to the cranking motor clutch, using three 6x60mm bolts. Pull the clutch assembly and idle gear free of the crankshaft.

Lift out the idle gear.

Crankcase Separation Single-Cylinder Engines

FIRST, THESE WORDS

A total of eight bolts secure the two halves of the crankcase together. The five bolts inside the starter clutch housing are steel bolts. The three bolts outside the housing are stainless steel bolts.

Remove the eight bolts securing the two halves of the crankcase together.

Install the crankcase splitting tool onto the aft crankcase half. Check to be sure the two supporting bolts are completely installed. Tighten the center bolt to separate the two halves of the crankcase.

After the crankcase halves have been separated, make a setup in an arbor press and press the crankshaft free of the aft crankcase half. Take precautions to **ENSURE** the crankshaft does not fall free to the floor.

If the forward crankshaft bearing is to be replaced, press the bearing free of the forward crankcase half. If the forward oil seal is to be replaced, pry the old seal free of the forward crankcase half.

If the rear crankshaft bearing is to be replaced, pull the bearing free of the aft crankcase half, using the special bearing puller. If the rear oil seal is to be replaced, pry the old seal free of the rear crankcase half.

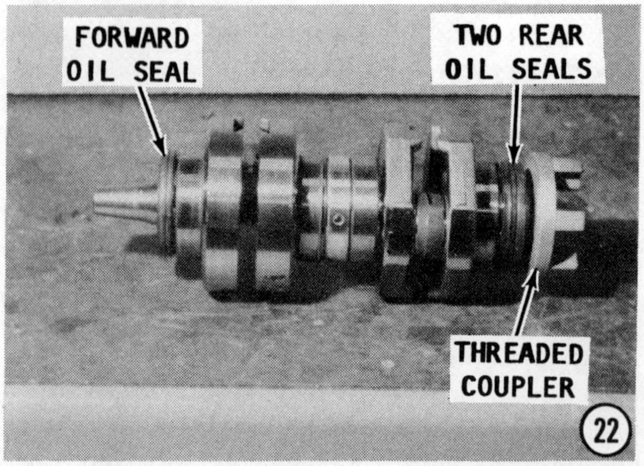

CLEANING AND INSPECTING

COMPLETE and **DETAILED** instructions for cleaning and inspecting **ALL** parts of the engine, including the block, will be found in Section 8-5, beginning on Page 8-46.

8-4 ASSEMBLING AND INSTALLATION

Detailed procedures are given to assemble virtually all parts of the engine, followed by instructions to install the engine.

The procedures are clearly identified for 2-cylinder engines and for single-cylinder engines.

Therefore, if certain parts were not removed or disassembled because the part was found to be fit for further service, simply skip the particular step involved and continue with the required tasks to return the engine to operating condition.

The following instructions pickup the work after all parts of the engine have been thoroughly cleaned and inspected according to the procedures outlined in Section 8-5.

SINGLE-CYLINDER ENGINES

The exploded drawing of a single-cylinder engine on Page 8-23 may be helpful during the assembling work.

Pack the lips of the forward and aft oil seals with Shell Alvania EP1 lubricant, or equivalent.

Install the aft oil seal into the aft crankcase half with the manufacturer's marks facing **UP**. Install the forward oil seals into the forward crankcase half, with the manufacturer's marks facing **DOWN**. The oil seals are correctly installed when the face of the seal is flush with the surface of the crankcase. After installation, the marks on both seals must face **FORWARD**.

Make a setup in an arbor press and install the forward main bearing into the forward crankcase half. Press the bearing into place until the outer race makes contact with the shoulder in the casting.

Press the rear main bearing onto the threaded end of the crankshaft, until it seats against the crankshaft journal.

Insert the crankshaft installing jig over the connecting rod between the crankshaft journals. Adjust the tool to fit "snug" in opening between the journals.

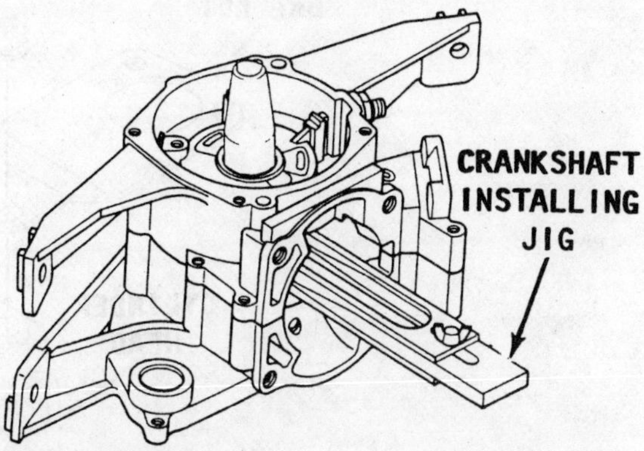

The 300 Series engine requires a special jig over the connecting rod in order to install the two crankcase halves together -- properly aligned.

Make a setup in an arbor press and press the crankshaft, with jig attached, into the forward crankcase half.

Apply a bead of Kawasaki Bond to the mating surface of the forward crankcase half.

Install the aft crankcase half over the forward crankcase half. The two locating pins must index into the matching holes in the opposite half. When the two crankcase halves are properly seated against each other, remove the installing jig.

Install and **HAND** tighten the five steel bolts inside the stator plate. Rotate the crankshaft by hand and check for binding or "rough" spots. If the crankshaft will not freely rotate or binding is felt, the crankshaft is not properly centered within the crankcase. Tap the end of the crankshaft with a soft head mallet to reposition the crankshaft, until it will rotate freely.

Tighten the five steel bolts, working in a criss-cross pattern, to a torque value of

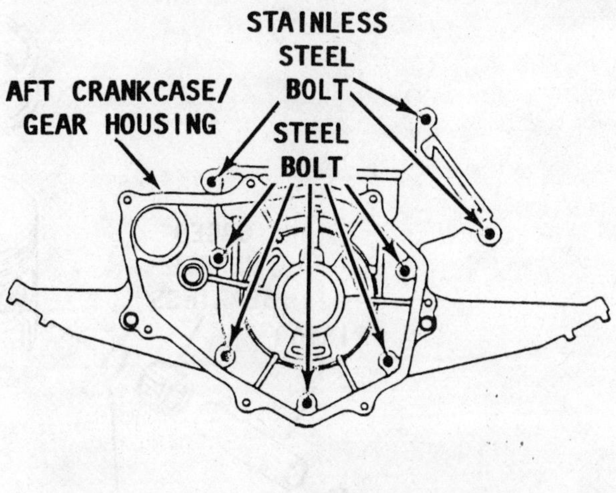

Location of the steel bolts and the stainless steel bolts used to secure the crankcase halves together.

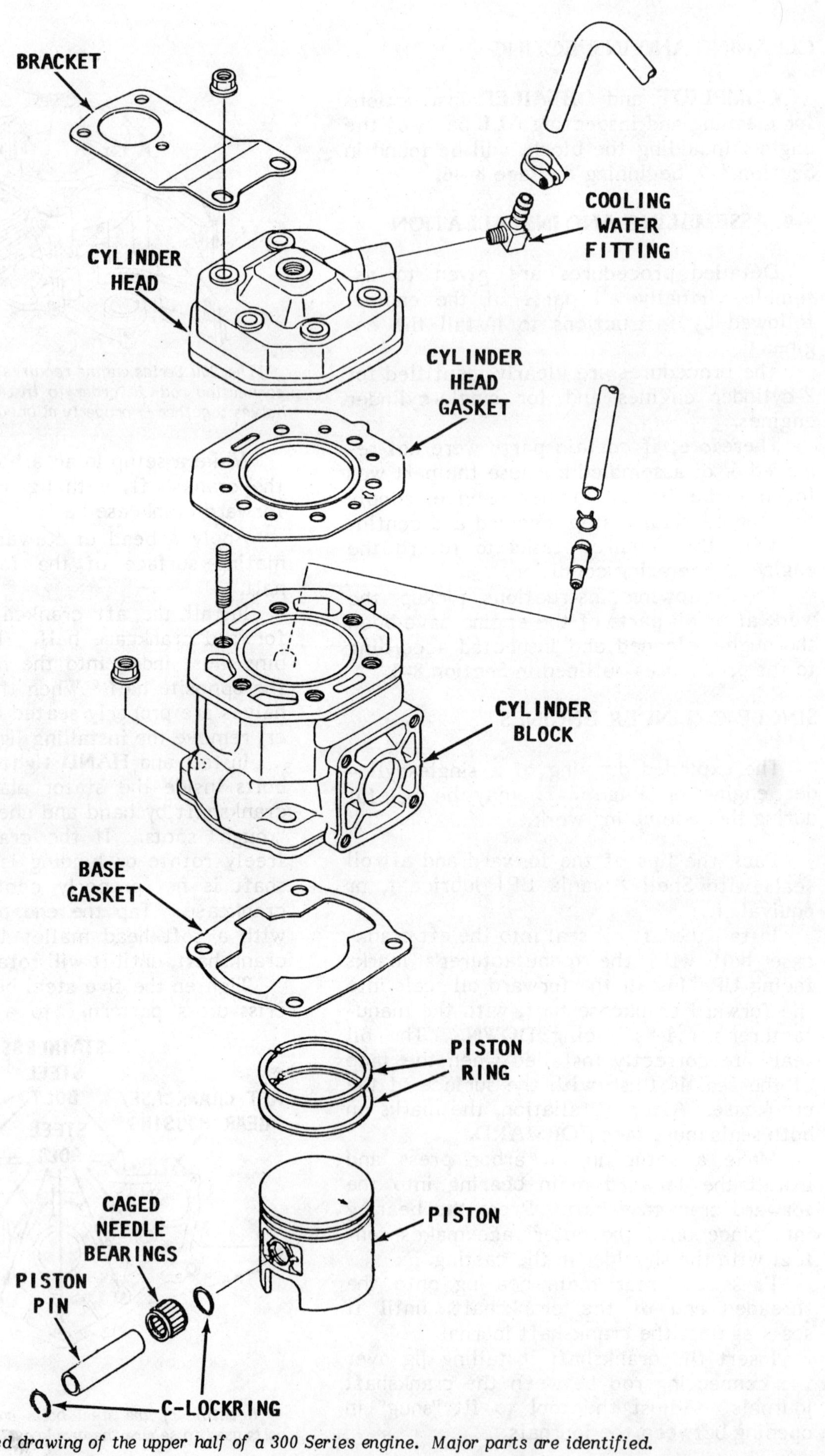

Exploded drawing of the upper half of a 300 Series engine. Major parts are identified.

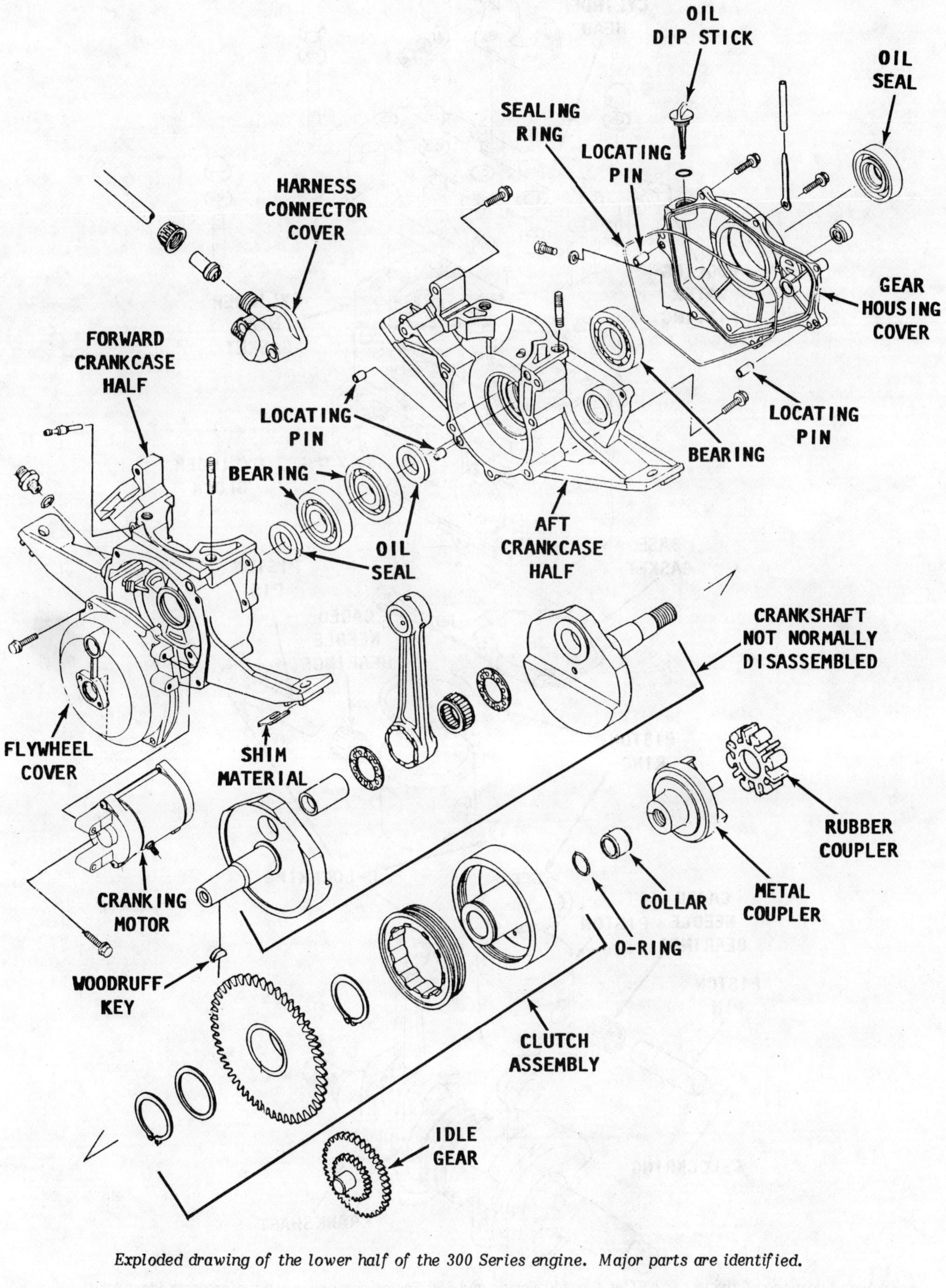

Exploded drawing of the lower half of the 300 Series engine. Major parts are identified.

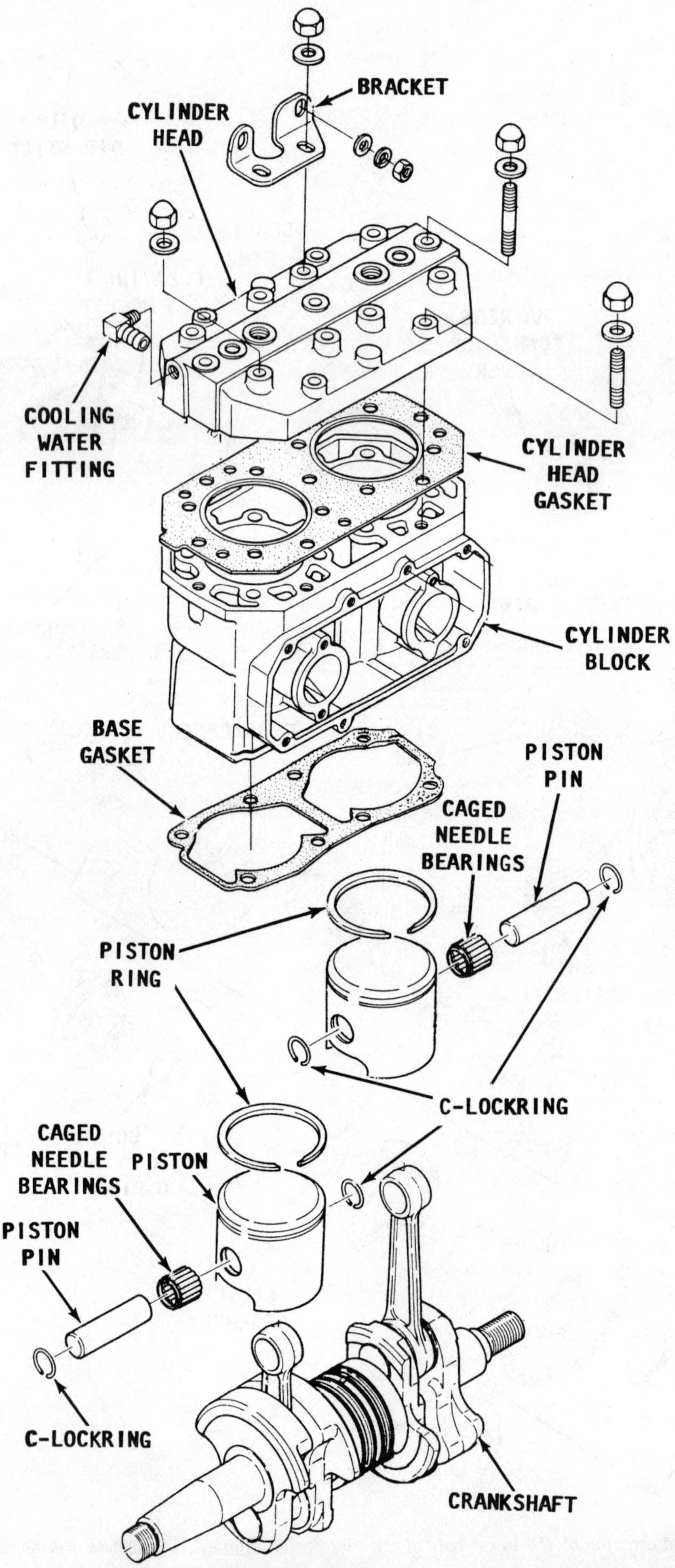

Exploded drawing of the upper half of the 440 Series and 550 Series engines, with major parts identified.

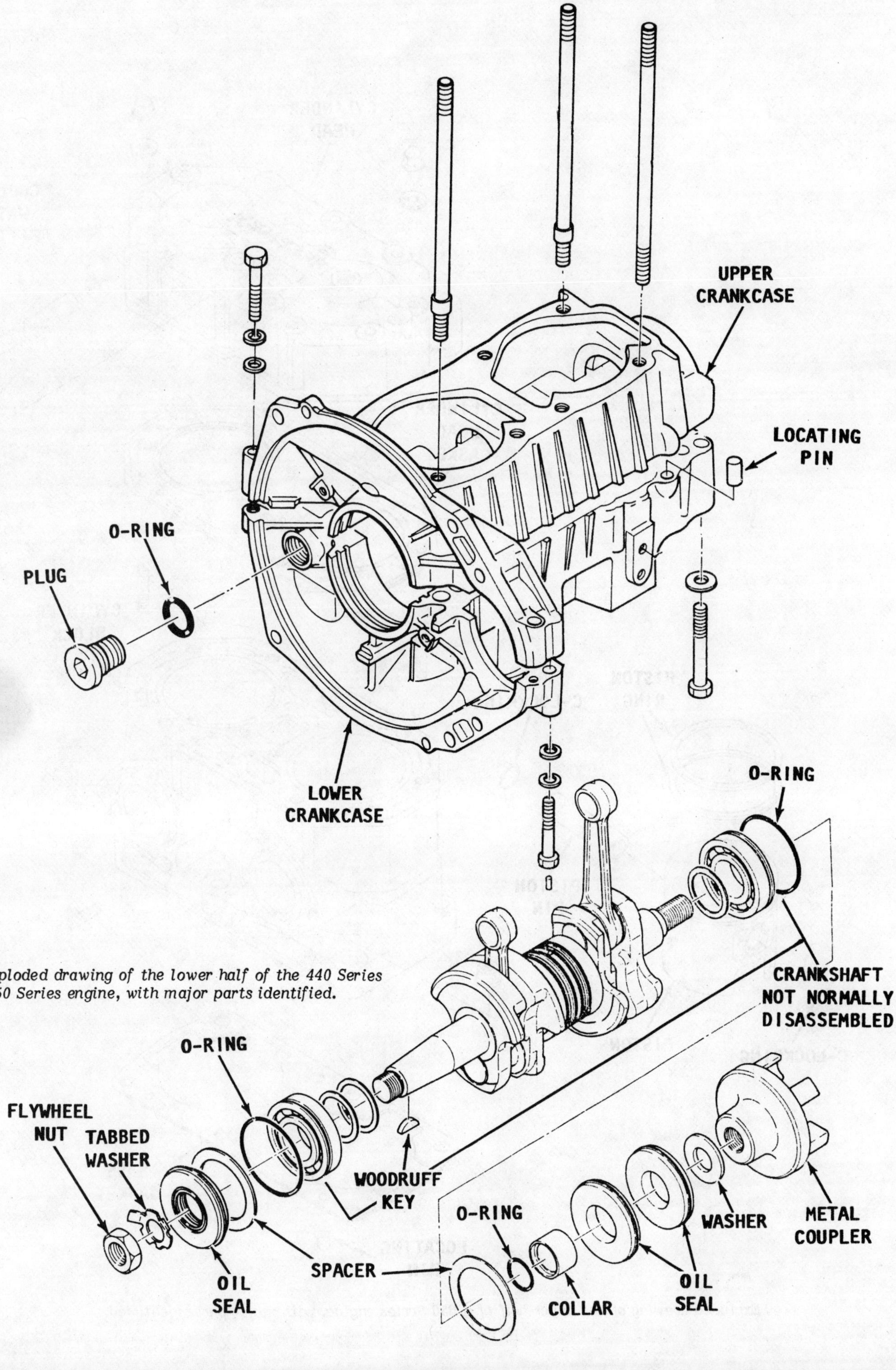

Exploded drawing of the lower half of the 440 Series and 550 Series engine, with major parts identified.

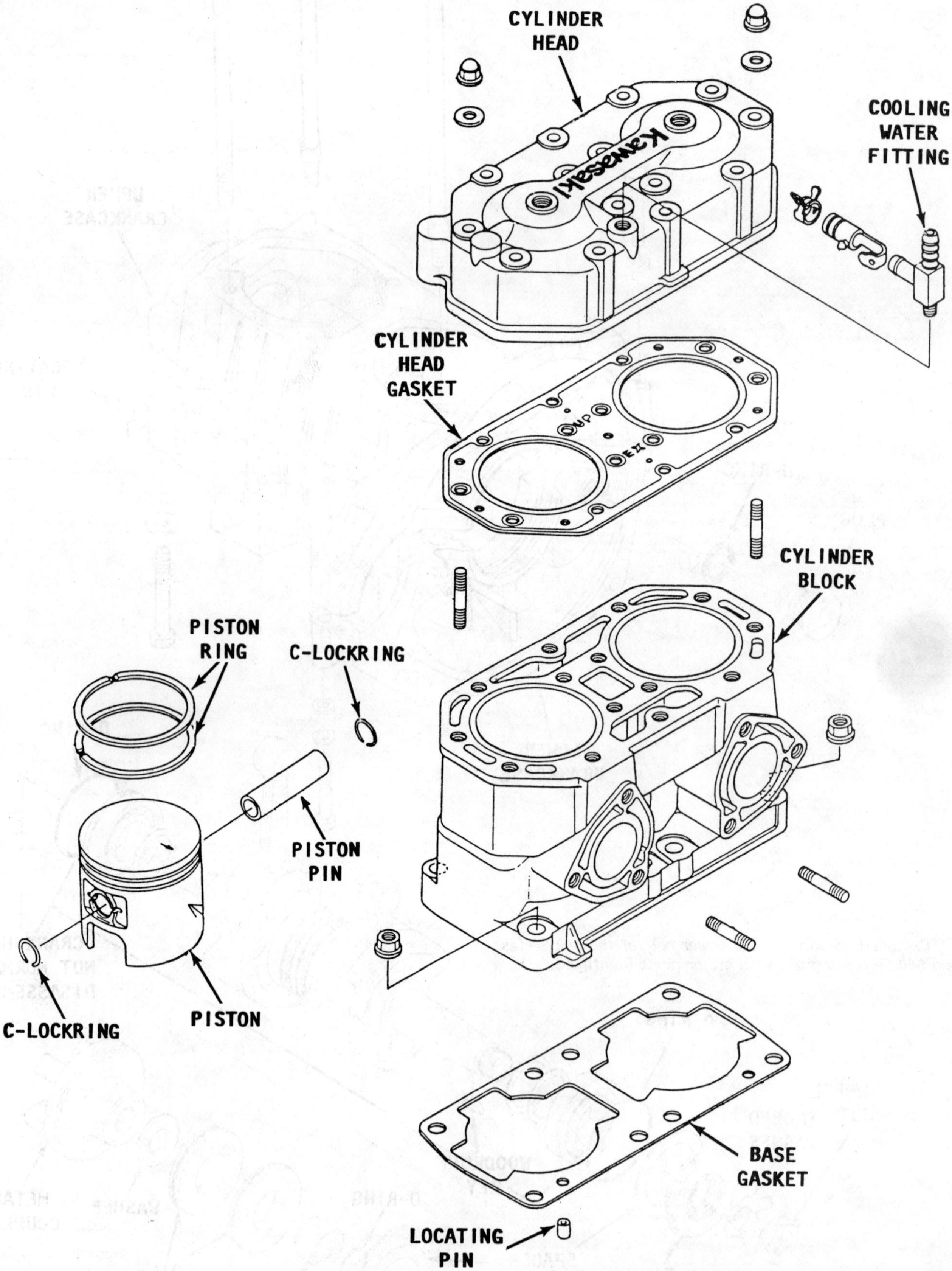

Exploded drawing of the upper half of a 650 Series engine, with major parts identified.

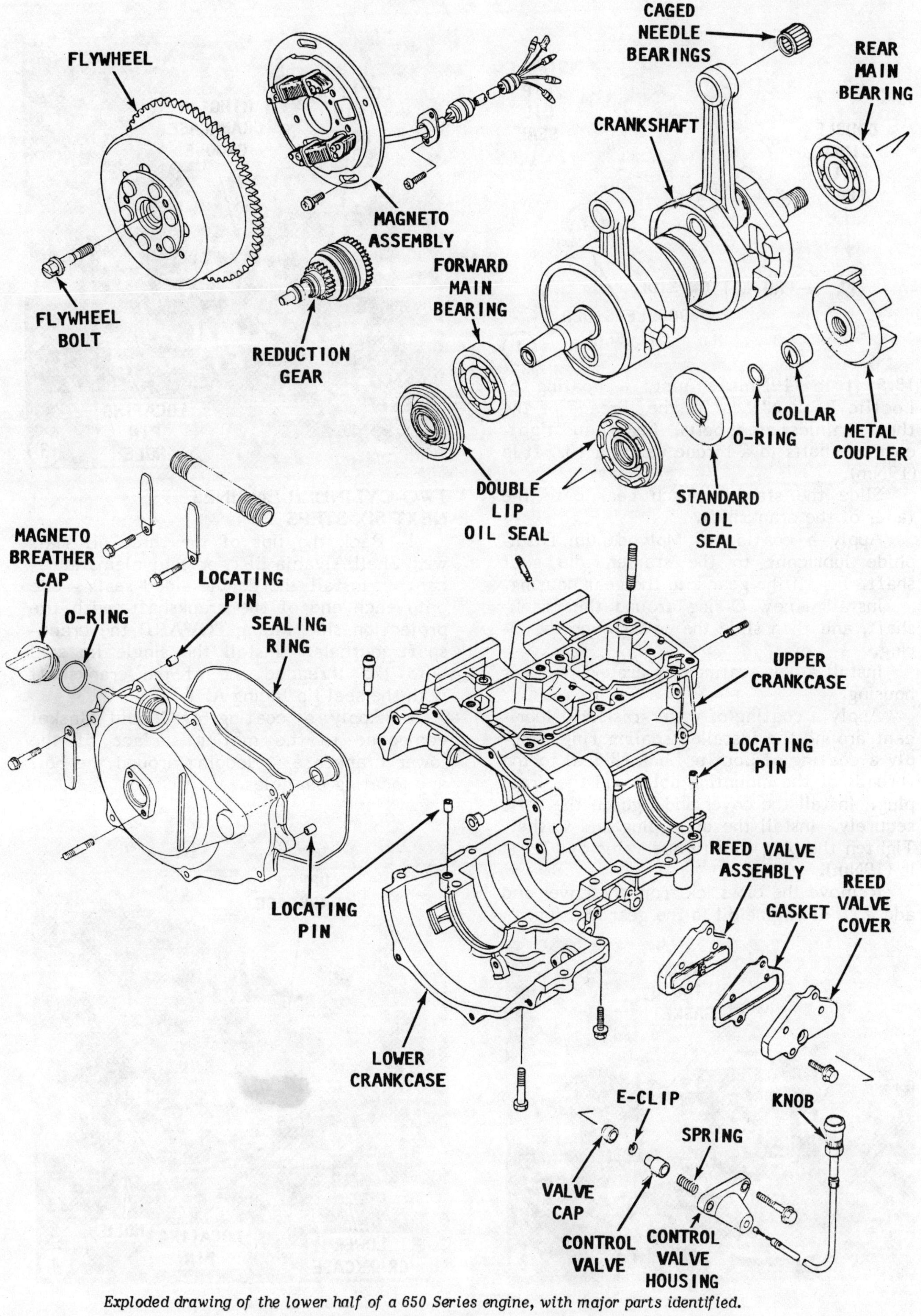

Exploded drawing of the lower half of a 650 Series engine, with major parts identified.

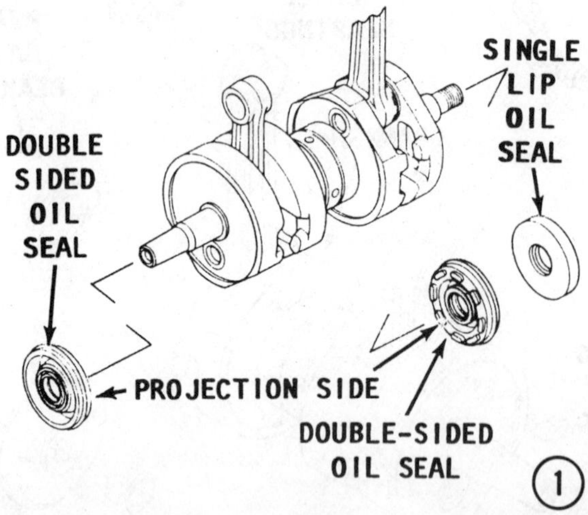

13.5 ft lb (19Nm). Apply a coating of Loctite Lock N' Seal to the threads of the three stainless steel bolts. Install and tighten these bolts to a torque value of 13.5 ft lb (19Nm).

Slide the starter clutch gear onto the taper of the crankshaft.

Apply a coating of Molybdenum Disulphide lubricant to the starter idle gear shaft. Install the gear into the gear housing.

Install a new O-ring around the crankshaft, and then slide the collar over the O-ring.

Install a new sealing ring around the gear housing.

Apply a coating of water-resistant lubricant around the installed sealing ring. Apply a coating of Loctite Lock N' Seal to the threads of the mounting bolts and the drain plug. Install the cover and tighten the bolts securely. Install the drain plug and washer. Tighten the plug to a torque value of 13.5 ft lb (19Nm).

Remove the dip stick from the cover and add 0.2L of engine oil to the gear housing.

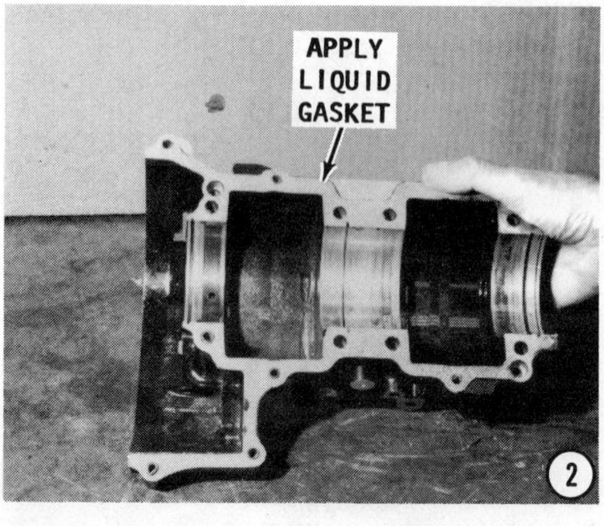

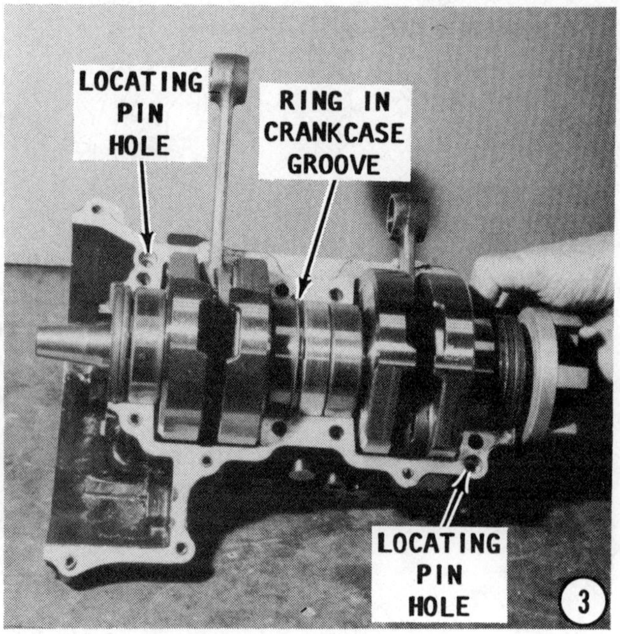

TWO-CYLINDER ENGINES NEXT SIX STEPS

1- Pack the lips of the three oil seals with Shell Alvania EP1, or equivalent, lubricant. Install the double sided seals --one onto each end of the crankshaft, with the projection side facing **TOWARD** the crankshaft journals. Install the single lip seal onto the threaded end of the crankshaft with the seal lip facing **AFT.**

2- Apply a coating of Liquid Gasket compound to the mating surface of the lower crankcase -- looping around the bolt and locating pin holes.

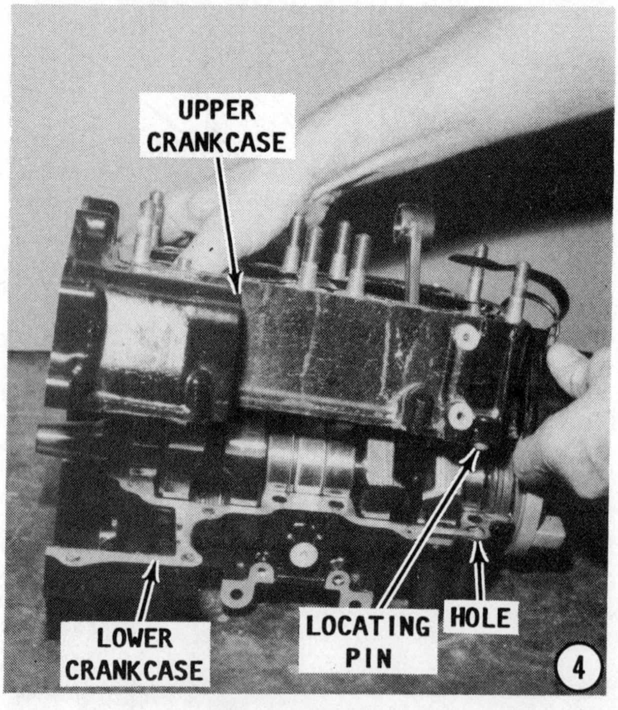

3- Seat the crankshaft assembly into the lower crankcase half and fit the labyrinth seal circlip into the groove in the crankcase. Check to be sure the locating pins are in place on one of the crankcase halves.

4- Lower the upper crankcase half over the crankshaft -- guiding the connecting rods through the two openings. Bring the two crankcase halves together with the locating pins indexed into the matching holes in the other half.

5- Apply a coating of non-permanent locking compound to the threads of the crankcase bolts.

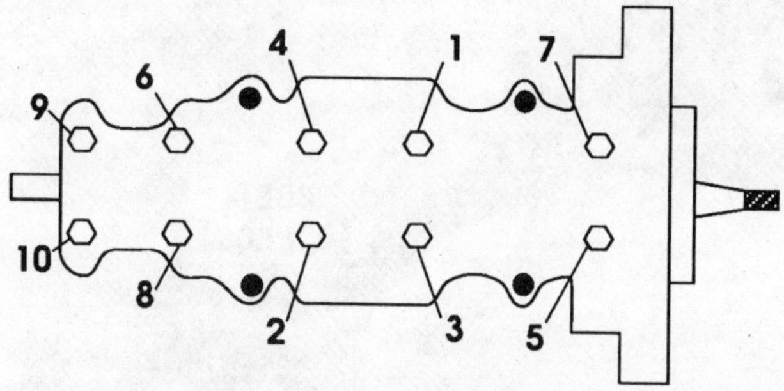

Crankcase tightening sequence for the 440 Series and the 550 Series engines.

CRITICAL BOLT WORDS

Refer to the accompanying illustrations for this step. Step 5 illustration applies to the 650 Series engines. The line drawing captioned illustration applies to the 440 Series and the 550 Series engines.

For the 650 Series -- bolt numbers 1 thru 8 are 8mm bolts and are to be tightened to a torque value of 18 ft lb (25Nm). Bolts numbered 9 thru 15 are 6mm bolts and are to be tightened to only 69 in lb (7.8Nm). These fifteen bolts do not have flat washers or lockwashers.

For the 440 Series and 550 Series -- bolt numbers 1 thru 10 are the large bolts with flat washers. These bolts are to be tightened to a torque value of 16 ft lb (22Nm). Bolts numbered 11 thru 13 are smaller with flat washers **AND** lockwashers. These bolts are to be tightened to only 52 in lb (6Nm).

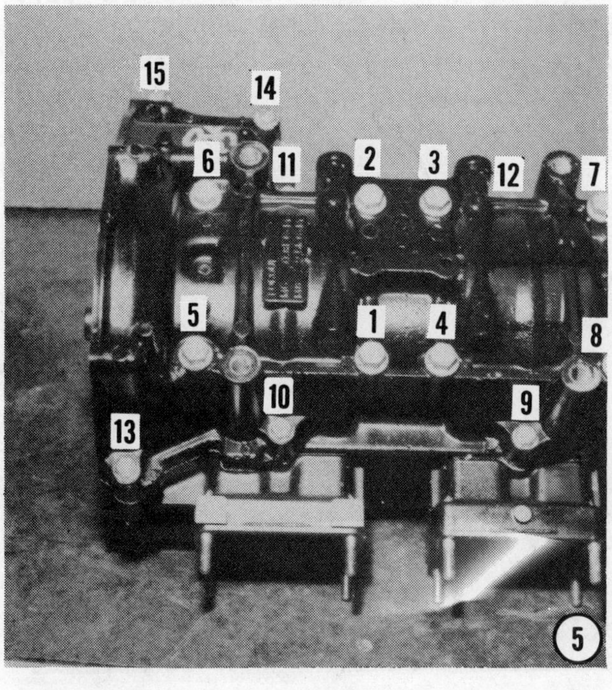

Pay attention -- to ensure the correct bolts are threaded into the proper openings.

Now, rotate the crankshaft by hand to be sure the crankshaft does not bind.

BAD NEWS

If binding is felt, it will be necessary to remove the crankcase and reseat the crankshaft. It will also be necessary to check the positioning of the labyrinth seal circlips and the bearing locating pins. If binding is still a problem after the crankcase has been installed a second time, the cause might very well be a broken piston ring.

Engine Bed Bracket Installation

6- With the engine upsidedown on the work surface, position the engine bed bracket over the crankcase. If servicing a 440 Series or 550 Series engine, the cutaway corners of the bracket **MUST** face **FORWARD.** If working on a 650 Series engine, the small notches at two corners, as shown,

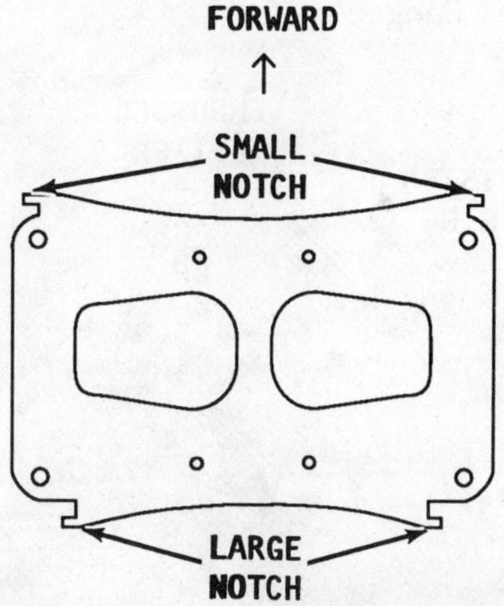

Simple line drawing to show the engine bed bracket with the large notches and the small notches, mentioned in the text.

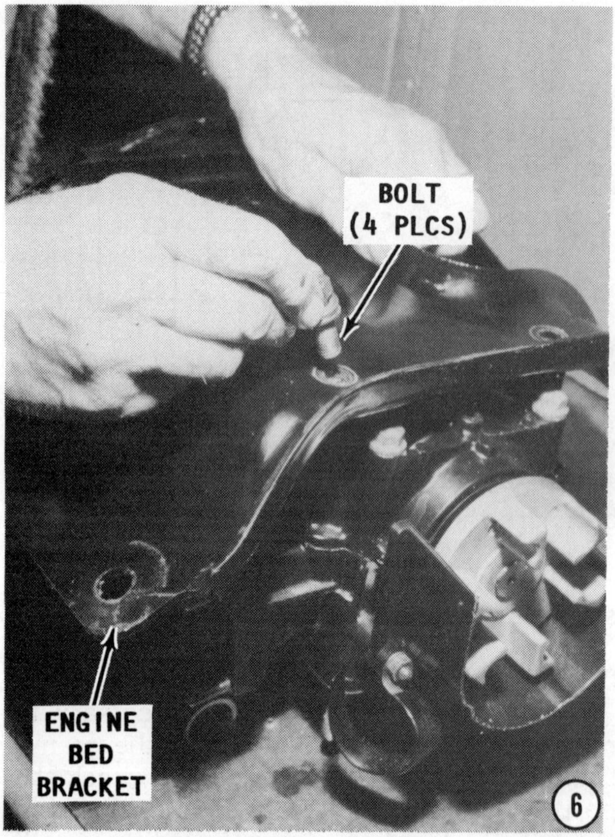

MUST face **FORWARD.** Apply a coating of non-permanent locking compound to the threads of the four mounting bolts. Install and tighten the bolts to a torque value of 35 ft lb (48Nm).

440 Series	35 ft lb (48Nm)
550 Series	35 ft lb (48Nm)
650 Series	27 ft lb (36Nm)

The 440 Series and 550 Series bolts have flat washers and lockwashers. The 650 Series uses flanged bolts.

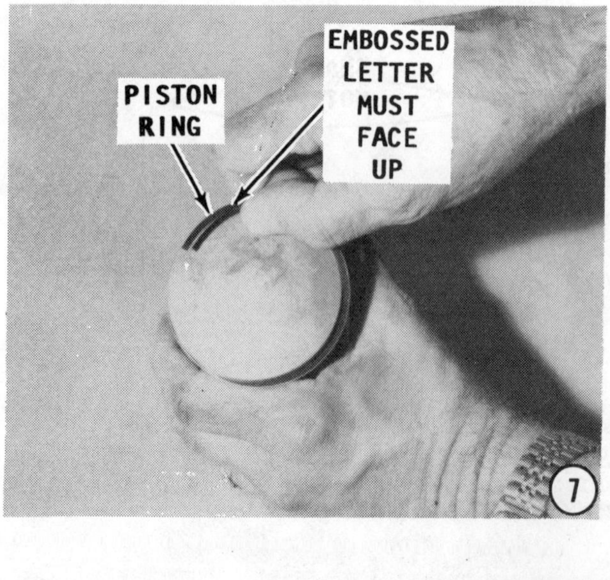

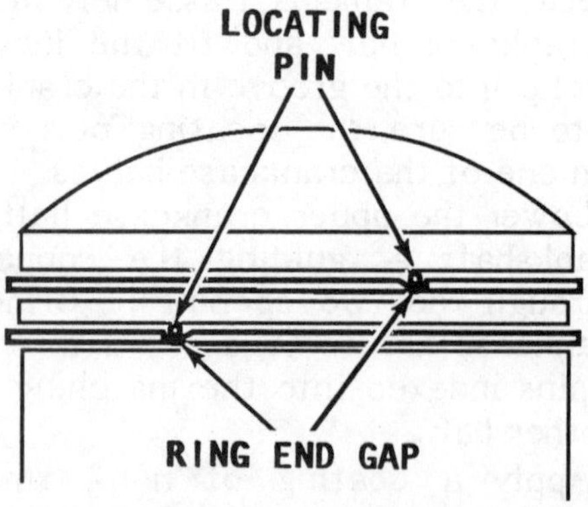

Align the ring end gap over the locating pin in each piston groove to prevent the ring from rotating in the groove. If the ring was free to rotate, an end might very well catch on the port opening and break.

Piston Installation
All Series Engines

7- Observe the embossed letter on each piston ring. The ring **MUST** be installed with this letter facing **UP.**

Install a new set of piston rings onto the piston/s. No special tool is necessary to install the piston rings. **HOWEVER,** take care to spread the ring only enough to clear the top of the piston. The rings are **EXTREMELY** brittle and will snap if spread beyond their limit. Align the ring end gap over the locating pin in each piston groove.

The purpose of the pin is to prevent the ring from rotating in the ring groove. The ring end gap must travel up and down a portion of the cylinder wall where there is no port opening. If the ring was free to rotate, an end might catch on the port opening and break the ring.

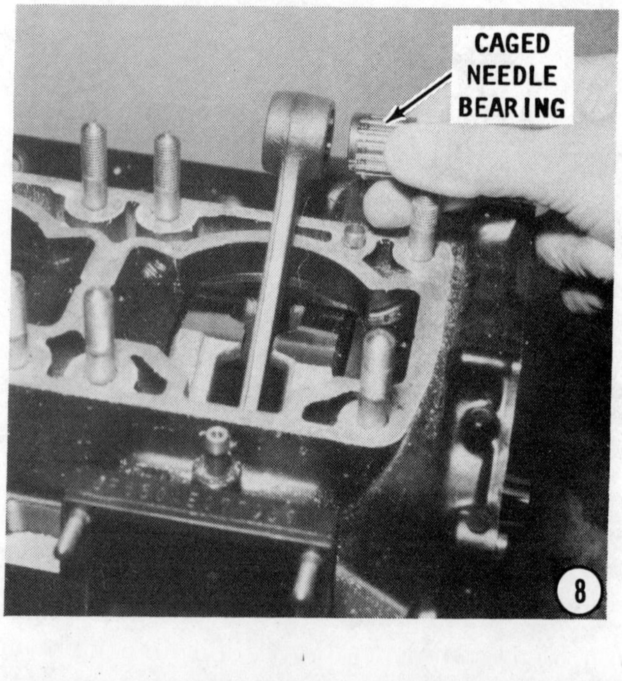

8- Dip the caged needle bearings into a container of engine oil, and then insert the cage into the small end of the connecting rod.

CRITICAL WORDS
An arrow is embossed on each piston crown. The direction of this arrow **MUST** be **TOWARD** the exhaust manifold -- port side.

9- Position the piston over the connecting rod with the arrow embossed on the crown facing **TOWARD** the exhaust manifold -- port side. Insert the piston pin into the piston and caged needle bearings. Center the pin in the piston.

SAFETY WORDS
The piston pin C-lockrings are made of spring steel and may slip out of the groove with considerable force. Therefore, warn other persons in the area and **WEAR** eye protection glasses while installing the piston pin lockrings in the next step.

10- Install a C-lockring at each end of the piston pin. The lockrings may be worked in by hand. Position the end gap away from the cutout on the piston opening, as shown in the accompanying line drawing.

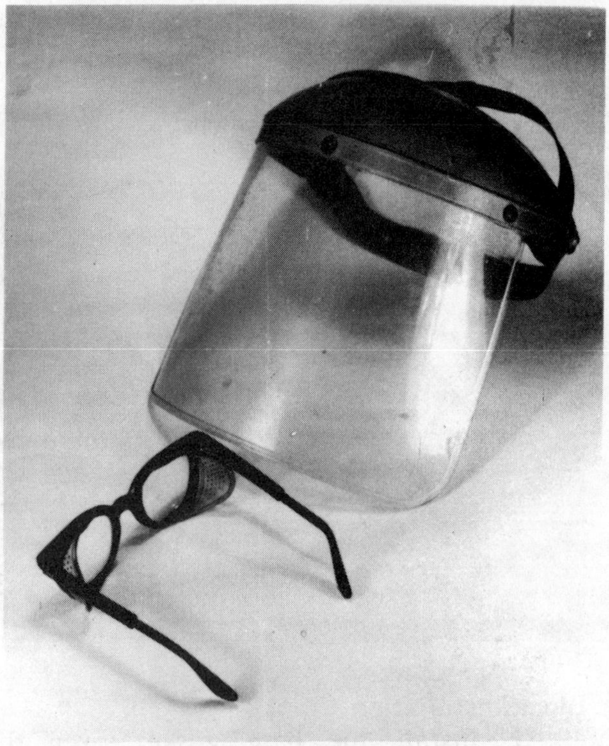

*C-lockrings are under tremendous tension during removal and installation, presenting a potential hazard. As an eye protection measure, safety glasses or a shield should **ALWAYS** be worn during work with such snap rings. Warn others in the area such work is in progress.*

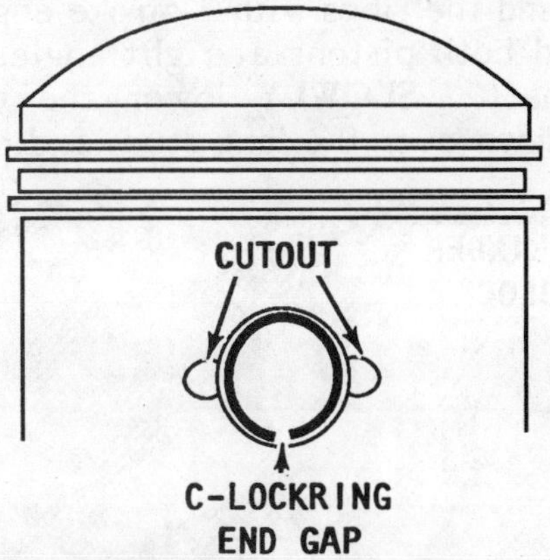

The C-lockring end gap must be positioned away from the two cutouts in the piston, as shown.

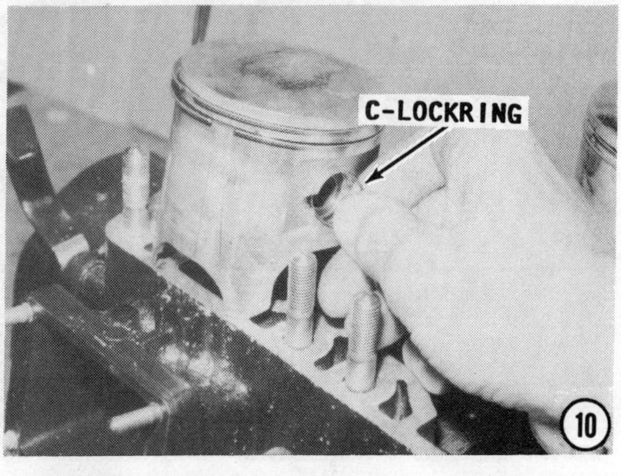

Block Installation

11- Install the base gasket over the crankcase studs. The manufacturer does not call for any type material be applied to either side of this gasket.

12- Coat the upper portion of each piston and the rings with 2-stroke engine oil.

Hold both pistons at right angles to the crankshaft. **SLOWLY** lower the cylinder block down over the long studs and onto the piston crown. Squeeze the rings inward around the locating pin and gently push the piston/s into the bore/s.

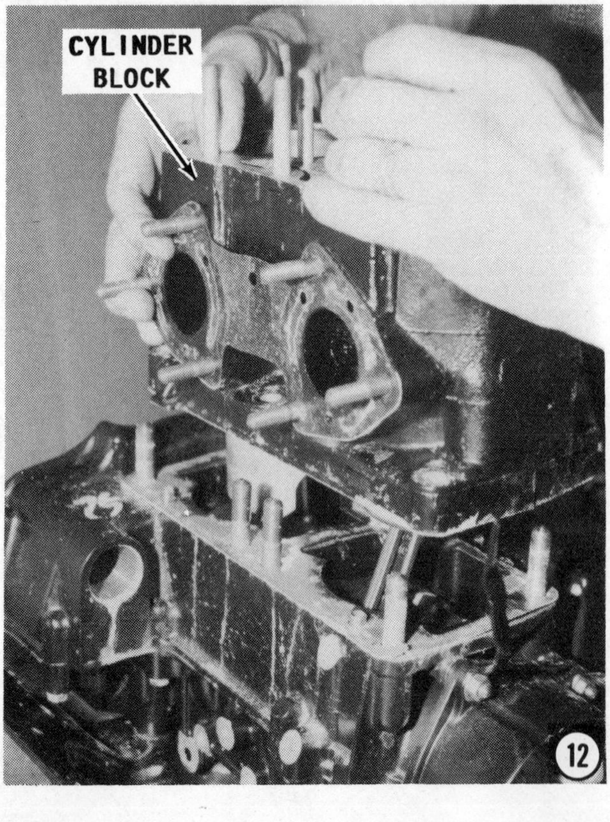

CRITICAL WORDS

If difficulty is experienced in fitting the piston/s into the cylinder/s, **DO NOT** force the piston. Such action might result in a broken piston ring. Make sure the ring end gap is aligned with the locating pin.

If more help is needed, obtain a large hose clamp and install the clamp around the piston to compress the rings. Try again -- it can be done.

13- Make a final check of the piston installation. The arrow **MUST** face **TOWARD** the exhaust side of the engine.

If servicing a 650 Series engine, install the nuts securing the cylinder block to the upper crankcase. Cross-tighten the nuts to a torque value of 25 ft lb (34Nm). For all other series engines, the block is secured between the head and the upper crankcase with the head nuts.

Reed Block Installation
650 Series Only

14- Observe the small hole in the long side of the reed block housing and the cutouts in the gaskets. Position a gasket on each side of the reed block housing. Slide the reed block housing over the mounting studs, with the hole at the bottom, as shown.

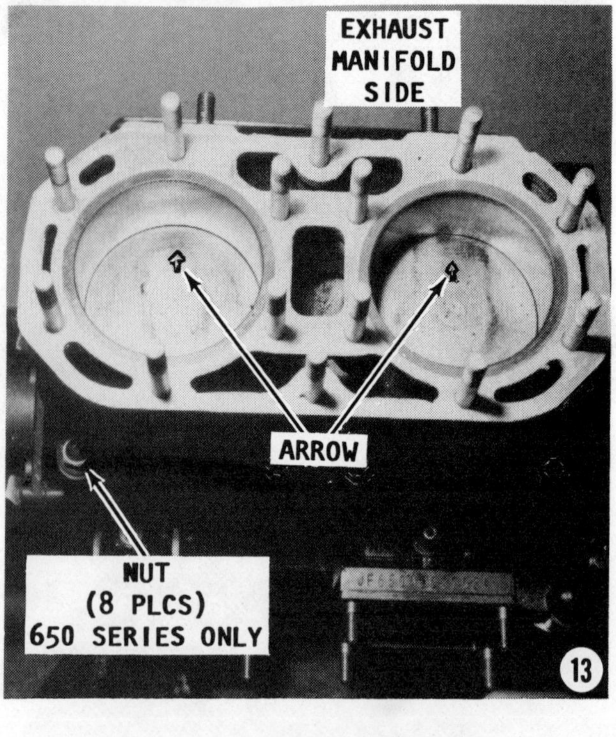

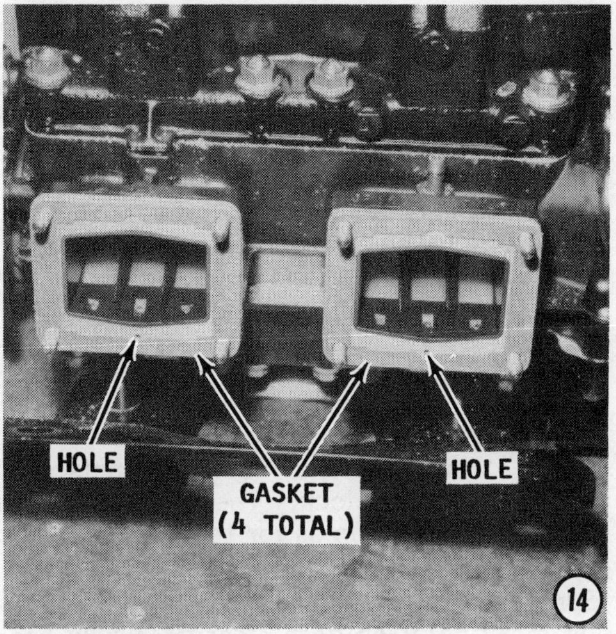

All Series Engines
Except 300 Series

15- Install the intake manifold and tighten the mounting nuts to a torque value of 69 in lb (7.8Nm).

Exhaust Manifold Installation
All Series Engines

16- Position a new exhaust manifold gasket onto the block.

17- Install the exhaust manifold to the engine block. Apply a coating of Loctite Lock N' Seal to the threads of the mounting studs or bolts. Studs are used on the 650 Series, but the 440 Series and the 550 Series engines have exhaust manifold mounting bolts. Tighten the stud nuts to a torque value of 14.5 ft lb (20Nm) or the bolts to a torque value of 52 in lb (6.0Nm).

Cylinder Head Installation

18- Observe the words **"UP"** and **"EX"** embossed on a new cylinder head gasket. The gasket **MUST** be installed with these words facing **UP** and with the **"EX"** toward the exhaust -- port side -- of the engine. Install the gasket as described.

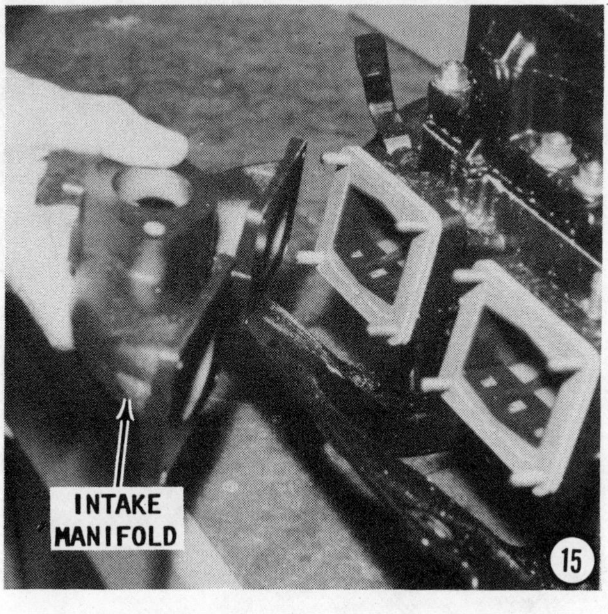

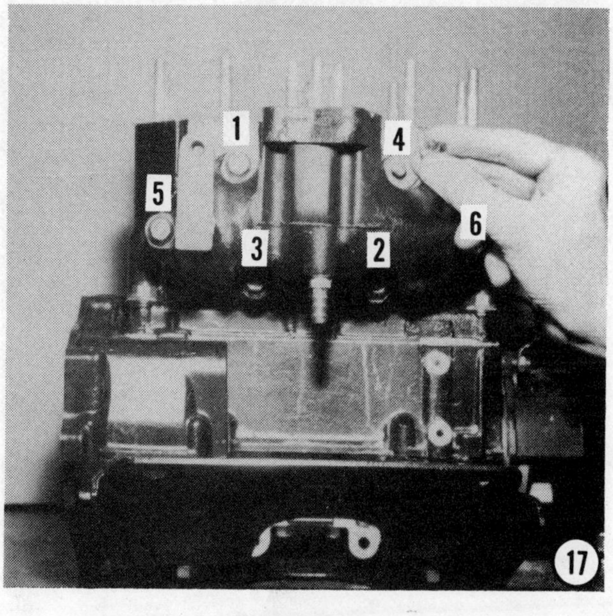

SPECIAL WORDS

The manufacturer recommends **NO** sealing agent be used on either side of the head gasket.

SEALING SURFACE WORDS
CRITICAL READING

Because of the high temperatures and pressures the sealing surfaces of the cylinder head and the block are the most prone to water leaks. No sealing agent is recommended **BECAUSE** it is almost impossible to apply an even coat of sealer. An even coat would be essential to ensure a air/water tight seal.

Some head gaskets are supplied with a "tacky" coating on both surfaces applied at the time of manufacture. This "tacky" substance will provide an even coating all around. Therefore, no further sealing agent is required.

HOWEVER, if a slight water leak should be noticed following completed assembly work and engine start up, **DO NOT** attempt to stop the leak by tightening the head bolts beyond the recommended torque value. Such action will only aggravate the problem and most likely distort the head.

FURTHERMORE, tightening the bolts, which are case hardened aluminum, may force the bolt beyond its elastic limit and cause the bolt to **FRACTURE. BAD NEWS,** very **BAD NEWS** indeed. A fractured bolt must usually be drilled out and the hole retapped to accommodate an oversize bolt, etc. Avoid such a situation.

Probable causes and remedies of a new head gasket leaking are:

a- Sealing surfaces not thoroughly cleaned of old gasket material. Disassemble and remove **ALL** traces of old gasket.

b- Damage to the machined surface of the head or the block. The remedy for this damage is the same as for the next case **"c"**.

c- Permanently distorted head or block. Spray a light **EVEN** coat of any type metallic spray paint on both sides of a new head gasket. Use only metallic paint -- any color will do. Regular spray paint does not have the particle content required to provide the extra sealing properties this procedure requires.

Assemble the block and head with the gasket while the paint is till **TACKY.** Install the head bolts and tighten in the recommended sequence and to the proper torque value and **NO** more!

Allow the paint to set for at least **24** hours before starting the engine. Consider this procedure as a temporary "band aid"

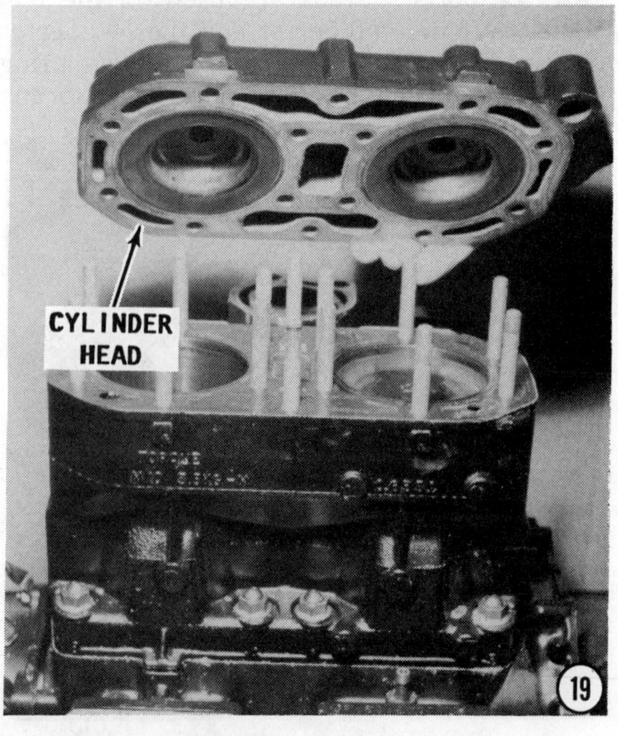

Cylinder head tightening sequence for the engines covered in this manual are almost always embossed on the head, as shown here. If for some reason they cannot be deciphered, the sequences are provided in the adjacent column.

type solution until a new head may be purchased or other permanent measures can be performed.

Under normal circumstances, if procedures have been followed to the letter, the head gasket will not leak.

19- Lower the head down over the mounting studs and onto the engine block.

20- Install the washers and nuts onto the studs. Tighten the nuts to the following

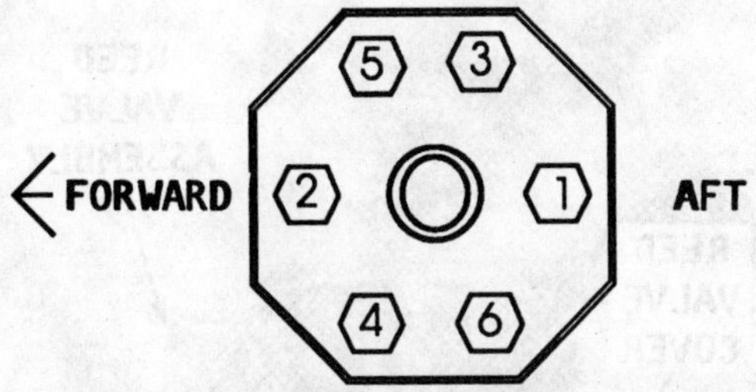

Cylinder head tightening sequence for the 300 Series engine.

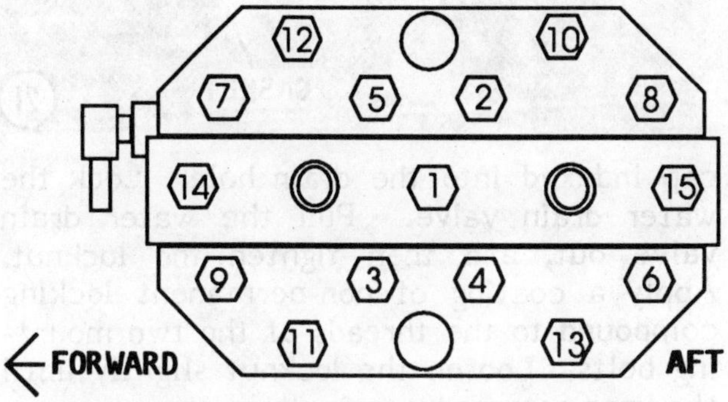

Cylinder head tightening sequence for the 440 and 550 Series engines.

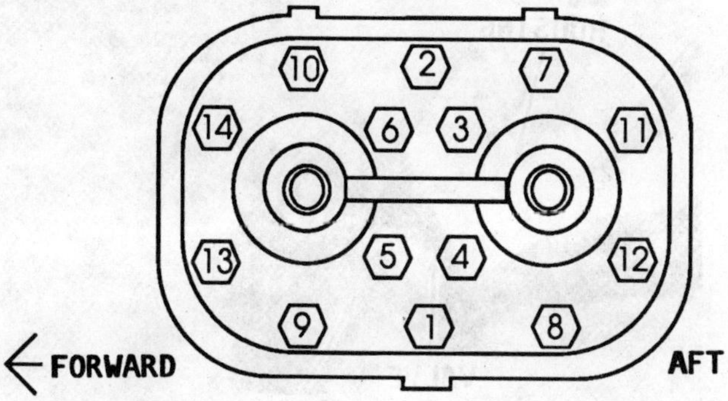

Cylinder head tightening sequence for the 650 Series engine.

torque values, in the sequence indicated in the accompanying illustration.

300 Series	23 ft lb (31Nm)
440 Series	16 ft lb (22Nm)
550 Series	16 ft lb (22Nm)
650 Series	22 ft lb (29Nm)

Water Drain Valve
650 Series Only

21- Position the reed valve cover over the reed valve assembly. Install and tighten the two securing bolts.

22- Install the control valve housing over the reed valve cover, with the valve

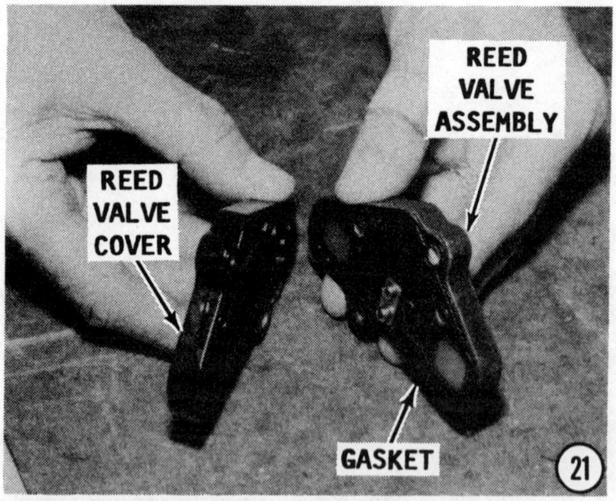

cap indexed into the drain hole. Lock the water drain valve. Pull the water drain valve out, and then tighten the locknut. Apply a coating of non-permanent locking compound to the threads of the two mounting bolts. Loosen the locknut slightly until the knob returns by spring tension.

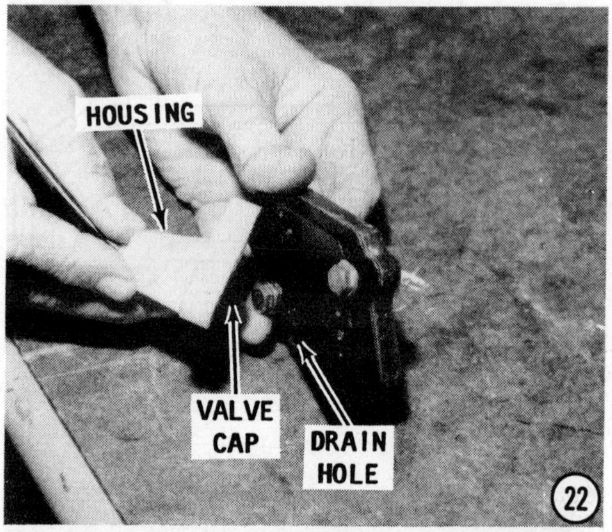

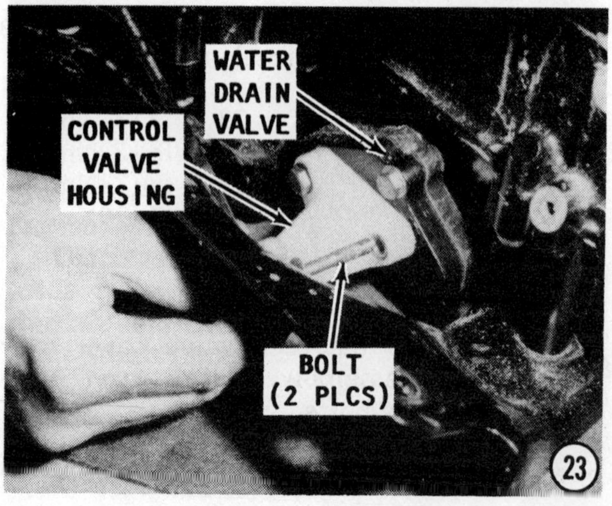

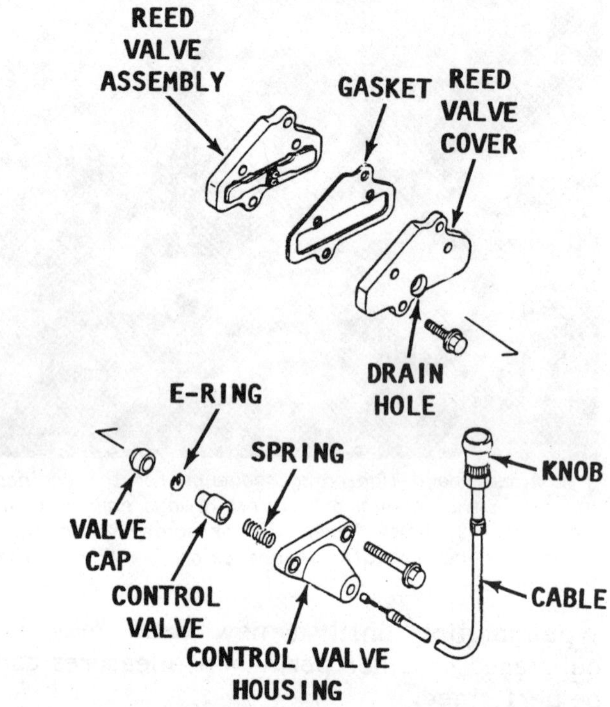

Exploded drawing of the water drain valve installed only on the 650 Series engines. Major parts are identified.

23- Install the water drain valve to the block and tighten the mounting bolts securely.

Flywheel Installation

24- Apply a coating of multi-purpose water resistant lubricant to the grommet in place on the stator harness. Feed the

harness connector through the opening in the stator plate and position the magneto assembly over the end of the crankshaft. Push the grommet into the opening and secure it in place with the two attaching screws. Tighten both screws securely.

Timing Marks

Identify the two marks made on the magneto assembly and the stator plate prior to removal. These marks serve to correctly position the magneto plate on the stator plate. The securing screws pass through elongated slots in the magneto assembly. Therefore, the assembly could be installed as much as 15° to its original position, drastically affecting the timing characteristics of the engine. If a **NEW** magneto assembly is being installed, match the new assembly to the old assembly and make a new mark as close to the original location as possible. The engine must be timed according to the procedures outlined in Section 7-7, of Chapter 7.

If **NO** timing marks were made prior to removal of the magneto assembly, as instructed in Disassembling, closely inspect the elongated slots in the magneto assembly and attempt to determine the original positioning of the screws. If no identifiable mark can be found, center the magneto assembly in the slot. This position may or may not be close to the correct position to permit the engine to start. Detailed timing procedure is presented in Section 7-7, of Chapter 7.

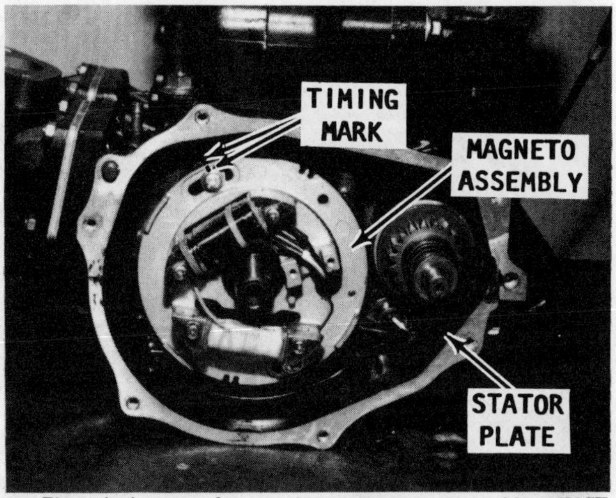

*The timing marks on the magneto assembly **MUST** align with the marks on the stator plate during installation.*

25- Apply a coating of Loctite #572, or equivalent, to the threads of the two magneto securing screws. Align the mark on the magneto assembly with the mark on the stator plate. Install and tighten the two screws securing the assembly to the stator plate **WITHOUT** disturbing the two aligned marks.

Flywheel Installation

FIRST, THESE WORDS

If the engine is to be timed, do not hesitate to install the flywheel using the Woodruff key and tightening the flywheel nut/bolt to the required torque value. The flywheel has cutouts affording access to the

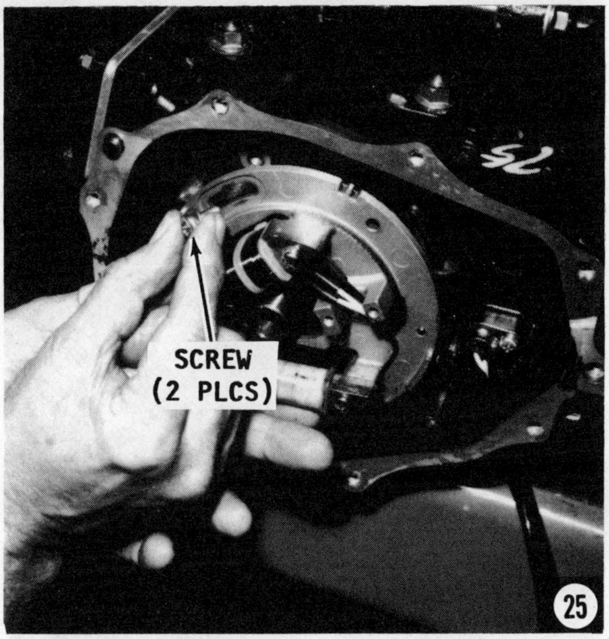

two magneto assembly securing screws. Therefore, the engine can be properly timed with the flywheel in place and secured.

650 Series Only

26- Slide the reduction gear over the splined armature shaft of the cranking motor. Apply a coating of multi-purpose water resistant lubricant to the reduction gear.

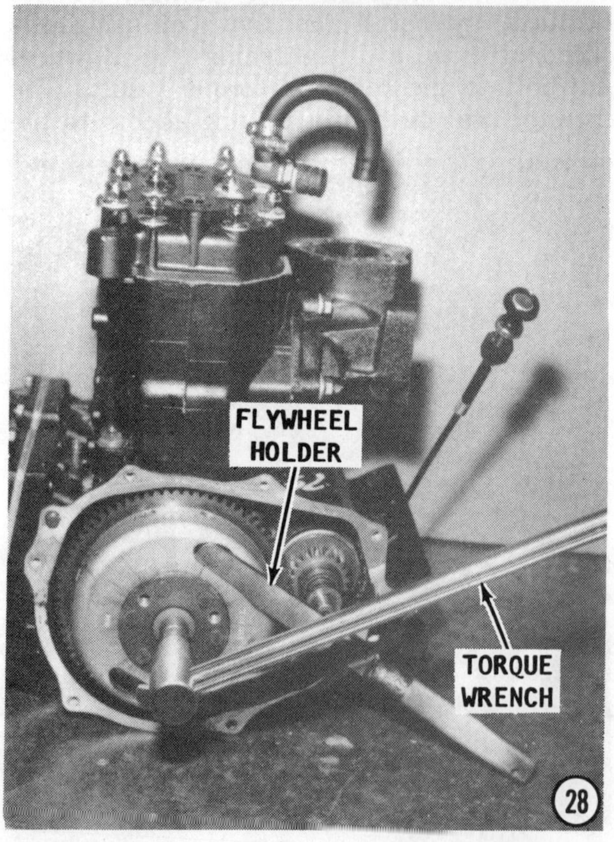

All Series

27- Apply just a "dab" of thick lubricant to the curved surface of the Woodruff key to hold it in place while the flywheel is being installed. Press the Woodruff key into place in the crankshaft recess. Clean any excess lubricant from the shaft to prevent the flywheel from "walking" during engine operation.

Check the flywheel magnets to ensure they are free of any metal particles. Double check the taper in the flywheel hub and the taper in the crankshaft to verify they are clean and contain no oil.

Now, slide the flywheel onto the crankshaft with the keyway in the flywheel aligned with the Woodruff key in place on the crankshaft. Rotate the flywheel **COUNTERCLOCKWISE** and at the same time check to be sure the flywheel does not contact any part of the magneto assembly or any wiring.

28- Thread the flywheel retaining device onto the crankshaft.

300 Series -- bolt and washer.
Late 300 Series -- flanged bolt.
440 Series -- tabbed lockwasher and nut.
550 Series -- tabbed lockwasher and nut.
650 Series -- regular washer and nut.

Use the special flywheel holding tool, or a spanner wrench, or the home made tool to prevent the flywheel from rotating while the flywheel nut/bolt is tightened to the following torque value for the series being serviced.

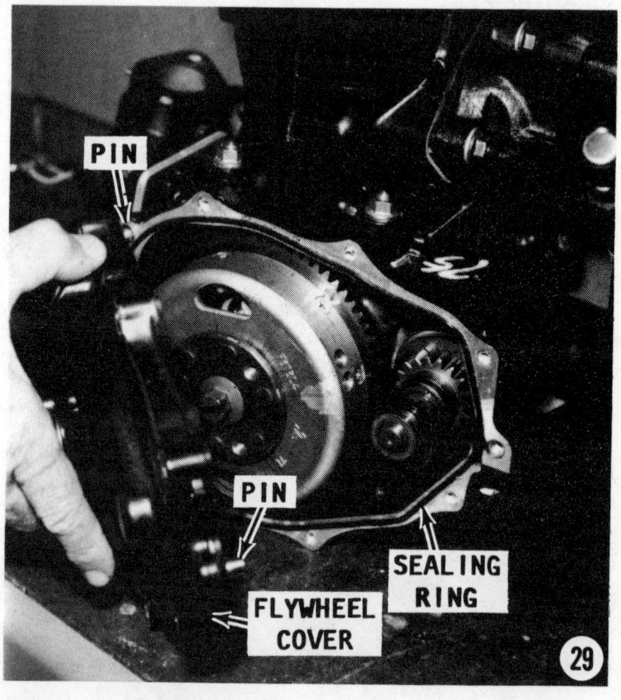

300 Series -- 29 ft lb (39Nm)
440 Series -- 115 ft lb (160Nm)
550 Series -- 115 ft lb (160Nm)
650 Series -- 72 ft lb (98Nm)

If servicing a 440 or 550 Series engine, use a hammer and punch to pry up the tabs on the tabbed lockwasher against the matching faces of the flywheel nut. If no tabs align, back off the nut slightly until the tabs align with a nut face.

29- Apply a coating of multi-purpose water resistant lubricant to a **NEW** sealing ring. Position the ring over the rim of the stator plate. Install the flywheel cover over the stator plate with the two pins -- one in the cover and the other in the plate indexed into their respective holes.

30- Apply a coating of Loctite Lock N' Seal to the threads of the attaching bolts, and then tighten the bolts in the sequence shown to a torque value of 12 ft lb (16Nm).

31- Install and tighten the spark plug/s to a torque value of 14 ft lb (20Nm). Secure the spark plug high tension leads/s to the spark plug/s.

Cranking Motor Installation
300 Series Engines

32- Clean the contact area between the cranking motor lugs and the flywheel cover

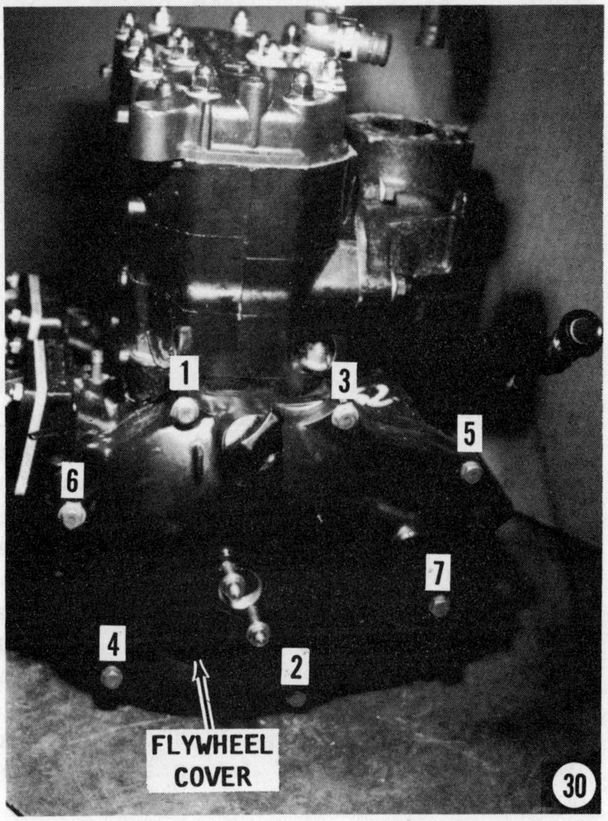

30

31

-- where the motor is grounded. Apply a coating of engine oil to the O-ring on the armature shaft. Insert the cranking motor into the rear gear housing. Check to be sure the splines on the armature index with the idle gear.

Install the two cranking motor mounting bolts and the battery ground cable onto the flywheel cover, as shown. Tighten the bolts securely.

Cranking Motor Installation
440 Series and 550 Series

33- Apply a coating of Loctite Superflex to the O-ring around the forward end cap. Apply engine oil to the pinion gear. Position the motor from rear of the engine forward against the flywheel housing. Apply a coating of Loctite Lock N' Seal to the threads of all four mounting bolts. Install the two

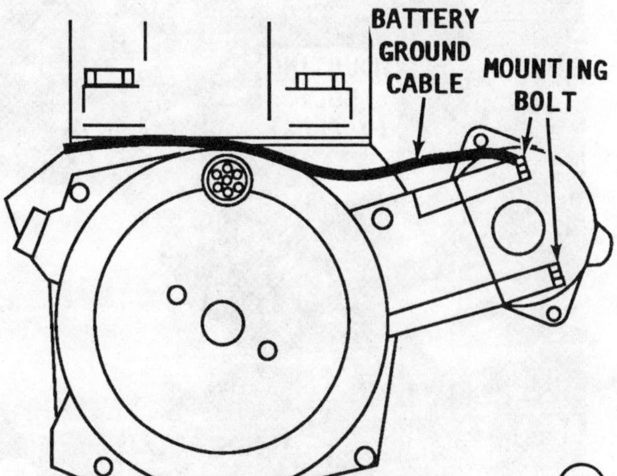

32

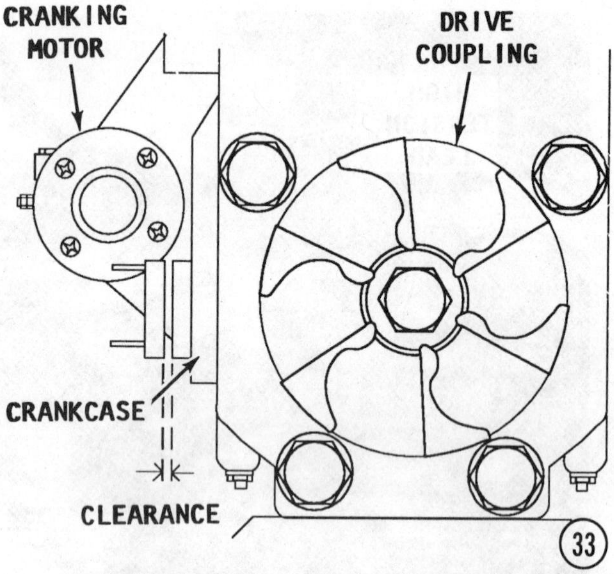

33

34

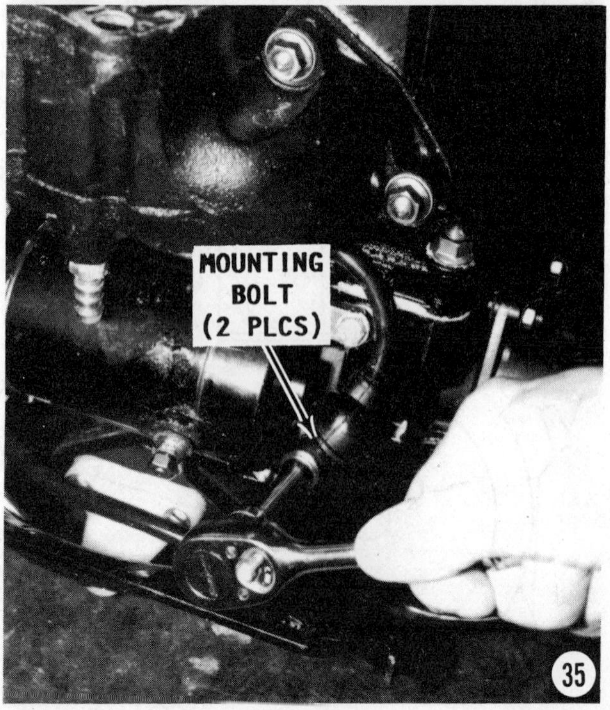

35

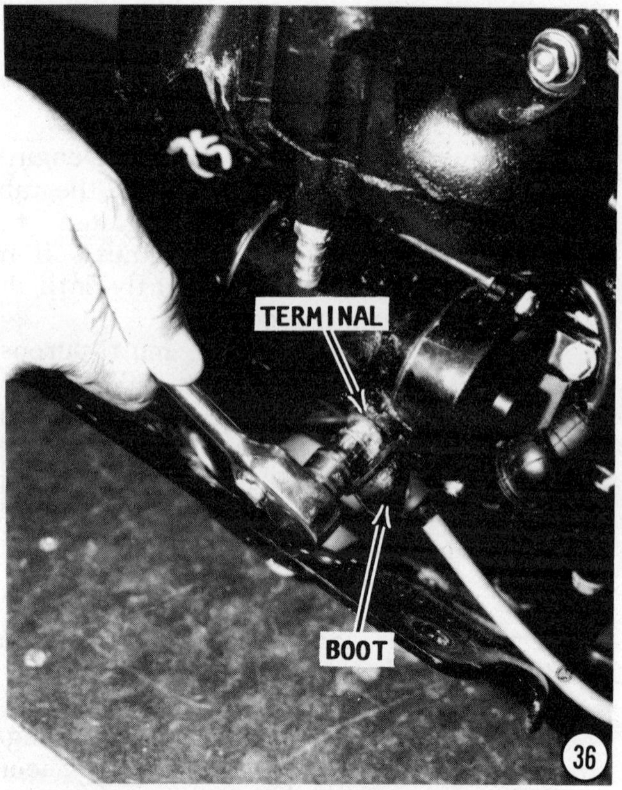

36

forward mounting bolts and tighten them to a torque value of 12 ft lb (16Nm).

Before installing the two aft mounting bolts, measure the distance between the rear mounting bracket and the crankcase. Fill this space with the original pieces of shim material or purchased pieces. Shim material is available in thicknesses of 0.4, 0.6, and 0.8mm. Install the two aft mounting bolts and tighten them to a torque value of 52 in lb (6Nm).

Cranking Motor Installation
650 Series

34- Clean the cranking motor lugs and the contact surface of the crankcase in the area where the motor is grounded. Apply a coating of engine oil to the large oil ring on the forward end cap. Position the motor against the crankcase and at the same time, insert the end of the armature shaft into the reduction gear inside the flywheel cover.

35- If the battery ground cable was originally connected to the crankcase with the lower mounting bolt, place the eye of the cable over the bolt and apply a non-permanent locking compound on both mounting bolts. Install and tighten the bolts securely.

36- Secure the large Red cable to the terminal on the motor frame and pull the boot over the connection.

ENGINE INSTALLATION

1- Insert the rubber coupler into the metal coupler on the jet pump driveshaft. Enlist the aid of an assistant and move the engine into the craft slightly forward of the engine mounts. Lower the engine slowly onto the rubber dampers and at the same time move the engine aft to engage the engine coupler with the rubber coupler on the jet pump driveshaft.

2- Apply a coating of Loctite #242 to the four engine mounting bolts. Temporarily thread the bolts -- just a few turns -- into the fixed engine rubber dampers.

SPECIAL WORDS

During disassembling, the instructions called for the shim material under each engine mount to be saved and identified to assist in engine alignment. If these groups of shim material are installed back in their original location, engine alignment should require very little adjustment.

Install the identified groups of shim material back into their original locations from which they were removed -- between the fixed engine rubber dampers and the engine bed bracket.

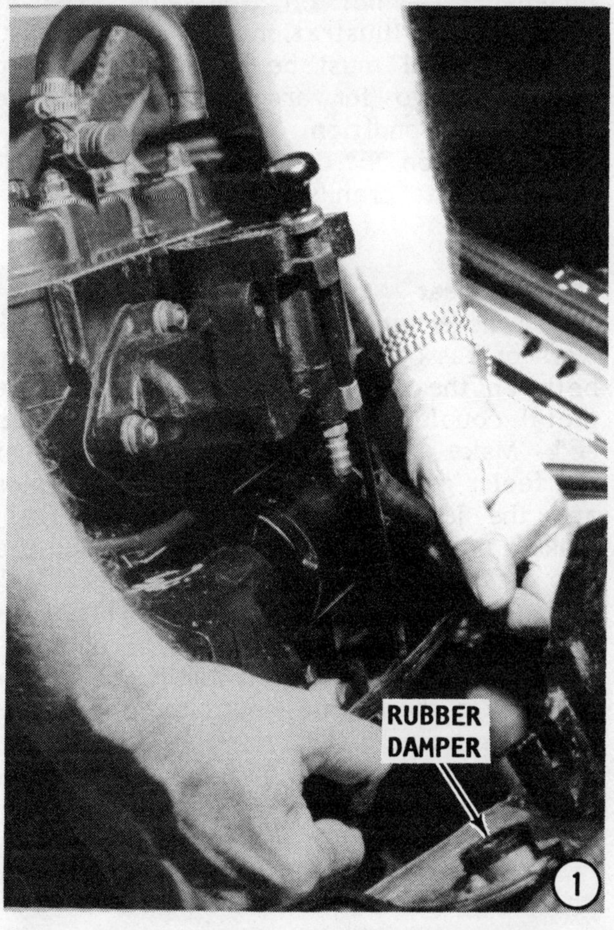

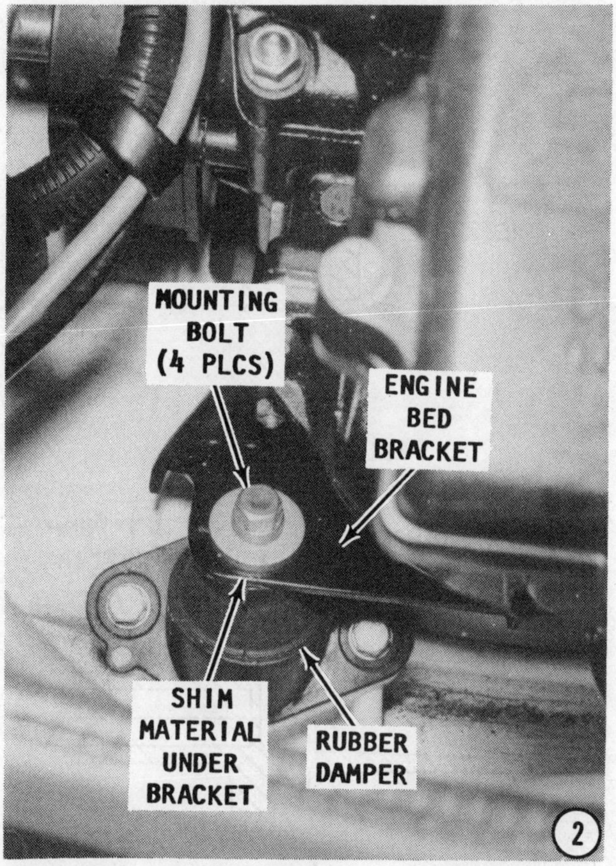

Engine Alignment

3- Several cross-section line drawings -- identified as "A" thru "E", and an exterior view "F", -- are included with the alignment procedures to indicate where measurements are to be taken; when proper alignment has been achieved; and when an adjustment is required.

Begin by placing a short straightedge across the two metal couplers, as indicated in illustration "A". Attempt to insert a

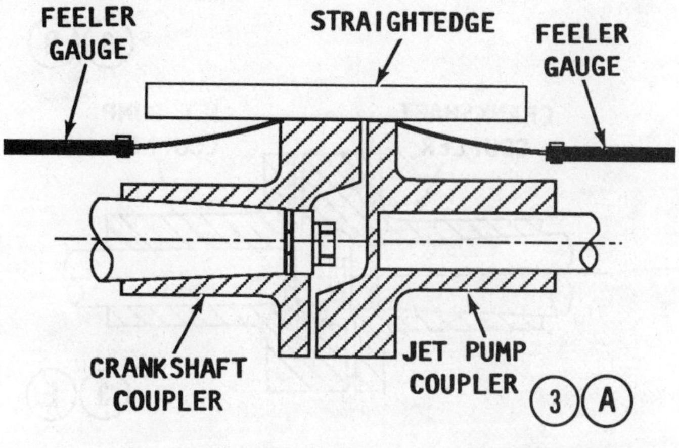

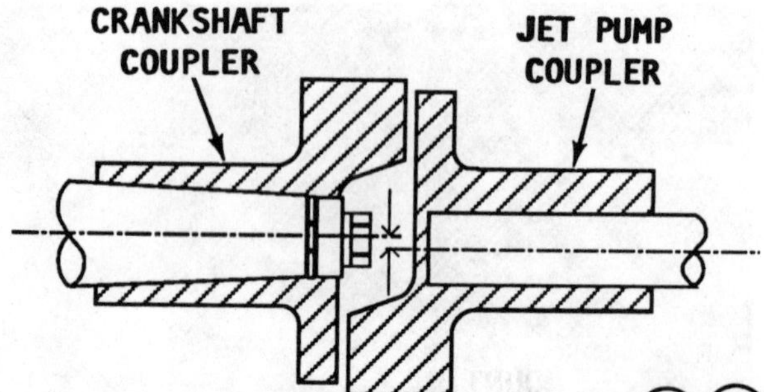

3 B

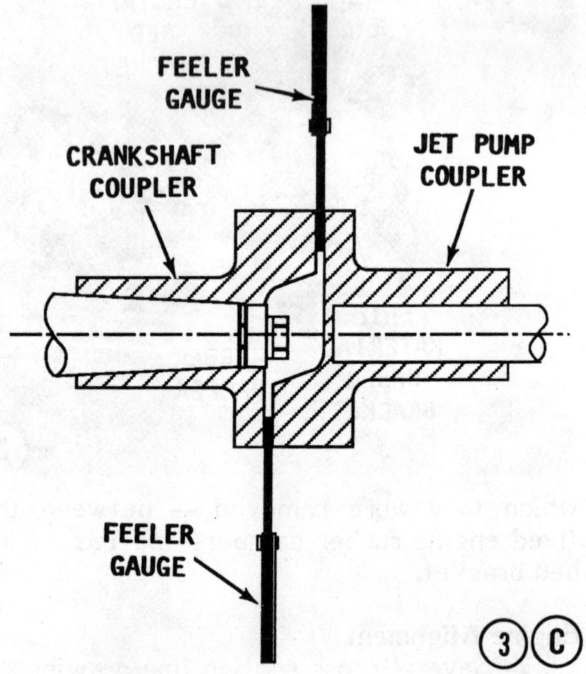

3 C

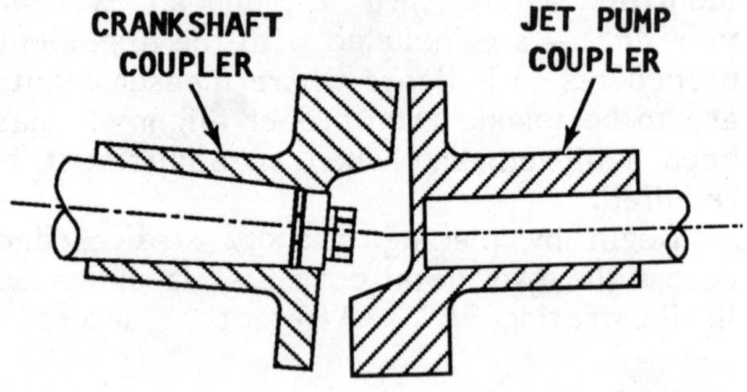

3 D

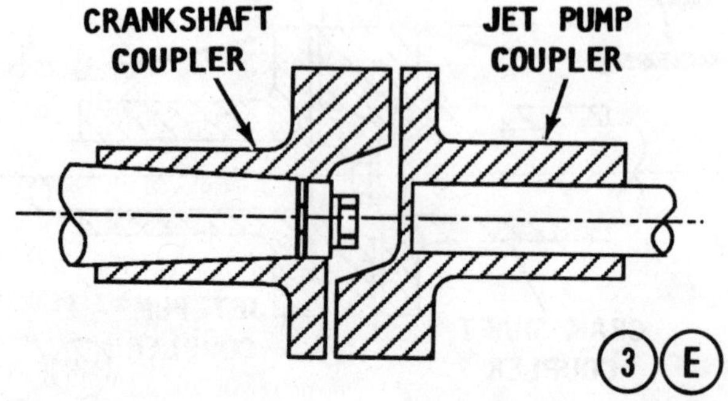

3 E

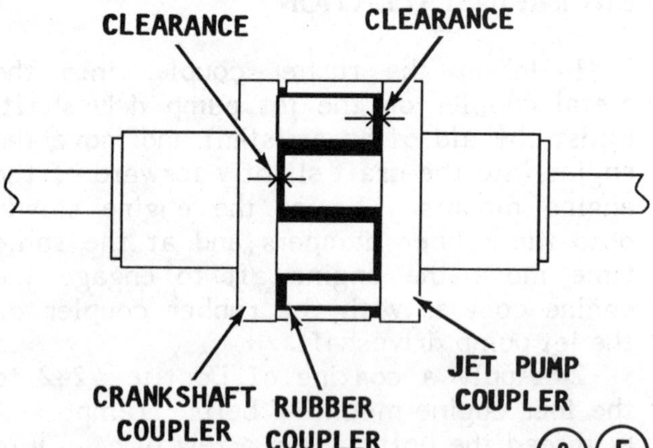

3 F

feeler gauge between the straightedge and the coupler on both sides. If the clearance exceeds 0.012" (0.3mm), an adjustment must be made to all four engine mounts to raise or lower the crankshaft. An extreme case is depicted in illustration "B".

Next, insert a feeler gauge between the flanges of both couplers, as indicated in illustration "C". Measure the clearance -- top and bottom --180° opposite. If the difference between the two measurements is more than 0.024" (0.6mm), the shafts are misaligned -- not on the same axis, as indicated in illustration "D". Therefore, shim material must be added or removed from the two forward engine mounts to correct the condition.

Illustration "E" depicts proper alignment of the engine crankshaft and the jet pump driveshaft.

Coupler Wear

To determine if the rubber coupler is worn or has hardened, insert a feeler gauge between the rubber coupler and the engine metal coupler, as indicated in illustration "F". Make a note of the clearance. Insert the feeler gauge between the rubber coupler and the jet pump metal coupler. Again make a note of the clearance. Now, add the two clearances. If the total exceeds 0.019" (0.5mm), the rubber coupler **MUST** be replaced.

Once proper alignment has been achieved, tighten the engine mounting bolts to the following torque value.

300 Series	27 ft lb (37Nm)
440 Series	27 ft lb (37Nm)
550 Series	27 ft lb (37Nm)
650 Series	16 ft lb (22Nm)

4- Position a new gasket on the exhaust manifold, and then ease the exhaust elbow into place.

5- Install and tighten the securing bolts to the following torque value.

300 Series	69 in lb (7.8Nm)
440 Series	12 ft lb (16Nm)
550 Series	12 ft lb (16Nm)
650 Series	36 ft lb (49Nm)

Install the water valve cable to the bracket on the exhaust elbow flange. Install the bolt securing the expansion chamber to the exhaust manifold. Tighten the bolt to a torque value of 14.5 ft lb (20Nm).

6- Install the bypass hose onto the fitting on the exhaust elbow.

Fuel Tank Installation

7- Check to be sure the mat is in position in the hull.

8- Lower the fuel tank into place and push the filler tube onto the fuel tank neck. Secure the tube to the neck with two hose clamps.

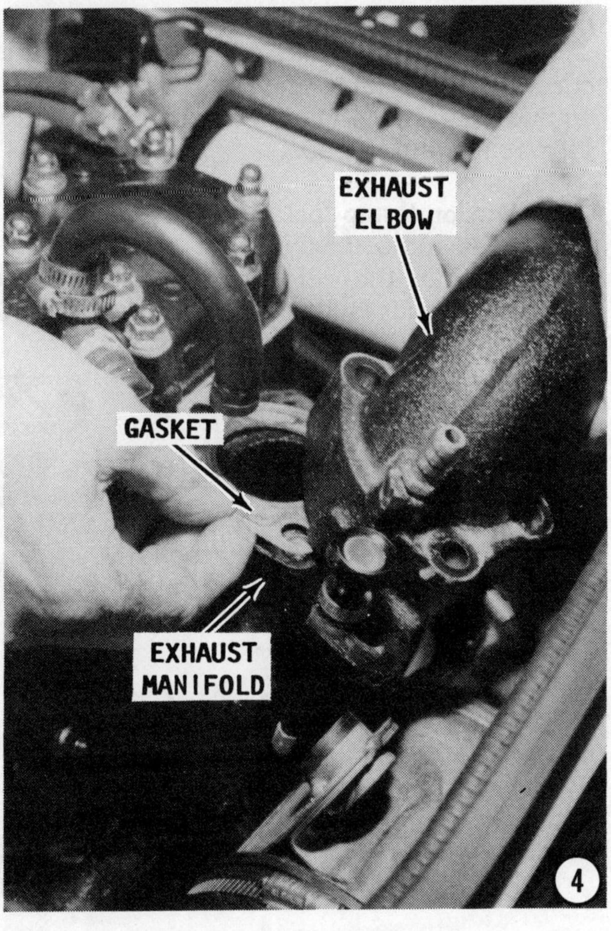

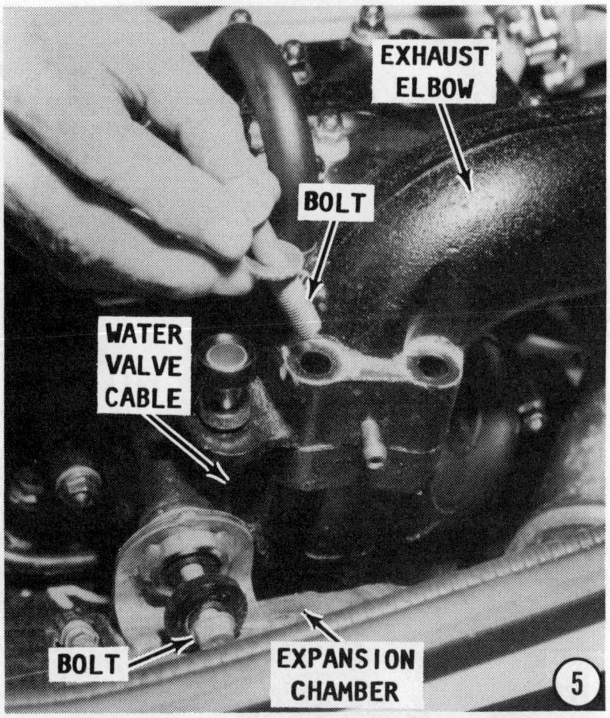

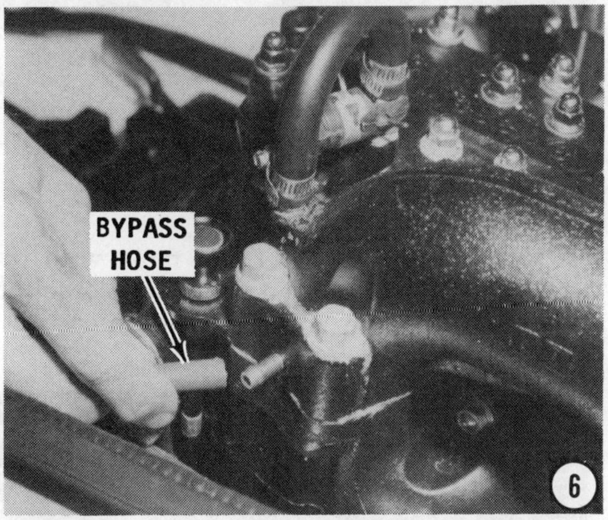

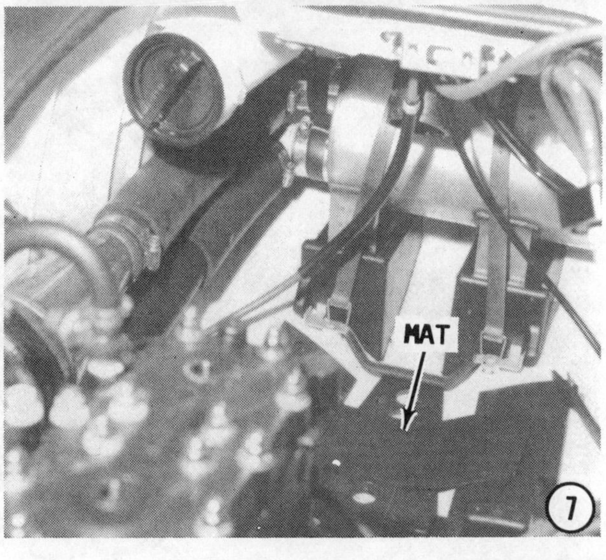

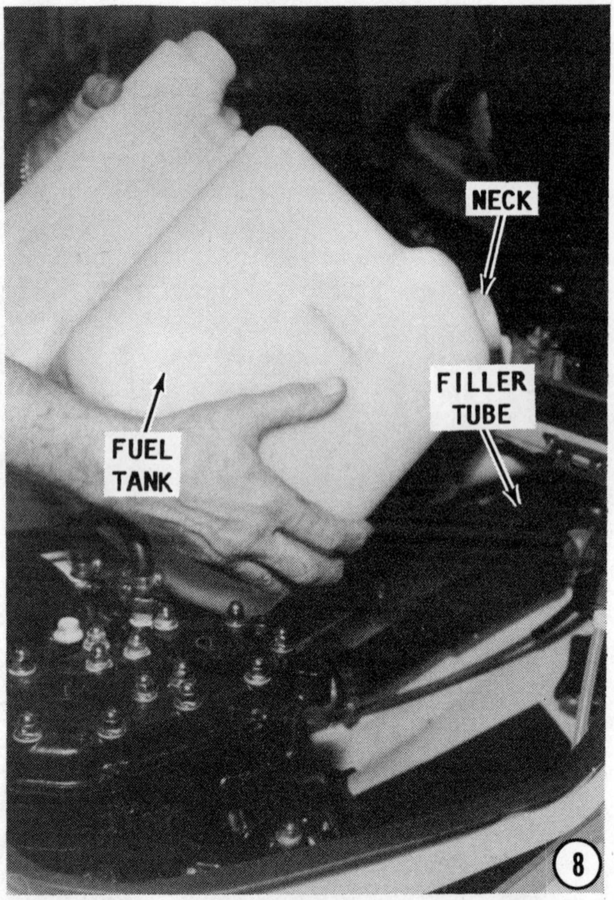

8

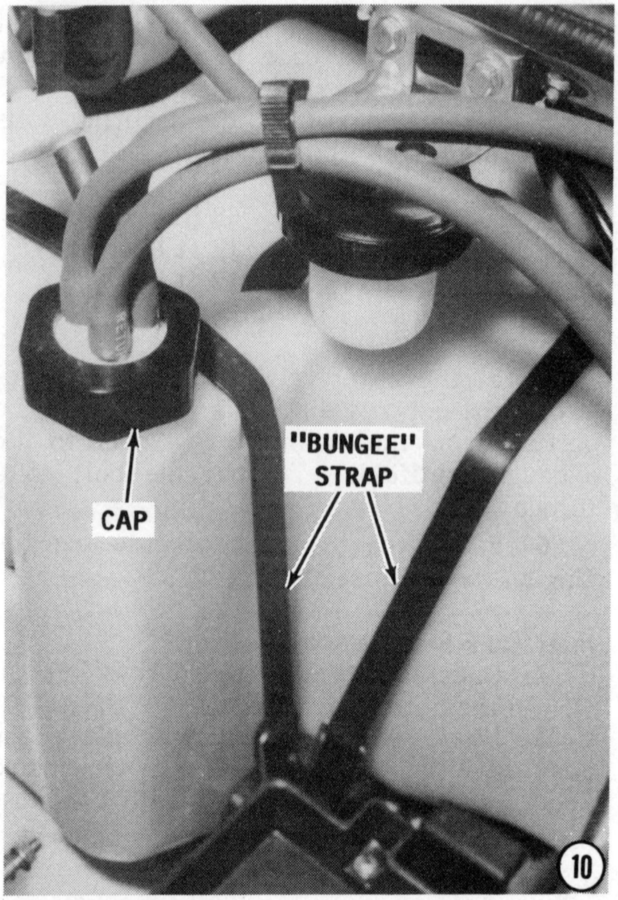

10

Oil Injection Tank Installation
300 Series and 650 Series

9- Move the oil injection tank, with the breather hose attached into the bracket. Check to be sure the rubber cushions surround the base of the tank.

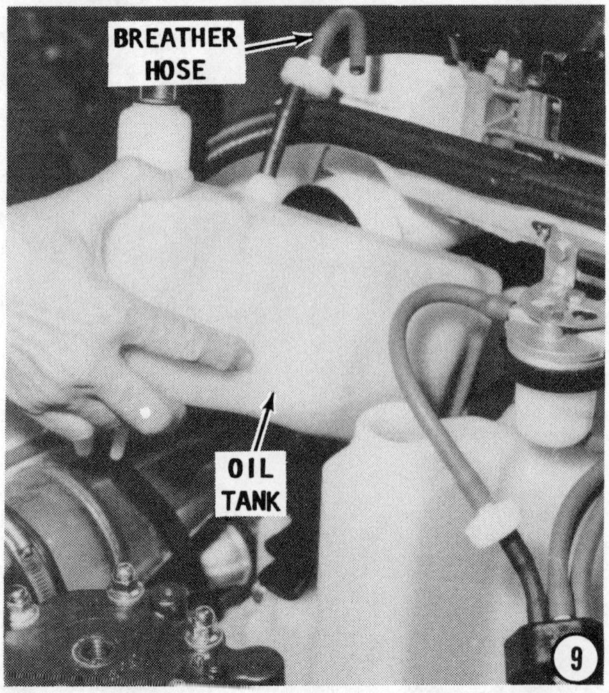

9

All Series Engines

10- Install the fuel tank filler cap. Secure the fuel tank and the oil tank -- if equipped -- with the two rubber "bungee" type restraining straps.

Oil Injection Pump Installation
300 Series and 650 Series

11- Slide the oil pump onto the two mounting studs with the oil pump shaft indexing into the crankshaft slot.

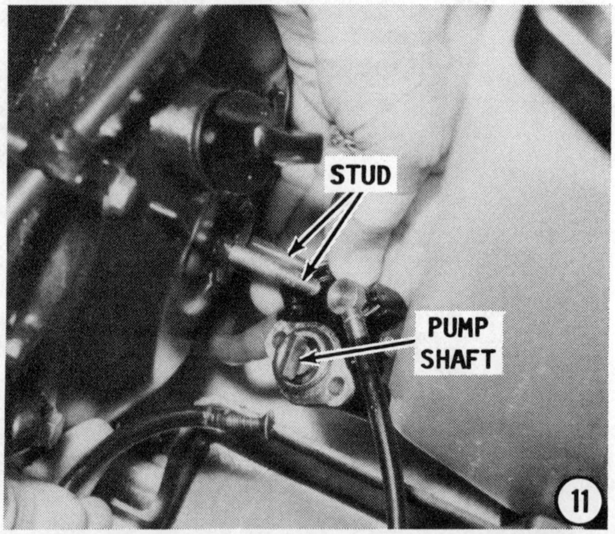

11

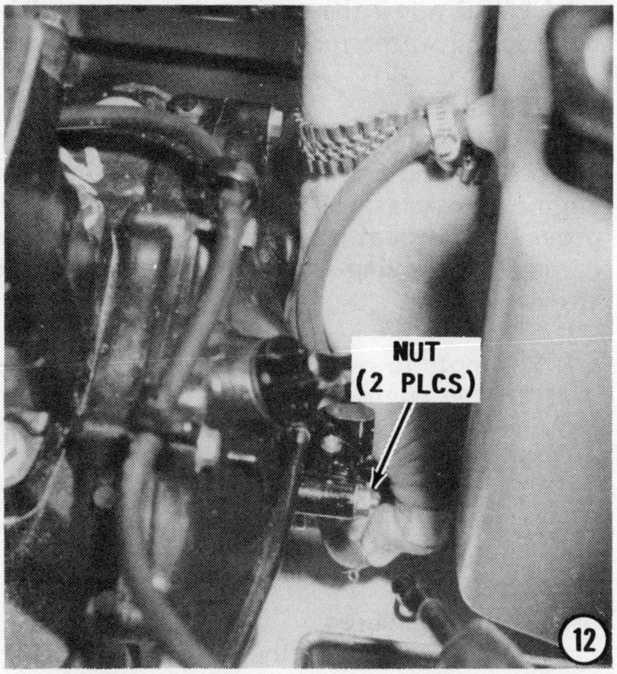

12- Secure the pump to the flywheel cover with the two nuts and washers. Tighten the nuts securely. Install the oil delivery hose to the fitting on the intake manifold.

Remove the screw from the end of the oil supply hose. Connect the hose to the inlet fitting on the pump. After all work has been completed and the engine is started, the oil injection **MUST** be "bled" -- purged of air -- according to the procedures outlined in Section 6-10, beginning on Page 6-49.

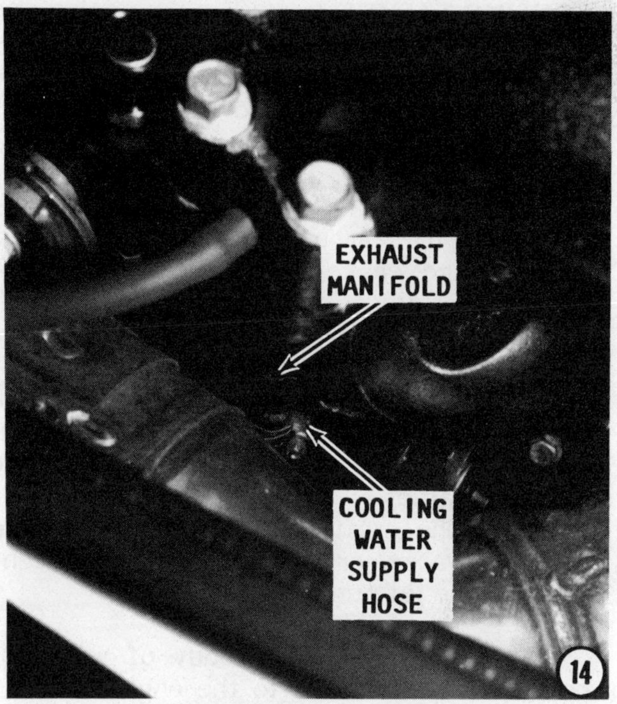

All Series Engines

13- Install the fuel pump pulse hose to the fitting on the intake manifold.

14- Secure the cooling water supply hose to the fitting on the underneath side of the exhaust manifold.

15- Secure the electrical leads to the engine block with the pliable metal clamps.

16- Tighten the hose clamps at the forward end of the expansion chamber. Tighten any and all other connections which may have been loosened.

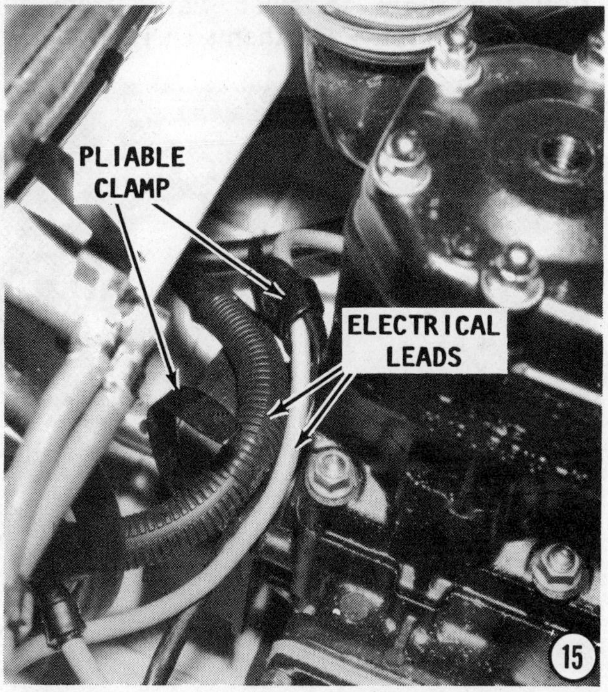

CLOSING TASKS

Position the cover over the coupler. Coat the threads of the attaching bolts with Loctite #242. Install and tighten the bolts securely.

All Series Engines

Install the carburetor, associated hoses, throttle and choke cables, see Chapter 6. Install the electrical box, see Chapter 7. Connect the stator harness lead at the electrical box. The oil injection system **MUST** be bled while operating the engine on a premix as directed in Section 6-10, beginning on Page 6-49. Connect the positive and negative cables to the battery.

Move the craft to a test tank, or to a body of water, or connect a flush attachment to the jet pump.

Move the craft to a body of water or connect a garden hose to the engine cooling water supply fitting or flush fitting on the cylinder head.

If a garden hose is used, start the engine and allow the rpm's to stabilize at idle speed **FOR JUST A FEW SECONDS**, and then turn the water on.

Adjust the water flow until a small trickle is discharged from the bypass outlet on the port side of the hull.

When the engine is to be shut down, turn the water off **FIRST** -- raise the aft portion of the hull -- **WHILE THE ENGINE IS OPERATING AT IDLE** -- "rev" the engine just a **COUPLE** times to clear water from the exhaust system -- and then shut it down.

NEVER allow the engine to operate without cooling water for more than 15 seconds.

CAUTION

Water must circulate through the jet pump -- to and from the engine, anytime the engine is operating. Circulating water will prevent overheating -- which could cause damage to moving engine parts and possible engine seizure.

Attempt to start and run the engine without the engine cover in place. This will provide the opportunity to check for water or fuel leaks.

After the engine is operating properly, install the engine cover. Follow the break-in procedures with a load on the jet pump.

Break-in Procedures

As soon as the engine starts, **CHECK** to be sure a fine stream of water is discharged from the exhaust relief hole.

DO NOT operate the engine at full throttle except for **VERY** short periods, until after 10 hours of operation as follows:

a- Operate at 1/2 throttle, approximately 2500 to 3500 rpm, for 2 hours.

b- Operate at any speed after 2 hours **BUT NOT** at sustained full throttle until another 8 hours of operation.

c- Mix gasoline and oil during the break-in period, total of 10 hours, at a ratio of 25:1.

d- While the engine is operating during the initial period, check the fuel, exhaust, and water systems for leaks.

8-5 CLEANING AND INSPECTING ALL ENGINES

The success of the overhaul work is largely dependent on how well the cleaning and inspecting tasks are completed. If some parts are not thoroughly cleaned, or if an unsatisfactory unit is allowed to be returned to service through negligent inspection, the time and expense involved in the work will not be justified with peak engine performance and long operating life. Therefore, the procedures in the following sections should be followed closely and the work performed with patience and attention to detail.

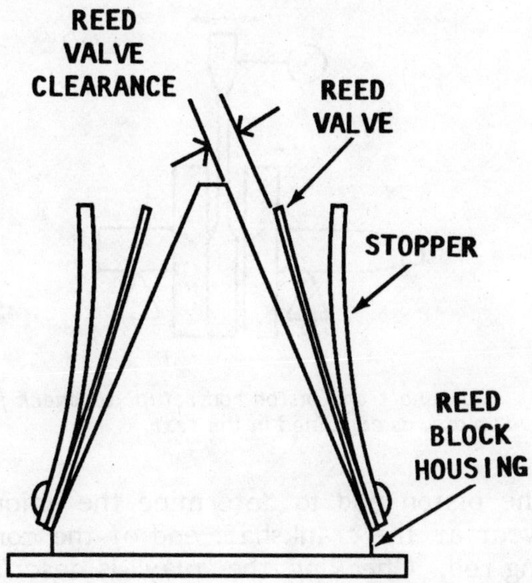

The maximum reed valve clearance is 0.008" (0.2mm), for all engines equipped with reed valves.

Reed Block Service
300 Series and 650 Series

Disassemble the reed block housing by first removing the screws securing the reed stoppers and reed petals to the housing. After the screws are removed, lift the stoppers and petals from the housing.

Clean the gasket surfaces of the housing. Check the surfaces for deep grooves, cracks or any distortion which could cause leakage.

Replace the reed block housing if it is damaged. The reed petals should fit flush against their seats, and not be preloaded against their seats or bend away from their seats.

Petal Clearance

The maximum petal clearance for all 300 Series and 650 Series engines is 0.008" (0.2mm).

If the reed petal is distorted, rarely, if ever, can the petal be successfully straightened. **THEREFORE,** it must be replaced.

Do not remove the reed valves unless they, or the stoppers, are to be replaced. The reeds **MUST** be replaced in sets.

Apply Loctite to the threads of the reed retaining screws. Tighten each screw gradually, starting in the center and working outwards across the reed block. Tighten the screws to a torque value of 3.5 ft lbs (5Nm).

TAKE CARE not to mar the special rubber coating on the reed block housing.

Crankshaft Service

Clean the crankshaft with solvent and wipe the journals dry with a lint free cloth. Inspect the main journals and connecting rod journals for cracks, scratches, grooves, or scores. Inspect the crankshaft oil seal surface for nicks, sharp edges or burrs which might damage the oil seal during installation or might cause premature seal wear. **ALWAYS** handle the crankshaft carefully to avoid damaging the highly finished journal surfaces. Blow out all oil passages with compressed air. The oil passageway leads from the rod to the main bearing journal. **TAKE CARE** not to blow dirt into the main bearing journal bore.

Inspect the threads at both ends for signs of abnormal wear. Check the crankshaft for runout by supporting it on two **"V"** blocks at the main bearing surfaces.

Install a dial indicator gauge above the main bearing journals. Rotate the crankshaft and measure the runout (or the out-of-round) and the taper at both ends (and in the center journal on all two-cylinder models).

The following crankshaft runout limits are recommended by the manufacturer.

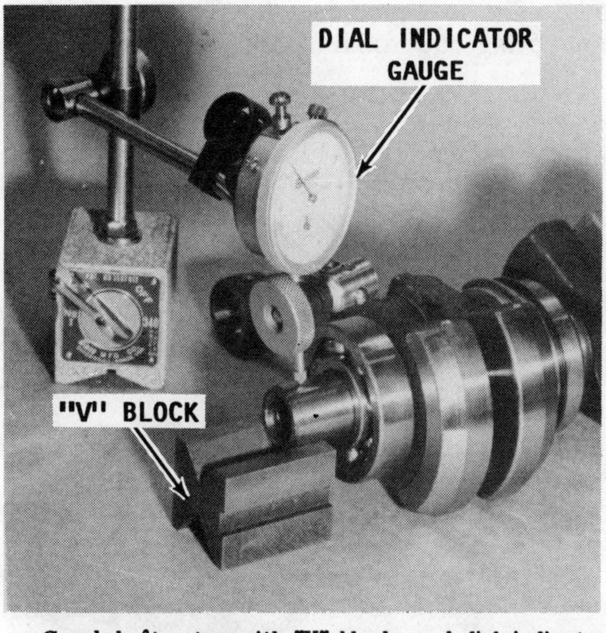

Crankshaft setup with "V" blocks and dial indicator to measure main bearing journals on an assembled two-stroke crankshaft. The ball bearings and connecting rods will not affect the readings.

CRANKSHAFT RUNOUT LIMITS

Series	Standard	Service Limit
300	0.0012" (0.03mm)	0.0031" (0.08mm)
440	0.0019" (0.05mm)	0.0031" (0.08mm)
550	0.0019" (0.05mm)	0.0031" (0.08mm)
650	0.0015" (0.04mm)	0.0039" (0.10mm)

If **"V"** blocks or a dial indicator are not available, a micrometer may be used to measure the diameter of the journal. Make a second measurement at right angles to the first. Check the difference between the first and second measurement for out-of-round condition. If the journals are tapered, ridged, or out-of-round by more than the specification allows, the journals should be reground, or the crankshaft replaced.

Any out-of-round or taper shortens bearing life. Good shop practice dictates new main bearings be installed with a new or reground crankshaft.

Normally the connecting rods would **ONLY** be pressed from the crankshaft if either the crankshaft and/or the connecting rods were to be replaced. Therefore, the connecting rod axial "play" is checked at the piston end to determine the amount of wear at the crankshaft end of the connecting rod. Checking the "play" is described in the following paragraph.

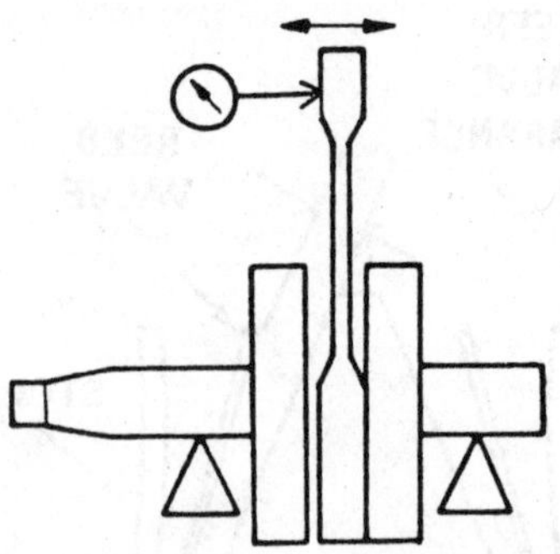

Crankshaft and piston rod setup to check for axial rod "play", as explained in the text.

Setup the crankshaft in the **"V"** blocks. Setup a dial indicator to touch the flat surface of the piston end of the rod. Now, hold the crankshaft steady in the **"V"** blocks and at the same time, rock the piston end of the rod along the same axis as the crankshaft. If the dial indicator needle moves through more than 0.08" (2mm) for all engines covered in this manual, the "play" is considered excessive. The rod must be pressed from the crankshaft, the journal

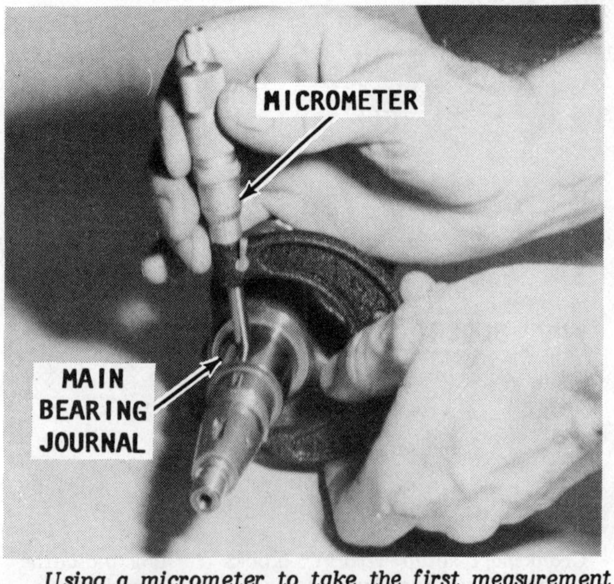

Using a micrometer to take the first measurement on a crankshaft main journal for out-of-round and taper.

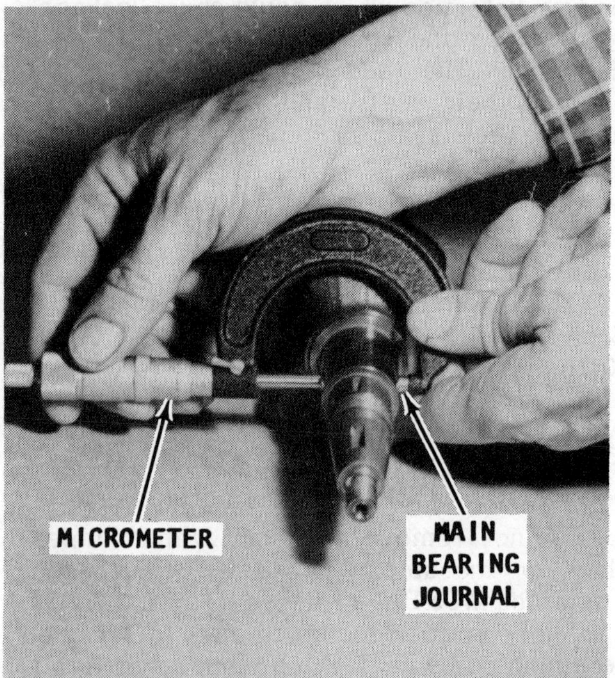

Taking a second measurement on a crankshaft main bearing journal at 90° to the first measurement.

checked and a determination made as to whether the crankshaft and/or the rod must be replaced. A new rod may be purchased, **BUT** it must be pressed onto the crankshaft throw with a hydraulic press.

To check the connecting rod side clearance at the crankshaft, first insert a feeler gauge between the connecting rod and the counterweight of the crankshaft. Acceptable clearances are as follows:

CONNECTING ROD CLEARANCE

Series	Standard	Service Limit
300	0.0177-0.0216" (0.45-0.55mm)	0.0314" (0.8mm)
440	0.0157-0.0196" (0.40-0.50mm)	0.0275" (0.7mm)
550	0.0157-0.0196" (0.40-0.50mm)	0.0275" (0.7mm)
650	0.0177-0.0216" (0.45-0.55mm)	0.0314" (0.8mm)

Inspect the crankshaft oil seal surfaces to be sure they are not grooved, pitted, or scratched. Replace the crankshaft if it is severely damaged or worn. Check all crankshaft bearing surfaces for rust, water marks, chatter marks, uneven wear or overheating. Clean the crankshaft surfaces with 320-grit carborundum cloth. **NEVER** spin-dry a crankshaft ball bearing with compressed air.

Clean the crankshaft and crankshaft ball bearing with solvent. Dry the parts, but not the ball bearing, with compressed air. Check the crankshaft surfaces a second time. Replace the crankshaft if the surfaces cannot be cleaned properly for satisfactory service. If the crankshaft is to be installed for service, lubricate the surfaces with light oil. **DO NOT** lubricate the crankshaft ball bearing at this time.

After the crankshaft has been cleaned, grasp the outer race of the crankshaft ball bearing installed on the lower end of the crankshaft, and attempt to work the race back-and-forth. There should not be excessive "play". A very slight amount of side "play" is acceptable because there is only about 0.001" (.025mm) clearance in the bearing.

Lubricate the ball bearing with light oil. Check the action of the bearing by rotating the outer bearing race. The bearing should have a smooth action and no rust stains. If the ball bearing sounds or feels rough or catches, the bearing should be removed and discarded.

Connecting Rod Service

Inspect the connecting rod bearings for rust or signs of bearing failure. **NEVER** intermix new and used bearings. If even one bearing in a set needs to be replaced, all bearings at that location **MUST** be replaced.

Clean the inside diameter of the piston pin end of the connecting rod with crocus cloth.

Clean the connecting rod **ONLY** enough to remove marks. **DO NOT** continue, once the marks have disappeared.

Assemble the piston end of the connecting rod with loose needle bearings, caged needle bearings, or no needle bearing, depending on the model being serviced. Insert the piston pin and check for vertical "play". The piston pin should have **NO** noticeable vertical "play".

If the pin is loose or there is vertical "play" check for and replace the worn part/s.

Inspect the piston pin and matching rod end for signs of heat discoloration. Overheating is identified as a bluish bearing surface color and is caused by inadequate lubrication or by operating the engine at excessive high rpm.

Piston Service

Inspect each piston for evidence of scoring, cracks, metal damage, cracked piston pin boss, or worn pin boss. Be especially critical during inspection if the craft and

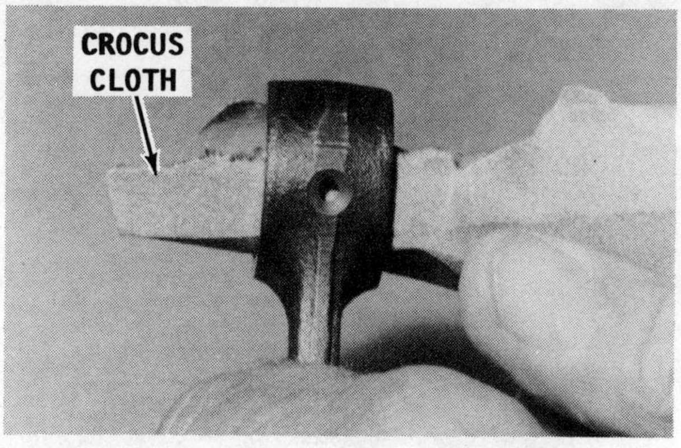

Cleaning the piston end of a connecting rod with crocus cloth.

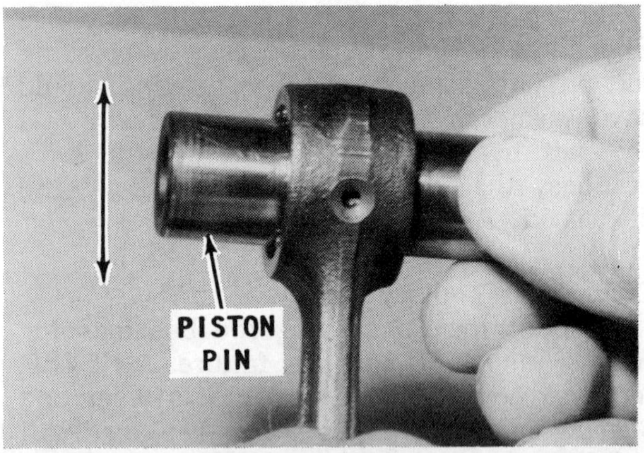

Checking the piston end of a connecting rod for vertical free "play" using a piston pin.

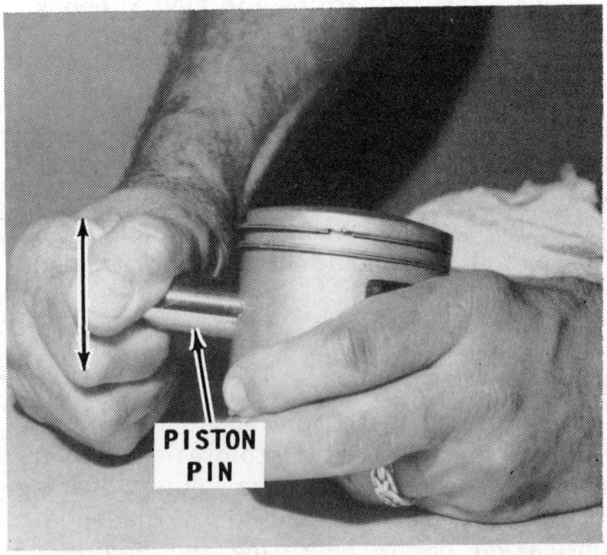

*Checking free "play" between the piston pin and the piston boss. There should be **NO** "play".*

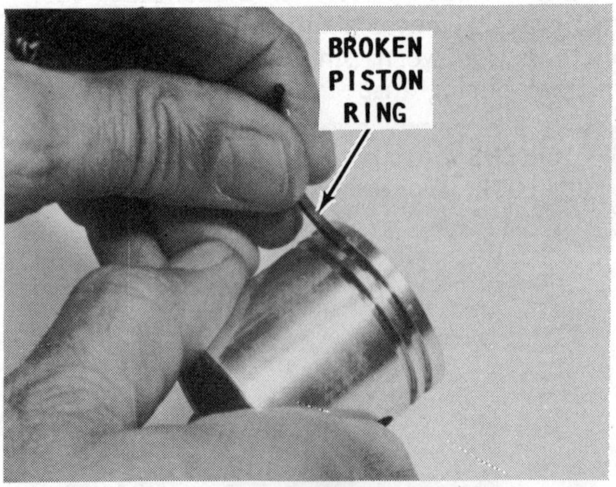

*Using part of a broken piston ring to clean the ring groove. Exercise **CARE** not to disturb the pin in each groove.*

It is believed, this crown seized with the cylinder wall when the unit was operated at high rpm and the timing was not adjusted properly. At the same instant, the rod apparently pulled the lower part of the piston downward, severing it from the crown.

engine has been submerged. If the piston pin is bent, the piston and pin **MUST** be replaced as a set for two reasons. First, a bent pin will damage the boss when it is removed. Secondly, a piston pin is not sold as a separate item.

Check the piston ring grooves for wear, burns distortion, or loose locating pins. During an overhaul, the rings should be replaced to ensure lasting repair and proper powerhead performance after the work is completed. Clean the piston dome, ring grooves and the piston skirt. Clean carbon deposits from the ring grooves using the recessed end of a broken piston ring.

NEVER use a rectangular ring to clean the groove for a tapered ring, or use a tapered ring to clean the groove for a rectangular ring.

Locating pins prevent the rings from rotating and catching on a port in the cylinder wall.

The pitted damage to this piston crown was probably caused by a broken piston ring working its way into the combustion chamber. The little "hills" then became "hot" spots on the crown, contributing to "dieseling" after the engine was shut down.

NEVER use an automotive-type ring groove cleaner, because such a tool may loosen the piston ring locating pins.

Clean carbon deposits from the top of the piston using a soft wire brush, carbon removal solution or by sand blasting. If a wire brush is used, **TAKE CARE** not to burr or round machined edges. Clean the piston skirt with crocus cloth.

The rings on this piston became stuck due to lack of adequate lubrication, incorrect timing, or overheating.

The rings on this piston were broken, possibly during installation, and then caused extensive damage to the piston. Ring parts found their way into the combustion chamber and caused damage to the piston crown.

Install the piston pin through the first boss only. Check for vertical free "play". There should be **NO** vertical free "play". The presence of "play" is an indication the piston boss is worn. The piston is manufactured from a softer material than the piston pin. Therefore, the piston boss will wear more quickly than the pin.

Excessive piston skirt wear **CANNOT** be visually detected. Therefore, good shop practice dictates, the piston skirt diameter be measured with a micrometer.

Piston diameters are are measured at a definite distance up from the bottom of the skirt at right angles to the piston pin axis. Size limits are as follows:

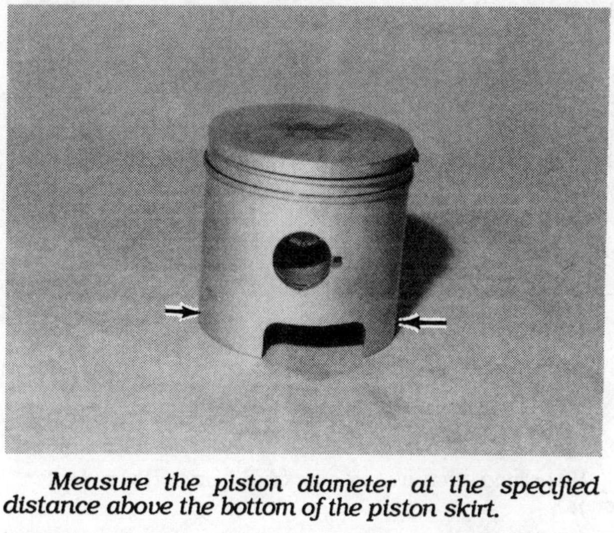

Measure the piston diameter at the specified distance above the bottom of the piston skirt.

STOCK PISTON DIAMETER

Series	Standard	Service Limit
300 3/4" Up	2.9906-2.9912" (75.96-75.97mm)	2.9846" (75.81mm)
440 3/16" Up	2.6689-2.6764" (67.79-67.98mm)	2.6689" (67.79mm)
550 3/16" Up	2.9496-2.9503" (74.92-74.94mm)	2.9441" (74.78mm)
650 3/4" Up	2.9905-2.9911" (75.96-75.97mm)	2.9846" (75.81mm)

Ring End Gap

Before the piston rings are installed onto the piston, the ring end gap clearance for each ring must be determined. The purpose of the piston rings is to prevent the blowby of gases in the combustion chamber. This cannot be achieved unless the correct oil film thickness is left on the cylinder wall.

This thin coating of oil acts as a seal between the cylinder wall and the face of the piston ring. An excessive end gap will allow blowby and the cylinder will lose compression. An inadequate end gap will scrape too much oil from the cylinder wall and

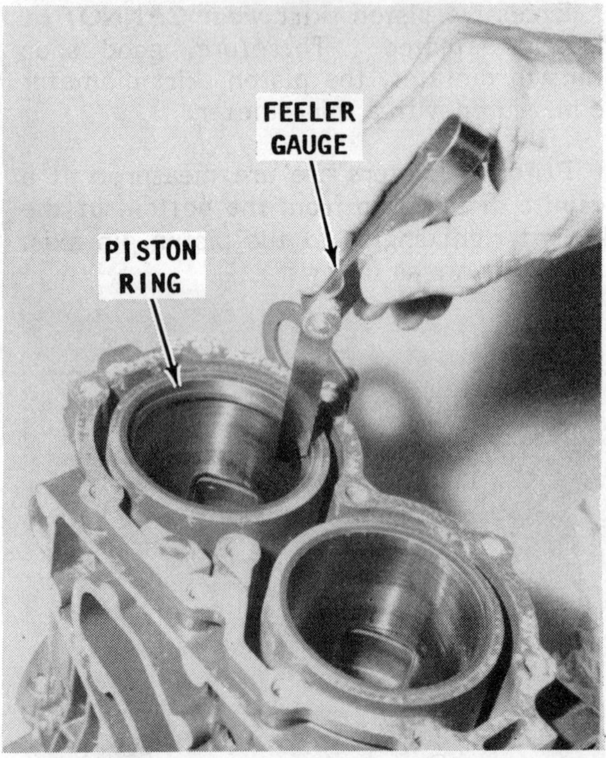

Measuring the piston ring end gap with a feeler gauge.

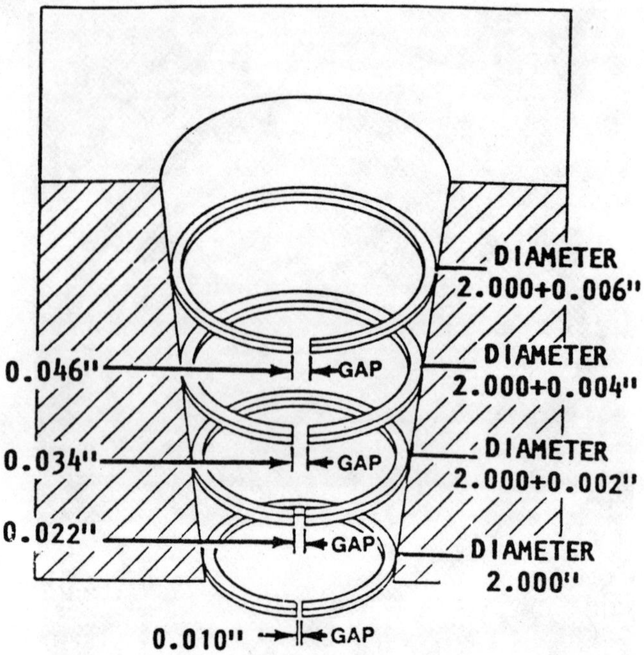

The cylinder taper drastically affects ring and end gap, as shown in this cross-section line drawing.

limit lubrication. Lack of adequate lubrication will cause excessive heat and wear.

IDEALLY the ring end gap measurement should be taken **AFTER** the cylinder bore has been measured for wear and taper **AND** after any corrective work, such as boring or honing, has been completed.

IF the ring end gap is measured with a taper to the cylinder wall, the diameter at the lower limit of ring travel will be smaller than the diameter at the top of the cylinder.

IF the ring is fitted to the upper part of a cylinder with a taper, the ring end gap will not be great enough at the lower limit of ring travel. Such a condition could result in a broken ring and/or damage to the cylinder wall and/or damage to the piston and/or damage to the cylinder head.

IF the cylinder is to be only honed, not bored, **OR** if only cleaned, not honed, the ring end gap should be measured at the lower limit of ring travel.

The ring should be inserted into the cylinder just above the ports to perform this measurement, using the piston crown. Be sure the ring is parallel to the surface of the block, then measure the ring end gap with a feeler gauge.

If the end gap is greater than the amount listed, replace the entire ring set.

If the end gap is less than 0.008-0.016" (0.20-0.40mm) for both top and bottom rings, carefully file the ends of the ring --

just a little at a time -- until the correct end gap is obtained.

Inspect the piston ring locating pins to be sure they are tight. There is one locating pin in each ring groove. If the locating pins are loose, the piston **MUST** be replaced.

Oversized Pistons and Rings

Scored cylinder blocks can be saved for further service by reboring and installing oversize pistons and piston rings. **HOWEVER,** if the scoring is over 0.0075" (0.13mm) deep, the block cannot be effectively rebored for continued use.

Check with the local dealer for oversize piston availability.

If oversize pistons are not available, the local marine shop may have the facilities to "knurl" the piston, making it larger.

Cylinder Block Service

Inspect the cylinder block and cylinder bores for cracks or other damage. Remove carbon with a fine wire brush on a shaft attached to an electric drill or use a carbon remover solution.

STOP: If the cylinder block is to be submerged in a carbon removal solution, the crankcase bleed system **MUST** be removed from the block to prevent damage to hoses and check valves.

Use an inside micrometer or telescopic gauge and micrometer to check the cylinders for wear. Check the bore for out-of-round and/or oversize bore. If the bore is tapered, out-of-round or worn more than the wear limit specified by the manufacturer, the cylinders should be rebored -- provided oversize pistons and rings are available.

Check with the dealer **PRIOR** to reboring. If oversize pistons and matching rings are not available, the block **MUST** be replaced.

GOOD WORDS

Oversize piston weight is approximately the same as a standard size piston. Therefore, it is **NOT** necessary to rebore all cylinders in a block just because one cylinder requires reboring.

Cylinder sleeves are an integral part of the die cast cylinder block and **CANNOT** be replaced. In other words, the cylinder cannot be "resleeved".

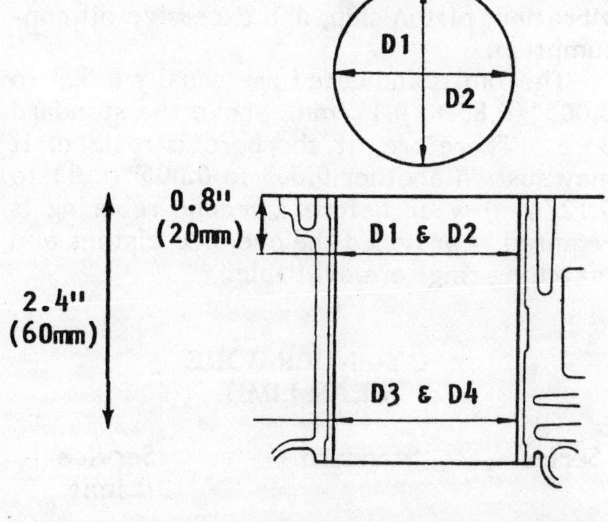

Top view diagram and cross-section of a typical cylinder to indicate where measurements are to be taken for wear limit, taper, and out-of-round limits.

Four inside cylinder bore measurements must be taken for each cylinder to determine an out-of-round condition, the maximum taper, and the maximum bore diameter.

In the accompanying illustration, measurements D1 and D2 are diameters measured at the top of the cylinder at right angles to each other. Measurements D3 and D4 are diameters measured from the top of the cylinder at right angles to each other.

Out-of-Round

Measure the cylinder diameter at D1 and D2. The manufacturer requires the difference between the two measurements should be less than 0.002" (0.050mm) for **ALL** models.

Maximum Taper

Measure the cylinder diameter at D1, D2, D3, and D4. Take the largest of the D1 or D2 measurements and subtract the smallest measurement at D3 or D4. The answer to the subtraction -- the cylinder taper -- should be less than 0.003" (0.08mm) for **ALL** models.

Bore Wear Limit

The maximum cylinder diameter D1, D2, D3, and D4 must not exceed the bore wear limits indicated in the following table **BEFORE** the bore is rebored for the **FIRST** time. These limits are only imposed on original parts because the sealing ability of the rings would be lost, resulting in power loss, increased engine noise, unnecessary vibration, piston slap, and excessive oil consumption.

The limits indicated are usually 0.003 to 0.005" (0.83 to 0.127mm) above the standard bore. Therefore, if the bore is resized, it may sustain another 0.003 to 0.005" (0.83 to 0.127mm) wear before a second reboring is required -- provided the oversize pistons and matching rings are available.

CYLINDER BORE WEAR LIMIT

Series	Standard	Service Limit
300	2.9941-2.9946" (76.05-76.06mm)	2.9961" (76.10mm)
440	2.6802-2.6808" (68.05-68.09mm)	2.7003" (68.59mm)
550	2.9557-2.9564" (75.07-75.09mm)	2.9759" (75.59mm)
650	2.9940-2.9946" (76.05-76.06mm)	2.9960" (76.10mm)

Piston Clearance

Piston clearance is the difference between a maximum piston diameter and a minimum cylinder bore diameter. If this clearance is excessive, the engine will develop the same symptoms as for excessive cylinder bore wear -- loss of ring sealing ability, loss of power, increased engine noise, unnecessary vibration, and excessive oil consumption.

Maximum piston diameter was described earlier in this section. Minimum cylinder bore diameter is usually determined by measurement D3 or D4 also described earlier in this section.

If the piston clearance exceeds the limits outlined in the following table, either the piston or the cylinder block **MUST** be replaced.

Calculate the piston clearance by subtracting the maximum piston skirt diameter from the maximum cylinder bore measurement and compare the results for the model being serviced.

PISTON CLEARANCE LIMITS

Series	Clearance
300	0.0033-0.0037" (0.084-0.094mm)
440	0.0037-0.0052" (0.09-0.13mm)
550	0.0053-0.0068" (0.13-0.17mm)
650	0.0033-0.0037" (0.084-0.094mm)

HONING CYLINDER WALLS

Hone the cylinder walls lightly to seat the new piston rings, as outlined in this section. If the cylinders have been scored, but are not out-of-round or the bore is

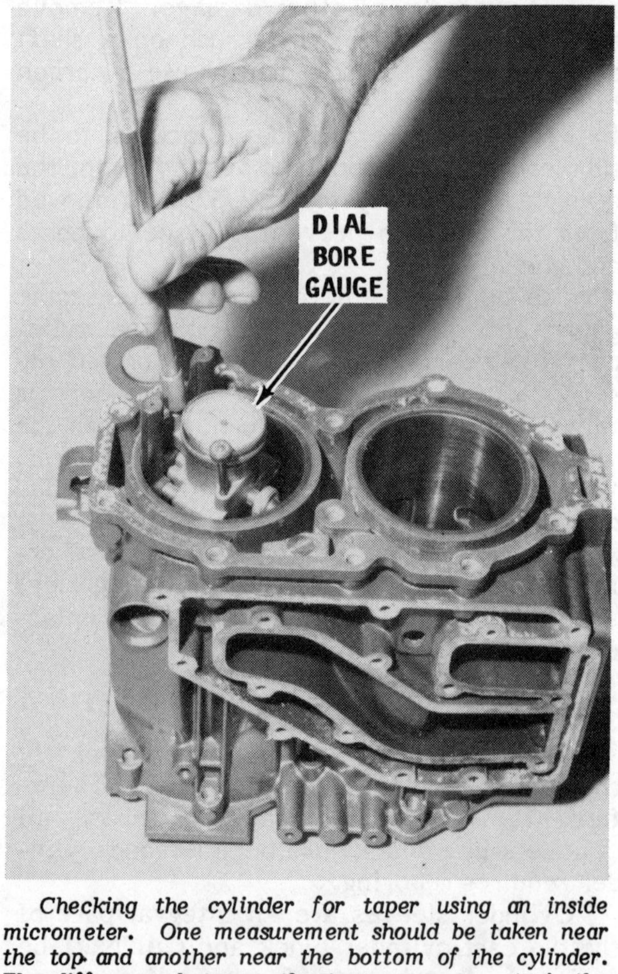

Checking the cylinder for taper using an inside micrometer. One measurement should be taken near the top and another near the bottom of the cylinder. The difference between the two measurements is the amount of taper.

rough, clean the surface of the cylinder with a cylinder hone as described in the following procedures.

SPECIAL WORDS

If overheating has occurred, check and resurface the spark plug end of the cylinder block, if necessary. This can be accomplished with 240-grit sandpaper and a small flat block of wood.

To ensure satisfactory engine performance and long life following the overhaul work, the honing work should be performed with patience, skill, and in the following sequence:

a- Follow the hone manufacturer's recommendations for use of the hone and for cleaning and lubricating during the honing operation. A "Christmas tree" hone may also be used.

b- Pump a continuous flow of honing oil into the work area. If pumping is not practical, use an oil can. Apply the oil generously and frequently on both the stones and work surface.

c- Begin the stroking at the smallest diameter. Maintain a firm stone pressure against the cylinder wall to assure fast stock removal and accurate results.

d- Expand the stones as necessary to compensate for stock removal and stone wear. The best cross-hatch pattern is obtained using a stroke rate of 30 complete cycles per minute. Again, use the honing oil generously.

e- Hone the cylinder walls **ONLY** enough to de-glaze the walls.

f- After the honing operation has been completed, clean the cylinder bores with hot water and detergent. Scrub the walls with a stiff bristle brush and rinse thoroughly with hot water. The cylinders **MUST** be thoroughly cleaned to prevent any abrasive material from remaining in the cylinder bore. Such material will cause rapid wear of new piston rings, the cylinder bore, and the bearings.

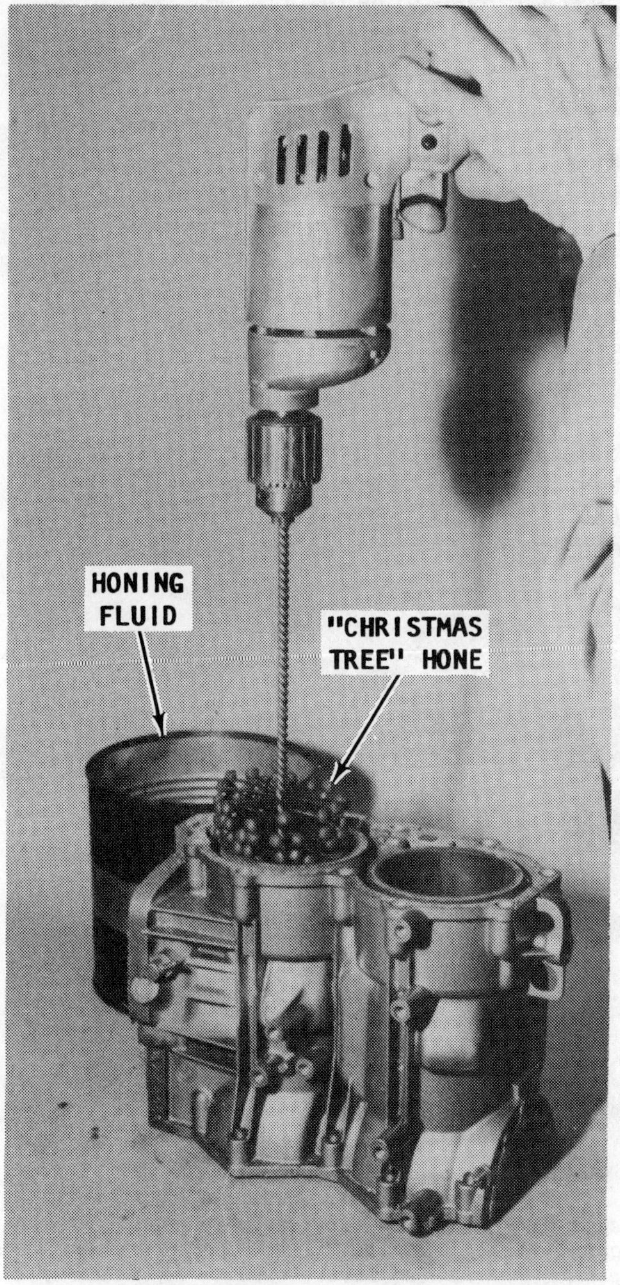

*Refinishing the cylinder wall using an electric drill and "Christmas tree" hone. **ALWAYS** keep the tool moving in long even strokes the entire cylinder depth. Use a continuous **LIBERAL** amount of honing fluid.*

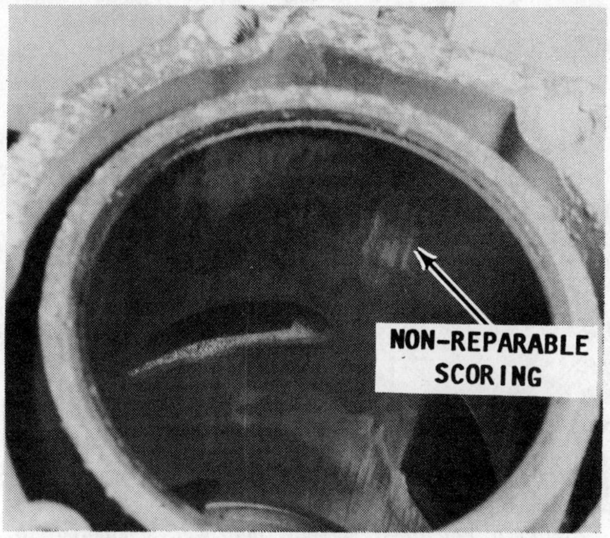

The walls of this cylinder were damaged beyond repair when a piston ring broke and worked its way into the combustion chamber.

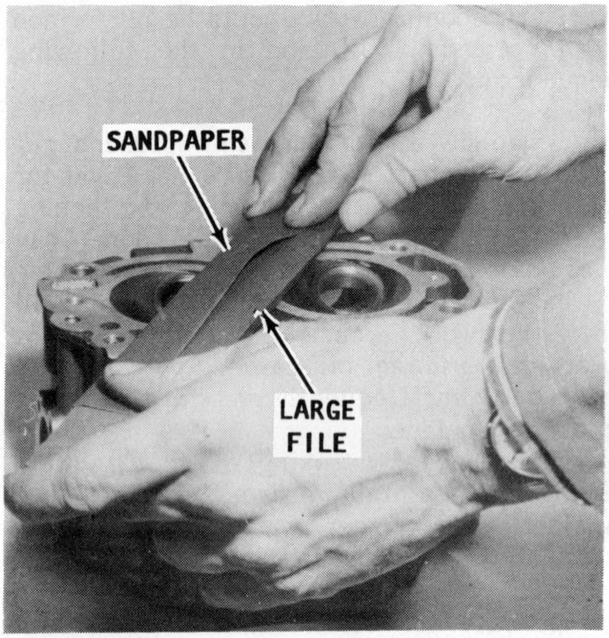

*When a suitable flat surface is not available, a **LARGE** flat file and piece of wet sandpaper may be used to resurface a head or cylinder block.*

g- After cleaning, swab the bores several times with engine oil and a clean cloth, and then wipe them dry with a clean cloth. **NEVER** use kerosene or gasoline to clean the cylinders.

h- Clean the remainder of the cylinder block to remove any excess material spread during the honing operation.

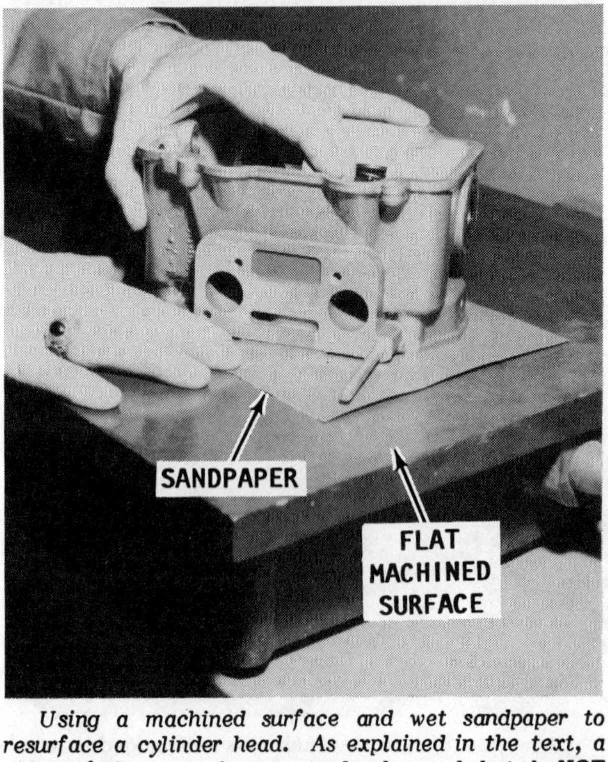

*Using a machined surface and wet sandpaper to resurface a cylinder head. As explained in the text, a piece of glass or mirror may also be used, but do **NOT** attempt this task on a workbench.*

Block and Cylinder Head Warpage

First, check to be sure all old gasket material has been removed from the contact surfaces of the block and the cylinder head. Clean both surfaces down to shiny metal, to ensure a true measurement.

Next, place a straight edge across the gasket surface. Check under the straight edge with a 0.004" (0.1mm) feeler gauge. Move the straight edge to at least eight different locations. If the feeler gauge can pass under the straight edge -- anywhere contact with the other is made -- the surface will have to be resurfaced.

The block or the cylinder head may be resurfaced by placing the warped surface on 400-600 grit **WET** sandpaper, with the sandpaper resting on a **FLAT MACHINED** surface. If a machined surface is not available a large piece of glass or mirror may be used. **DO NOT** attempt to use a workbench or similar surface for this task. A workbench is never perfectly flat and the block or cylinder head will pickup the imperfections of the surface and the warpage will be made worse.

Sand -- work -- the warped surface on the wet sandpaper using large figure **"8"** motions. Rotate the block, or head, through 180° (turn it end-for-end) and spend an equal amount of time in each position to

This cylinder head was removed from the same engine as the pitted piston shown on Page 8-51. In order to return this head to a serviceable condition, the combustion chamber surface must be refinished to a smooth surface using a stone. If left in its present condition, the sharp corners of the gouges will become "hot" spots resulting in pre-ignition and possible "dieseling" after the engine is shut down.

avoid removing too much material from one side.

If a suitable flat surface is not available, the next best method is to wrap 400-600 grit wet sandpaper around a **LARGE** file. Draw the file as evenly as possible in one sweep across the surface. Do not file in one place. Draw the file in many, many directions to get as even a finish as possible.

As the work moves along, check the progress with the straight edge and feeler gauge. Once the 0.004" (0.1mm) feeler gauge will no longer slide under the straight edge, consider the work completed.

If the warpage cannot be reduced using one of the described methods, the block or cylinder head should be replaced.

ONE LAST CHANCE

If the warpage cannot be reduced and it is not possible to obtain new items -- and the warped part must be assembled for further use -- there is a strong possibility of a water leak at the head gasket. In an effort to prevent a water leak, follow the instructions outlined in Assembly on Page 8-34.

8-6 SEALANTS, ADHESIVES, LUBRICANTS, AND FUEL STABILIZERS

It is common practice for the larger manufacturers of personal watercraft to

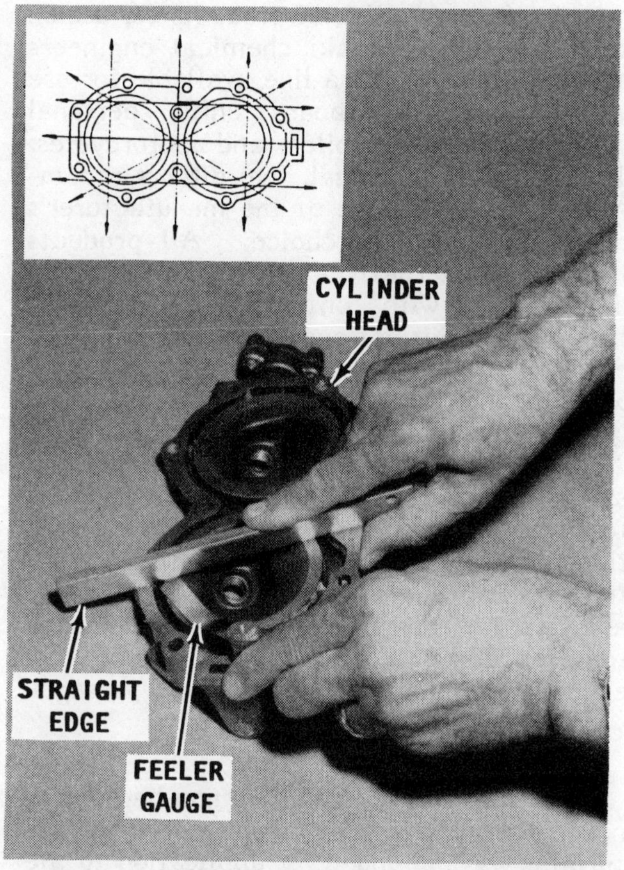

Using a straight edge and feeler gauge to check the cylinder head surface for warpage. Move the straight edge to at least eight different positions, as indicated in the superimposed drawing.

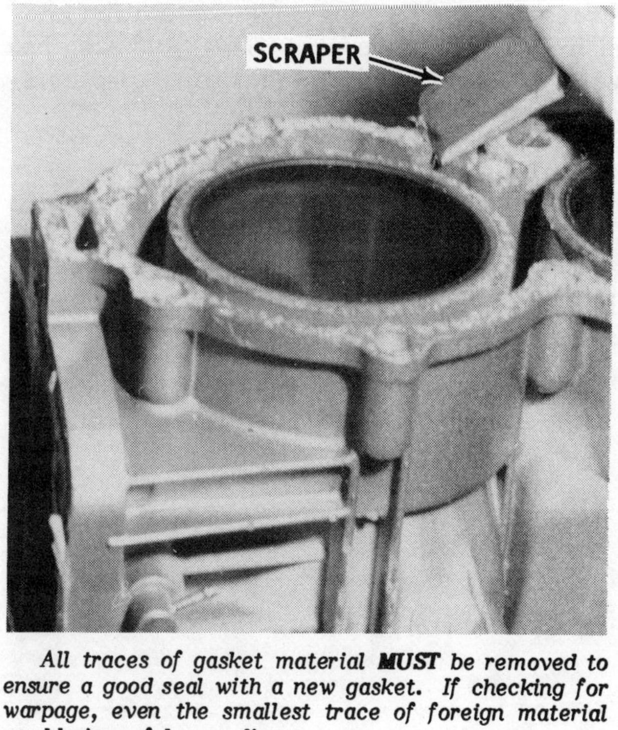

*All traces of gasket material **MUST** be removed to ensure a good seal with a new gasket. If checking for warpage, even the smallest trace of foreign material would give a false reading.*

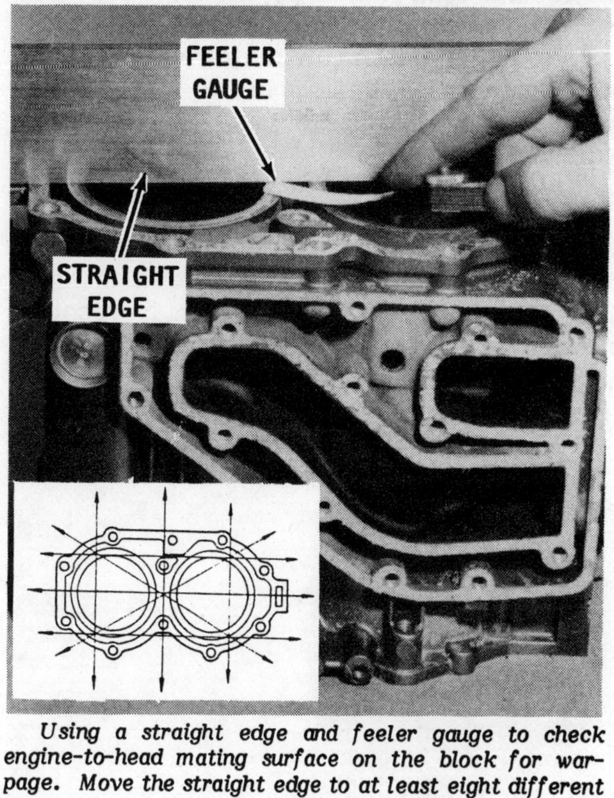

Using a straight edge and feeler gauge to check engine-to-head mating surface on the block for warpage. Move the straight edge to at least eight different positions, as indicated in the superimposed drawing.

market their own line of products for use on their craft. Kawasaki chemical engineers have developed such a line available for use with Kawasaki outboard units, personal watercraft, snowmobiles, and motorcycles. Throughout this manual, the authors recommend the application of the manufacturer's products as a first choice. All products listed are alternatives of equal value, and may be used with confidence if the manufacturer's line is not available.

Sealants and Adhesives

Four sealants are recommended and they are **NOT** interchangeable. Each is designed to perform under a different set of conditions. Follow the directions on the package for cleaning and preparing surfaces. Sealants and adhesives **MUST** be applied **ONLY** to clean and dry parts. Apply sparingly -- excessive amounts may block oil passageways and cause serious damage.

Loctite Lock N' Seal is a non-hardening, non-permanent locking compound. This material is recommended for application to the threads of load bearing fasteners. Loctite helps prevent loosening of the bolt due to vibration, thread wear, and corrosion.

Loctite Stud N' Bearing Mount compound is also a non-hardening, non-permanent locking compound. This material is recommended for application to the threads of load bearing fasteners submerged under water.

Loctite Superflex is a water resistant silicone sealer which provides a very effective flexible seal and is able to withstand high temperatures. This sealer is recommended for use where gaskets are required next to a metal surface at high temperature, for example, on the exhaust manifold cover gaskets. Loctite Superflex is blue in color.

Kawasaki Bond is a non-hardening material and is recommended for metal-to-metal joints such as the crankcase halves. This substance is highly resistant to oil and gasoline.

Lubricants

Different lubricants are recommended in the lubrication procedures presented in this manual. These lubricants are **NOT** interchangeable, each is designed to perform under varying conditions.

Shell Alvania EP1 is a general marine lubricant, chemically formulated to resist salt water. This lubricant is recommended for application to bearings, bushings, and oil seals.

Kawasaki Lubricant is a two-stroke engine oil. It has a petroleum base, and is considered a clean burning lubricant. Yamalube reduces carbon deposits and ensures maximum protection against engine wear. This lubricant also contains an ashless detergent to keep piston rings "free". Oil additives are usually not recommended by the manufacturer, and in some cases the use of such a substance may invalidate the warranty.

Fuel Stabilizer

"Sta-Bil" Fuel Conditioner and Stabilizer is recommended during engine operation and during the storage period. This fluid absorbs water in the fuel system and protects against corrosion.

If used during operation, this fuel additive will prevent the formation of gum and varnish deposits and greatly extend the period between required carburetor overhauls.

When added to the fuel during storage, the additive will prevent the fuel from "souring" for up to twelve full months.

9 ELECTRICAL

9-1 INTRODUCTION

The battery, cranking system, starter interlock, stop "kill" switch, electric bilge pump and fan, and temperature warning system, are all considered subsystems of the electrical system. Each of these areas will be covered in this chapter beginning with the battery. The charging system is considered a subsystem of ignition and is covered in detail in Chapter 7.

9-2 BATTERIES

The battery is one of the most important parts of the electrical system. Because of its job and the consequences, (failure to perform when needed), the best advice is to purchase a well-known brand, with an extended warranty period, from a reputable dealer.

Personal Watercraft Batteries

The battery developed for use in a personal watercraft is the same kind used for many years on motorcycles. Personal watercraft batteries are required to perform under the most rigorous conditions.

Personal watercraft batteries have a much heavier exterior case than the usual automobile battery to withstand the violent pounding and shocks imposed on it as the craft moves through rough water. The battery case must also be vibration and seepage resistant. Therefore, a personal watercraft battery should always be the best the owner can afford.

The caps of marine batteries are "spill proof" to prevent acid from spilling into the bilges when the craft is subjected to violent maneuvers. These special caps are designed to vent gases regardless of the battery position, even when upside down.

Because of these features, the personal watercraft battery will recover from a low charge condition and give satisfactory service over a much longer period of time.

Battery Construction

A battery consists of a number of positive and negative plates immersed in a solution of diluted sulfuric acid. The plates contain dissimilar active materials and are kept apart by separators. The plates are grouped into what are termed elements. Plate straps on top of each element connect all of the positive plates and all of the negative plates into groups.

The battery is divided into cells which hold a number of the elements apart from

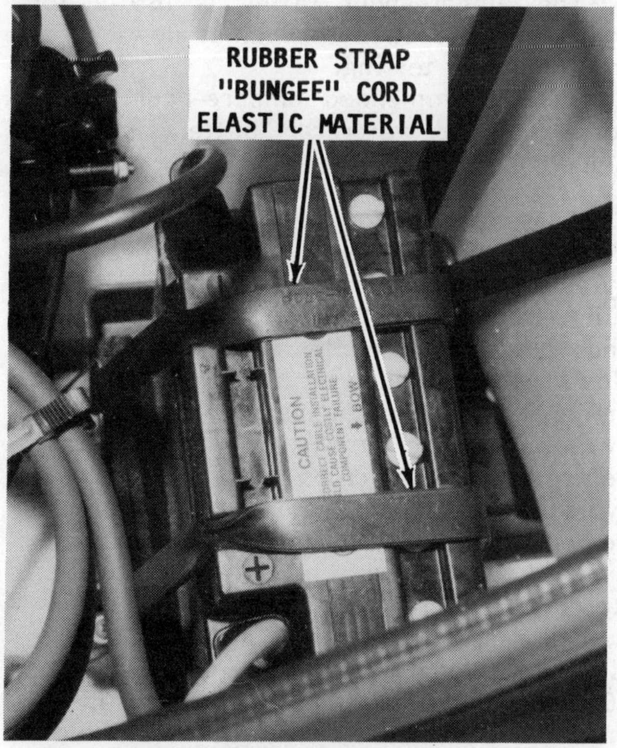

On a Personal Watercraft, two heavy-duty elastic, rubber straps, or "bungee" type cords do an excellent job of holding the battery in place.

the others. The entire arrangement is contained within a hard plastic case. The top is a one-piece cover and contains the filler caps for each cell. The terminal posts protrude through the top where the battery connections for the craft are made. Each of the cells is connected to its neighbor in a positive-to-negative manner with a heavy strap called the cell connector.

Battery Ratings

Four different methods are used to measure and indicate battery electrical capacity:

1- Ampere-hour rating
2- Cold cranking performance
3- Reserve capacity
4- Watt-hour rating

Ampere-Hour

The ampere-hour rating of a battery refers to the battery's ability to provide a set amount of amperes for a given amount of time under test conditions at a constant temperature of 80° (27°C). Amperes x hours equals ampere-hour rating. Therefore, if the battery is capable of supplying 3 amperes of current for 5 consecutive hours, the battery is rated as a 15 ampere-hour battery.

The ampere-hour rating is useful for some service operations, such as slow charging or battery testing.

The recommended ampere-hour rating for a personal watercraft battery is 15-19 ampere-hours.

Cold Cranking Performance

Cold cranking performance is measured by first cooling a fully charged battery to 0°F (-18°C), and then testing it for 5 seconds to determine the maximum current flow. In this manner the cold cranking amperes rating is the number of amperes available to be drawn from the battery before the voltage drops below 9.8 volts. A typical cold cranking ampere rating for a personal watercraft battery is 240 amps.

Reserve Capacity

The reserve capacity of a battery is considered the length of time -- in minutes --at 80°F (27°C) a 25 ampere current can be maintained before the voltage drops below 10.5 volts. This test is intended to provide an approximation of how long the engine, including electrical accessories such as bilge pump, could operate satisfactorily if the stator assembly or charging coil did not produce sufficient current. A typical rating is 25 minutes.

Watt-Hour

Watt-hour is a very useful rating of battery power. It is determined by multiplying the number of ampere hours times the voltage. Therefore, a 12-volt battery rated at 16 ampere-hours would be rated at 192 watt-hours (16 x 12 = 192).

If possible, the new battery should have a power rating equal to or higher than the unit it is replacing.

Battery Installation

Every battery installed in a personal watercraft must be secured in a well-protected ventilated area. If the battery area lacks adequate ventilation, hydrogen gas which is given off during charging could become very explosive. This is especially true if the gas is concentrated and confined.

Battery Service

The battery requires periodic servicing and a definite maintenance program will ensure extended life. If the battery should

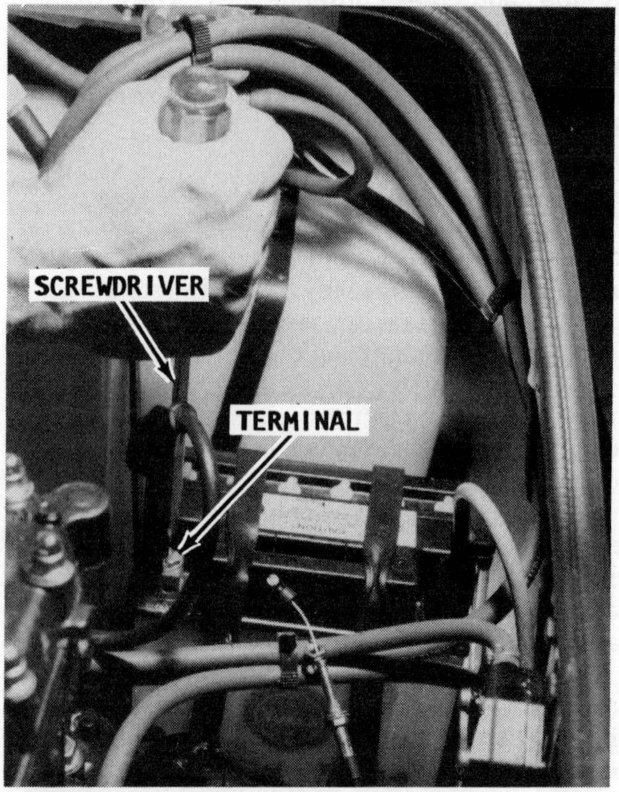

The battery cables can be removed using a slotted screwdriver or socket.

test satisfactorily, but still fails to perform properly, one of five problems could be the cause.

1- Slow speed engine operation for long periods of time resulting in an undercharged condition.

2- A defect in the charging system. A faulty stator assembly or charging coil, defective rectifier, or high resistance somewhere in the system could cause the battery to become undercharged.

3- Failure to maintain the battery in good order. This might include a low level of electrolyte in the cells; loose or dirty cable connections at the battery terminals; or possibly an excessively dirty battery top.

Remove the battery from the engine compartment for service. Inspect and service the battery, cables and connections. Check for signs of corrosion. Inspect the battery case for cracks or bulges, dirt, acid, and electrolyte leakage.

Make sure the breather pipe is attached to the battery and not pinched by any part of the engine compartment.

Cleaning

Dirt and corrosion should be cleaned from the battery just as soon as it is discovered. Any accumulation of acid film or dirt will permit current to flow between the terminals. Such a current flow will drain the battery over a period of time. In time an acid film on top of the battery will rot the two rubber hold down straps.

Clean the exterior of the battery with a solution of diluted ammonia or a soda solution to neutralize any acid which may be present. Flush the cleaning solution off with clean water. **TAKE CARE** to prevent any of the neutralizing solution from entering the cells, by keeping the caps tight.

A poor contact at the terminals will add resistance to the charging circuit. This resistance will cause the voltage regulator to register a fully charged battery, and thus cut down on the stator assembly or charging coil output adding to the low battery charge problem.

Scrape the battery posts clean with a suitable tool or with a stiff wire brush. Clean the inside of the cable clamps to be sure they do not cause any resistance in the circuit.

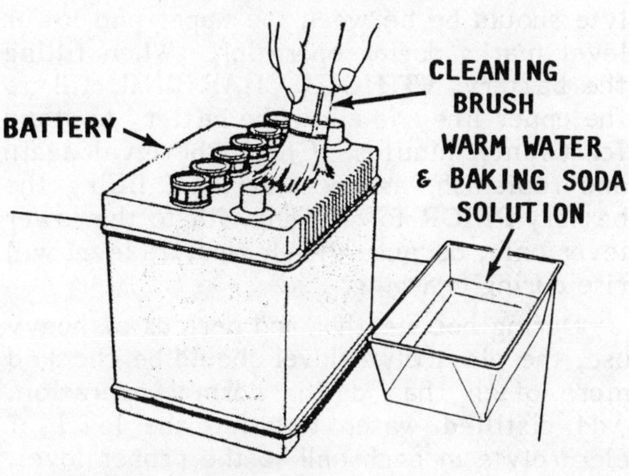

The battery top should be kept clean. A water and baking soda solution applied with a brush will neutralize battery electrolyte and clean the top nicely.

Battery Sulfation

Battery sulfation can be identified by the formation of white crystals forming on top of exposed battery plates. This usually occurs when a battery has been stored in a discharged condition over a period of time and the electrolyte has evaporated leaving the tops of the plates exposed to the air. This condition may also occur if the battery is not serviced at regular intervals during the season and the electrolyte not replenished when low.

Once white crystals form, the plates are permanently damaged and the battery will never hold a charge. It must be replaced.

Filling the Battery

Remove each filler cap using a pair of pliers. Fill each cell to the proper level with distilled water. The level of electro-

*The battery **MUST** be filled correctly. The proper procedure for this task is explained in the text.*

lyte should be between the upper and lower level marks during operation. When filling the battery, **WITHOUT CHARGING,** fill to the upper line and allow the battery to stand for twenty minutes. Check the level again and replenish, as necessary. If filling the battery **PRIOR** to charging, fill to the lower lever only, because the electrolyte level will rise during charging.

During hot weather and periods of heavy use, the electrolyte level should be checked more often than during normal operation. Add distilled water to bring the level of electrolyte in each cell to the proper level. **TAKE CARE** not to overfill, because adding an excessive amount of water will cause loss of electrolyte and any loss will result in poor performance, short battery life, and will contribute quickly to corrosion. **NEVER** add electrolyte from another battery. **NEVER** add sulphuric acid because such acid would make the electrolyte too stong and permanently damage the battery. **NEVER** add ordinary tap water. The minerals in tap water will shorten battery life.

Battery Testing

A hydrometer is a device to measure the percentage of sulphuric acid in the battery electrolyte in terms of specific gravity. When the condition of the battery drops from fully charged to discharged, the acid leaves the solution and enters the plates, causing the specific gravity of the electrolyte to drop.

Use a temperature corrected hydrometer to test the specific gravity of the electrolyte. At 20°C (68°F) the hydrometer reading should be 1.20-1.26 on the scale.

It may not be common knowledge, but hydrometer floats are calibrated for use at 80°F (27°C). If the hydrometer is used at any other temperature, hotter or colder, a correction factor must be applied. Remember, a liquid will expand if it is heated and will contract if cooled. Such expansion and contraction will cause a definite change in the specific gravity of the liquid, in this case the electrolyte.

A quality hydrometer will have a thermometer/temperature correction table in the lower portion, as shown in the accompanying illustration. By knowing the air temperature around the battery and from the table, a correction factor may be applied to the specific gravity reading of the

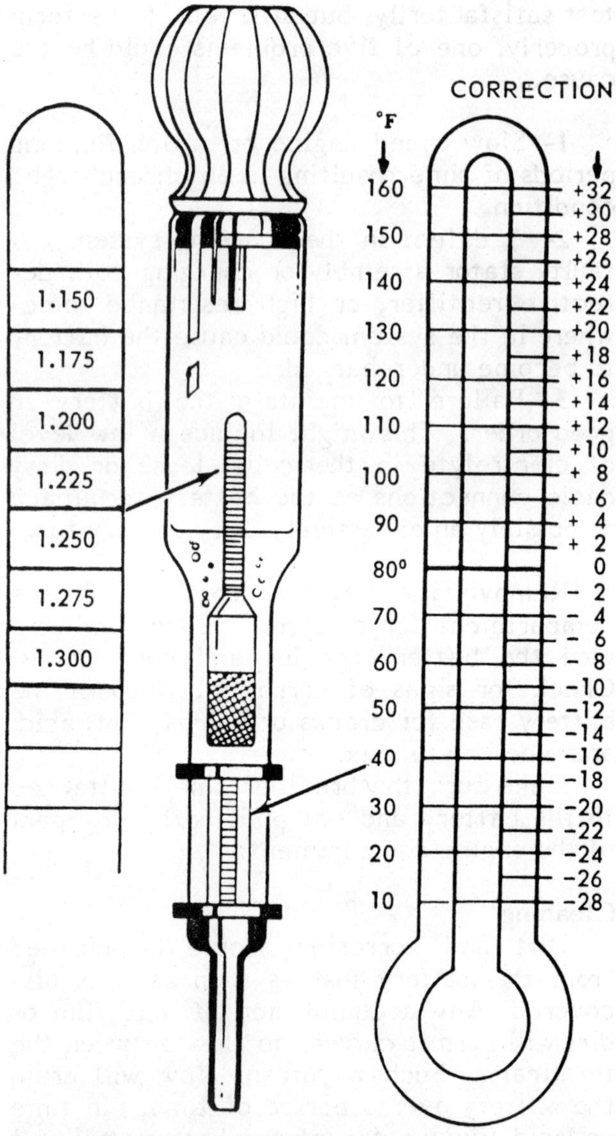

A check of the electrolyte in the battery should be on the maintenance schedule for any craft. A hydrometer reading of 1.300, or in the green band, indicates the battery is in satisfactory condition. If the reading is 1.150 or in the red band, the battery must be charged. Observe the six safety points listed in the text when using a hydrometer.

hydrometer float. In this manner, an accurate determination may be made as to the condition of the battery.

The following six points should be observed when using a hydrometer.

1- NEVER attempt to take a reading immediately after adding water to the battery. Allow at least 1/4 hour of charging at the recommended rate to thoroughly mix the electrolyte with the new distilled water. This time will also allow for the necessary gases to be created.

2- ALWAYS be sure the hydrometer is clean inside and out as a precaution against contaminating the electrolyte.

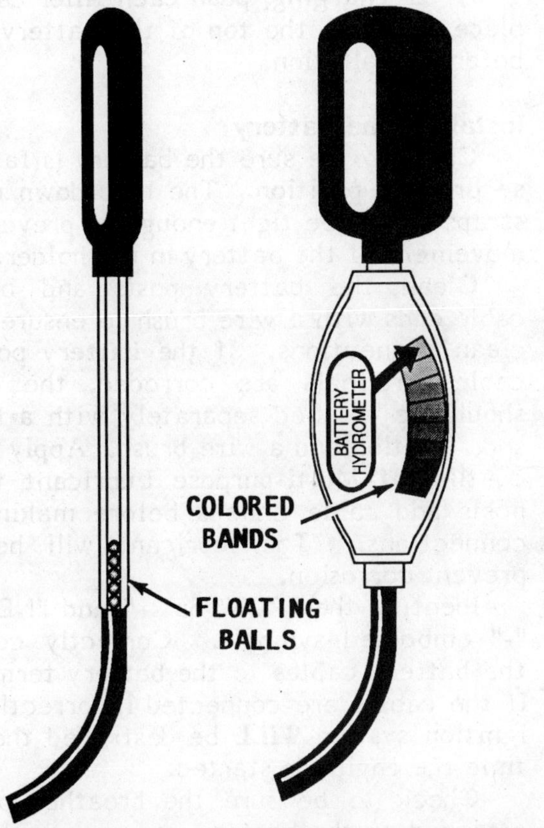

Two types of battery hydrometers, slightly different than the "old style". Their action is explained in the text.

3- If a thermometer is an integral part of the hydrometer, draw liquid into it several times to ensure the correct temperature before taking a reading.

4- BE SURE to hold the hydrometer vertically and suck up liquid only until the float is free and floating.

5- ALWAYS hold the hydrometer at eye level and take the reading at the surface of the liquid with the float free and floating.

Disregard the light curvature appearing where the liquid rises against the float stem. This phenomenon is due to surface tension.

6- DO NOT drop any of the battery fluid on the boat or on your clothing, because it is extremely caustic. Use water and baking soda to neutralize any battery liquid that does accidentally drop.

After withdrawing electrolyte from the battery cell until the float is barely free, note the level of the liquid inside the hydrometer.

Reading the Hydrometer

Different manufacturers of battery hydrometers use different methods of displaying specific gravity readings. Three methods are used:

Red, white, and green bands
Small floating balls
Graduated scale

Colored Band Hydrometer

When using a battery hydrometer with colored bands of red, white, and green, the electrolyte level is interpreted as follows:

Draw electrolyte from the battery cell until the float is barely free, note the level of the liquid inside the hydrometer.

If the level is within the green band range for all cells, the condition of the battery is satisfactory. If the level is within the white band for all cells, the battery is in fair condition.

If the level is within the green or white band for all cells except one, which registers in the red, the cell is shorted internally. No amount of charging will bring the battery back to satisfactory condition.

If the level in all cells is about the same, even if it falls in the red band, the battery may be recharged and returned to service. If the level fails to rise above the red band after charging, the only solution is to replace the battery.

Small Float Ball Hydrometer

If using a battery hydrometer with small floating balls, the electrolyte level is interpreted as follows:

If four or five balls are floating on top of the electrolyte drawn into the hydrometer, the battery is 100% charged. If three balls are floating, the battery is 75% charged. If two balls are floating, the battery is 50% charged. If only one ball is floating the battery is 25% charged. If no balls float, the battery is 0% charged.

If the same number of balls float for each cell, the condition of the battery is satisfactory.

If all the cells but one float balls, the cell is shorted internally. No amount of charging will bring the battery back to satisfactory condition.

If the number of floated balls for all cells is about the same, even if it is only one or two, the battery may be recharged and returned to service. If the number of balls fails to increase after charging, the only solution is to replace the battery.

Graduated Scale Hydrometer

If using a battery hydrometer with a graduated scale, note the level of the liquid inside the hydrometer. If the level is between the 1.280 and 1.265 mark, the battery is 100% charged. If the level is closer to the 1.210 mark, the battery is 75% charged. If the level is closer to the 1.160 mark, the battery is 50% charged. If it is closer to the 1.120 mark the battery is only 25% charged. If the level is below 1.100 the battery is at 0% charge.

If the level is the same and above 1.160 for all cells, the condition of the battery is satisfactory. If the level is the same and above 1.120 for all cells, the battery is in fair condition.

If the level is the same for all cells except one, which registers low on the scale, the cell is shorted internally. No amount of charging will bring the battery back to satisfactory condition.

If the level in all cells is about the same, even if it registers low on the scale, the battery may be recharged and returned to service. If the level fails to rise above 1.210 after charging, the only solution is to replace the battery.

Charging the Battery

If it is necessary to charge the battery, leave the filler caps lightly resting on the cell openings to allow the gases to escape.

The charging current should not exceed 1.9 amps for 10 hours.

NEVER use a fast charger, such as those found at automotive service stations, to charge a personal watercraft battery. These chargers operate at too high an output, ideal for automotive batteries, but very damaging to personal watercraft batteries. If such a battery is fast charged, excessive heat will be generated which will warp the plates and cause internal shorting. The plates will also shed active material which will accumulate at the bottom of the battery and cause internal shorting.

If, under the recommended charging procedures, the temperature of the electrolyte rises above 115°F (45°C) during charging, reduce the charging rate and increase the charging time proportionately. If possible, stop, allow the battery to cool and then resume charging for the recommended length of time.

After charging, push each filler cap into place and wipe the top of the battery clean before installation.

Installing the Battery

Check to be sure the battery is fastened securely in position. The hold-down rubber straps should be tight enough to prevent any movement of the battery in the holder.

Clean the battery posts and battery cable ends with a wire brush to ensure good, clean connections. If the battery posts or cable terminals are corroded, the cables should be cleaned separately with a baking soda solution and a wire brush. Apply a thin coating of Multi-purpose Lubricant to the posts and cable clamps before making the connections. The lubricant will help to prevent corrosion.

Identify the "POS" or "+" and "NEG" or "-" embossed symbols. Correctly connect the battery cables to the battery terminals. If the cables are connected incorrectly, the ignition system **WILL** be destroyed the first time the engine is started.

Check to be sure the breather pipe is attached to the battery and not pinched by any part of the engine compartment.

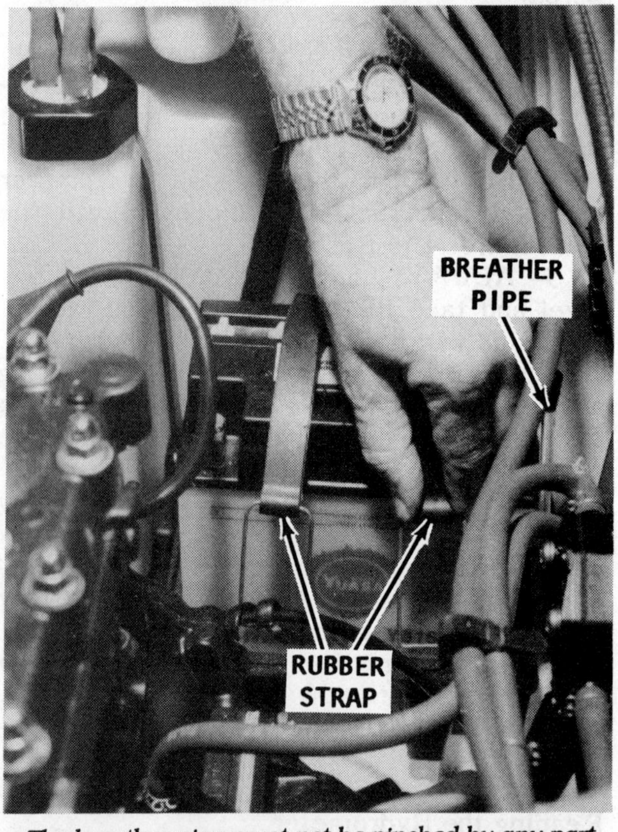

The breather pipe must not be pinched by any part of the engine compartment after the holding straps are secured.

Jumper Cables

If a booster battery is used for starting an engine, the jumper cables must be connected correctly and in the proper sequence to prevent damage to either battery, or to the diodes in the rectifier/regulator.

ALWAYS connect a cable from the positive terminal of the dead battery to the positive terminal of the good battery **FIRST.** **NEXT,** connect one end of the other cable to the negative terminal of the good battery and the other end to a good ground on the engine. **DO NOT** connect the negative jumper from the good battery to the negative terminal of the low battery. Such action will almost always cause a spark which could ignite gases escaping through the vent holes in the battery filler caps. Igniting the gases may result in an explosion destroying the battery and causing severe personal **INJURY.**

By making the negative (ground) connection on the engine, if an arc is created, it will not be near the battery.

DISCONNECT the battery ground cable before replacing a stator assembly or charging coil, or before connecting any type of meter to the ignition system.

If it is necessary to use a trickle charger on a dead battery, **ALWAYS** disconnect one of the cables from the battery **FIRST,** to prevent burning out the diodes in the regulator/rectifier.

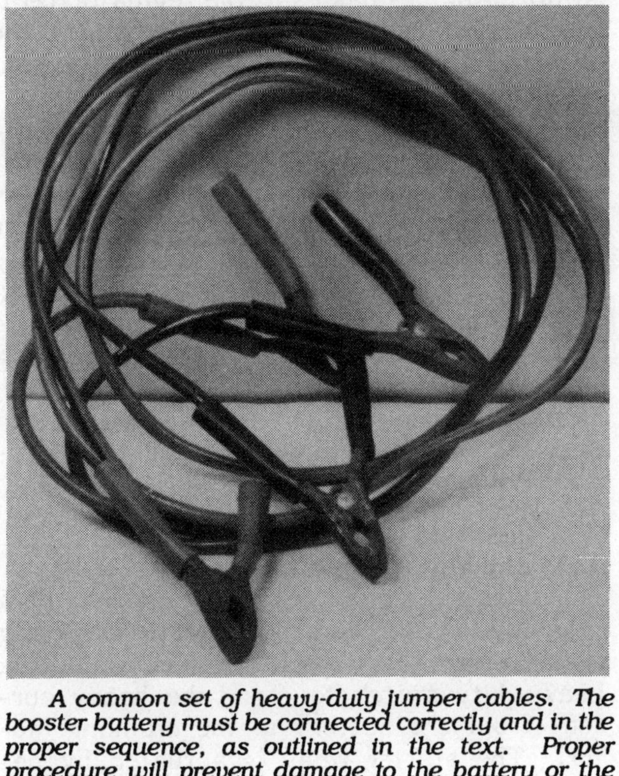

A common set of heavy-duty jumper cables. The booster battery must be connected correctly and in the proper sequence, as outlined in the text. Proper procedure will prevent damage to the battery or the diodes in the rectifier/regulator.

NEVER use a trickle charger as a booster to start the engine because the diodes will be **DAMAGED.**

Storage

If the craft is to be laid up for the winter or for more than a few weeks, special attention must be given to the battery to prevent complete discharge or possible damage to the terminals and wiring. Before putting the craft in storage, disconnect and remove the battery. Clean the battery thoroughly of any dirt or corrosion, and then charge it to full specific gravity reading. After the battery is fully charged, store the battery in a clean cool dry place where it will not be damaged or knocked over, preferably on a couple blocks of wood. Storing the battery up off the deck, will permit air to circulate freely around and under the battery and will help to prevent condensation.

NEVER store the battery with anything on top of it or cover the battery in such a manner as to prevent air from circulating around the filler caps. All batteries, both new and old, will discharge during periods of storage, more so if they are hot than if they remain cool. Therefore, the electrolyte level and the specific gravity should be checked at regular intervals. A drop in the specific gravity reading is cause to charge the battery back to a full reading.

One of the largest personal watercraft battery manufacturers recommends that a battery in storage should be checked every two weeks and charged if necessary. The electrolyte level can also be replenished at this time to prevent sulfation occurring.

In cold climates, care should be exercised in selecting the battery storage area. A fully charged battery will freeze at about 60° below zero. A discharged battery, almost dead, will have ice forming at about 19° above zero.

9-3 TACHOMETER

An accurate tachometer can be connected to any engine. Such an instrument provides an indication of engine speed in revolutions per minute (rpm). This is accomplished by measuring the number of electrical pulses per minute generated in the primary circuit of the ignition system.

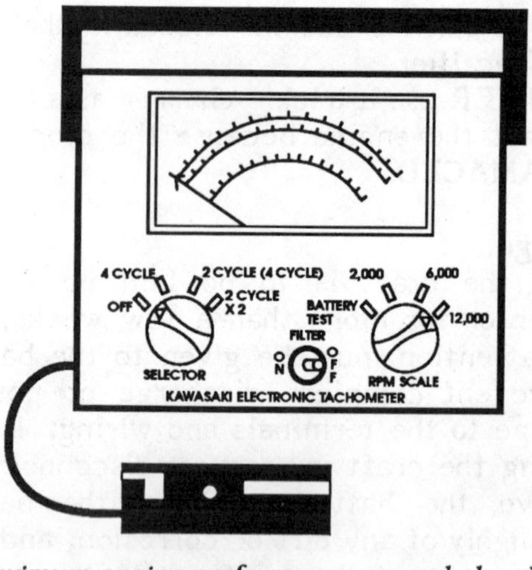

Maximum engine performance can only be obtained through proper tuning using a tachometer.

The meter readings range from 0 to 6,000 rpm, in increments of 100. Tachometers have solid-state electronic circuits which eliminates the need for relays or batteries and contributes to their accuracy. The electronic parts of the tachometer susceptible to moisture are coated to prolong their life.

The manufacturer recommends the use of a Kawasaki Electronic Tachometer. Instructions on how to connect the leads are provided with the meter. However, as a general rule, on a two stroke engine with CDI type ignition, the tachometer leads are usually connected across the charging coil leads -- refer to Chapter 7 for charging coil lead color identification. The color code differs for each model.

Components of the ignition circuit and the charging circuit are housed and protected by the flywheel.

9-4 ELECTRICAL SYSTEM GENERAL INFORMATION

The electrical system consists of three separate circuits:

Charging circuit
Cranking motor circuit
Ignition circuit

Charging Circuit

Components of the charging circuit include:

Permanent magnets attached to the inside perimeter of the flywheel.
A charging coil installed on the stator plate.
A rectifier/regulator located elsewhere on the engine.
An external battery
Necessary wiring to connect it altogether.

The negative side of the rectifier/regulator is grounded. The positive side of the rectifier/regulator is connected to the battery. The negative side of the battery is connected to a good ground on the engine.

The alternating current generated in the stator windings passes to the regulator/rectifier. The rectifier/regulator changes the alternating current (AC) to direct current (DC) to charge the 12-volt battery.

The stator is located behind and protected by, the flywheel. Therefore, the stator, including the charging coil, seldom causes problems in the charging circuit. Most problems in the charging circuit can usually be traced to the rectifier/regulator or to the battery. If either the stator or the rectifier/regulator fails the troubleshooting tests, the defective unit cannot be repaired, it **MUST** be replaced.

See Chapter 7 to service and test components in the charging system.

Cranking Motor Circuit

The cranking motor circuit consists of a cranking motor and a starter-engaging mechanism. A starter relay is used as a heavy-duty switch to carry the heavy current from the battery to the cranking motor. The starter relay is actuated by depressing the **START** button.

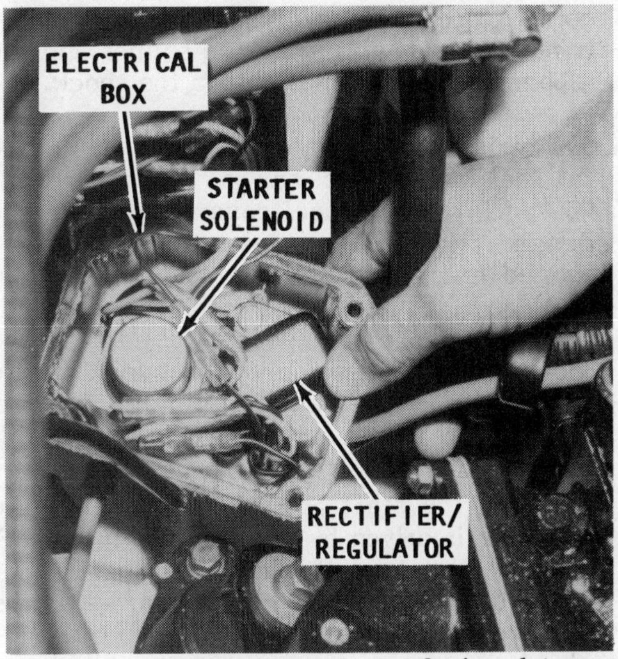

The starter solenoid and the rectifier/regulator are housed inside the electrical box.

Ignition Circuit

The ignition circuit is covered extensively in Chapter 7, Ignition.

9-5 CRANKING MOTOR CIRCUIT SERVICE

DESCRIPTION

As the name implies, the sole purpose of the cranking motor circuit is to control operation of the cranking motor to crank the engine until the air/fuel mixture ignites and the flywheel develops enough torque to continue under its own power. The circuit includes a solenoid or magnetic switch to connect or disconnect the motor from the battery. The operator controls the switch with a push button.

The cranking motor is a series wound electric motor which draws heavy current from the battery. It is designed to be used only for short periods of time to crank the engine for starting. To prevent overheating the motor, cranking should not be continued for more than 30-seconds without allowing the motor time to cool -- say at least three minutes. Actually, this time can be spent in making preliminary checks to determine why the engine fails to start.

Theory of Operation

On Kawasaki Personal Watercraft, three different methods are used to transmit power from the cranking motor to the engine crankshaft through a combination of meshed gears or clutch mechanisms.

300 Series

The 300 Series has a splined armature shaft in constant mesh with an idle gear. The cranking motor drives this idle gear. A second idle gear on the same shaft as the first is in constant mesh with a one-way clutch on the engine crankshaft. The accompanying simplified functional line drawing illustrates this arrangement. The one-way clutch consists of an eccentric cam mechanism. Power is transmitted from the one way clutch to the cam roller, the outer race, and finally to the crankshaft.

On a cranking motor equipped with a one-way clutch: The armature shaft is in constant mesh with an idle gear. The idle gear rotates on a short shaft housed in the flywheel cover. Another smaller idle gear also rotates on the same shaft. This smaller gear is in constant mesh with the one-way clutch gear on the flywheel.

When the engine starts to crank, the one-way clutch rotates and the internal cams rise up and lock the oneway clutch to the outer race. The rotation is therefore transmitted to the crankshaft. When the engine starts and the start button is released, engine rpm rises above cranking rpm. The internal cams fall away from the outer race. As soon as the start button is released, current no longer flows to the cranking motor. Therefore, the cranking motor, the two idle gears and the one-way clutch no longer rotate.

440 and 550 Series

The 440 and 550 Series are equipped with a Bendix drive on the end of the armature shaft. When the motor is operated, the pinion gear moves forward and meshes with teeth on the flywheel ring gear, as depicted in the accompanying illustration. A rubber cushion is built into the Bendix drive to absorb the shock when the pinion gear meshes with the flywheel ring gear.

After the engine starts, the pinion gear is driven faster than the armature shaft, and as a result, it spins out of mesh with the flywheel. The parts of the drive **MUST** be properly assembled for efficient operation.

If the drive is removed for cleaning, **TAKE CARE** to assemble the parts in the correct order. If the screw shaft assembly is reversed, it will strike the splines and the rubber cushion will not absorb the shock.

650 Series

The 650 Series has a reduction gear housed in the flywheel cover containing the Bendix drive. The armature shaft has a splined end which meshes inside the reduction gear. When the cranking motor is operated, the pinion gear -- a part of the reduction gear assembly -- moves forward and meshes with the teeth on the flywheel ring gear. The simple line drawing on this page illustrates this arrangement.

When the start button is depressed, the idle gear is driven by the cranking motor. A one-way clutch on the idle gear shaft slides forward, engaging the pinion gear -- at the forward end of the shaft -- with the flywheel ring gear.

After the engine starts and the start button is released, engine rpm rises above cranking rpm. The rollers inside the clutch move in a different direction at increased speed and no longer contact the outer race of the one-way clutch. Therefore, the one-way clutch ceases to rotate and slides back on the shaft disengaging the pinion gear from the flywheel ring gear.

Cranking Motor Noises

The sound of the motor during cranking is a good indication of how the cranking motor is operating -- properly or not. Naturally, temperature conditions will affect the speed at which the cranking motor is able to crank the engine. The speed of cranking a cold engine will be much slower than when cranking a warm engine. An experienced operator will learn to recognize the favorable sounds of the engine cranking under various conditions.

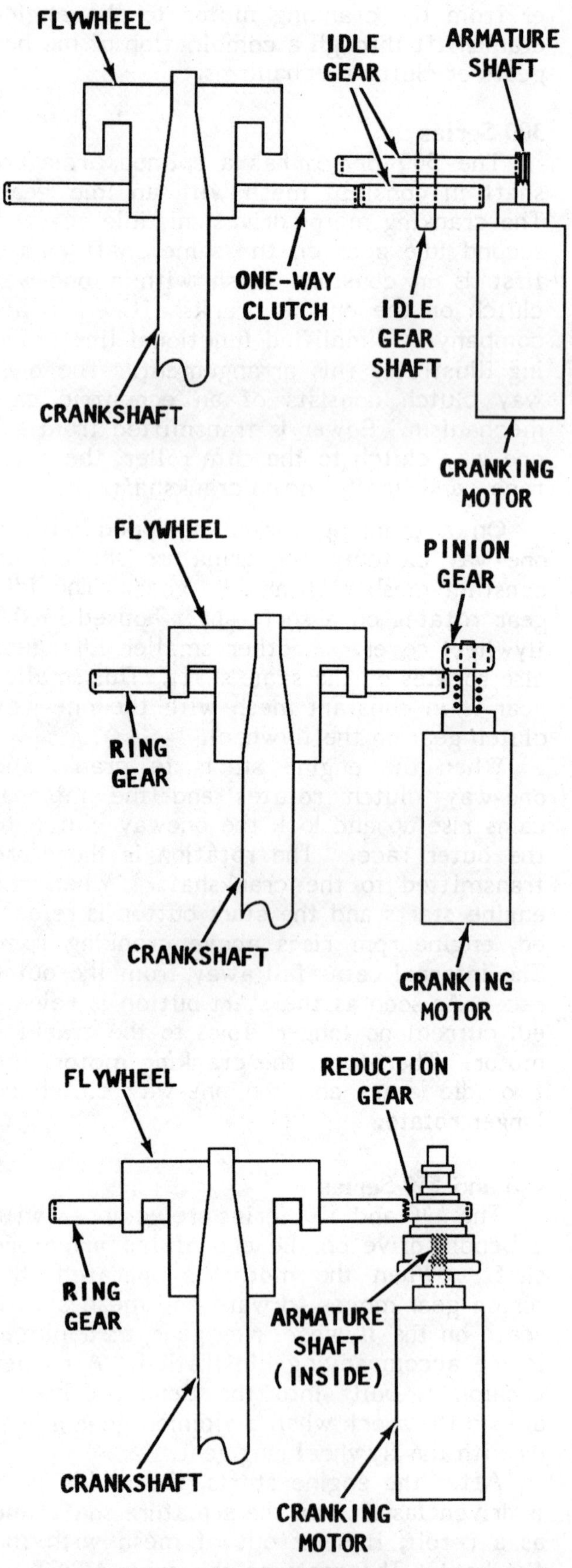

Line drawing to depict the three power train methods for transferring energy from the cranking motor to the engine flywheel. The 300 Series (top), the 440 and 550 Series (center), and the 650 Series (bottom), are covered in this section.

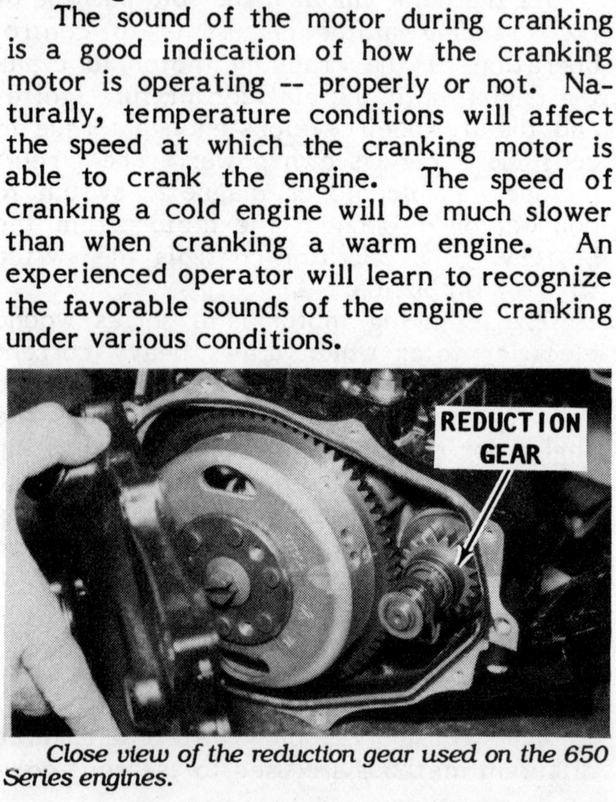

Close view of the reduction gear used on the 650 Series engines.

Faulty Symptoms

If the cranking motor spins, but fails to crank the engine, the cause is usually a corroded or gummy Bendix drive. The drive should be removed, cleaned, and given an inspection.

If the cranking motor cranks the engine too slowly, the following are possible causes and the corrective actions that may be taken:

a- Battery charge is low. Charge the battery to full capacity.

b- High resistance connections at the battery, solenoid, or motor. Clean and tighten all connections.

c- Undersize battery cables. Replace cables with sufficient size.

Maintenance

The cranking motor does not require periodic maintenance or lubrication, except for lubrication of the pinion gear. If the motor fails to perform properly, the checks outlined in the previous paragraph should be performed.

Naturally, the motor will have to be removed if the corrective actions outlined under **Faulty Symptoms** above, does not restore the motor to satisfactory operation.

Otherwise, the cranking motor is not normally removed from the engine until it fails.

CRANKING MOTOR TROUBLESHOOTING

Before wasting too much time troubleshooting the cranking motor circuit, the following checks should be made. Many times, the problem will be corrected.

a- Battery fully charged.
b- Main fuse is "good" (not blown).
c- All electrical connections clean and tight.
d- Wiring in good condition -- insulation not worn or frayed.

The following troubleshooting procedures are presented in a logical sequence.

Do not operate the cranking motor for more than 15 seconds. Prolonged cranking motor operation will cause overheating and damage the motor.

After each test, allow the cranking motor to cool for a minute or so.

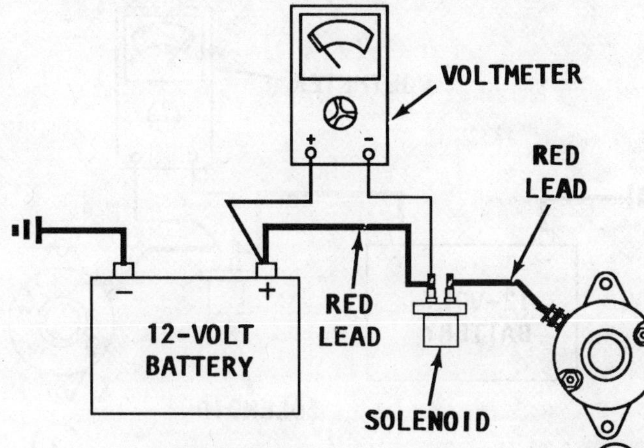

1

Never depress the **START** button to activate the cranking motor while the engine is operating. Such action will damage the pinion and/or flywheel gears.

Cranking Circuit Tests

1- Remove both spark plug leads from the plugs and ground the ends to the engine to prevent accidental engine start. Keep the leads grounded for the following tests.

Obtain a voltmeter and select the Vx1 DC scale. Observe the starter solenoid. Both large terminals have Red leads connected. One terminal has the battery cable connected and the other terminal has the lead from the cranking motor connected. Make contact with the negative Black meter lead to the terminal on the starter solenoid with the battery cable connected.

Make contact with the positive Red meter lead to the positive battery terminal. The voltmeter should register less then 0.25V DC. If the reading exceeds 0.25V DC, clean the connections at both ends of the

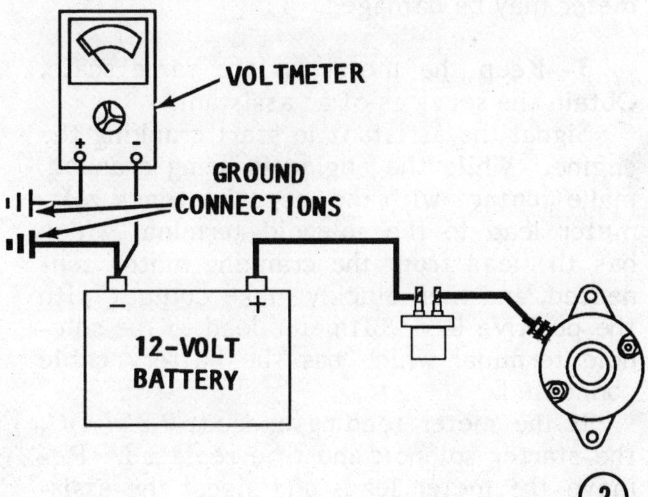

2

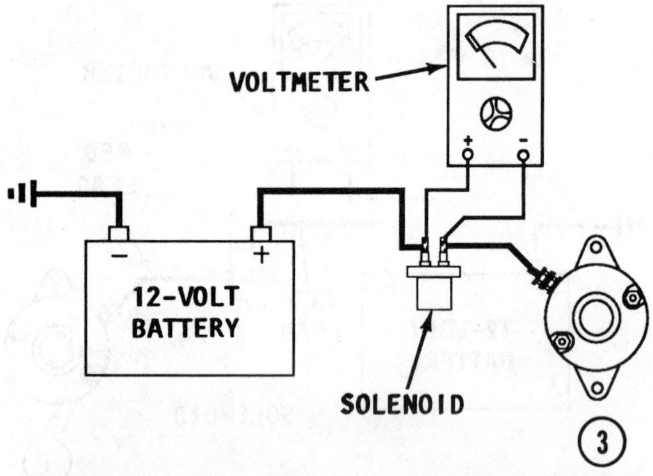

positive battery cable and repeat the test. If the reading does not change, replace the cable.

2- Keep the meter on the same scale. Make contact with the negative Black voltmeter lead to the negative terminal on the battery. Make contact with the positive Red voltmeter lead to a good engine ground. The voltmeter should register less then 0.25V DC. If the reading exceeds 0.25V DC, clean the connections at both ends of the negative battery cable and repeat the test. If the reading does not change, replace the cable.

CRITICAL WORDS

During the following tests -- Step 3 and Step 4, both meter leads must be placed on the solenoid terminals **WHILE** the engine is being cranked. **ALSO,** both meter leads must be removed from the solenoid terminals **WHILE** the engine is being cranked and **BEFORE** the engine has stopped cranking. If these precautions are not followed, the voltmeter may be damaged.

3- Keep the meter on the same scale. Obtain the services of an assistant.

Signal the assistant to start cranking the engine. While the engine is being cranked, make contact with the negative Black voltmeter lead to the solenoid terminal which has the lead from the cranking motor connected, and then quickly make contact with the positive Red voltmeter lead to the solenoid terminal which has the battery cable connected.

If the meter reading exceeds 0.25V DC, the starter solenoid must be replaced. Remove the meter leads and signal the assistant to cease cranking.

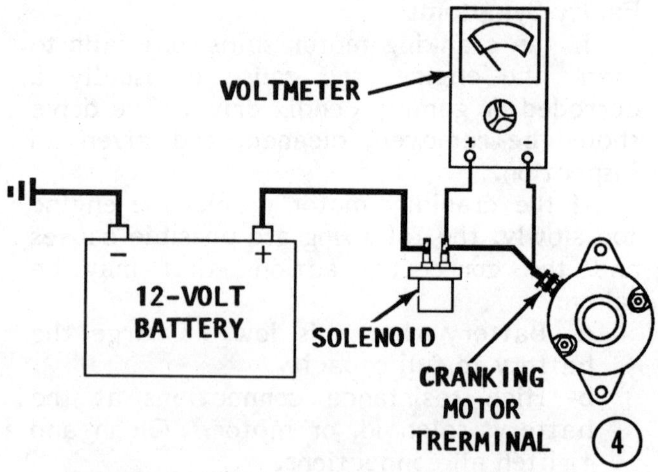

4- Keep the meter on the same scale and the assistant at the **START** button. Observe the same precautions as before in Step 3. While the engine is being cranked, make contact with the negative Black voltmeter lead to terminal on the cranking motor, and then quickly make contact with the positive Red voltmeter lead to the solenoid terminal which has the Red cranking motor connected.

If the meter reading exceeds 0.25V DC, the connections at both ends of the solenoid-to-cranking motor lead should be cleaned and the test repeated. If the reading does not change, the lead must be replaced.

CRANKING MOTOR RELAY REMOVAL FOR TESTING

Description

The cranking motor relay is actually a switch between the battery and the engine. The switch cannot be serviced. Therefore, if testing indicates the switch to be faulty, it **MUST** be replaced. The relay is housed inside the electrical case. The case must be removed from the engine to gain access to the relay. The relay must be removed from the case for testing purposes.

Electric Case Removal Models 300, 440 and 550

1- Disconnect the battery ground cable from the battery. Remove the spark plug lead from the spark plug. Remove the two bolts securing the connector cover to the electric case. Pull off the cover. Unplug the 4-prong connector and the four free leads at their quick disconnect fittings.

Pull back the two boots over the large cables at the cranking motor relay. Discon-

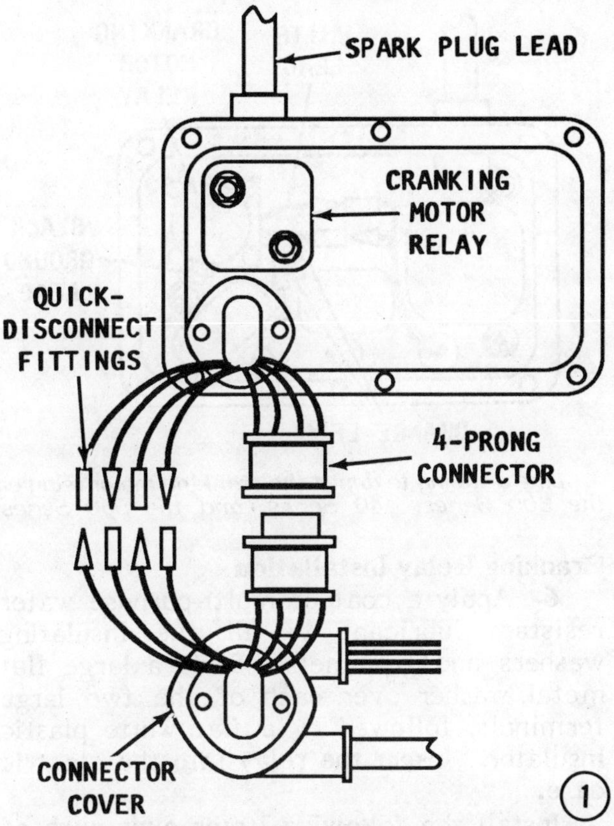

nect both large cables from the relay terminals. Remove the six bolts securing the electrical box to the engine compartment bulkhead. Remove the two bolts securing the two halves of the case together, and then open the case.

Electric Case Removal 650 Series Only

2- Disconnect the battery ground cable from the battery. Remove the spark plug leads from the spark plugs. Pull back the

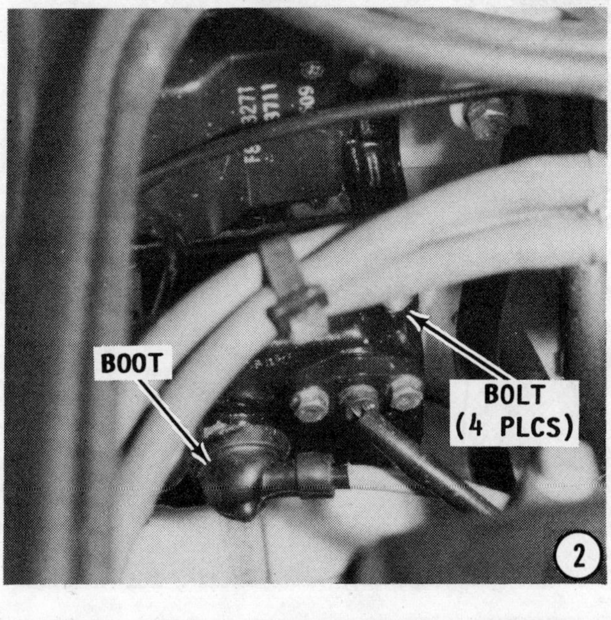

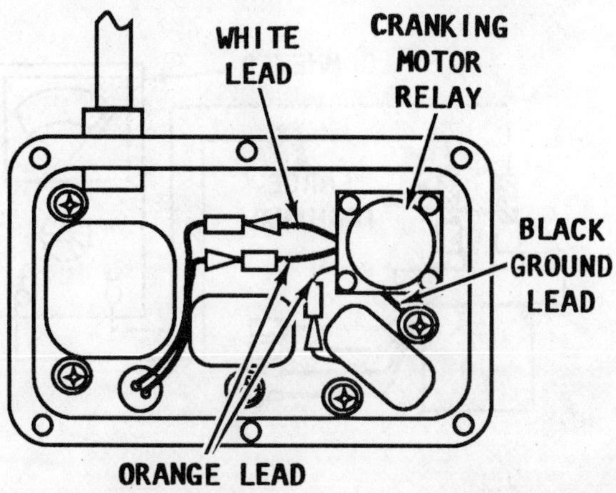

Line drawing to depict the cranking motor relay on the 300 Series, 440 Series, and the 550 Series.

large boot over both large cables at the cranking motor relay. Disconnect both cables from the relay terminals. Remove the four bolts securing the cover to the electrical box on the engine compartment bulkhead. Note there are two different bolt sizes.

Cranking Motor Relay Removal All Series

3- Inside the case -- disconnect the ground switch from the relay at the regulator/rectifier. Disconnect the two Orange leads and the White lead at their quick-disconnect fittings -- if working on a Jet Mate, disconnect the Yellow/Red, Red and Black leads. Outside the case --remove the nuts from the two large relay terminals. Pull the relay out of the case, taking care to observe the arrangement of washers and insulating grommets.

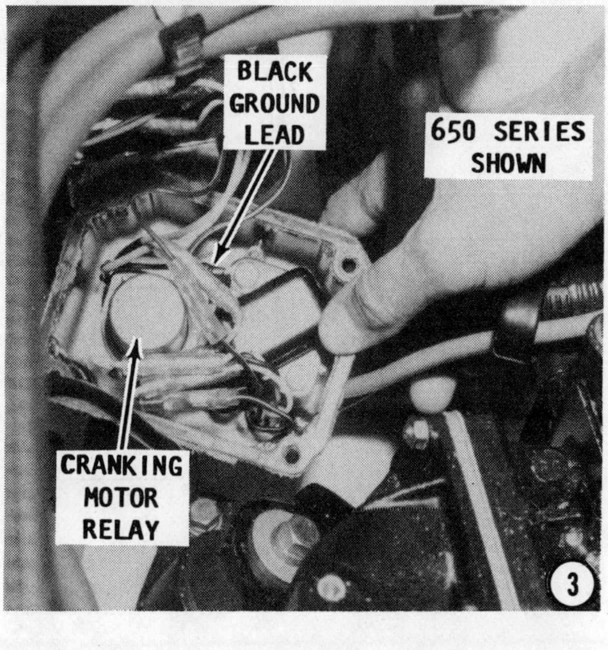

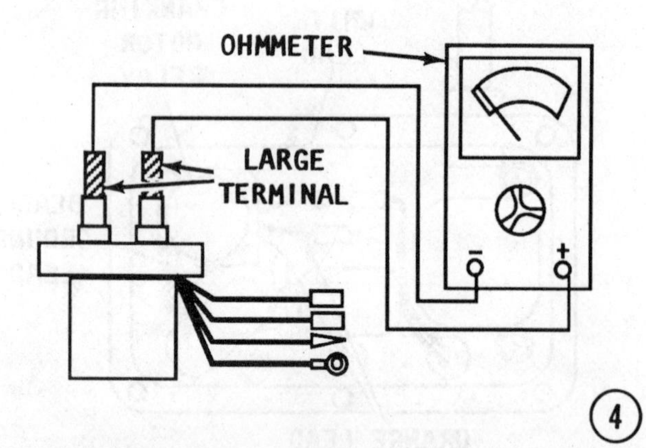

Cranking Relay Testing

4- Select the Rx1 ohm scale on an ohmmeter. Make contact with the two meter leads across the two large relay terminals. The meter should read an infinite resistance. If the meter shows a resistance of less than infinity, the relay is defective and must be replaced.

5- Keep the ohmmeter on the same scale. Obtain a 12V battery. Connect the negative battery terminal to the small Black ground lead from the relay. Connect the positive battery terminal to the small White lead from the relay. If working on a JetMate, connect the positive battery terminal to the Yellow/Red relay lead. All models -- make contact with the two meter leads across the two large relay terminals. If the relay emits a "click" and the ohmmeter registers zero ohms, the relay is in good condition. If no "click" is heard, or if the meter registers a high or infinite resistance, the relay is defective and must be replaced.

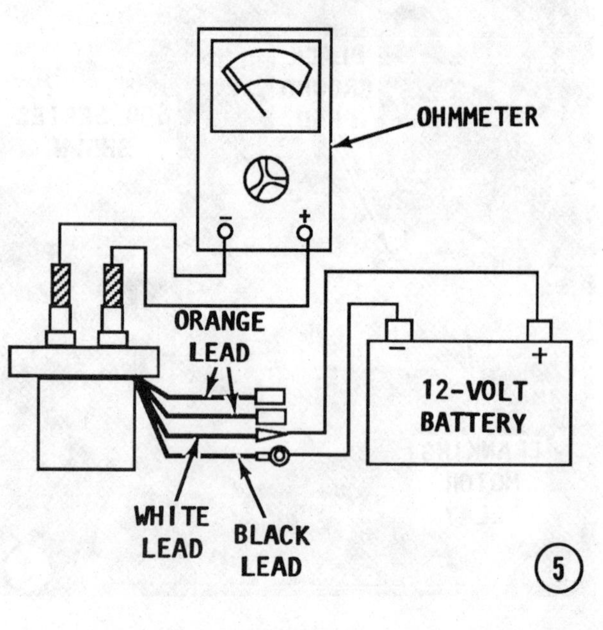

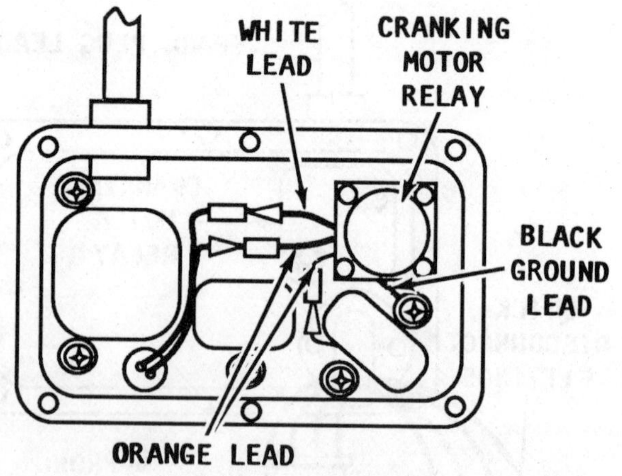

Line drawing to depict the cranking motor relay on the 300 Series, 440 Series, and the 550 Series.

Cranking Relay Installation

6- Apply a coat of multi-purpose water resistant lubricant to all the insulating washers and grommets. Slide a large flat metal washer over each of the two large terminals, followed by a flat white plastic insulator. Insert the relay into the electric case.

Install the following items over each of the relay terminals protruding from the case, in the manner shown: large Black plastic insulator, large White plastic insulator, small Black plastic insulator, small metal washer and finally, the locknut. Tighten both locknuts securely. Connect the Orange and White leads -- or if working on a JetMate, the Yellow/Red, Red and Black leads, matching color-to-color and secure the relay Black ground lead under the rectifier/regulator mounting screw.

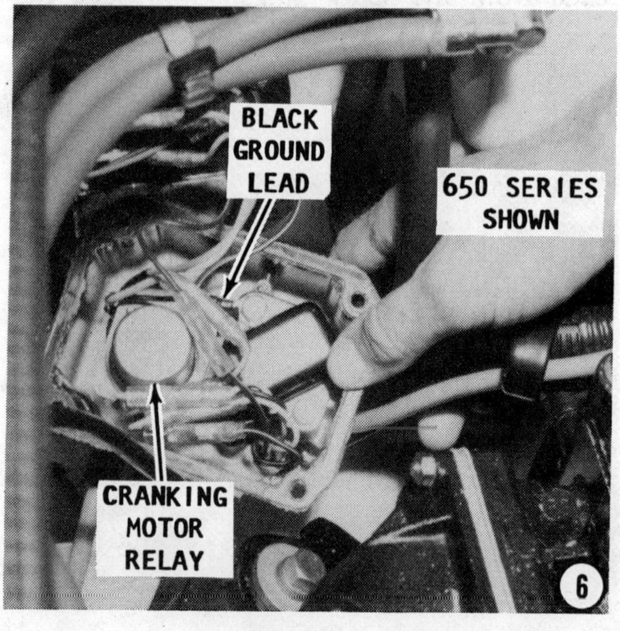

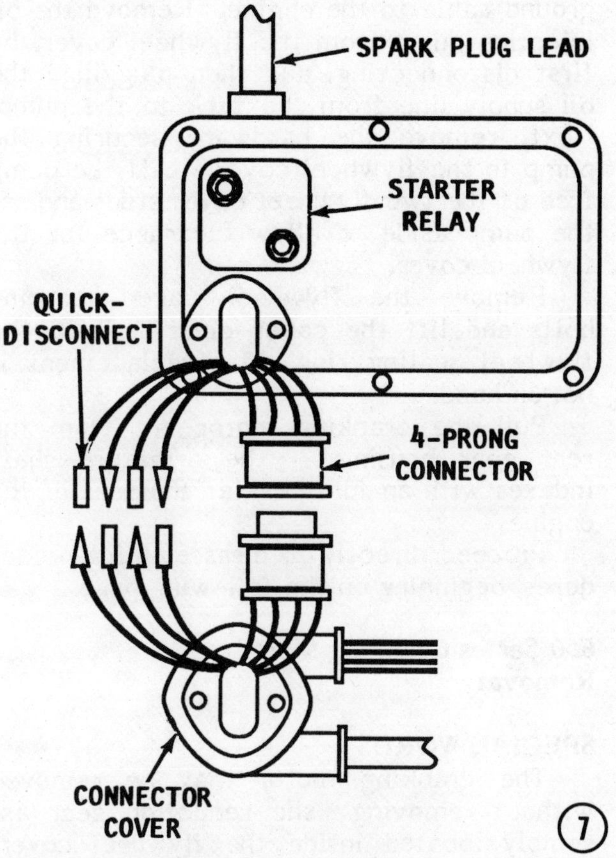

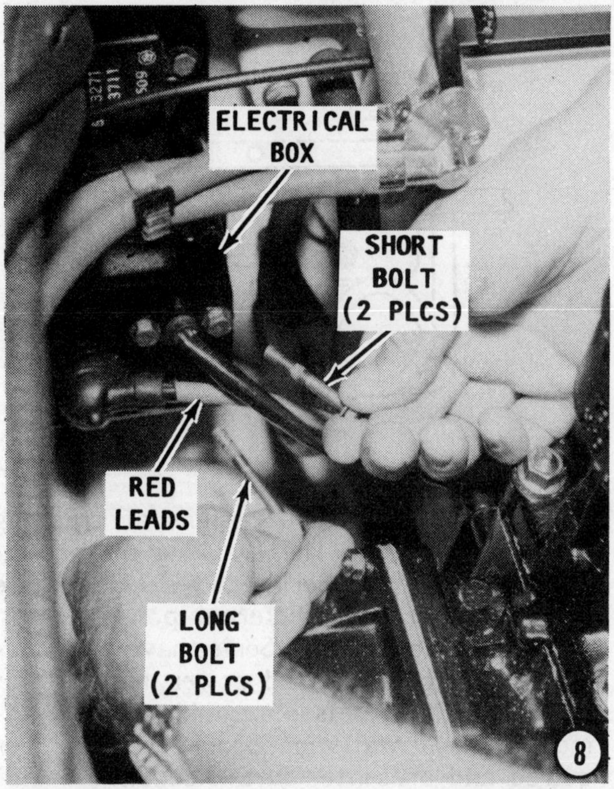

Electrical Box (Case) Installation 300, 440, and 550 Series

7- Secure the halves of the case together with the two bolts and washers. Position the case against the engine compartment bulkhead and secure it in place with the six bolts. Connect the two large cables to the relay terminals -- one from the battery and one to the cranking motor. Cover the terminals with the rubber boots -- one on each cable.

Connect the two halves of the 4-prong connector. Connect the four free leads, matching color-to-color, at their quick-disconnect fittings. Install the connector cover over the electrical box and secure it in place with the two bolts.

Install the spark plug leads and connect the battery ground cable.

Electrical Box (Case) Installation 650 Series Only

8- Install the cover over the electrical case mounted on the engine compartment bulkhead. Secure the cover to the box with the four bolts and washers. Notice there are two different length bolts, as shown. Connect the two large cables to the relay terminals -- one from the battery and one to the cranking motor. Cover the terminals with the single large boot.

Install the spark plug leads and connect the battery ground cable.

9-6 CRANKING MOTOR SERVICE

Description

Marine cranking motors are very similar in construction and operation to the units used in the automotive industry.

NEVER operate a cranking motor for more than 30-seconds without allowing it to cool for at least three minutes. Continuous operation without the cooling period can cause serious damage to the cranking motor.

Three types of power train arrangements are used on the cranking motors covered in this manual. Stating this fact in other terms -- Kawasaki uses three methods of transferring energy from the cranking motor to the engine flywheel.

Method one -- Series 300 type uses a one-way clutch.
Method two -- Series 440 and Series 550 type -- uses a pinion gear.
Method three -- Series 650 type -- uses a reduction gear.

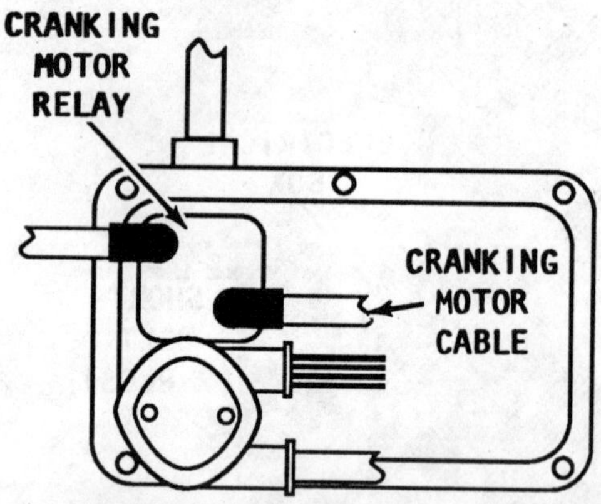

①

However, only two designs of cranking motors are used. The cranking motors installed on the 300 Series and 650 Series are identical, but with different power trains. The 440 Series and 550 Series have a different cranking motor and power train. This means -- three sets of cranking motor removal and installation procedures are required and presented in this section. After the cranking motor is on the bench, only two sets of disassembling and assembling procedures are necessary.

Follow only the set of instructions for the Kawasaki Series being serviced.

300 Model Cranking Motor Removal

1- Disconnect the cranking motor cable from the lower left relay terminal on the electric case.

2- Remove the two cranking motor mounting bolts from the flywheel cover. One of these bolts also secures the battery ground cable to the engine. Remove the oil injection pump from the flywheel cover, by first disconnecting, and then plugging the oil supply line from the tank to the pump. Next, remove the hardware securing the pump to the flywheel cover. Lift the pump free of the two flywheel cover studs and set the pump aside to allow clearance for the flywheel cover.

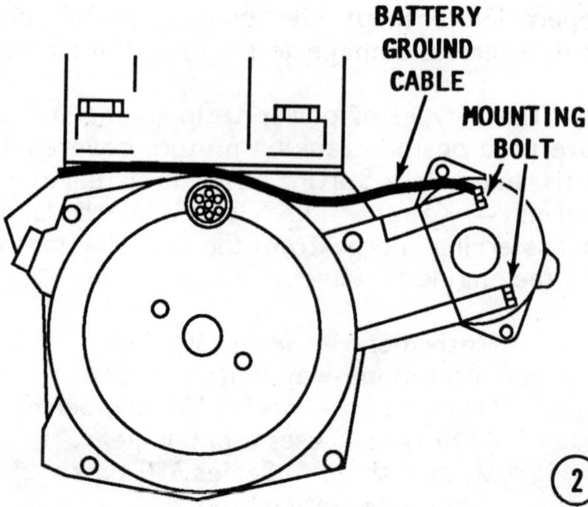

②

Remove the flywheel cover securing bolts and lift the cover clear. **SAVE** the flywheel sealing ring, if a replacement is not on hand.

Pull the cranking motor out from the rear gear housing -- the armature shaft indexes with an idle gear at the rear of the engine.

Proceed directly to disassembling procedures, beginning on the following page.

650 Series Cranking Motor Removal

SPECIAL WORDS

The cranking motor may be removed without removing the reduction gear assembly located inside the flywheel cover. However, if the reduction gear needs service, see Chapter 7 for removal of the flywheel cover, flywheel, and reduction gear.

The exhaust pipe and exhaust manifold may have to be removed on some 650 Series.

①

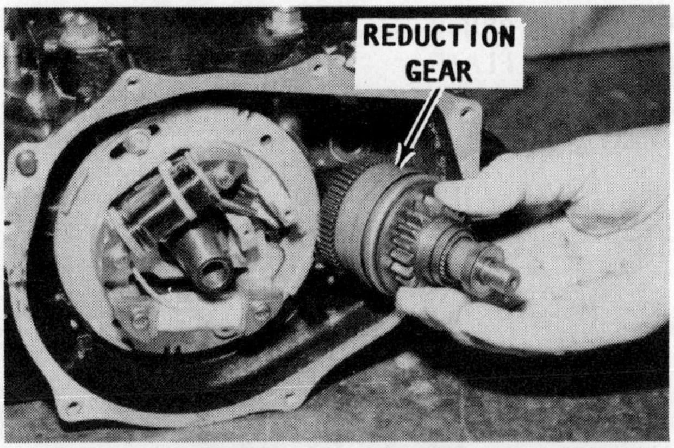

The cranking motor on a 650 Series engine can be removed without disturbing the reduction gear. However, if the reduction gear must be replaced, the flywheel must be removed, see Chapter 7.

1- Disconnect the negative cable from the battery.

2- Push back the protective boot from the terminal on the cranking motor and remove the large Red cable.

3- Remove the two mounting bolts securing the cranking motor to the crankcase. On some 650 Series, the lower mounting bolt also secures the battery ground cable to the crankcase.

4- Ease the armature shaft back out of the reduction gear inside the flywheel cover.

300 and 650 Series Cranking Motor Disassembling

1- Remove and discard the O-ring around the armature shaft. Make a mark

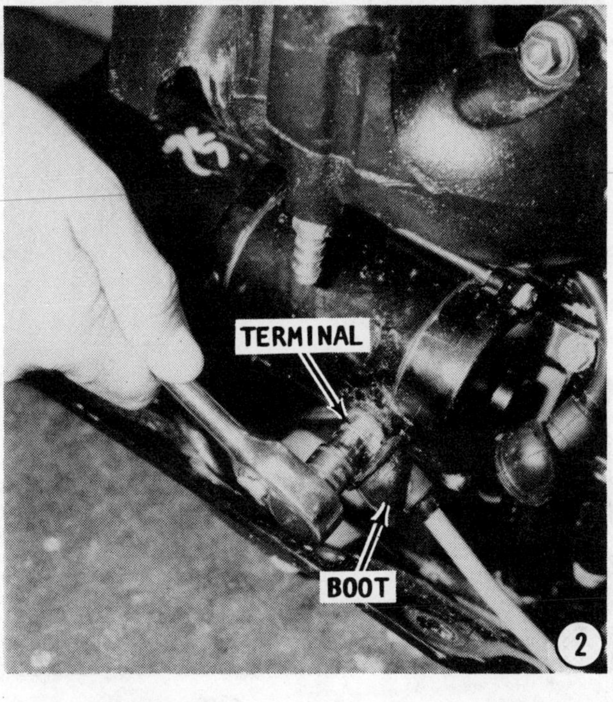

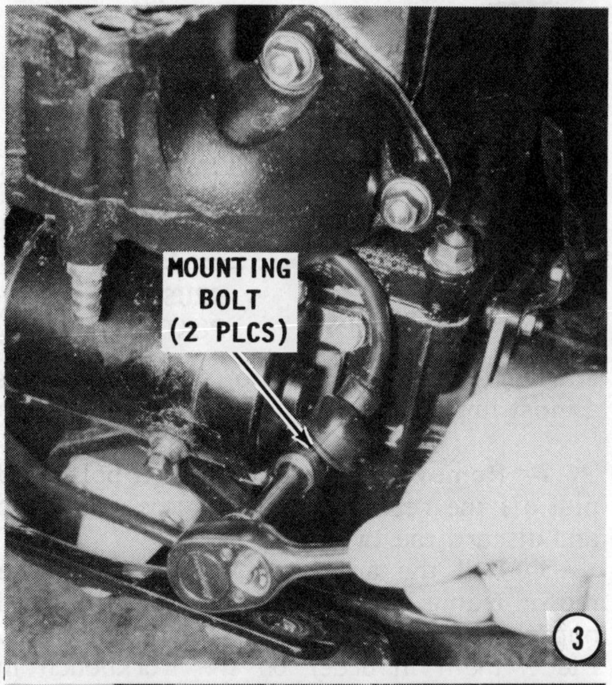

across the mating surface of the front cover and motor frame and another mark across the mating surface of the motor frame and the rear cover, as an aid during assembling.

SPECIAL WORDS

Without some type of alignment marks, the task of inserting the thru bolts and

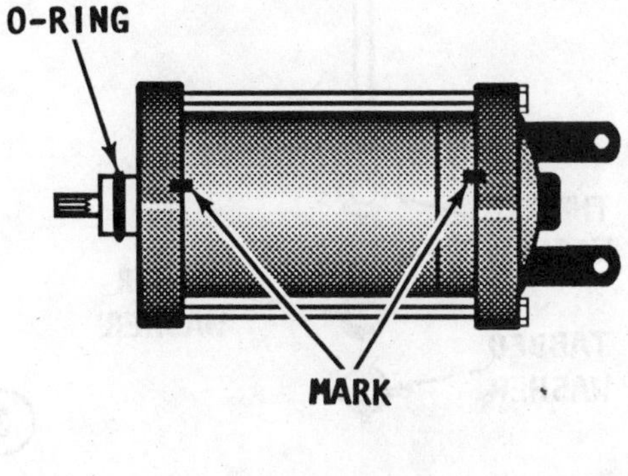

1

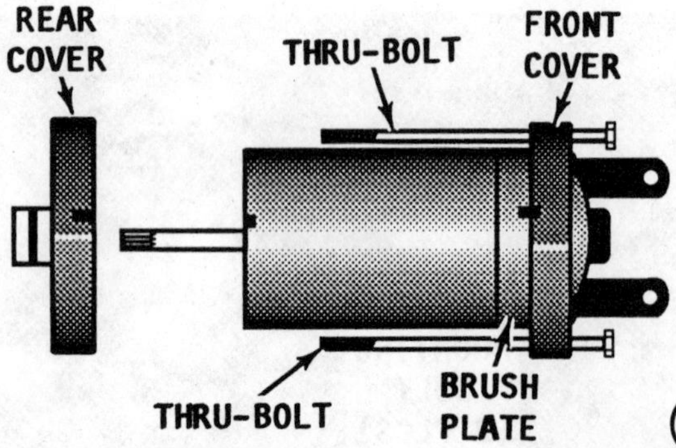

2

starting the threads into the end cap is almost impossible.

2- Remove the two long thru bolts and pull off the rear and front covers. Remove and discard the two O-rings.

3- Pull the armature shaft free of the motor frame. Make a note of the type and number of washers at each end of the shaft. Place them in order on the workbench in order of removal as an assist in later assembly. Incorrect placement of these washers will cause the armature to be misaligned with the brushes and the field magnets during assembling.

4- Identify the two positive brushes. These brushes have insulated leads between the carbon brush and the positive terminal in the motor frame. Lift off the retaining

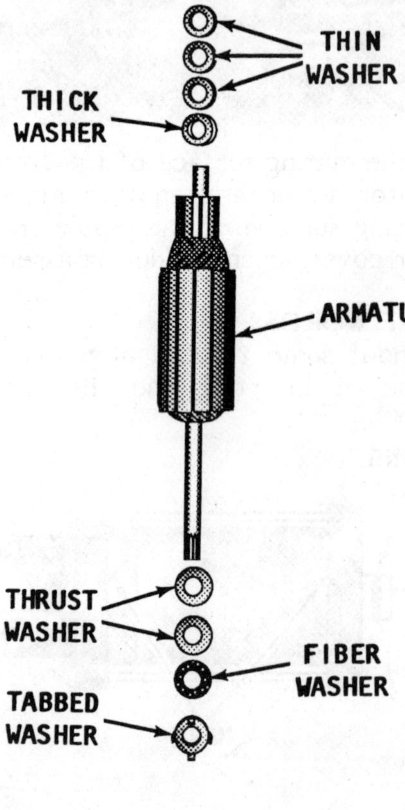

3

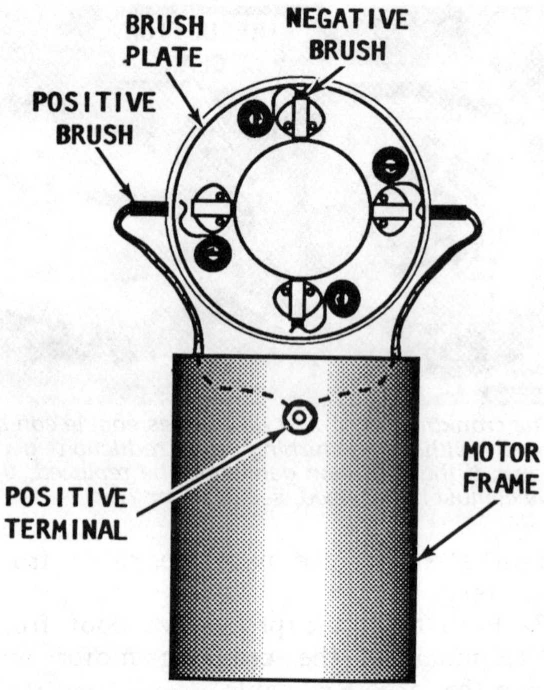

4

springs and pull out the carbon brushes from their holders to release them from the brush plate. Slide the brush assembly from the armature shaft.

Proceed to Cleaning and Inspecting on Page 9-20.

440 Series and 550 Series Cranking Motor Removal

1- Disconnect the battery ground cable at the battery.

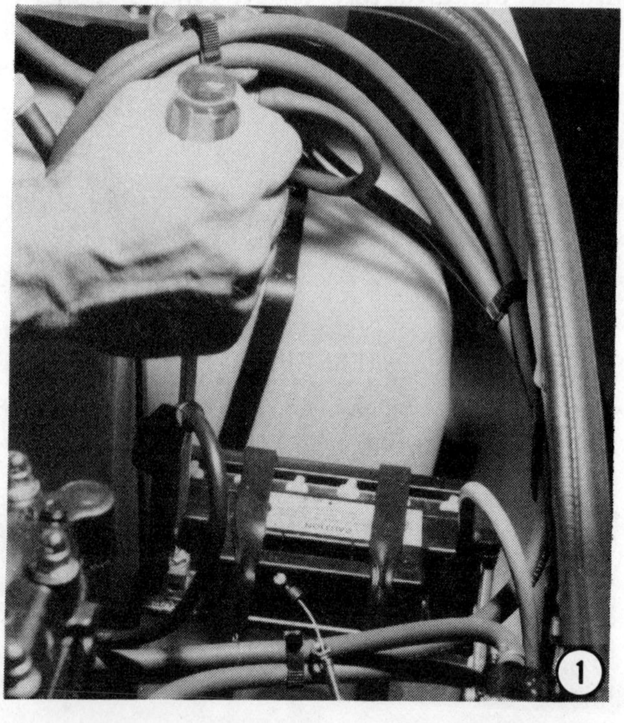

1

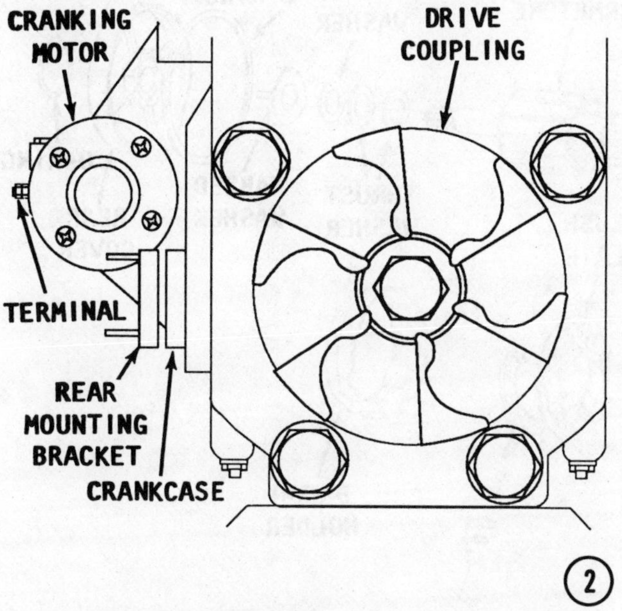

2- Remove the two rear cranking motor mounting bolts. Remove and **SAVE** any shim material found between the rear mounting bracket and the crankcase.

Remove the two front cranking motor mounting bolts.

Slide the cranking motor aft -- toward the rear of the engine. Disconnect the large Red cable from the terminal on the motor.

Armature Shaft End Play

Before disassembling the cranking motor, check the armature end play. Clamp the motor horizontally in a vise equipped with soft jaws. Make a set up with a dial indicator against the armature end. Zero the dial indicator. Now, push the armature shaft in and out. At the same time, observe the deflection of the dial indicator needle.

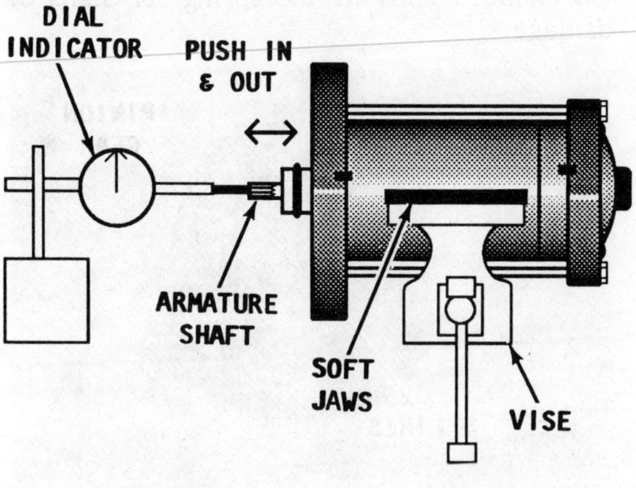

Setup to measure the armature end "play", as explained in the text.

The needle should move through a total range of 0.05-0.30mm. If the needle moves through more than 1.00mm, make a note to add thrust washers from the rear of the armature shaft, as required. If there is no needle movement, remove one of the thrust washers from the rear of the armature shaft.

440 and 550 Series
Cranking Motor Disassembling

1- With the cranking motor on a work surface, pry the snap ring out of its groove and free of the armature shaft.

After the snap ring is free, slide the collar, spring, and pinion gear free of the armature. Remove the large **O**-ring around the upper end cap. Discard this **O**-ring **ONLY** if a replacement is available.

2- Make a mark on both end caps and matching marks on the field frame assembly as an aid to alignment and installation of the thru bolts, during assembling.

SPECIAL WORDS

Without some type of alignment marks, the task of inserting the thru bolts and starting the threads into the end cap is almost impossible.

Remove the two thru bolts and the two small brush plate retaining screws from the rear end cap. Observe the four small **O**-rings around these bolts and screws. Take care not to lose them. If these **O**-rings are

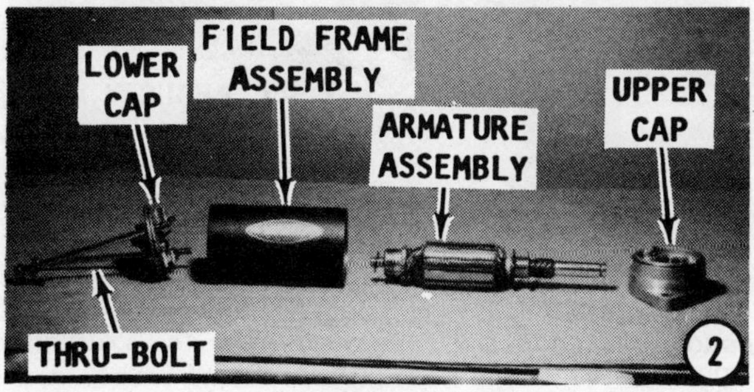

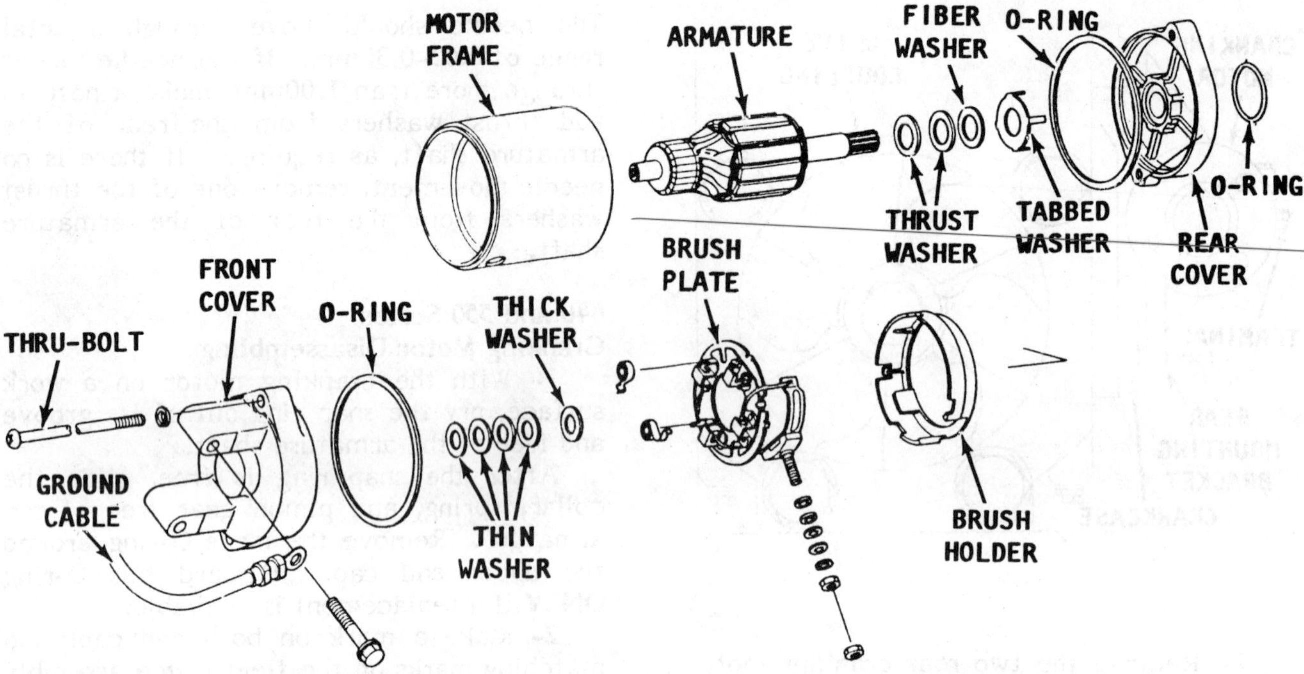

Exploded drawing of a 300 Series cranking motor, with major parts identified.

no longer fit for service they must be replaced. Remove both end caps. If the end caps are stuck, tap around the end cap circumference with a soft head mallet. Use a small screwdriver and pry off the curved springs from the positive brush holders.

Pull out the two carbon brushes and slide the brush plate free of the armature. Note the number of washers on each end of the armature. Incorrect placement of these washers will cause the armature to be misaligned with the brushes and the field magnets during assembling.

Remove the **O**-ring around each end cap. Discard this **O**-ring **ONLY** if a replacement is available.

The internal splines can only be inspected after the pinion gear has been removed from the cranking motor.

CLEANING AND INSPECTING

Pinion Gear Assembly 440 and 550 Series Only

Inspect the pinion gear teeth for chips, cracks, or a broken tooth. Check the splines inside the pinion gear for burrs and to be sure the gear moves freely on the armature shaft.

BAD NEWS

This type cranking motor pinion gear assembly cannot be repaired if the unit is defective. Replacement of the pinion gear assembly -- as a unit -- is the only answer.

Check to be sure the return spring is flexible and has not become distorted. Check both ends of the spring for signs of damage.

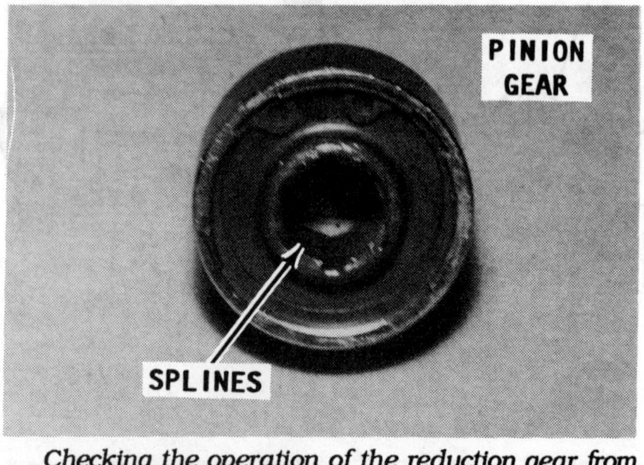

Checking the operation of the reduction gear from a 650 Series engine.

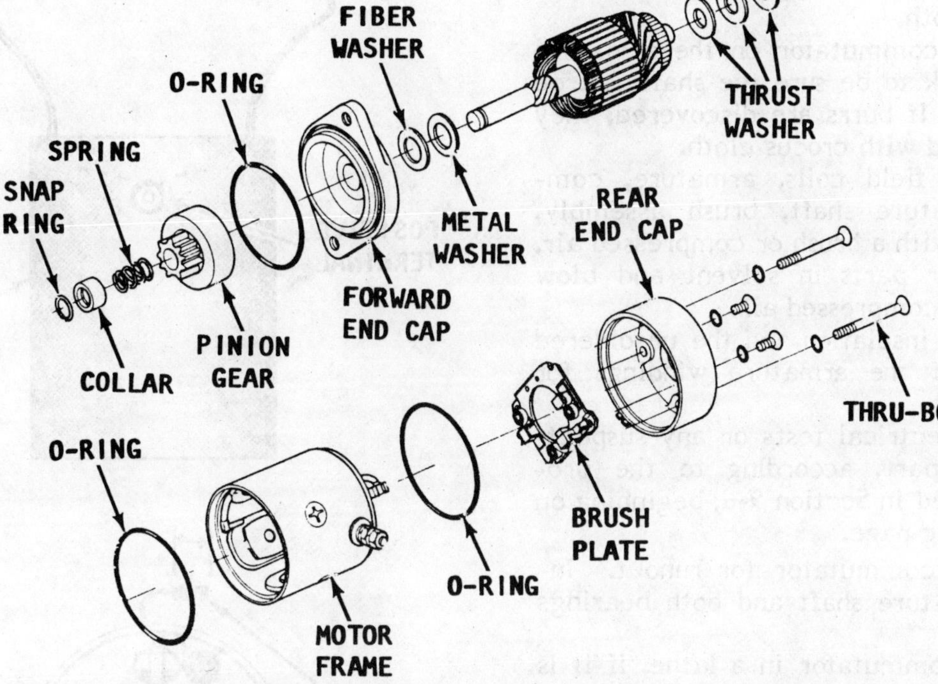

Exploded drawing of the cranking motor installed on 440 Series and 550 Series engines, with major parts identified.

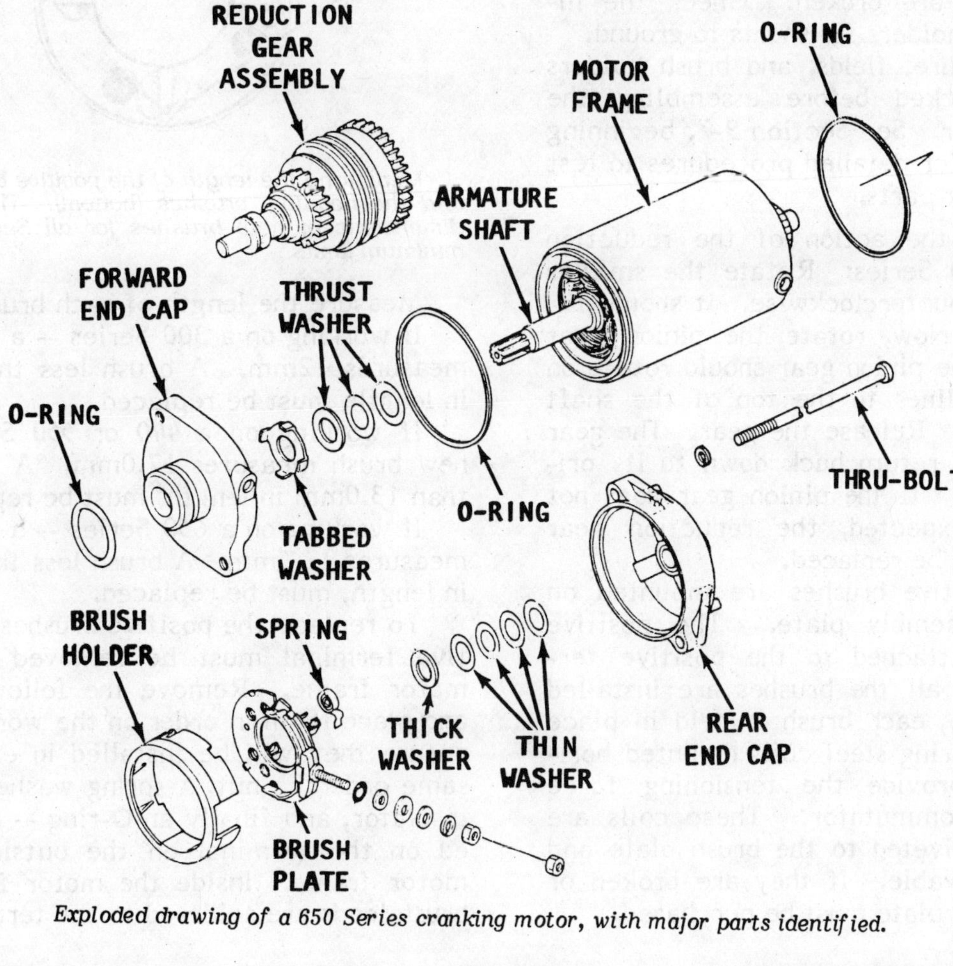

Exploded drawing of a 650 Series cranking motor, with major parts identified.

All Series

Clean the armature shaft and check to be sure the shaft is free of any burrs. If burrs are discovered, they may be removed with crocus cloth.

Clean the commutator on the armature shaft and check to be sure the shaft is free of any burrs. If burrs are discovered, they can be removed with crocus cloth.

Clean the field coils, armature, commutator, armature shaft, brush assembly, and end caps with a brush or compressed air. Wash all other parts in solvent and blow them dry with compressed air.

Inspect the insulation and the unsoldered connections of the armature windings for breaks or burns.

Perform electrical tests on any suspected defective part, according to the procedures outlined in Section 9-6, beginning on on the following page.

Check the commutator for runout. Inspect the armature shaft and both bearings for scoring.

Turn the commutator in a lathe, if it is out-of-round by more than 0.005" (0.13mm).

Check the springs in the brush holder to be sure none are broken. Check the insulated brush holders for shorts to ground.

The armature, fields, and brush holders must be checked before assembling the cranking motor. See Section 9-7, beginning on Page 9-23 for detailed procedures to test cranking motor parts.

To check the action of the reduction gear on a 650 Series: Rotate the smaller pinion gear counterclockwise. It should rotate freely. Now, rotate the pinion gear clockwise. The pinion gear should rotate on the curved splines to the top of the shaft and then stop. Release the gear. The gear should quickly return back down to its original position. If the pinion gear does not function as expected, the reduction gear assembly must be replaced.

The negative brushes are mounted on the brush assembly plate. The positive brushes are attached to the positive terminal. When all the brushes are installed onto the plate, each brush is held in place with wound spring steel coils mounted horizontally to provide the tensioning force against the commutator. These coils are permanently riveted to the brush plate and are not removable. If they are broken or missing, a new plate must be purchased.

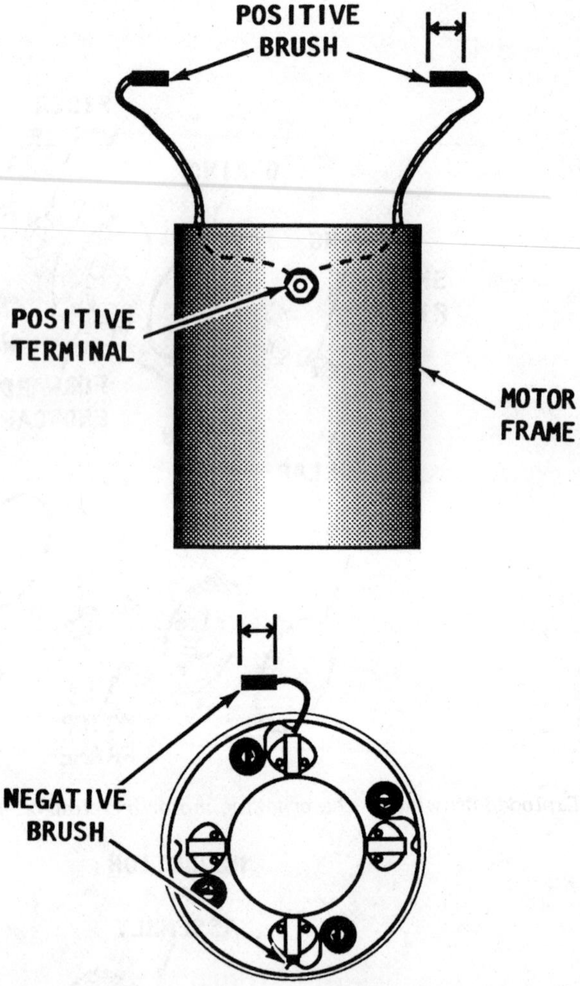

Measuring the length of the positive brushes (top), and the negative brushes (bottom). The text gives dimensions of new brushes for all Series and the minimum limits.

Measure the length of each brush.

If working on a 300 Series -- a new brush measures 12mm. A brush less than 8.5mm in length, must be replaced.

If working on a 440 or 550 Series -- a new brush measures 17.0mm. A brush less than 13.0mm in length, must be replaced.

If working on a 650 Series -- a new brush measures 12.5mm. A brush less than 6.5mm in length, must be replaced.

To replace the positive brushes, the positive terminal must be removed from the motor frame. Remove, the following parts and place them in order on the workbench to ensure they will be installed in exactly the same order: a nut, a spring washer, a large insulator, and finally an O-ring -- all installed on the terminal on the outside of the motor frame. Inside the motor frame, the brush leads are soldered to the terminal.

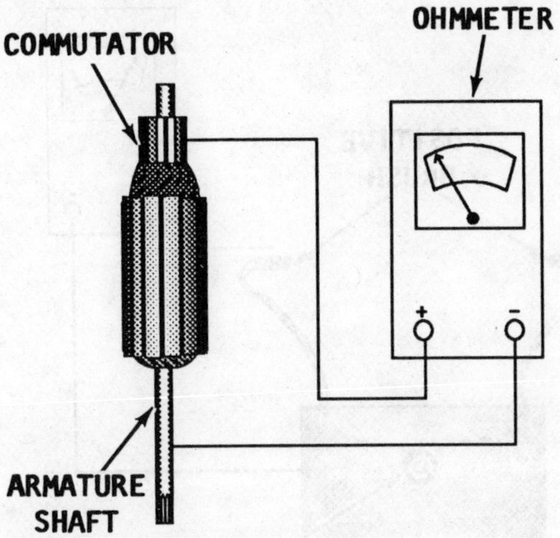

(1)

9-7 TESTING CRANKING MOTOR PARTS

SPECIAL WORDS

Most marine shops and all electric motor rebuild shops will test an armature for a very modest charge. If the armature has a short, it **MUST** be replaced.

Check the armature for a short circuit by placing it on a growler and holding a hack saw blade over the armature core while the armature is rotated. If the saw blade vibrates, the armature is shorted. Clean between the armature bars, and then check again on the growler. If the saw blade still vibrates, the armature must be replaced. Occasionally carbon dust from the brushes will short the armature. Therefore, blow the slots in the armature clean with compressed air.

1- Make contact with one probe of a test light or ohmmeter on the armature core or shaft. Make contact with the other probe on the commutator. If the light comes on,

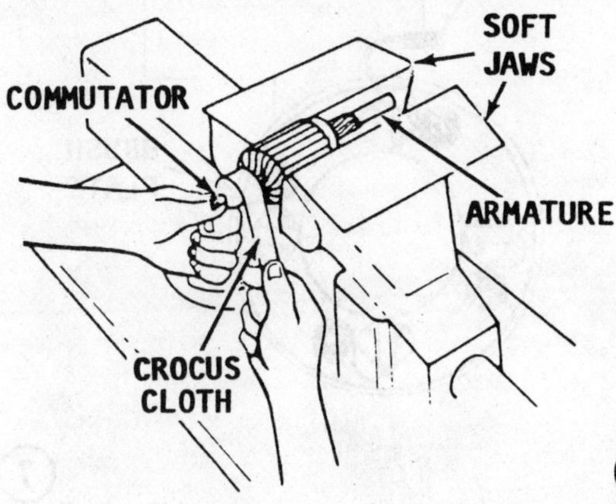

(2)

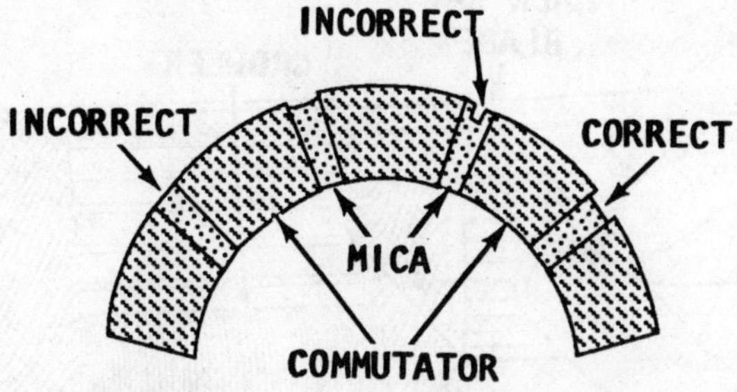

Simple line drawing to illustrate correct and incorrect mica cutting on a commutator.

or the meter indicates continuity, the armature is grounded and must be replaced.

Turning the Commutator

2- True the commutator, if necessary, in a lathe. **NEVER** undercut the mica because the brushes are harder than the insulation. Undercut the insulation between the commutator bars 1/32" (0.80mm) to the full width of the insulation and flat at the bottom. A triangular groove is not satisfactory. After the undercutting work is completed, clean out the slots carefully to remove dirt and copper dust. Sand the commutator lightly with No. 500 or No. 600 sandpaper to remove any burrs left from the undercutting.

3- Test light probes, placed on any two commutator bars, should light and indicate continuity.

4- Check the armature a second time on the growler for possible short circuits.

Positive Brush

5- Obtain an ohmmeter. Make contact with one lead of the ohmmeter to the positive brush. Make contact with the other

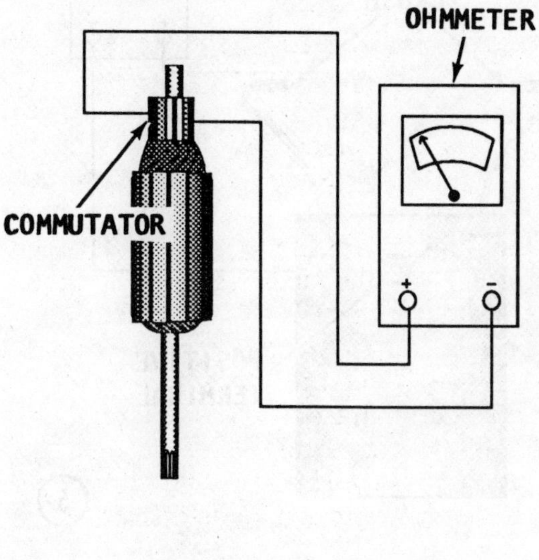

(3)

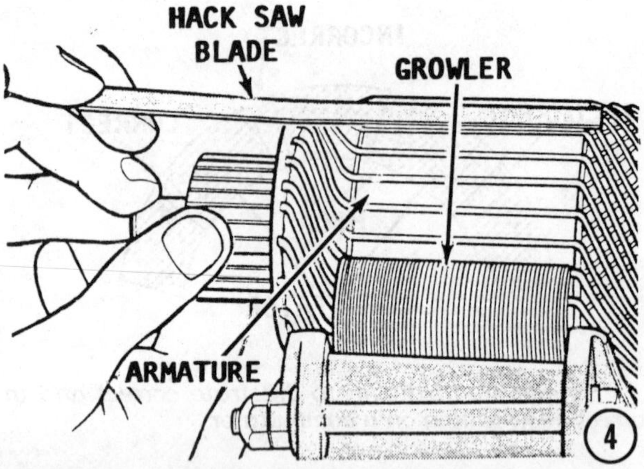

test lead to the end cap/motor frame. The ohmmeter should indicate **NO** continuity. If the meter indicates continuity, the positive lead is shorted to the motor frame and must be replaced.

6- Make contact with one ohmmeter lead to a positive brush. Make contact with the other ohmmeter lead to the positive terminal. The meter should register continuity. Repeat this test for the other positive brush. If the meter registers no continuity in either test, the positive brushes have an open in the circuit and must be replaced.

Negative Brush

7- Make contact with one ohmmeter lead to the negative brush. Make contact with the other test lead to the brush plate. The ohmmeter should indicate continuity. If the meter indicates no continuity, there is an open in the negative lead or the lead is not grounded properly to the brush plate.

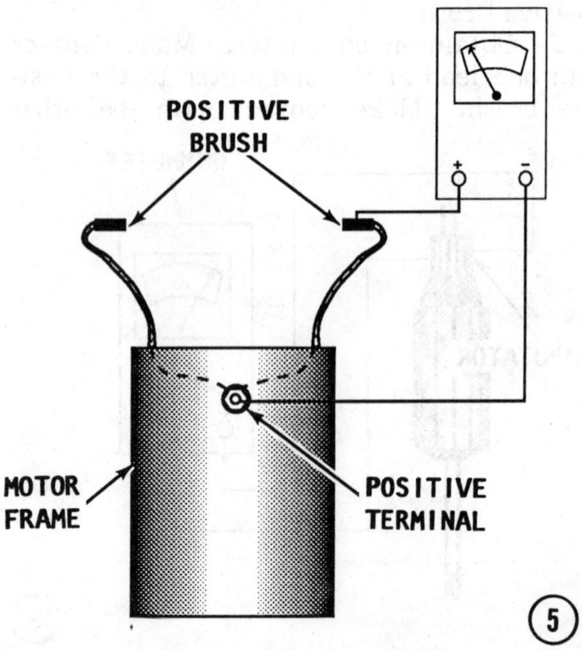

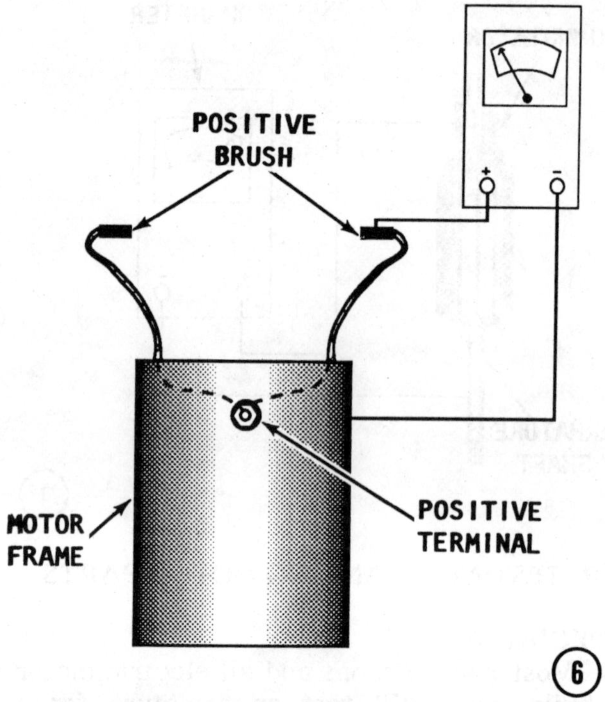

8- Make contact with the ohmmeter leads across the positive and negative brush holders on the brush plate. The meter should register no continuity. Repeat the test for the other pair of positive and negative brush holders. If the meter registers continuity in either test, the positive brush holders are shorted to the plate and the plate must be replaced.

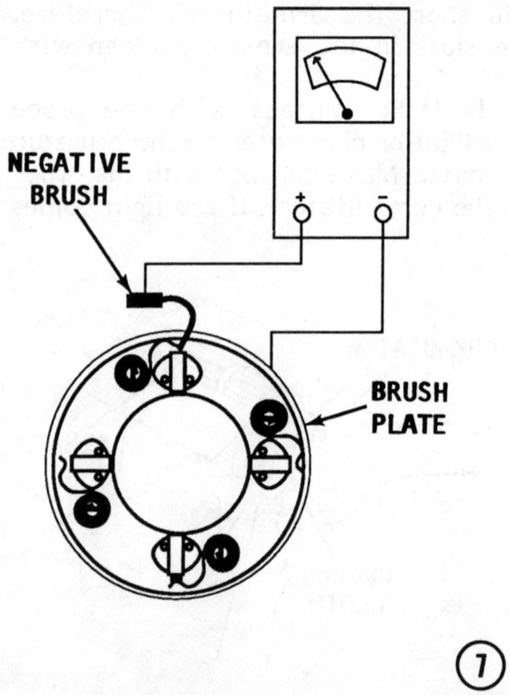

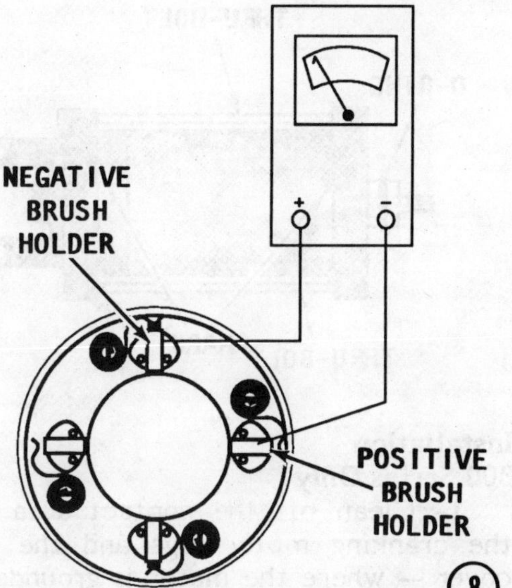

8

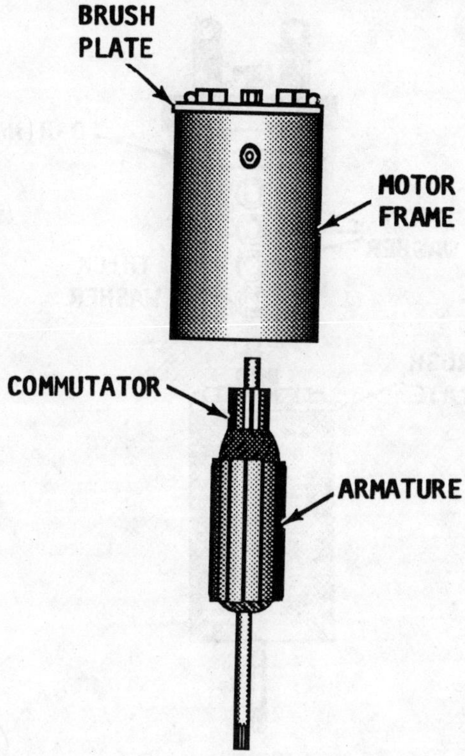

2

9-8 CRANKING MOTOR ASSEMBLING

Assembling
300 and 650 Series

1- If the positive brushes were removed or replaced, apply Loctite Superflex to the bolt hole in the motor frame. Insert the terminal, with positive brushes attached, through the bolt hole from the inside. Install a washer and an insulator over the terminal and then secure it in place with a locknut.

2- Install the brush plate over the motor frame, with the two positive brush leads indexed into the two slots in the plate. Install the positive brushes into their respective holders. Insert the armature shaft into the other end of the motor frame. If necessary, push back the brushes to allow the plate to slide over the commutator.

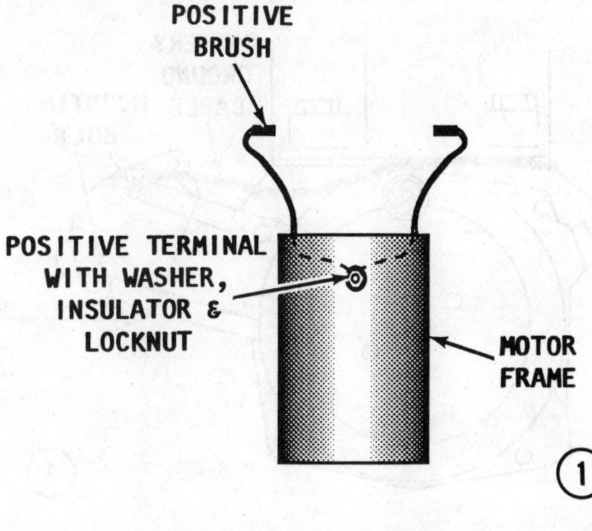

1

3- Check each of the four brush springs for tension. Rotate the armature shaft. The brushes must stay in contact with the commutator surface without binding.

4- Install the thick washer, followed by the thin washers over the commutator end of the armature. Install a **NEW** O-ring around the motor frame, and then install the front cover, aligning the indexing mark made during disassembling. Hold all the pieces of the motor together until the long thru-bolts are installed in a later step.

5- Install the two thrust washers, the fiber washer and the tabbed washer -- with tabs facing up, over the other end of the armature. Install a **NEW** O-ring around the motor frame, followed by the rear cover, aligning the marks made during disassembling. The tabs on the tabbed washer must

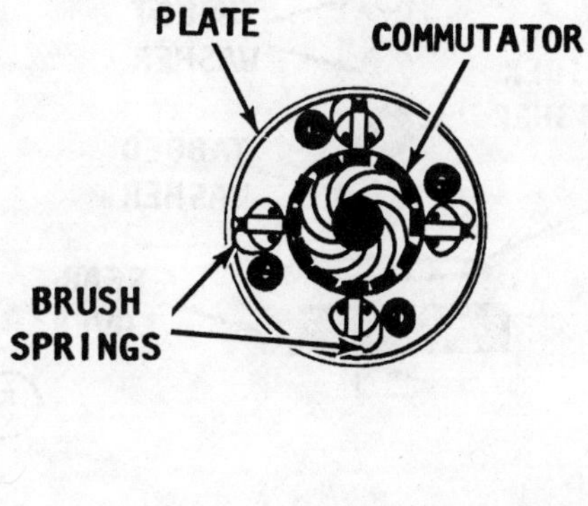

3

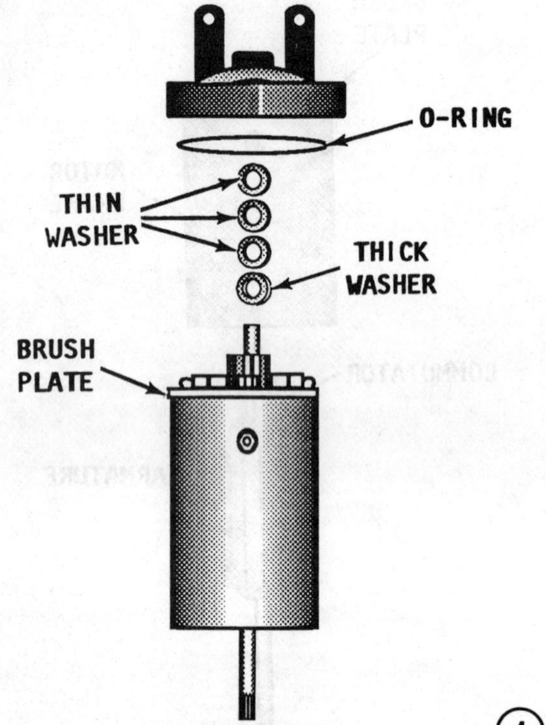

4

index into the four slots cut into the hub of the rear cover.

6- Install the two thru-bolts and tighten them securely. Install a new O-ring around the outside of the rear cover.

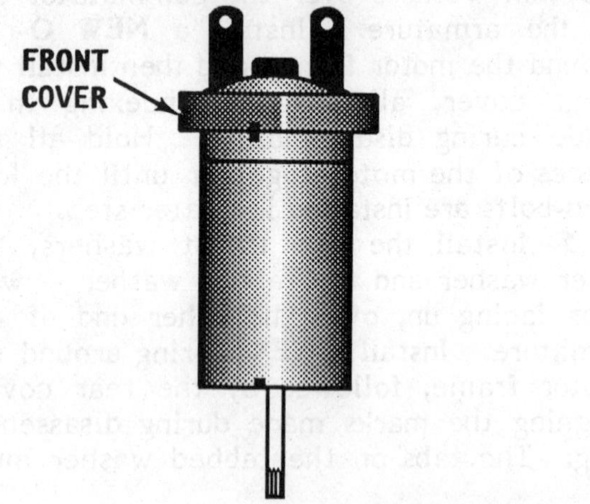

5

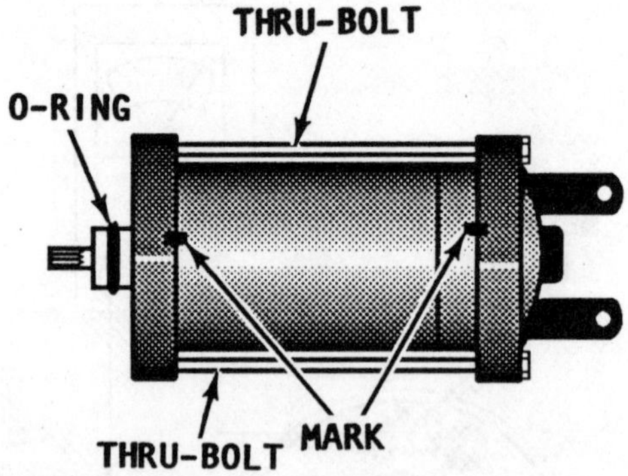

6

Installation
300 Series Only

1- Clean off the contact area between the cranking motor lugs and the flywheel cover -- where the motor is grounded. Coat the O-ring on the armature shaft with engine oil. Insert the cranking motor into the rear gear housing. Check to be sure the splines on the armature shaft index with the idle gear.

Install a new flywheel cover ring, and then install the flywheel cover. Tighten the attaching hardware securely.

Install the oil injection pump, with the attaching hardware. Unplug, and then connect the oil supply line, disconnected during disassembling.

CRITICAL WORDS

When the oil injection system was opened to remove the pump, air entered the system. This air **MUST** be purged -- "bled" before the engine is started. See Chapter 6, Section 6-10, beginning on Page 6-49.

Apply a coat of Loctite Lock N' Seal to the threads of both attaching bolts. Position the battery ground cable over one of the mounting holes as shown. Secure the

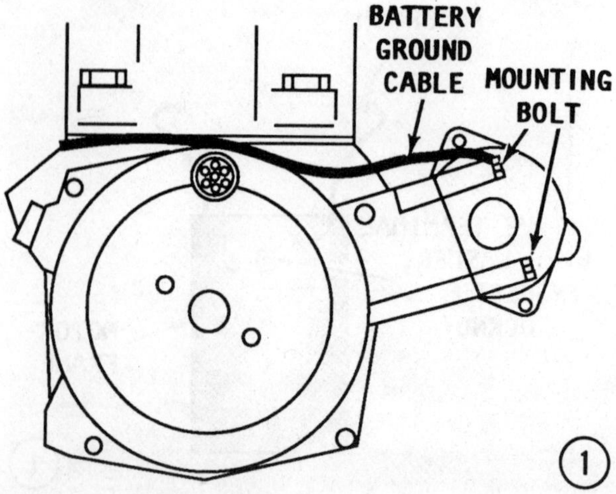

1

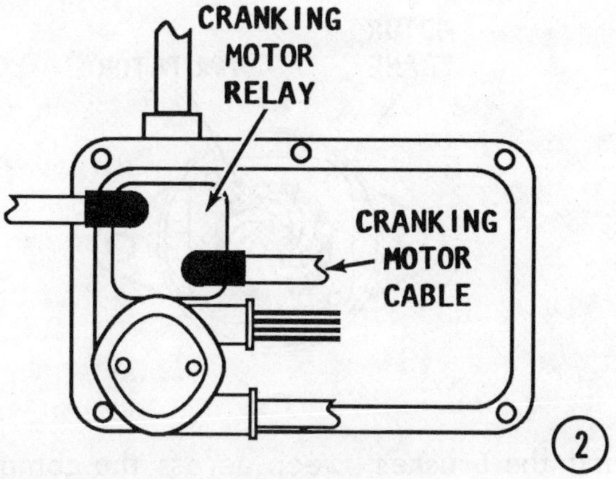

cranking motor to the flywheel cover with the two attaching bolts. Tighten the bolts securely.

2- Connect the cranking motor cable to the lower left relay terminal on the electric case.

Installation
650 Series Only

1- Clean the cranking motor lugs and the contact surface of the crankcase in the area where the motor is grounded. Coat the large O-ring on the forward end cap with engine oil. Position the motor against the crankcase and at the same time insert the end of the armature shaft into the reduction gear inside the flywheel cover.

2- If the battery ground cable was originally connected to the crankcase with the lower mounting bolt, then place the eye of the cable over the bolt and apply a non-permanent locking substance on both mounting bolts. Install and tighten the bolts securely.

3- Secure the cranking motor cable to the terminal on the motor frame. Pull the boot over the connection.

4- Connect the negative cable to the battery.

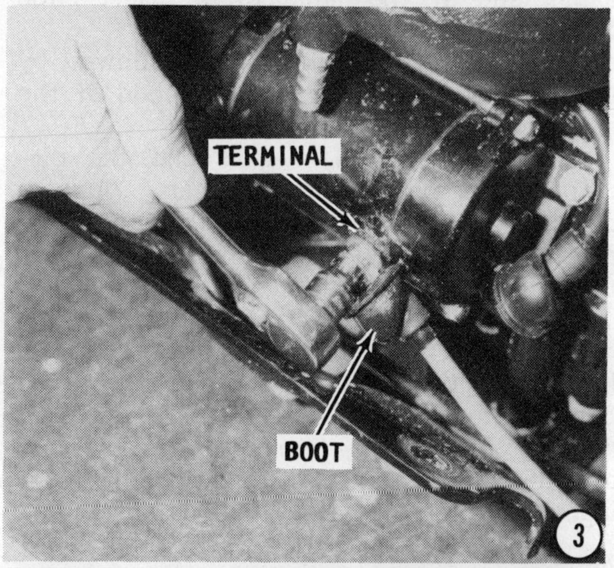

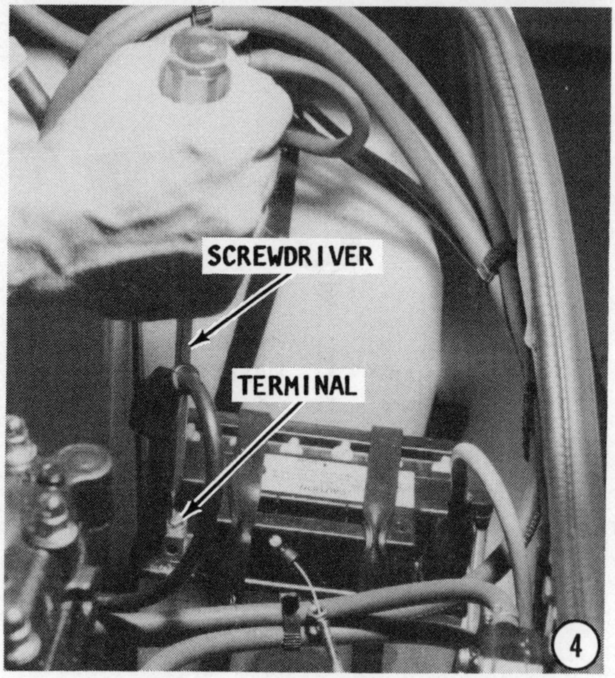

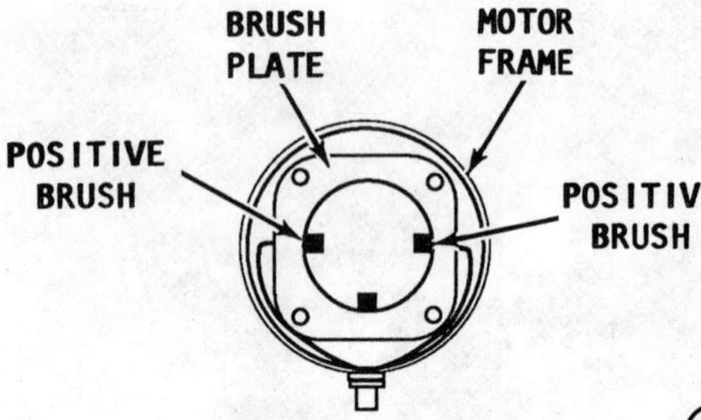

1

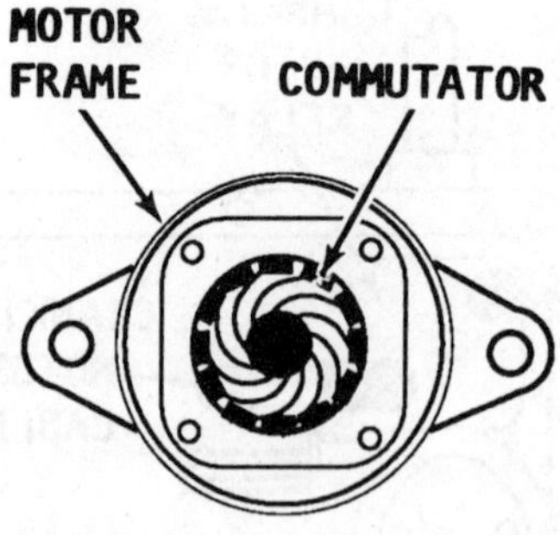

3

Cranking Motor Assembling 440 and 550 Series

1- Position the two positive brushes outside the motor frame. Place the brush plate inside the frame. Push the tensioning spring aside, and at the same time, insert each brush into its holder.

2- Slide the metal washer, followed by the fiber washer onto the spiral end of the armature shaft.

Stretch a new O-ring around the circumference of the front end cap. Apply a coating of water resistant lubricant to the armature shaft and install the end cap over the spiral end of the armature.

3- Push the commutator end of the armature shaft through the motor frame. Hold back the three brushes to allow the commutator to pass through. **TAKE CARE** not to distort the brush springs. Once assembled, rotate the shaft inside the bushing and check to be sure there is no binding and the brushes sweep across the commutator smoothly.

4- Install the original number of thrust washers over the rear of the armature shaft. If the armature end play was measured and found it be out of specifications, then adjust the number/thickness of thrust washers accordingly.

Apply a thin coating of Loctite Superflex, or equivalent waterproof anti-sieze lubricant, onto the bushing in the end cap.

Stretch a new O-ring around the circumference of the rear end cap and install it over the brush plate. **TAKE CARE** not to lose the brushes from the commutator. Seat the end cap into the motor frame with the marks made prior to disassembly aligned and the bolt and screw holes in the rear end cap aligned with those in the brush plate.

Make sure each of the two thru bolts and two screws have O-rings in good condition. Install the two brush plate securing screws and tighten them securely.

Check the alignment of the rear end cap with the marks made during disassembly.

Now, hold it all together and slide the two thru-bolts through the rear end cap, through the motor frame, and then thread them into the forward end cap.

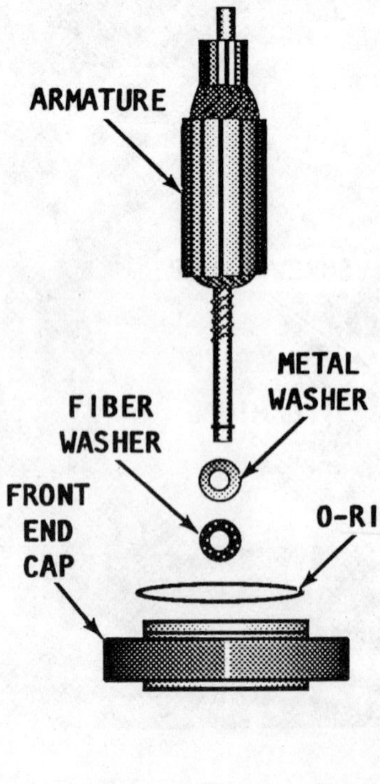

2

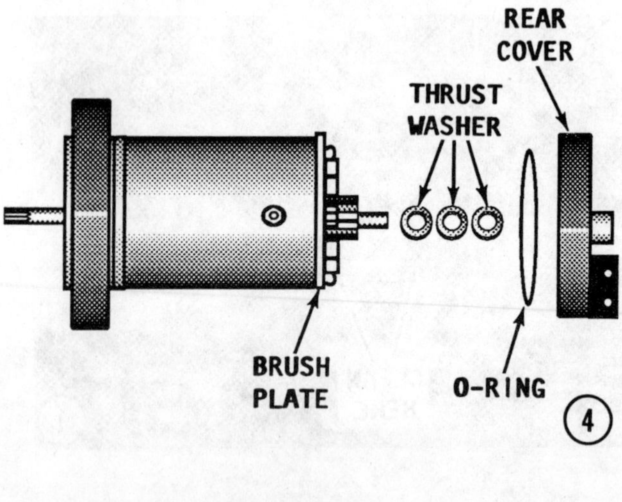

4

SPECIAL WORDS

Installing the two thru- bolts is not as simple a task as it sounds. The reason: as the bolts pass the magnets inside the motor frame, the magnets will attract the bolts, "pulling" them out of line. With a bit of patience, the bolts can be made to align with the holes in the upper end cap and the threads started.

After the bolts are successfully threaded into the end cap, tighten them securely.

Position the large O-ring into the groove around the upper end cap. Install the following components onto the armature shaft in the sequence given: first, the pinion gear, followed by the spring, the collar, and finally the snap ring. It will be necessary to hold the collar down tight against the spring while the snap ring is installed. The ring must be crimped around the groove in the shaft with a pair of pliers. When properly installed the collar will completely hide the snap ring.

Cranking Motor Installation
440 and 550 Series

1- Apply a coating of Loctite Superflex to the O-ring around the forward end cap. Apply engine oil to the pinion gear. Connect the Red lead to the cranking motor terminal. Position the motor from the rear of the engine up against the rear flywheel housing. Apply Loctite Lock N' Seal to the threads of all four mounting bolts. Install the front two mounting bolts and tighten them to a torque value of 12 ft lb (16Nm).

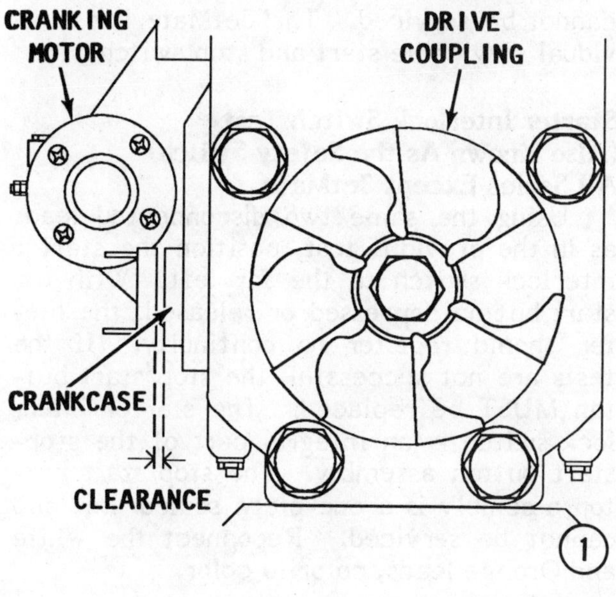

Before installing the two rear mounting bolts, measure the distance between the rear mounting bracket and the crankcase. Fill this space with the original pieces of shim material or purchased pieces. Shim material is available in thicknesses of 0.4, 0.6 and 0.8mm. Install the two rear mounting bolts and tighten them to a torque value of 52 in lb (6Nm).

2- Connect the battery ground cable to the battery.

All Series
Check Completed Work

Secure the craft in a test tank or body of water. If a body of water or tank is not possible, connect a flush attachment and hose to the jet pump.

CAUTION

Water must circulate through the jet pump -- to and from the engine, anytime the engine is operating. Circulating water will prevent overheating -- which could cause damage to moving engine parts and possible engine seizure.

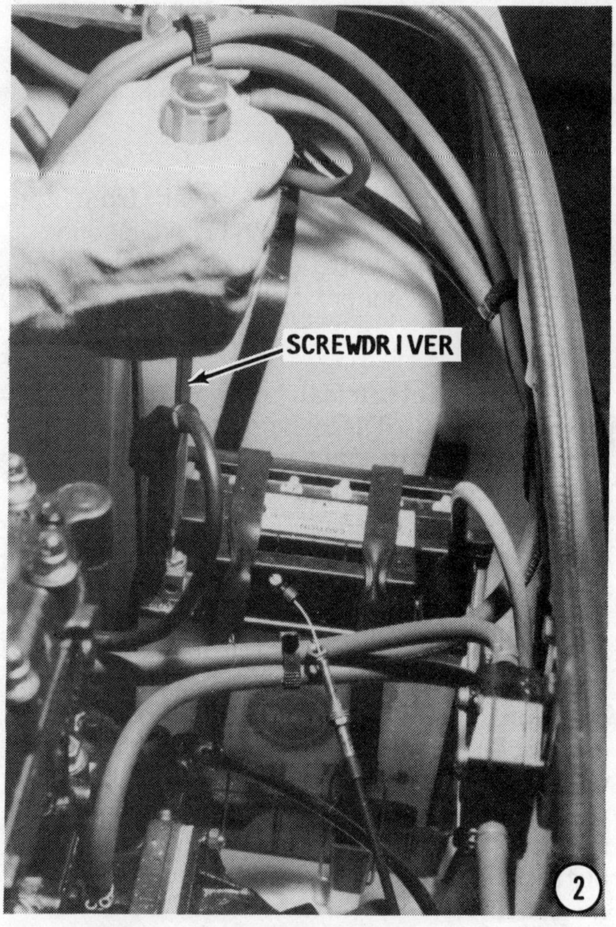

NEVER, AGAIN, NEVER operate the engine at high speed with a flush device attached. The engine, operating at high speed with such a device attached, would **RUNAWAY** from lack of a load on the impeller, causing extensive damage.

Crank the engine with the cranking motor and start the unit. Shut the engine down and start it several times to check operation of the cranking motor.

9-9 TESTING OTHER ELECTRICAL COMPONENTS

This short section provides testing procedures for other electrical parts installed on the engine. If a unit fails the testing, the faulty part **MUST** be replaced. In most cases, removal and installation is through attaching hardware.

Colored Lead Identification

Test leads are identified by their color to aid in locating the correct lead. However, the manufacturer has been known to change lead colors. Therefore, identify the component and consult the wiring diagram for the correct component connections if colors vary.

Start Button Test

All Models Except JetMate

Remove the connector cover from the electrical box, see Page 9-12. Disconnect the Orange and White start switch leads at their quick-disconnect fittings.

JetMate

Open the electrical box and remove the grommet cap. Disconnect the Red/Purple and Green/White start switch leads at their quick-disconnect fitting.

All Series

Obtain an ohmmeter. Select the Rx1 ohm scale. Connect the ohmmeter leads across the two disconnected leads from the connector cover or from the control stick. Position the starter interlock switch (if equipped), to the far right -- the small lever under the start button, on the right. With the start button depressed, the meter should register almost zero ohms. With the start button released, the meter should register no continuity. Both tests **MUST** be successful. If the tests are not successful, the start button **MUST** be replaced. On all models except the JetMate, the stop/start button assembly is a one piece sealed unit and cannot be serviced. The JetMate has individual servicable start and stop switches.

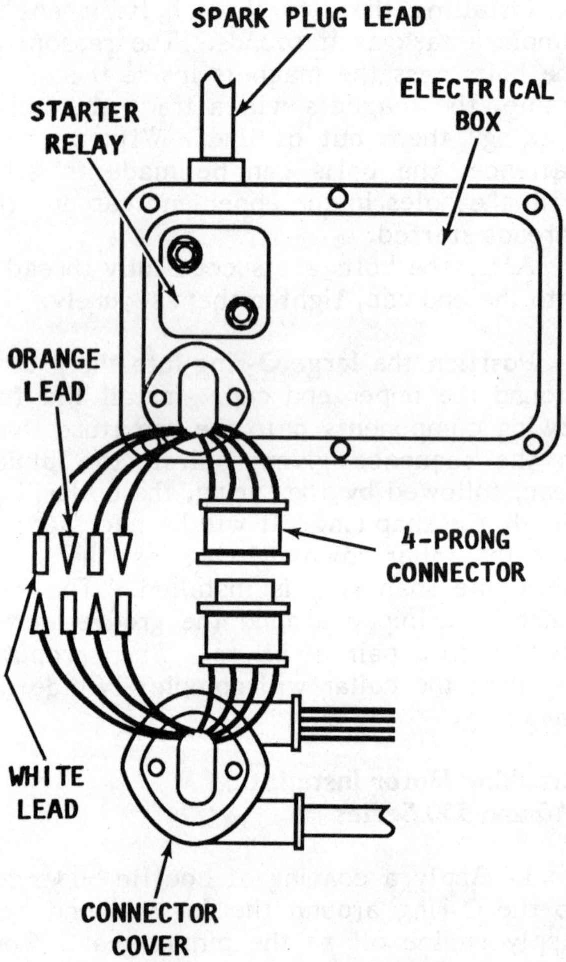

Line drawing of a typical electrical box to illustrate the connections for resistance tests outlined in this section.

Starter Interlock Switch Test (Also Known As the Safety Switch)

All Series Except JetMate

Using the same two disconnected leads as in the previous test, position the starter interlock switch to the far left. With the start button depressed or released, the meter should register no continuity. If the tests are not successful, the stop/start button **MUST** be replaced. The starter interlock switch is an integral part of the stop-start button assembly. The stop/start button assembly is a one piece sealed unit and cannot be serviced. Reconnect the White and Orange leads, color to color.

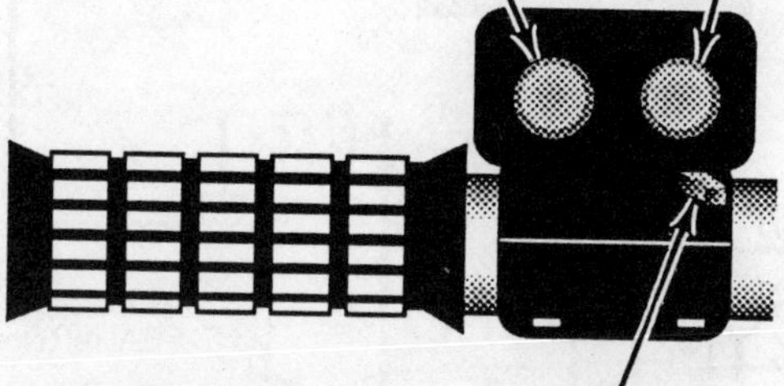

Tests for the stop, start, and the starter interlock switch are given on this page.

Starter Interlock Switch Test
(Also Known As the Safety Switch)
JetMate only

The quick-disconnect fittings for the starter lockout switch are located under the control panel. Disconnect the Green/White and the Black/Yellow leads from the switch. Obtain an ohmmeter and select the Rx1 ohm scale. With the shift lever in the engine starting position, the meter should register almost zero ohms. With the shift lever in the Forward or Reverse position, the meter should register no continuity.

If the tests are not successful, the starter interlock switch must be replaced. The switch assembly is a one piece sealed unit and cannot be serviced. Reconnect the Green/White and the Black/Yellow leads, color to color.

Stop Switch Test
All Series Except JetMate and SC

Remove the connector cover from the electric case, see Page 9-12. Disconnect the Blue and Black stop switch leads at their quick disconnect fittings.
Obtain an ohmmeter. Select the Rx1 ohm scale. Connect the ohmmeter leads across the two disconnected leads from the connector cover. Position the starter interlock switch to the far right. With the stop button depressed, the meter should register almost zero ohms. With the stop button released, the ohmmeter should register no continuity.

Both tests **MUST** be successful. If the tests are not successful, the stop button **MUST** be replaced. The stop/start button assembly is a one piece sealed unit and cannot be serviced.

JetMate and SC

These Series may have an engine kill switch on the control panel and an engine stop switch on the control stick. To test the engine stop switch: Open the electric case and remove the grommet cap. Identify the White and Black stop switch leads. Disconnect the two leads at their quick disconnect fittings. Obtain and ohmmeter and select the Rx1 ohm scale. Connect the meter leads across the disconnected switch leads. With the safety lanyard in place behind the kill switch and the stop button depressed, the meter should register almost zero ohms. With the stop button released, the ohmmeter should register no continuity.

Both tests **MUST** be successful. If the tests are not successful, the stop button **MUST** be replaced. Remove the top of the control stick, disconnect the White and Black leads at their quick disconnect fittings and remove the switch.

To test the engine "kill" switch: Locate the White and Black leads from the "kill" switch behind the control panel. Disconnect both leads at their quick-disconnect fittings. Select the Rx1 scale on the ohmmeter. Make contact with the Red meter lead to the White switch lead and the negative meter lead to the Black switch lead. With the lanyard in place behind the switch, the meter should register no continuity. With the lanyard removed from the switch, the meter should register almost zero ohms.

These readings may seem incorrect. However, bear in mind that the "kill" switch leads are connected directly to the CDI Igniter. The leads are not directly part of the ignition circuit, which would ground the ignition if the lanyard was removed from the switch.

Stop Switch Relay Test
If Equipped

The stop switch relay is an electronic solid state device to prevent the engine from restarting after the stop switch has been released. If an engine starts up again immediately after the stop switch has been released, the stop switch relay is defective and must be replaced. A defective relay will allow the engine to restart due to the momentum of the flywheel.

POSITIVE METER LEAD CONNECTION

NEGATIVE METER LEAD CONNECTION	BLACK/WHITE	BLUE	BLACK	BROWN
BLACK/WHITE		SOME READING	∞ NO READING	∞ NO READING
BLUE	SOME READING		∞ NO READING	∞ NO READING
BLACK	∞ NO READING	∞ NO READING		SOME READING
BROWN	∞ NO READING	∞ NO READING	SOME READING	

Simple chart to visually indicate the stop switch relay tests, listed in the text.

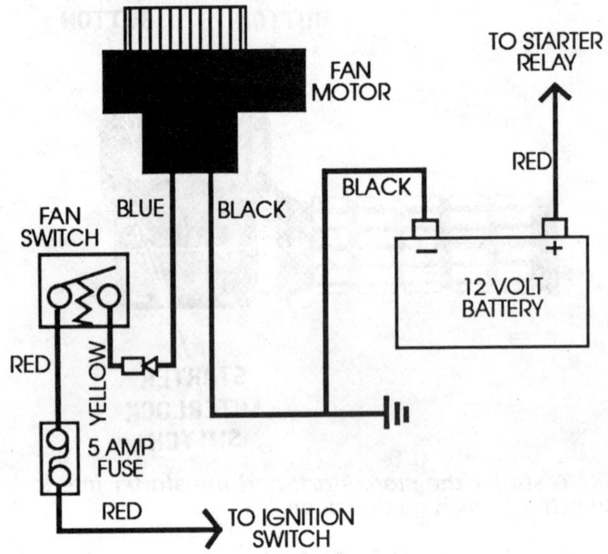

Functional diagram of the blower fan circuit installed on some 650 Series units.

On most models, the stop switch relay is located inside the electrical box and has Black, Brown, Blue, and Black/White leads. Open the electrical box, see Page 9-12. Disconnect the four leads to the stop switch relay at the quick-disconnect fittings. Obtain an ohmmeter. Select the Rx100 ohm scale. The manufacturer does not give specific resistance values between different colored leads. There is either "no" reading, or "some" reading on this resistance scale. Observe the correct meter lead connections as shown in the accompanying table. If any of the test values are not as expected, replace the relay.

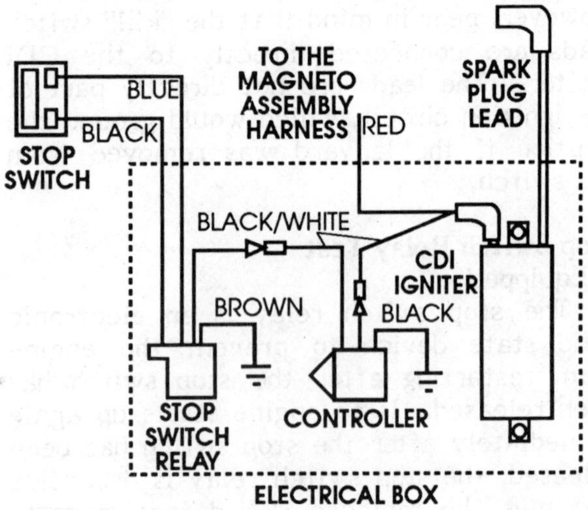

Simple functional diagram of the stop switch circuit for the 300 Series.

Testing Controller 300 Series Only

In addition to the stop switch relay, the stop switch circuit on the 300 Series has an additional relay called the "controller", connected between the stop switch relay and the CDI igniter. This controller is housed inside the electrical box with two Black/-White leads and a Black ground lead attached.

To test the controller, obtain an ohmmeter and select the Rx1K (1,000) ohm scale. Disconnect the controller leads at the two quick-disconnect fittings.

Measure the resistance between the two Black/White leads -- either meter lead to either Black/White lead. Reverse the meter leads and measure the resistance again. Now, for one of the tests the meter should indicate 3-4K ohms. For the other test, the meter should indicate infinity. Since both controller leads are the same color -- it cannot be stated which test will indicate the ohms and which will indicate infinity.

If the meter readings are not as expected, the controller relay must be replaced.

Electric Bilge Pump Test Some 650 Series Only

To test the electric bilge pump: Obtain a fully charged 12 volt battery and two leads. Disconnect the two pump leads --the Black ground lead and the Brown lead from the bilge pump switch. Connect the battery

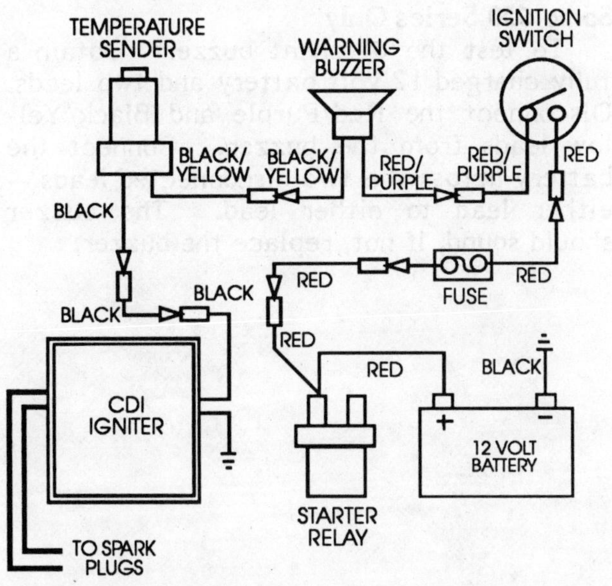

Functional diagram of the temperature warning system circuit installed on some 650 Series craft.

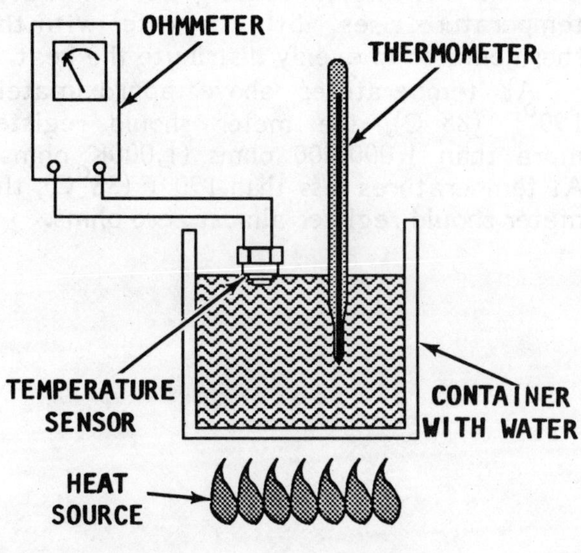

Testing the operation of the temperature sensor, as explained in the text.

across the two disconnected leads --positive battery terminal to the Brown disconnected lead and negative battery terminal to the Black ground lead. Check to be sure adequate water is in the bilge area for the pump to function. The pump should pump out the water from the bilge, if not, replace the pump.

Blower Fan Test
Some 650 Series Only

If the fan does not operate, the fault could lie in one or more of three areas:

Fan 5 amp fuse
Fan switch
Fan motor

To inspect the fuse, remove the fuse from the fuse holder. If the fuse is suspect, replace it with a new one.

To test the fan switch, obtain an ohmmeter and select the Rx1 ohm scale. Connect the meter leads across the Red and Yellow leads to the fan switch. With the switch in the **OFF** position, the meter should register no continuity. With the switch in the **ON** position, the meter should register almost zero ohms. Replace the switch if the results are not as expected.

To test the fan motor: Obtain a fully charged 12 volt battery and two leads. Disconnect the two fan motor leads -- the Black ground lead and the Blue lead from the fan switch. Connect the battery across the two disconnected leads -- positive battery terminal to the Blue disconnected lead and negative battery terminal to the Black ground lead. The fan should operate. If not, replace the fan.

Temperature Warning System
Some 650 Series Only

The temperature sender is located inline with the by-pass hose to measure the temperature of the cooling water. If the cooling water temperature rises above 203°F (95°C), the sender sends a signal to the warning buzzer to alert the operator of the craft. If the warning buzzer sounds, shut down the engine immediately to prevent damage from overheating.

To test operation of the temperature sender, disconnect the Black/Yellow and Black leads at their quick-disconnect fittings. Remove the sender from the by-pass hose. Obtain an ohmmeter and select the Rx1000 (Rx1K) ohm scale. Connect the meter leads across the sender leads. Heat a pan of water on the stove and have a thermometer handy to measure the water temperature. Submerge only the sensor end in the water, as shown, without the sensor touching the sides or the bottom of the pan. Observe the meter reading as the water temperature rises. Stir the water with the thermometer to evenly distribute the heat.

At temperatures above approximately 190°F (88°C), the meter should register more than 1,000,000 ohms (1,000K ohms). At temperatures less than 190°F (88°C), the meter should register almost zero ohms.

Overheat Buzzer Test
Some 650 Series Only

To test the overheat buzzer: Obtain a fully charged 12 volt battery and two leads. Disconnect the Red/Purple and Black/Yellow leads from the buzzer. Connect the battery across the two disconnected leads -- either lead to either lead. The buzzer should sound, if not, replace the buzzer.

10
JET PUMP

10-1 INTRODUCTION

Model Identification and Chapter Coverage

Kawasaki personal watercraft use both axial and mixed flow jet pumps. The 300 Series, 440 Series, and 650 Series use the axial flow pump and the 550 Series uses the mixed flow pump.

Two different designs of axial flow pumps are used. The axial flow pump on the 300 Series and 440 Series has a one-piece impeller/driveshaft -- supported by two ball bearings at the front coupler and a bushing type bearing in the pump case. The axial pump on the 650 Series, including the Jet-Mate, has a two-piece impeller shaft -- actually a long drive shaft and a short impeller shaft -- each shaft with two ball bearings at each end.

The engineers at Kawasaki provide impeller-to-pump case clearances for **ALL** craft equipped with axial or mixed flow jet pumps. Therefore, a special section -- 10-2 -- deals with impeller-to-pump case clearance. Shimming procedures for the mixed flow pump installed on the 550 Series is covered in Section 10-5. No shimming is required on the axial flow pumps, but impeller-to-pump case clearances **MUST** be observed.

One set of service procedures covering the two designs of axial flow pumps **AND** the mixed flow pump is presented in Section 10-4. Where differences occur between the designs or the type flow pump, these differences, will be clearly identified.

The bearing housing supporting the driveshaft at the bulkhead is covered in Section 10-6.

Engine crankshaft-to-impeller shaft alignment is covered in the engine chapter in Section 8-41.

SPECIAL WORDS ON JET PUMPS

The axial flow pump is installed on one of the larger horsepower engines. These engines are capable of delivering higher engine/pump shaft rpms, thus moving the craft at greater speeds.

Both the mixed flow and axial flow pumps have the same basic components.

Pump parts are **NOT** interchangeable, because many of the castings are of unique design even though they share the same function and are similar in appearance. The main difference between the two pumps, regarding maintenance, concerns positioning the impeller on the impeller shaft.

The mixed flow pump is equipped with shim material to position the impeller at a specific distance from the pump case.

The axial flow pump does not have shim material to position the impeller.

Neither the axial or the mixed flow pump has a removable liner, just a solid aluminum casting. Therefore, if the impeller-to-pump case clearance becomes exces-

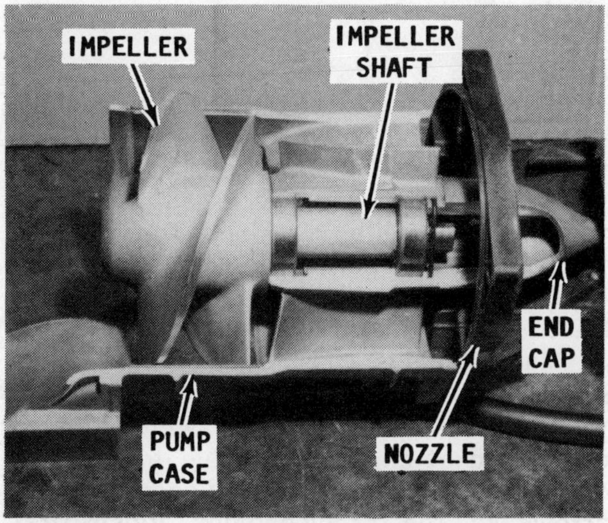

Cutaway of a 650 Series axial flow jet pump, with a few major parts identified.

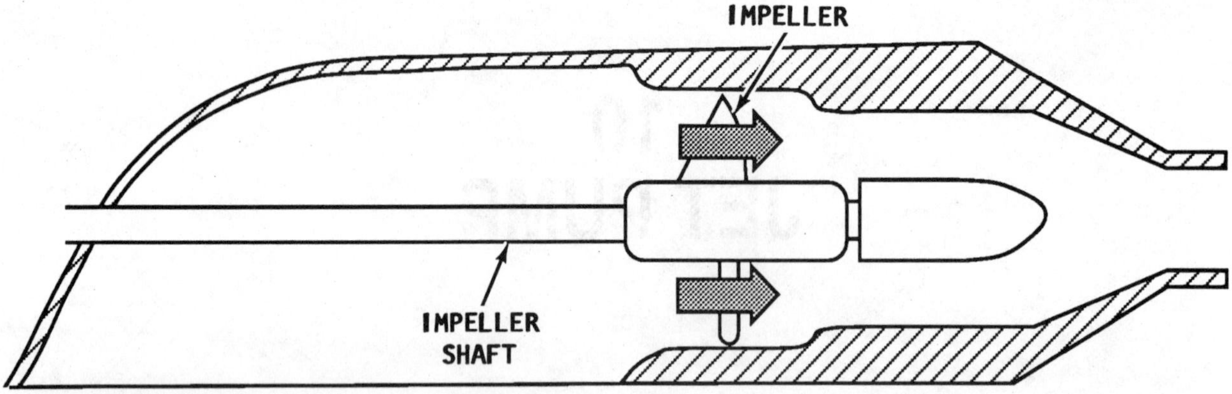

Simple cross-section line drawing to depict water flow through an axial flow jet pump. The water passes through parallel to the axis of impeller rotation.

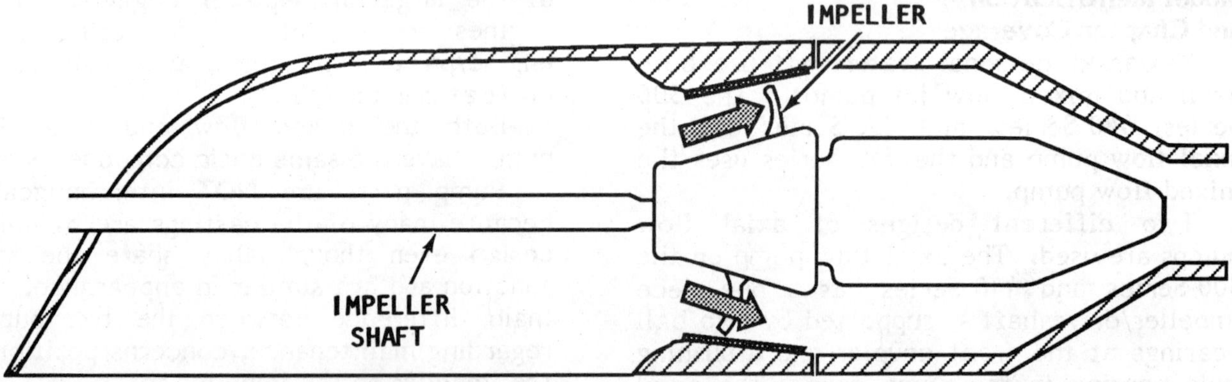

Line drawing to illustrate water flow through a mixed flow jet pump. Notice how the water is directed upward at an angle to the axis of impeller rotation.

sive, and cannot be corrected by the addition of shim material -- on the mixed flow pump -- or the pump case is deeply grooved, the case **MUST** be replaced.

On older models, before 1983, the pump case was equipped with a non-removable liner. The liner was never sold as a replacement part.

Axial Flow

Water in an axial pump moves on a single axis, as depicted in the adjacent illustration, thus the word "axial". In simple terms, water is ingested and discharged parallel to the axis of impeller rotation. The inside of the pump case is parallel to the water flow.

Mixed Flow

Many experts consider the mixed flow system to be the most efficient for personal watercraft application. The water flow is discharged at an inclined angle to the axis of impeller rotation, as indicated by the arrows in the accompanying illustration. The impeller blades are inclined in such a manner to cause the water to be thrown off at a 45 degree angle to the impeller shaft within the pump case.

From the pump case, the water then flows through the guide vane casting and forced out through the outlet nozzle and steering nozzle. The vanes in the guide vane casting help straighten the water flow through the nozzle. The inside of the pump case is inclined at an angle to the water flow. Therefore, this design pump incorporates shim material to space the impeller blades at a specified distance from the pump case. Mixed flow pump efficiency may decrease as much as 30% if the impeller-to-pump case clearance increases by only 0.04" (1mm).

Axial or Mixed?

Axial and mixed flow pumps are intended for different applications. One is not necessarily better than the other. The axial pump can be operated at higher engine rpm and is more suitable for racing. A mixed flow pump adds thrust at low engine rpm due to the pump design. Therefore, the mixed flow pump performs more efficiently at low engine rpm. This type pump is more suitable for a craft equipped to carry extra weight or intended to pull a water skier.

A jet pump seldom requires service except for the following:

Impeller-to-pump case clearance becomes excessive.
Impeller is damaged.
The oil seal around the impeller shaft fails -- allowing water to enter hull.
Silicone seal around the perimeter of the pump case allows the pump to ingest air.

10-2 IMPELLERS

Impellers provide thrust using a combination of water flow and water pressure in a closed environment.

The blades of an impeller overlap, thus water is trapped and forced through the impeller into the pump case and outlet nozzle. The water moving past the impeller resists cavitation because it is under pressure. Blades in the pump case are angled exactly opposite to those of the impeller. These blades redirect the water to establish a concentrated flow through the nozzle.

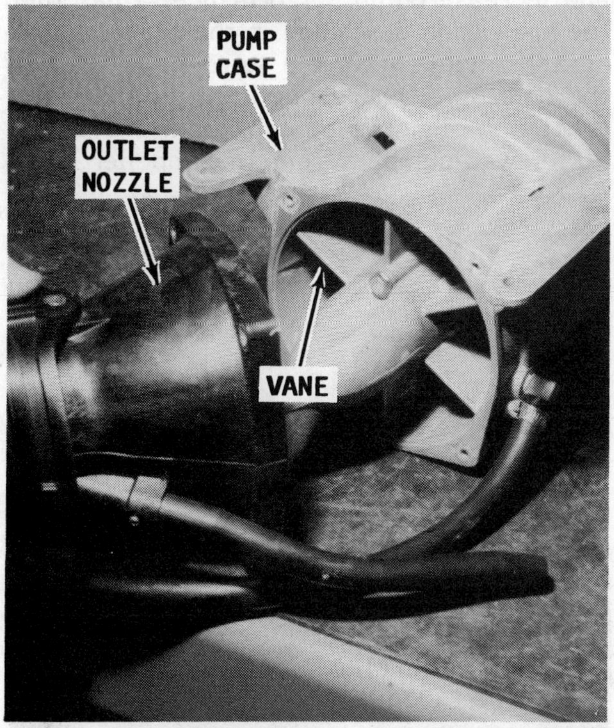

The vanes in the pump case straighten water flow and the conical shape of the outlet nozzle increases flow velocity.

The manufacturer recommends a radius of about 1/64" (0.3-0.5mm) for the leading edge of the impeller blade. A sharper radius will result in cavitation and a greater radius will reduce pump efficiency.

Cavitation Burns

Cavitation burns are the worst enemy of an impeller. These burns literally "eat away" material from the impeller, leaving holes and weakening the impeller structure. In extreme cases entire blades have been known to "depart" from the impeller hub because the cavitation burns at the base of the blade were so severe.

Cavitation burns are the result of imperfections (damage) on the impeller blades or air mixed with the water flow.

Cavitation burns may be caused by:

Wave jumping - air is sucked into the pump as the craft leaves the surface of the water.

Worn seals between the impeller and intake housings -- allowing air to enter.

Leading edge of the impeller is damaged.

To explain exactly what causes cavitation burns, first, let us examine the last cause. If the impeller leading edge is damaged and "mushrooms" over the blade, a low pressure area will form under the mushroomed lip.

Under atmospheric pressure -- at sea level -- water boils at 212°F. Anytime there is an imperfection on the surface of an impeller blade and water passes over that imperfection, the water pressure is lowered. When the pressure is lowered, water will boil at a much lower temperature. Therefore, air bubbles will form in the boiling water under the mushroomed lip. The bubbles will then creep down the surface of the blade and accumulate at the impeller hub. A high pressure area is formed at the base of the blade. Here, the air bubbles will collapse and reform back into water with a release of energy. This energy is absorbed by the impeller and the end result is impeller material is "eaten away".

The impeller should be dressed to a slightly rounded edge. It is impossible to

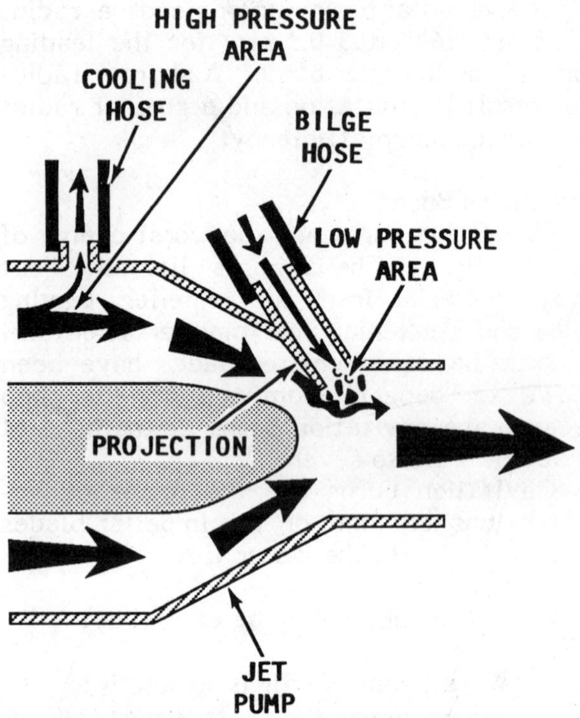

The direction of water flow through the cooling hose and the bilge hose is opposite. The cooling hose routes water from the pump to the engine. Water is "vacuumed" from the bilge through the bilge hose to the pump where it is forced out through the nozzle.

achieve a perfectly straight sharp edge, as one side will always be convex and the other side concave. The concave side will form an area of low pressure and encourage cavitation as described above.

The other two causes of cavitation -- defective seals and wave jumping -- occurs when air is sucked in with the water flow. The same principles apply, as just described, in the previous paragraphs. Air bubbles will creep down the surface of the blade and accumulate at the impeller hub. A high pressure area is formed at the base of the blade. Here, the air bubbles will collapse and reform back into water with a release of energy. As explained earlier, this energy is absorbed by the impeller and the end result -- impeller material is "eaten away".

Cooling Water and Bilge Hoses

On most pumps, two hoses are attached to the jet pump housing. One hose channels some of the water flowing through the pump to the exhaust manifold -- the hottest part of the engine -- to cool the block during operation. The other hose siphons water from the bottom of the hull, and is referred to as the bilge system. The cooling hose has water flowing **FROM** the pump. The bilge hose has water flowing **TO** the pump. Both hoses attach to the pump housing in a similar manner.

What determines the direction of water flow?

The answer to this question is in a simple explanation of high and low pressure areas along the inside surface of the housing.

Close inspection of the area inside the housing at the cooling water hose fitting reveals a smooth rounded shoulder with no obstruction or obstacle to impede the flow of water. A high pressure area develops here and draws the water down the hose attached to the fitting.

Further inspection of the area inside the housing at the bilge hose fitting reveals a small protrusion around the opening. This protrusion causes disturbance to the water flow and creates an area of low pressure. This low pressure area will have the effect of emptying air and water from the hose aft with the impeller water flow -- similar to the action of a vacuum cleaner, as depicted in the accompanying illustration. If the other end of the hose is submerged in water inside the hull, the water will be "vacuumed" out. In this manner, the bilge system drains the bilge.

Many an owner -- with good intentions -- thinking his action would smooth water flow and increase pump performance -- has filed the protrusion described in the previous paragraph smooth with the surrounding area of the housing. During operation of the water craft, the bilge system worked in **REVERSE.** Water was pumped from the impeller hous-

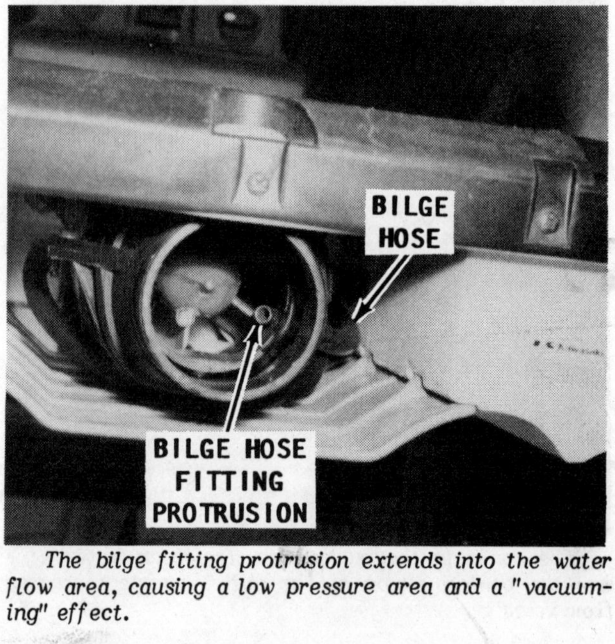

The bilge fitting protrusion extends into the water flow area, causing a low pressure area and a "vacuuming" effect.

ing into the bilge area, quickly filling the bilge and possibly sinking the craft before the rider had time to realize what was happening.

10-3 IMPELLER-TO-PUMP CASE CLEARANCE -- ALL AXIAL AND MIXED FLOW JET PUMPS

SPECIAL WORDS

The following procedure may be performed with the watercraft elevated on saw horses enabling a person to work underneath, or with the watercraft on its side allowing access to the rock grate and impeller.

If the impeller-to-pump case clearance is to be determined with the watercraft raised on saw horses, the battery may be left secured in the craft.

If the watercraft is to be positioned on its side, first disconnect the electrical leads at both battery terminals, and then remove the battery from the watercraft.

Remove the three bolts and washers securing the rock grate to the jet pump. Insert a feeler gauge between each impeller blade and the pump case and determine the clearance. Make a note of all three clearances, one for each blade. Calculate the average of the three clearances.

On a new craft the clearance will be what the manufacturer refers to as the "standard" clearance -- the first column. The second column is the "service limit". This value is the maximum allowable clearance.

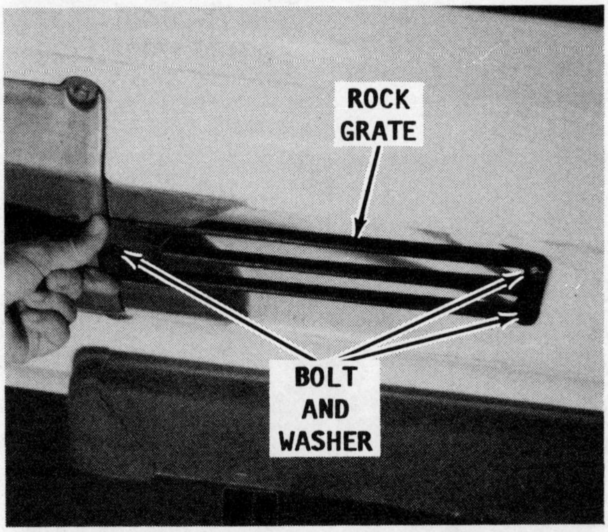

Removing the rock grate to gain access to the jet pump in order to measure clearance between the impeller and the pump case, as explained in the text.

IMPELLER-TO-PUMP CASE CLEARANCE

Series	Standard Clearance	Service Limit
300	0.0138-0.0177" (0.35-0.45mm)	0.0350" (0.87mm)
440	0.0138-0.0177" (0.35-0.45mm)	0.0350" (0.87mm)
550	0.0079-0.0118" (0.20-0.30mm)	0.0236" 0.60mm)
All 650 Including JetMate	0.0079-0.0118" (0.20-0.30mm)	0.0236" (0.60mm)

Axial Flow Pump
300 Series, 440 Series, and 650 Series

If the clearance is less then the maximum value, no action is required.

If the clearance is more than the service limit, inspect the condition of the pump case. If the pump case has scratches deeper than 0.04" (1mm), replace the pump case. If the pump case is satisfactory, the problem must be with the impeller. Visually inspect the impeller for nicks, scratches, pitting or a "mushroomed" edge. If the cause of the excessive clearance cannot be determined visually, the pump must be disassembled and the pump case measured and compared with specifications.

The jet pump must be removed from the craft to perform this work.

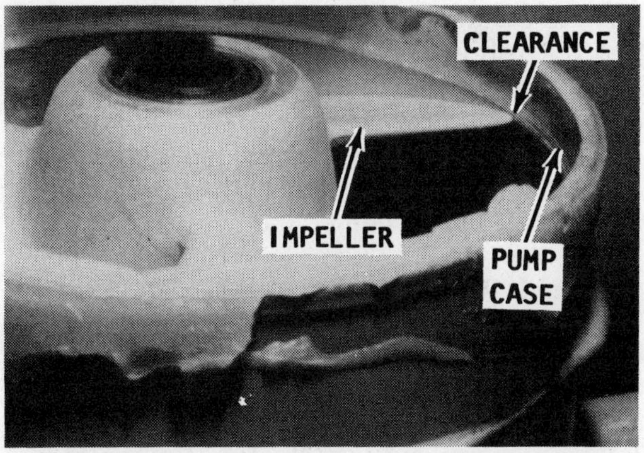

On both axial and mixed flow Kawasaki jet pumps, a feeler gauge is inserted between the impeller blade and the pump case to measure clearance.

Mixed Flow Pump
550 Series

If the clearance is less then the maximum value, no adjustment is required.

If the clearance is greater than the maximum value, and the pump case is in good condition, the impeller clearance must be adjusted. Shim material must be added between the pump case and the impeller. The addition of 0.04" (1mm) of shim material will decrease the impeller clearance by 0.005" (0.144mm).

Mixed flow pump efficiency decreases by as much as 30% when the blade to pump case clearance increases by only 0.04" (1mm).

If the clearance is greater than the maximum value, but the pump case has scratches deeper than 0.04" (1mm), replace the pump case.

The jet pump must be removed from the craft to add shim material behind the impeller. Procedures for pump removal are presented in the next section -- 10-4. The impeller-to-pump case clearance is **NOT** used to determine the total amount or the additional amount of shim material used behind the impeller. The thickness of shim material used depends entirely on the height of the end cap shoulder and the depth to which the bearings seat inside the guide vane casting.

A sample calculation and a bearing shim material chart are provided in the assembling procedures to correctly space the impeller on the pump shaft for the specified impeller-to-pump case clearance.

10-4 JET PUMP SERVICE

SPECIAL WORDS ON ACCOMPANYING ILLUSTRATIONS

The following series of illustrations were taken during service of an axial flow pump. The mixed flow pump differs slightly in appearance, and has some additional components. Where differences occur, these differences have been clearly identified.

SPECIAL WORDS ON SPECIAL TOOLS

The following list of special tools are required in order to perform some of the necessary work on the jet pump. Some tasks become extremely difficult, if not impossible, without the special tools or some type of equivalent. Manufacturers' part numbers are given. In specific steps of the procedures -- whenever possible -- a special effort has been made to give some type of equivalent or homemade tool which may be substituted. However, the cost of a special tool may be offset by money saved if an expensive part is not damaged.

SPECIAL TOOLS FOR JET PUMP SERVICE

300 Series and 440 Series
Axial Flow Pump

Seal and Bearing Remover	P/N 57001-1058
Seal Installation Kit	P/N W56019-004
Bearing Driver Set	P/N 57001-1129
Drive Shaft Holder	P/N W56019-003
Impeller Wrench	P/N 57001-1229
Adaptor	P/N 57001-1231

550 Series
Mixed Flow Pump

Impeller Shaft Holder	P/N K61079-007
Pump Seal Protector	P/N 57001-3003
Seal Installation Kit	P/N W56019-004
Seal Installation Adaptor	P/N 57001-3004

650 Series
Including JetMate
Axial Flow Pump

Impeller Wrench	P/N 57001-1228
Oil Seal and Bearing Remover	P/N 57001-1058
Bearing Driver Set	P/N 57001-1129

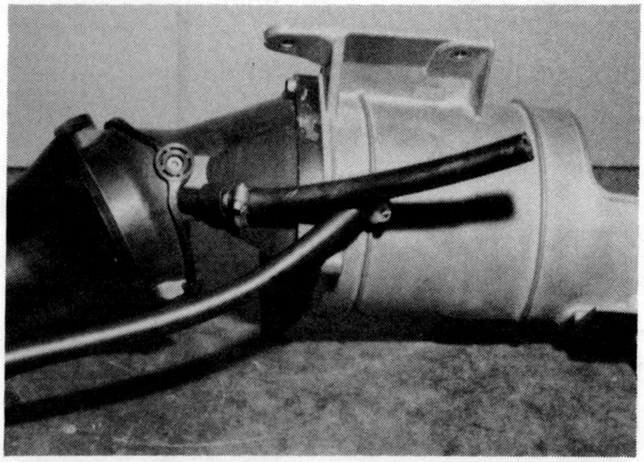

A jet pump is almost impossible to service -- disassemble and assemble -- without the use of some special tools.

REMOVAL

Disconnect both battery cables from the battery and remove the battery from the craft.

CAUTION

The battery **MUST** be removed to avoid spilling electrolyte into the bilge through the battery vent holes when the craft is tilted to one side.

Rotate the fuel control valve, if equipped, to the **"OFF"** position. Drain or remove the fuel tank. If working on a 300 Series or 650 Series engine, drain the oil from the oil injection tank. Place an old blanket or equivalent on the workshop floor to cushion the edges of the craft and prevent the craft from sliding while the work progresses. Rotate the craft on the blanket to expose the rock grate, and then block the hull to hold the desired position.

GOOD WORDS

Before starting service work the impeller-to-pump case clearance should be checked, as described in the previous section -- 10-2.

1- If the clearance is checked, the rock grate will have been removed. If the clearance was not checked -- remove the three bolts and washers securing the grate to the hull.

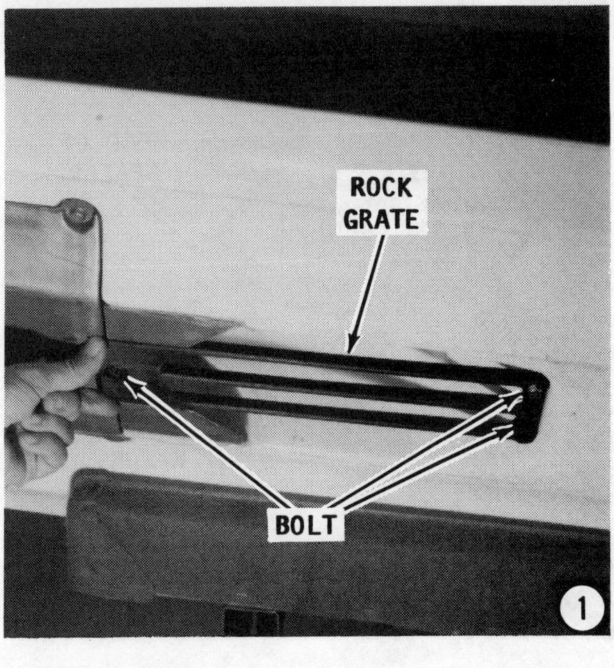

2- Remove the bolts and washers securing the cover over the jet pump.

300 Series	Four bolts
440 Series	Four bolts
550 Series	Four bolts
JetMate	Four bolts
650 Series	Six bolts

SPECIAL WORDS
300SX SERIES ONLY

Loosen the clamp around the large exhaust duct at the fitting on the hull, and slide the duct free of the fitting.

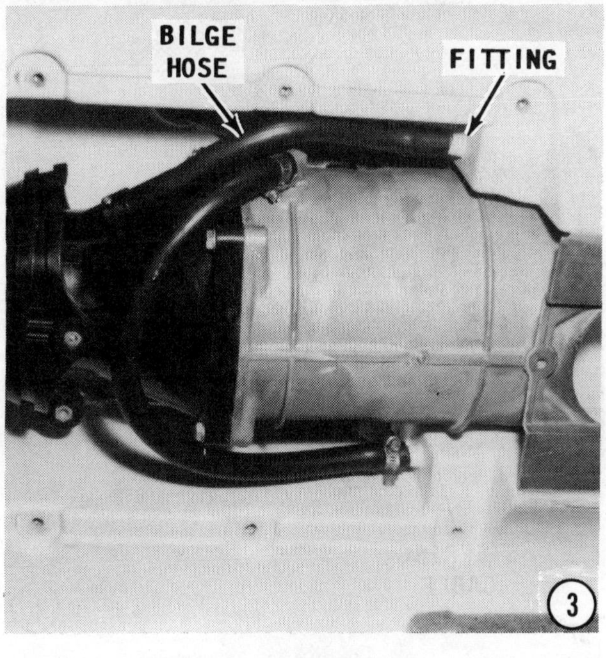

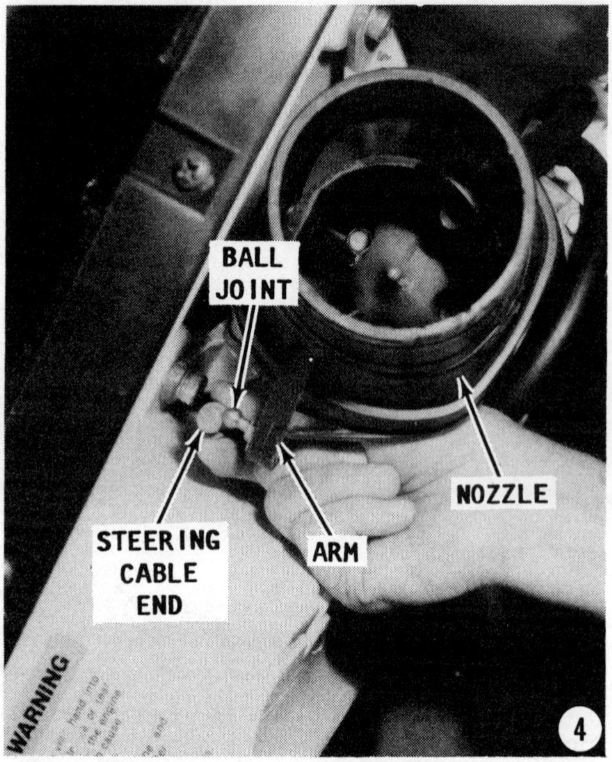

3- Loosen the hose clamp around the bilge hose fitting on the craft hull. Work the hose free of the fitting.

4- Pull back the steering cable end and disengage the cable end from the ball joint on the steering nozzle arm.

5- Loosen the hose clamp around the cooling hose fitting on the pump case. Work the hose free of the fitting.

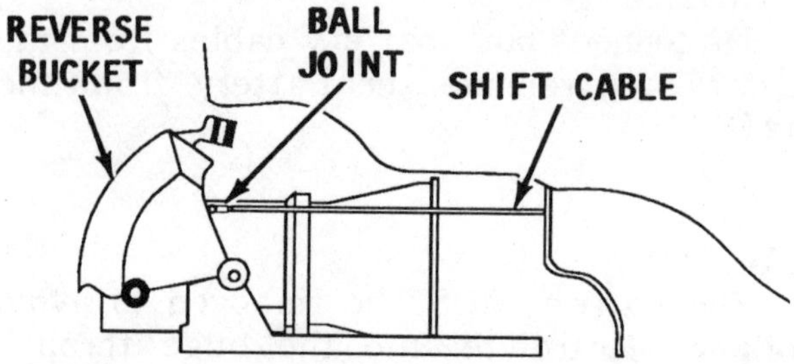

Simple line drawing to show the relationship of the reverse bucket, ball joint, and the shift cable on a JetMate or SC Series watercraft.

SPECIAL WORDS JETMATE AND SC SERIES

Pull back the reverse cable end and disengage the cable from the ball joint on the reverse bucket.

All Models

The jet pump is now "disconnected" and ready to be pulled from the craft.

6- Obtain a razor knife and cut into the silicone seal around the perimeter of the pump case casting. This is not an easy task, especially if working with the sealant installed at the factory. The entire perimeter must be cut several times -- many times -- before the pump will come free of the hull.

7- Using the correct size allen wrench, remove the four allen head bolts securing the jet pump to the craft hull.

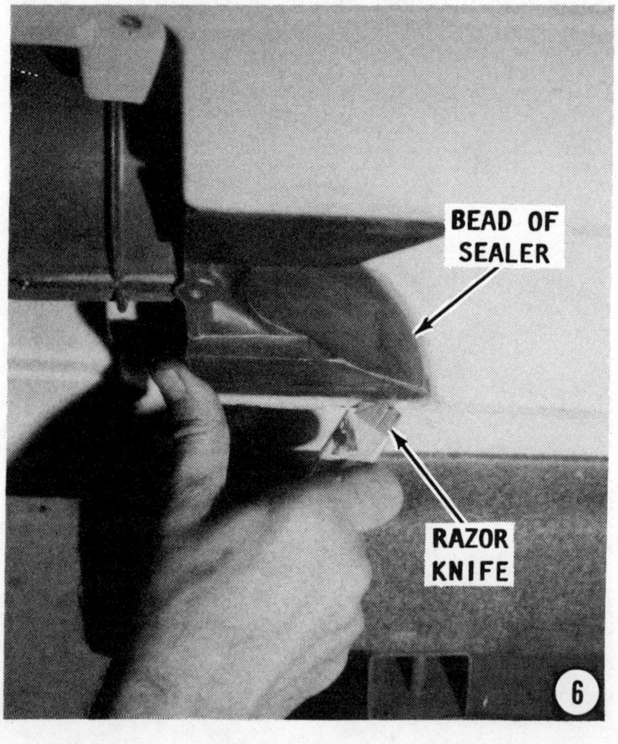

SPECIAL WORDS ON SHIM MATERIAL AND CRANKSHAFT-TO-IMPELLER SHAFT ALIGNMENT

Observe the location and thickness of any shim material placed between the pump case ears and the craft hull. Be sure to identify each of the four stacks to **ENSURE** each stack will be installed back in its original location.

Unless the coupler showed signs of wear, the impeller shaft was properly aligned with the crankshaft. Therefore, it is crucial to install the four stacks of shim material to keep the same alignment.

If the coupler shows signs of wear, as evidenced by the edges of the "petals" being worn away by the fingers of the metal coupler, the impeller shaft was misaligned with the crankshaft. Care must be exercised to achieve proper alignment when a new coupler is installed.

The axial flow pump installed on all 650 Series units, including the JetMate, has a long driveshaft with external splines at one end and a metal coupler at the other end. The coupler "links" the engine crankshaft with the impeller shaft.

The axial flow pump installed on the 300 Series and 440 Series units, and the mixed flow pump installed on the 550 units have the same long shaft. However, on these

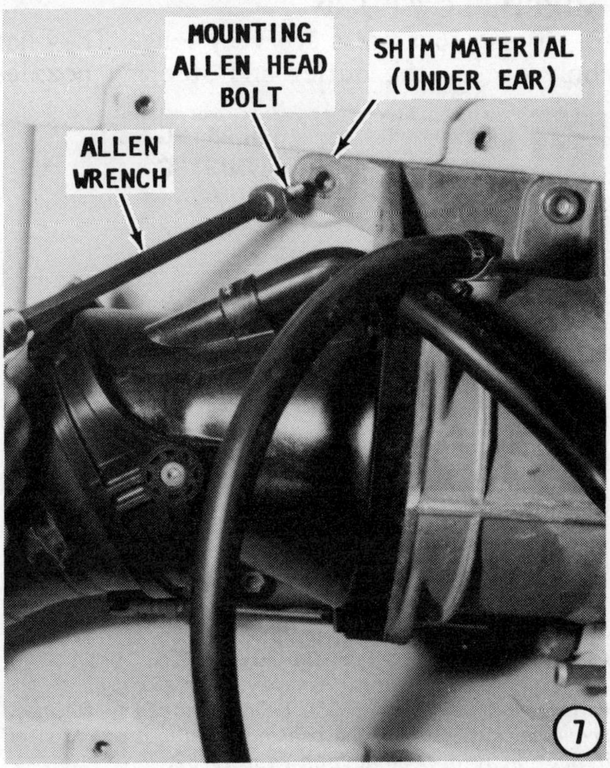

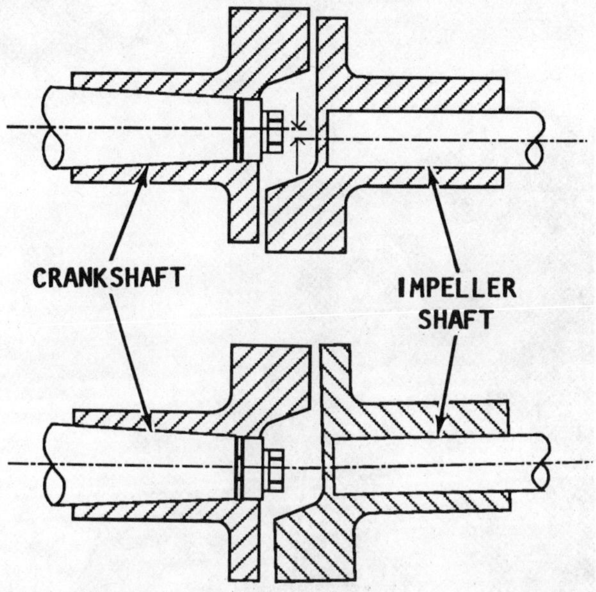

Two types of crankshaft/impeller shaft misalignment contributing to rubber coupler wear.

series, the shaft is refered to as the impeller shaft, because it is the only shaft in the pump.

The external splines of the long shaft mesh with the internal splines inside the forward half of the impeller. The metal coupler indexes into a rubber coupler. The crankshaft is also equipped with a metal coupler which indexes into the forward face of the rubber coupler. Therefore, any flexing -- due to misalignment of the drive line -- takes place at the rubber coupler --and is

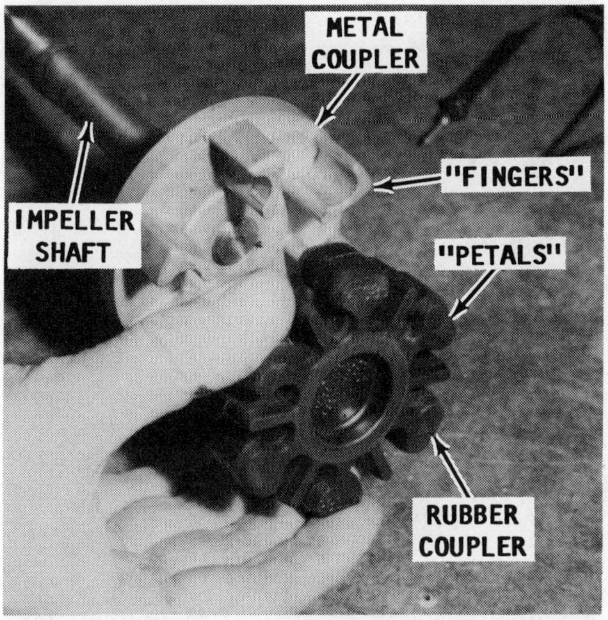

A used rubber coupler still in good condition because the engine crankshaft was properly aligned with the jet pump impeller shaft.

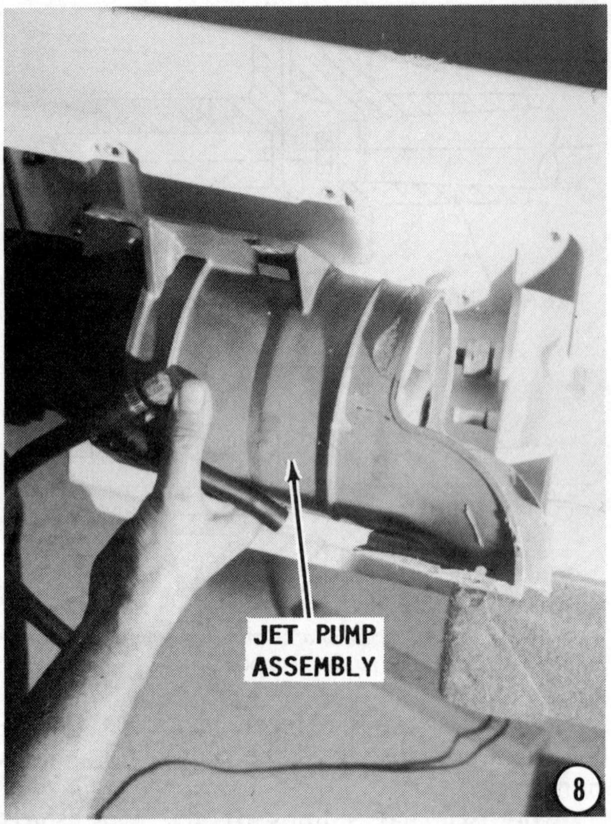

evidenced by wear described earlier in this step.

Alignment of the engine crankshaft-to-impeller shaft is covered on Page 8-41.

8- Slide the jet pump aft off the impeller/driveshaft and then ease the pump out and clear of the craft.

If working on a 300 Series or 440 Series pump: Insert the pin in the adaptor into the hole in the driveshaft. Clamp the driveshaft holder tool around the adaptor and shaft. Insert the coupling holder into the metal coupler and unscrew the coupler from the threaded end of the driveshaft.

9- If the engine has been removed: Pull the driveshaft (650 Series), or impeller shaft (all other Series), straight out of the bearing housing. Remove the rubber coupler from the metal coupler on the shaft end.

The work continues with the jet pump on a suitable work surface.

GOOD WORDS ON THE BEARING HOUSING

If the only work to be performed is to replace or service the bearing housing, the housing is now freely accessible with the pump moved out of the way. If no work is to be performed on the jet pump, proceed to Section 10-4 to service the bearing housing.

DISASSEMBLING

1- Remove the four bolts securing the outlet nozzle and the steering nozzle to the pump case.

SPECIAL WORDS 550 SERIES MIXED FLOW PUMP

These four bolts are very long. They not only secure the outlet and steering nozzles

The bearing housing is only accessible after the engine and jet pump have been removed. Detailed procedures to service the bearing housing are presented in Section 10-6, beginning on Page 10-32.

to the pump case, but also pass though the guide vane casting, between the outlet nozzle and the pump case.

Look for and **SAVE** the single locating pin at the contact surfaces of the pump case and guide vane casting. Another pin is located between the guide vane casting and the outlet nozzle. These pins may be left on either surface when the pump is split.

2- Separate the outlet nozzle -- with steering nozzle attached -- from the pump case, or guide vane casting (550 Series only). The two surfaces are probably stuck together, because they should have been assembled with a sealing agent. Tap lightly all around the perimeter of the outlet nozzle with a soft head mallet. **DO NOT PRY** between the two surfaces. If a deep enough gouge is made, air will be injested into the pump and result in a substantial loss of performance.

Clean all traces of sealant from both contact surfaces.

650 Series
Including JetMate

3- Remove the two bolts and washers securing the end cap to the pump case and remove the cap. If a replacement is available, remove and discard the O-ring around the cap. Remove the old lubricant and wipe the cap clean.

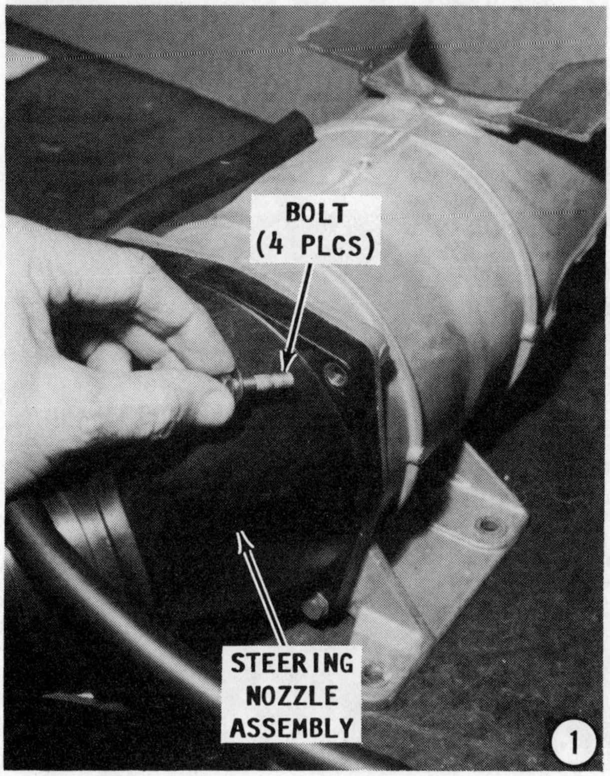

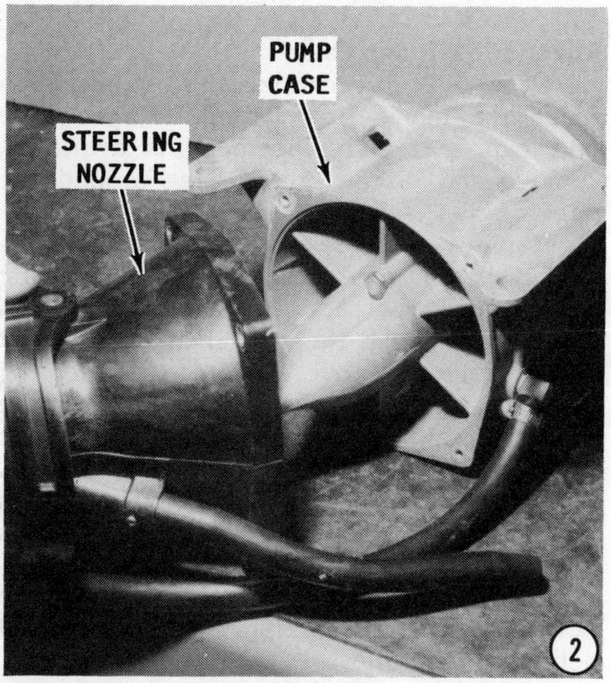

300 Series
and 440 Series Only

These axial pumps do not have a removable end cap. It is cast into the pump case. However, the lubrication fitting at the end of the end cap should be removed at this time to prevent hydraulic lock when the pump case is installed on the driveshaft.

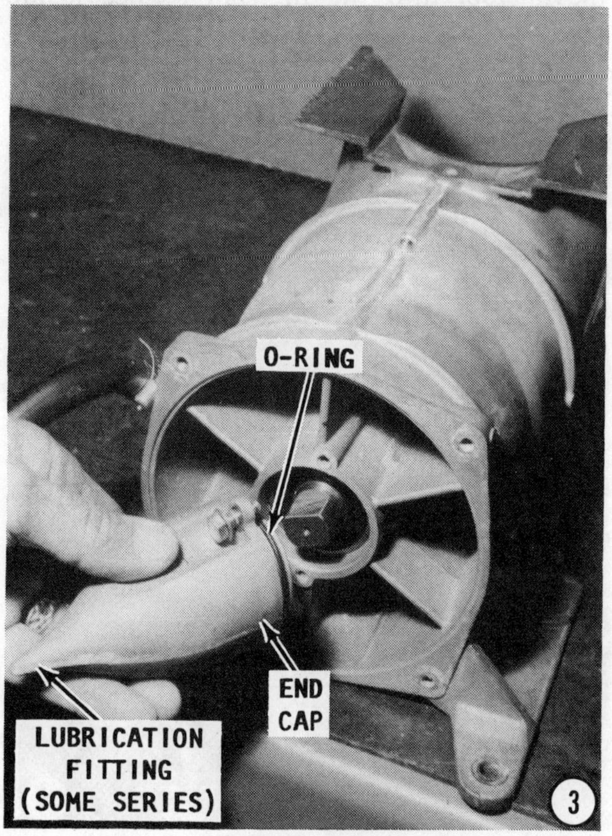

550 Series
Mixed Flow Pump Only

Using the correct size allen wrench, remove the lubrication fitting cover on the end cap. If a replacement is available, remove and discard the O-ring around the cap.

Remove the three bolts and washers securing the end cap to the guide vane casting and remove the cap. Remove and discard the O-ring around the cap, if a replacement is available. Remove the White plastic plug over the end cap. Remove and **SAVE** the pieces of shim material between the plastic plug and the bearing on the impeller shaft. The pieces of shim material found in this location are a "constant" thickness and do not affect the impeller-to-pump case clearance. Be sure to save them and keep them **SEPARATE** from the shim material located behind the impeller.

Impeller Removal
650 Series
Including JetMate

4- Clean the slotted end of the impeller shaft of any lubricant remaining after the end cap was removed. Clamp this end of the impeller shaft into a vise equipped with soft jaws. Insert the special impeller removal tool into the internal splines inside the impeller. With the correct size wrench on the tool, loosen the impeller from the threads on the impeller shaft.

5- Remove the impeller from the impeller shaft. If the oil seal at the forward face of the impeller needs to be replaced, pry it out with a screwdriver.

Impeller Shaft Removal
650 Series
Including JetMate

6- Pull out the impeller shaft from the rear of the pump case. If replacements are available, remove and discard the two O-rings around the shaft.

Impeller Removal
550 Series
Mixed Flow Pump Only

Clean the nut cast into the impeller of any lubricant remaining after the end cap was removed. Position the correct size open end wrench over the impeller nut. Insert the special impeller removal tool into the internal splines inside the impeller. Loosen the impeller from the threads on the impeller shaft.

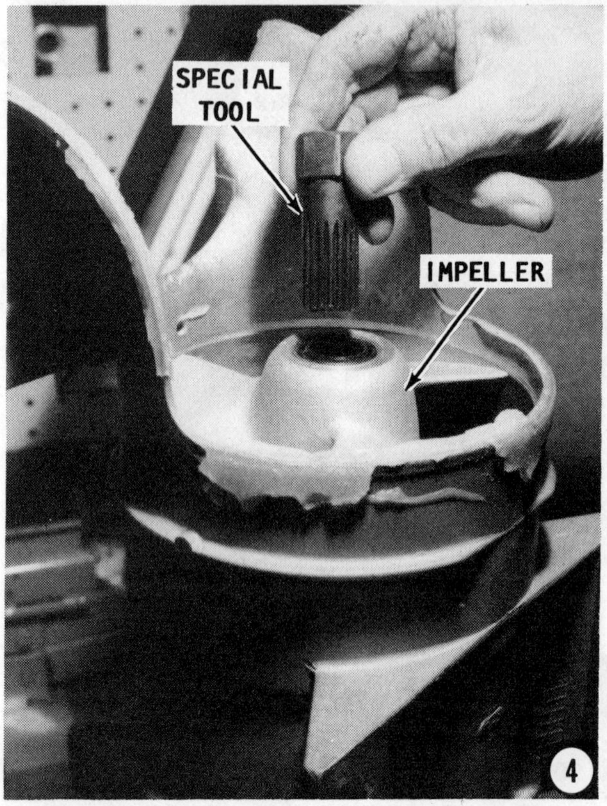

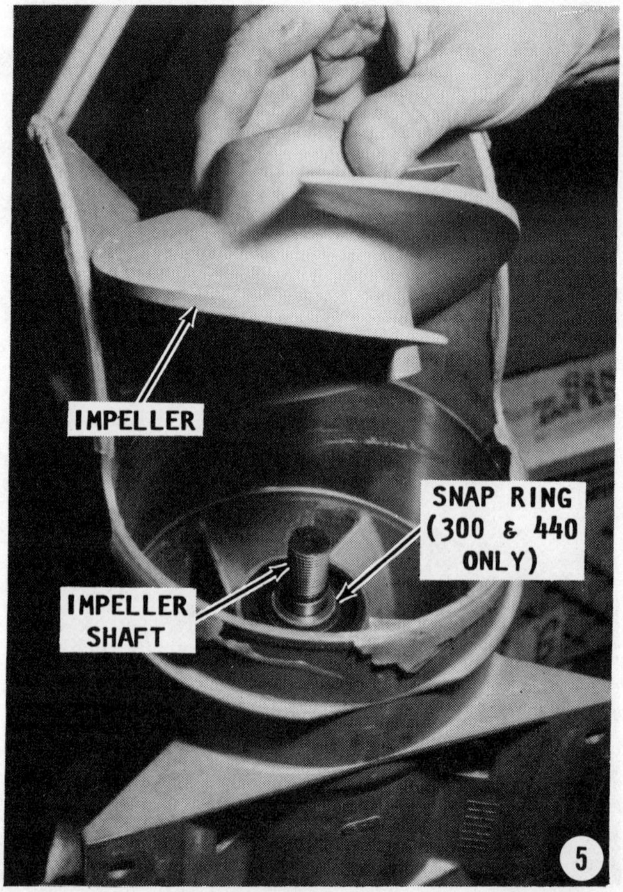

**CRITICAL WORDS
550 SERIES
MANUFACTURED BEFORE 1991**

The impeller has **LEFT-HAND** threads.

Units manufactured after 1991 have **RIGHT-HAND** threads, as do all the axial jet pumps.

**SPECIAL WORDS
550 SERIES ONLY**

These mixed flow pumps have pieces of shim material between the impeller and the oil seal in the guide vane casting. **SAVE** the shim material for later installation. The thickness of shim material is critical to properly position the impeller with a specific clearance from the pump case. If the clearance was correct no adjustment is required and the original thickness of shim material may be installed. If the clearance was excessive, the thickness of shim material will need to be increased.

Remove the impeller from the impeller shaft. Pull out the impeller shaft from the guide vane casting. If a replacement is available, remove and discard the small **O**-ring at the threaded end of the shaft.

**650 Series
Including JetMate**

7- Lift out the short bushing from the forward end of the pump case.

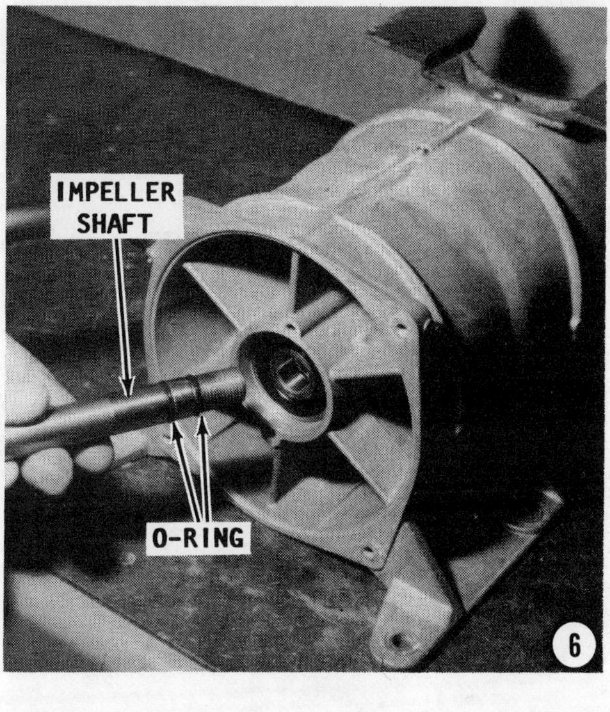

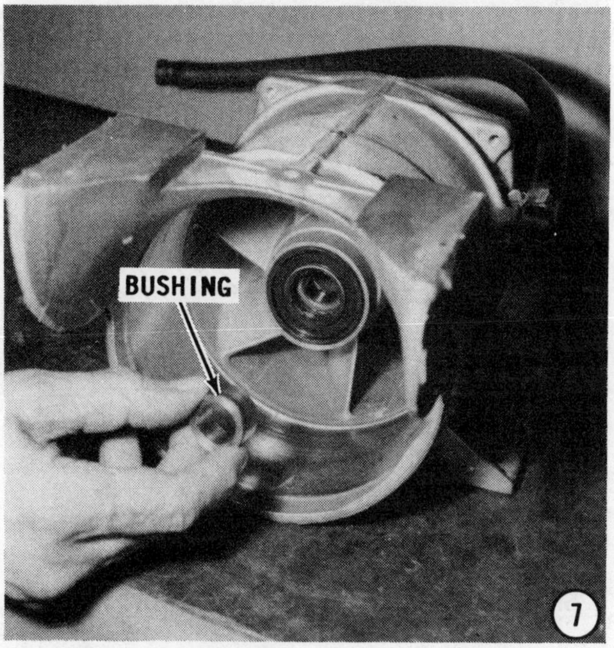

**Pump Case Oil Seal
and Bushing Removal
300 Series and 440 Series**

SAFETY WORDS

The snap ring mentioned in the following step is under tremendous pressure when installed. Therefore, wear a pair of safety glasses and exercise extra **CARE** when re-

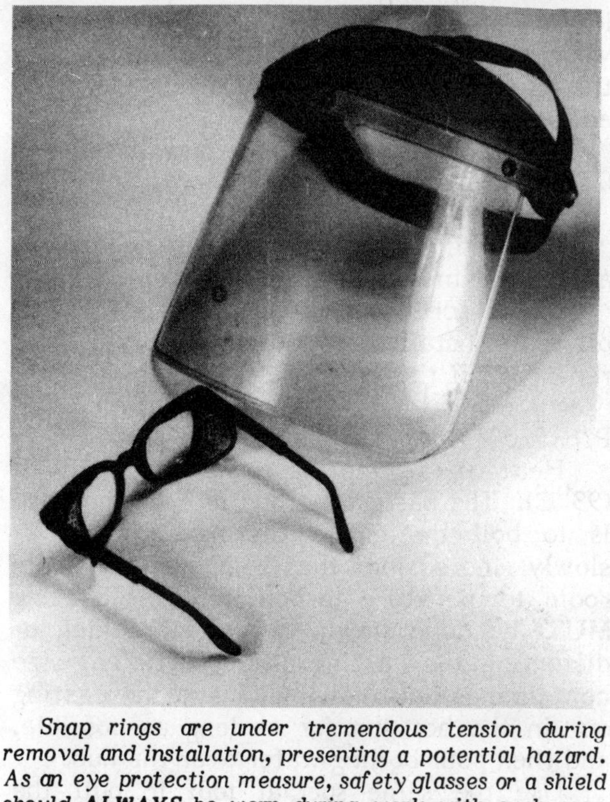

*Snap rings are under tremendous tension during removal and installation, presenting a potential hazard. As an eye protection measure, safety glasses or a shield should **ALWAYS** be worn during work with such snap rings. Warn others in the area such work is in progress.*

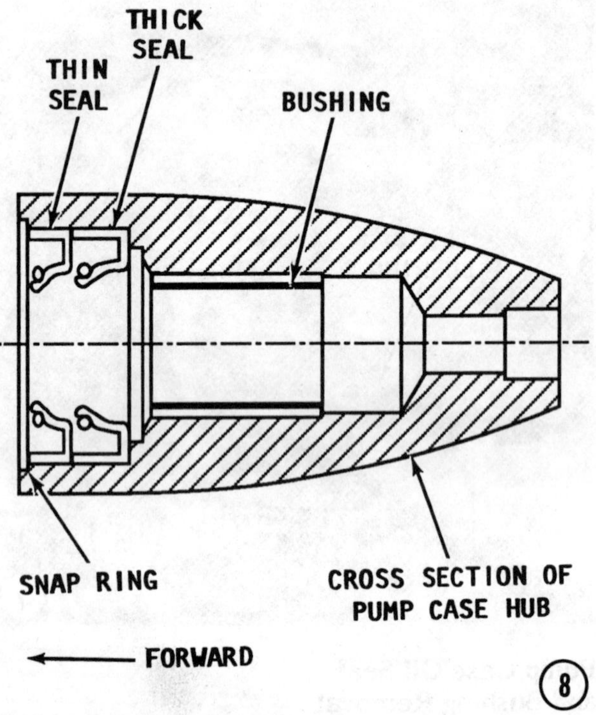

moving the snap ring to prevent injury to self or others in the area.

8- Use a pair of internal snap ring pliers and remove the snap ring from the aft hub of the pump case.

If the bushing inside the pump case is no longer fit for service -- the two oil seals were probably the cause of the failure and the oil seals and the bushing must be replaced.

Standard diameter for a new bushing is 0.7874" (20.0mm).

Service limit is 0.7953" (20.2mm).

A telescopic micrometer is needed to accurately measure this dimension.

Use a screwdriver and pry out the two oil seals installed behind the snap ring just removed. If the seals are frozen to the hub, use the seal and bearing remover tool P/N 57001-1058 to remove them.

Heat the pump case to about 200°F (93°C). The easiest method to heat the case is to boil the entire casting in water -- slowly increasing the temperature from room temperature to boiling. This heating **MUST** be done slowly to avoid cracking or distorting the casting. If a sufficient size container is not available, leave the casting out in the hot sun for as long as possible, and then pour boiling water over the hub.

Use the same special tool to pull the bushing from the pump case.

Pump Case Oil Seal and Bearing Removal 650 Series Including JetMate

9- Use a screwdriver and pry out the two oil seals at the forward end of the pump case. If the seals are frozen to the hub, use

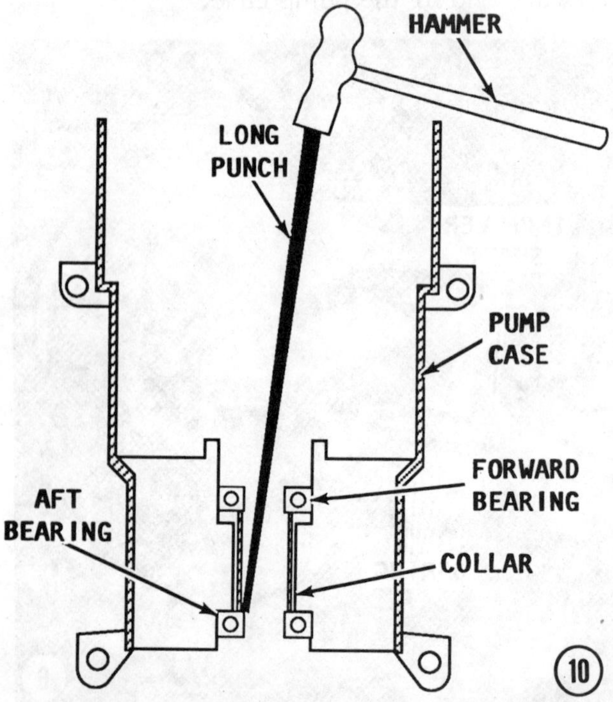

the seal and bearing remover tool P/N 57001-1058 to remove them.

10- Use a slim long punch and tap out the aft bearing first. Next, remove the spacer, and then tap out the forward bearing.

Impeller Shaft Removal
550 Series
Mixed Flow Pump Only

11- If the two oil seals in the forward face of the guide vane casting are still satisfactory, obtain an oil seal protector to remove the impeller shaft. However, if the seals have failed, a protector is not necessary. Nevertheless, a seal protector **MUST** be used when installating the impeller shaft.

If a small split occurs in the lip of one of these oil seals, water will find its way -- under pressure -- from the impeller past the two seals. This water will destroy the two bearings supporting the impeller shaft and the pump will seize.

Slide the seal protector tool over the pump shaft threads. If this work is performed without the use of the protector, the threads on the pump shaft may damage the oil seal lips inside the guide vane casting. Use a soft head mallet and tap the shaft free of the casting.

12- If the seals are to be removed, use a screwdriver and pry out the two oil seals from the forward face of the guide vane casting.

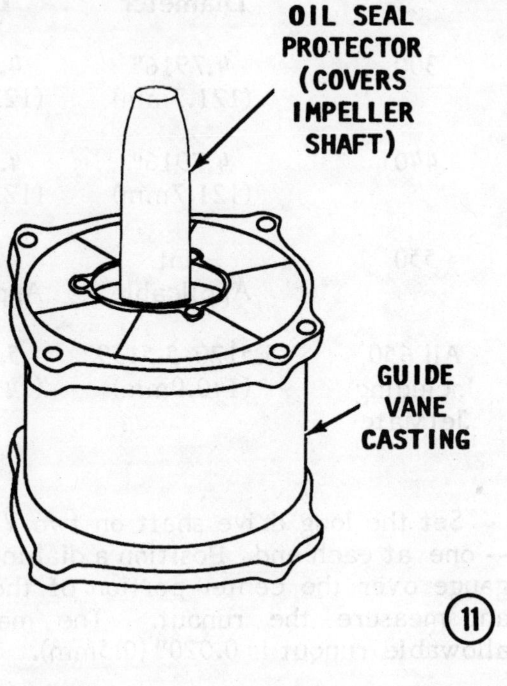

11

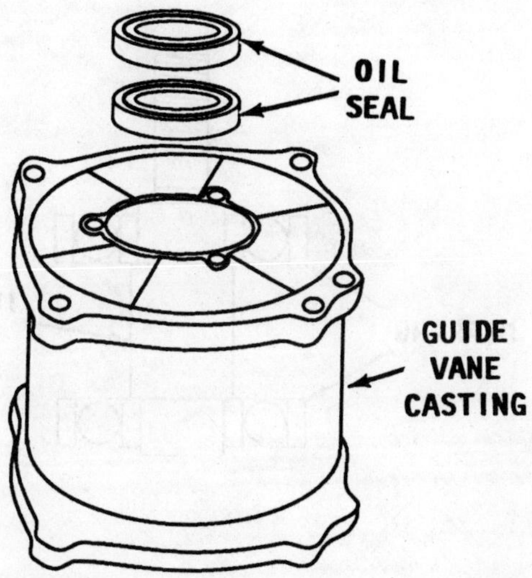

12

Impeller Shaft Disassembling
550 Series
Mixed Flow Pump Only

13- Insert a slim long punch into the open (threaded), end of the impeller shaft. Tap on the punch to remove the shaft plug from the other end of the shaft.

14- Make a setup in an arbor press, and carefully press the impeller shaft free of the two roller bearings -- one at each end of the shaft. Take care not to allow the shaft to free fall to the floor.

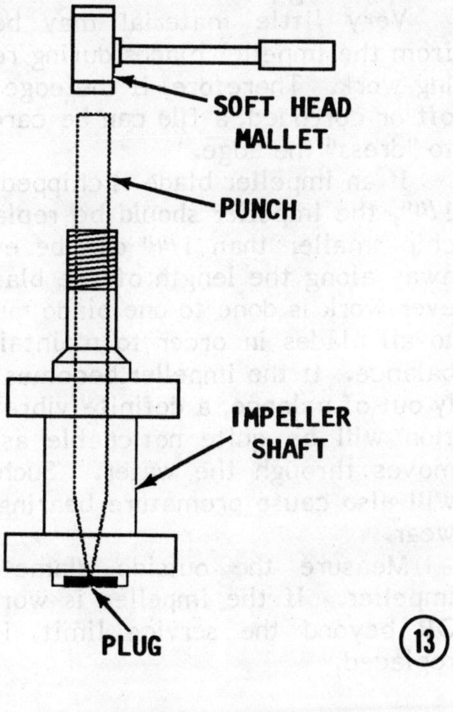

13

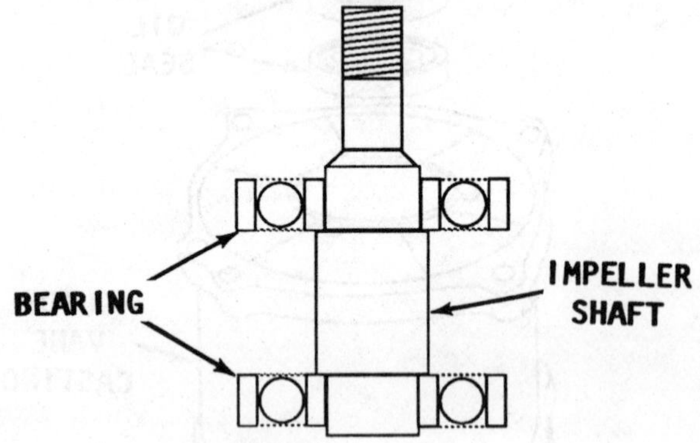

14

CLEANING AND INSPECTING

Inspect the leading edge of each impeller blade. Ideally, the edge should be rounded to a 1/32" radius, **NOT** a knife-edge! The cross-section shape of an impeller blade is similar to that of an airplane wing.

If the craft is consistently operated in shallow water with a sandy or "pebbly" bottom, the leading edge of the impeller blades will become rounded or will become chipped with missing pieces.

Very little material may be removed from the impeller blades during reconditioning work. Therefore, if the edge is rounded off or corroded, a file can be carefully used to "dress" the edge.

If an impeller blade is chipped more than 1/4", the impeller should be replaced. Any chip smaller than 1/4" can be evenly filed away along the length of the blade. Whatever work is done to one blade must be done to all blades in order to maintain impeller balance. If the impeller becomes excessively out of balance, a definite vibration condition will be quite noticeable as the craft moves through the water. Such vibration will also cause premature bearing and shaft wear.

Measure the outside diameter of the impeller. If the impeller is worn close to **OR** beyond the service limit, it must be replaced.

IMPELLER OUTSIDE DIAMETER

Series	Standard Diameter	Service Limit
300	4.7640" (121.0mm)	4.7246" (120.0mm)
440	4.7640" (121.0mm)	4.7246" (120.0mm)
550	Not Applicable	Not Applicable
All 650 Including JetMate	5.4921-5.4960" (139.5-139.6mm)	5.4527" (138.5mm)

Inspect the interior of the pump case. If the case surface has deep scratches, the pump case must be replaced. Measure the inside diameter of the pump case. If the diameter is greater than the service limit -- the case should be replaced.

PUMP CASE INSIDE DIAMETER

Series	Standard Diameter	Service Limit
300	4.7916" (121.7mm)	4.8309" (122.7mm)
440	4.7916" (121.7mm)	4.8309" (122.7mm)
550	Not Applicable	Not Applicable
All 650 Including JetMate	5.5120-5.5159" (140.0mm)	5.5553" (141.1mm)

Set the long drive shaft on two V-blocks -- one at each end. Position a dial indicator gauge over the center portion of the shaft and measure the runout. The maximum allowable runout is 0.020" (0.5mm).

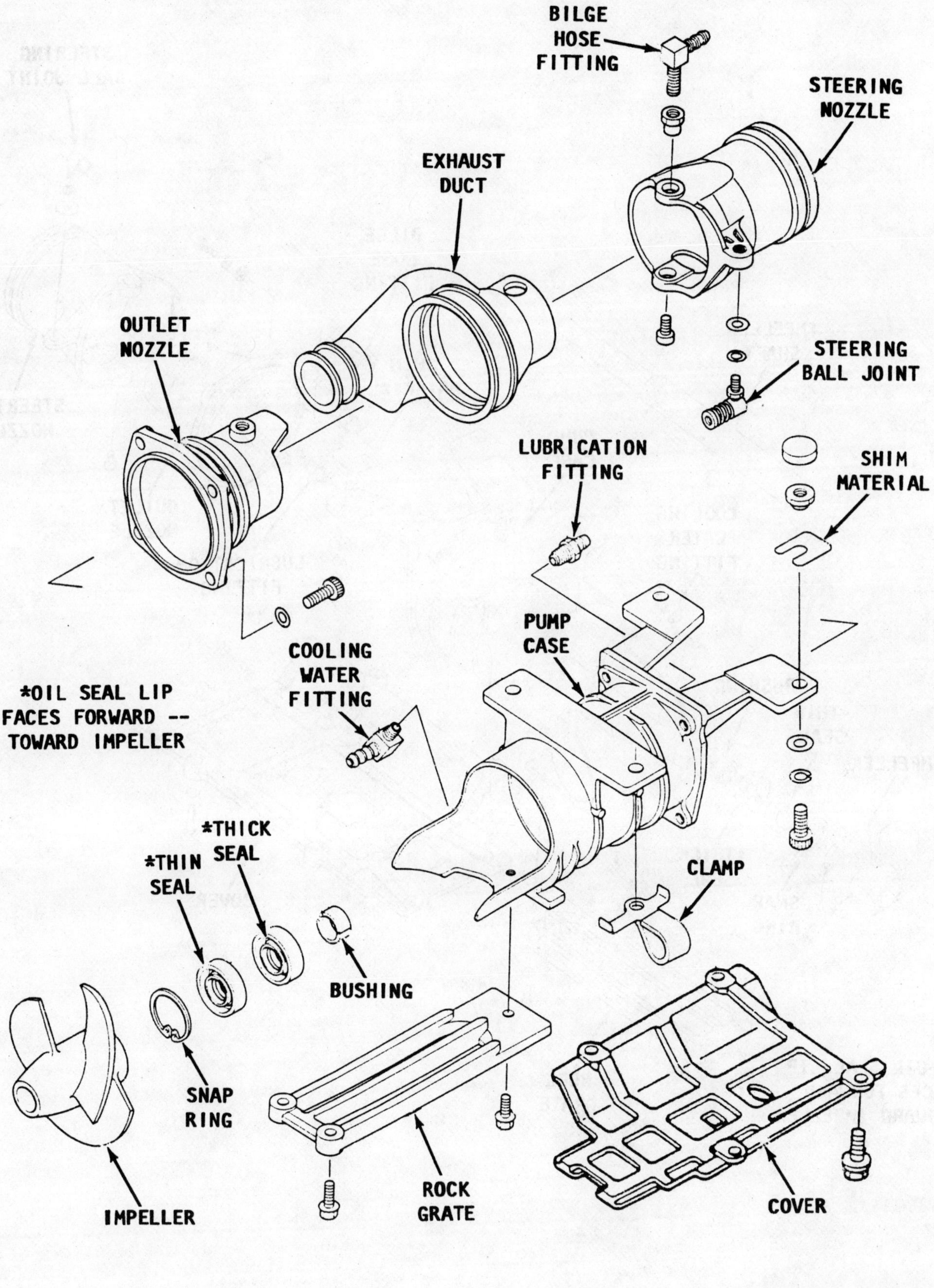

Exploded drawing of the 300SX axial flow pump, with major parts identified.

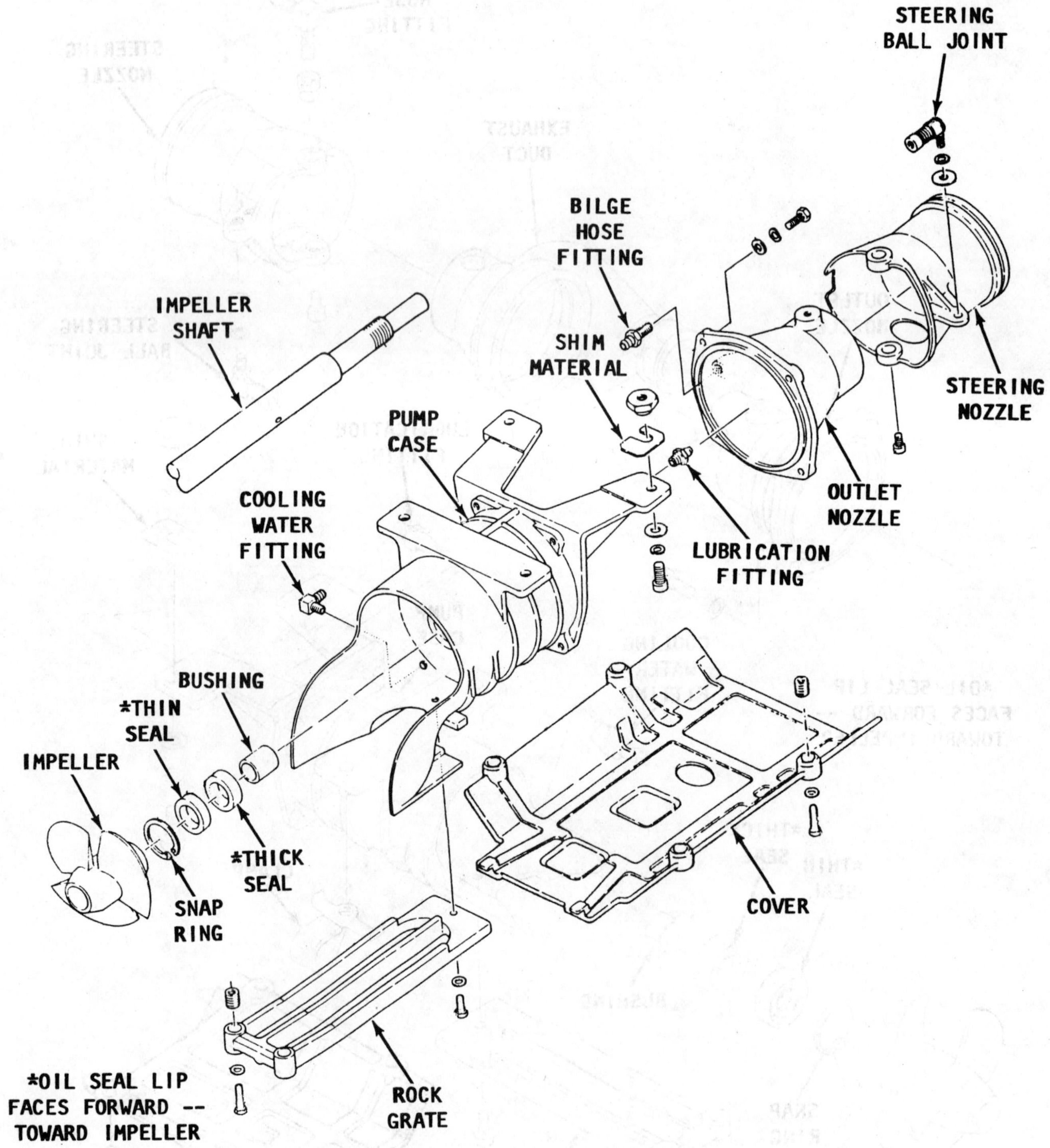

Exploded drawing of the axial flow pump used on the 300 Series (except the 300SX Series), and the 440 Series watercraft. Major parts have been identified.

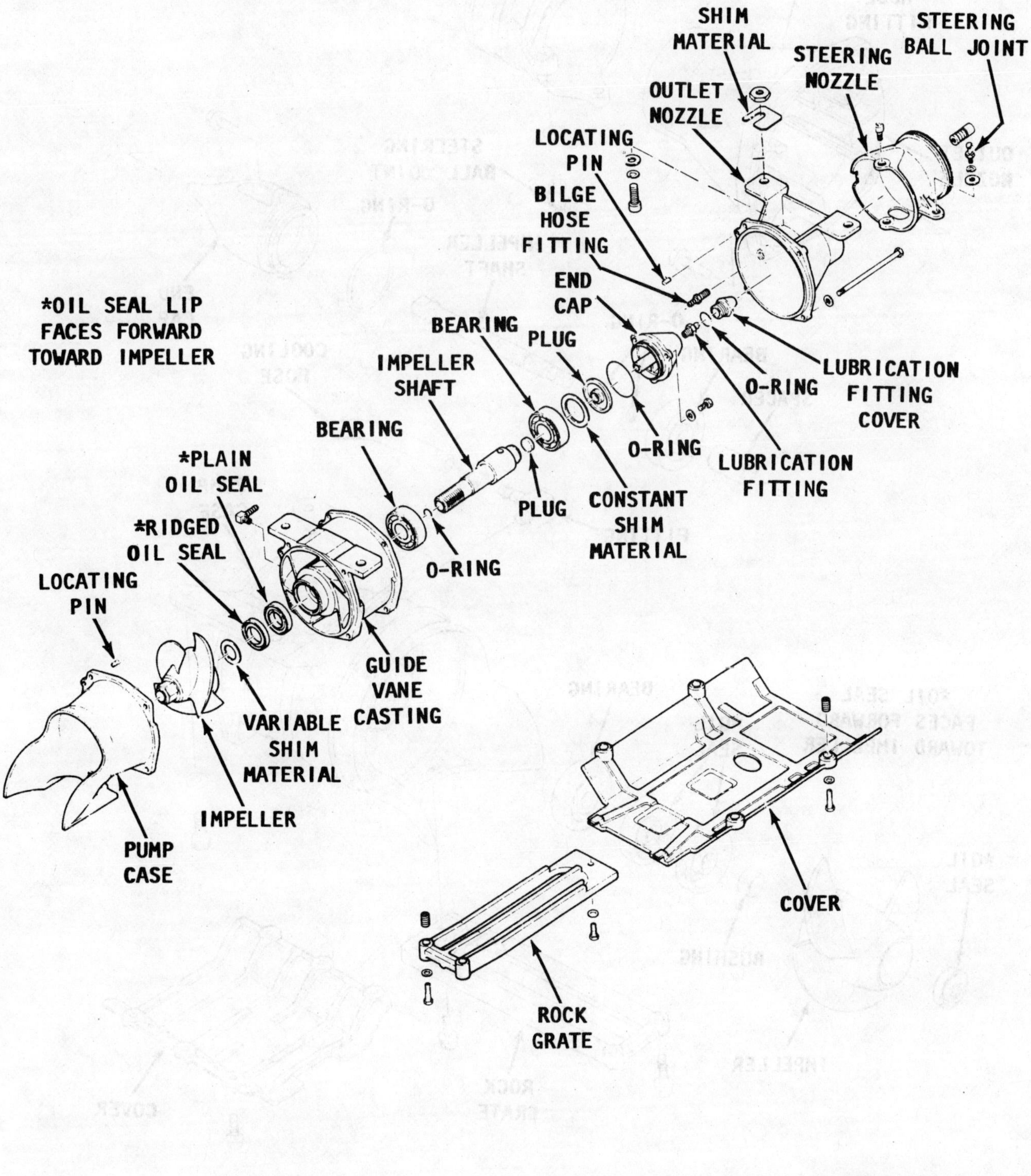

Exploded drawing of the mixed flow jet pump installed on the 550 Series watercraft. Major parts have been identified.

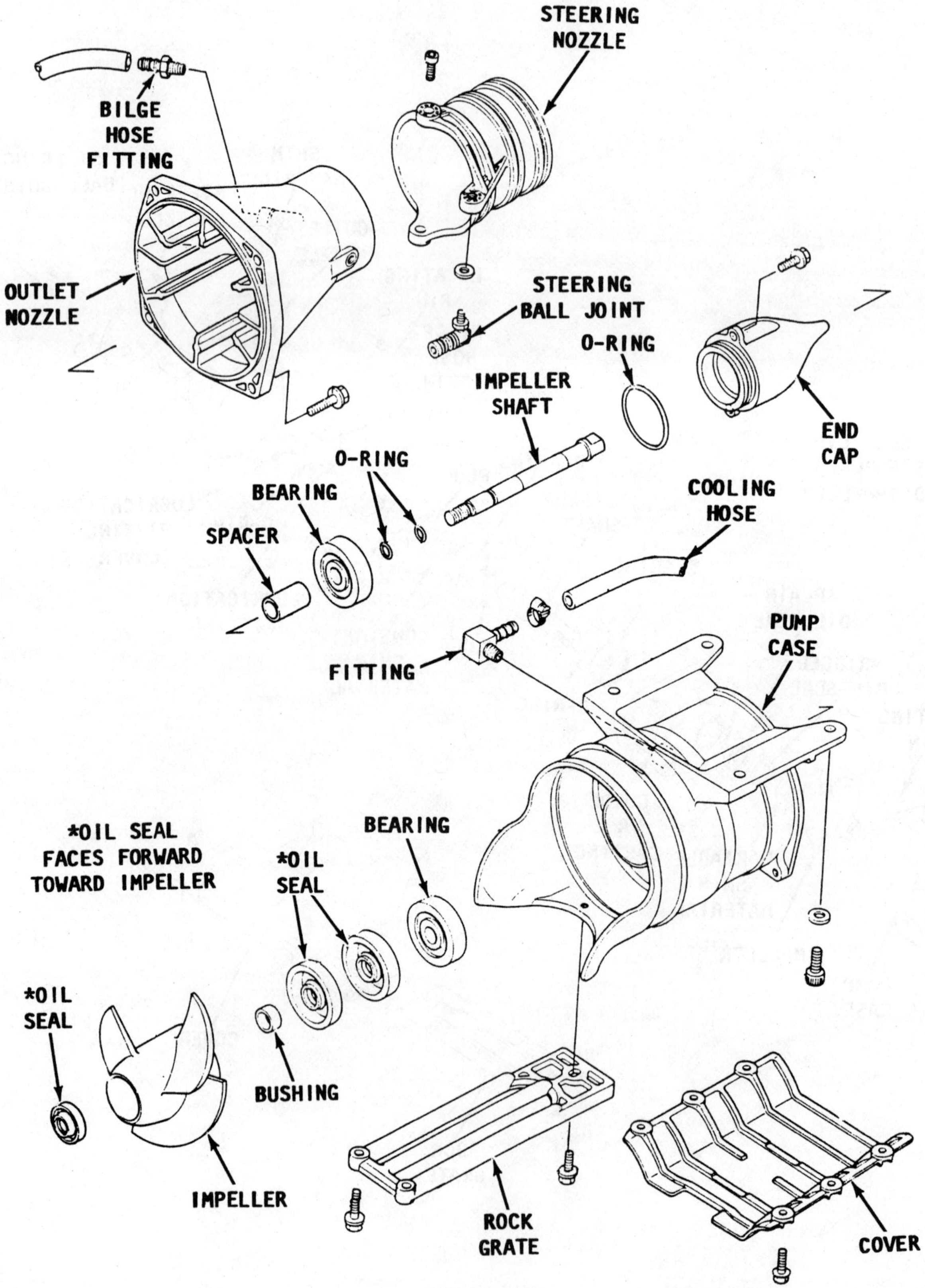

Exploded drawing of the axial flow jet pump used on all 650 Series watercraft, ***EXCEPT*** *the JetMate and SC Series. Major parts have been identified.*

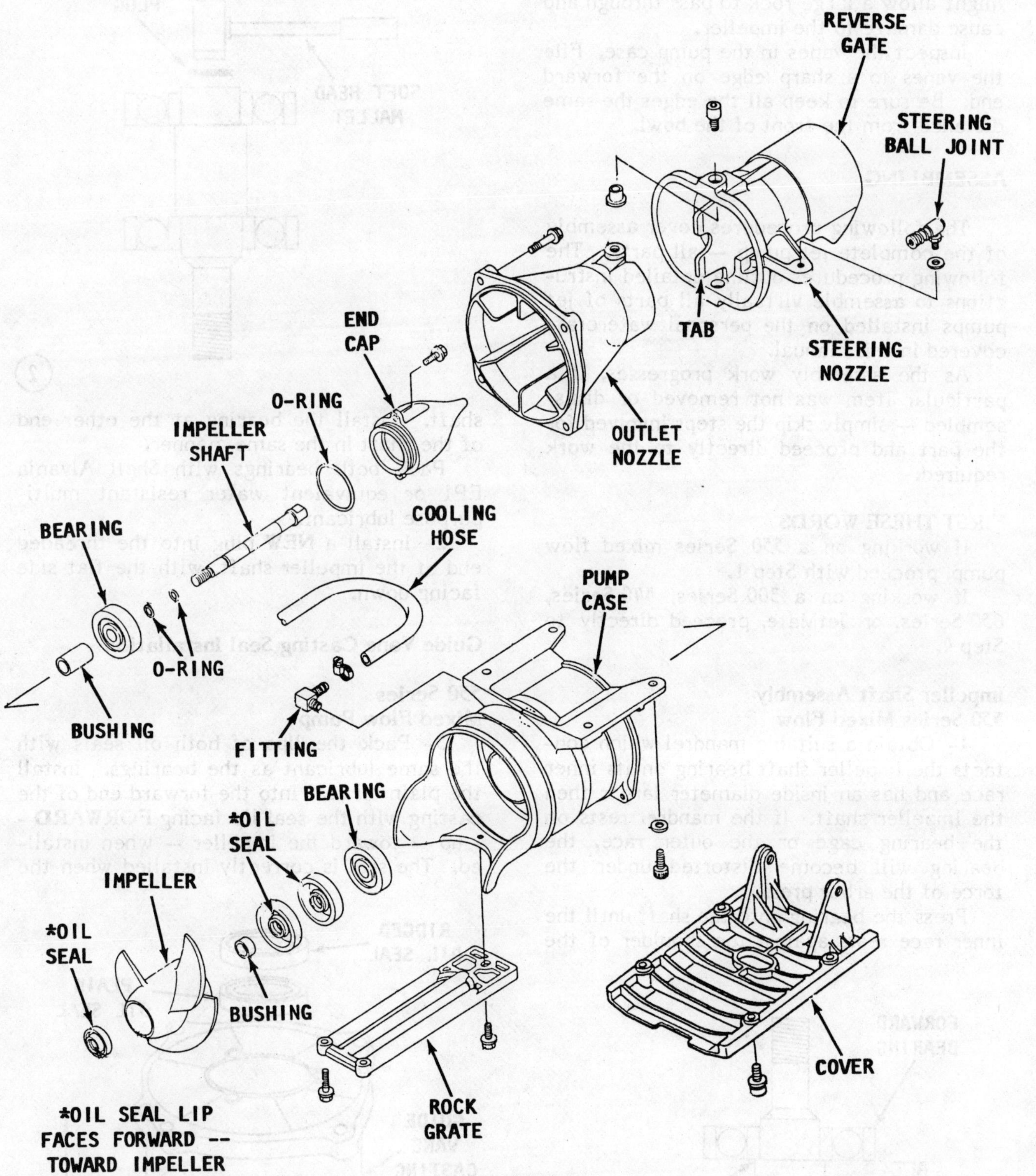

Exploded drawing of the axial flow jet pump with reverse gate installed on the JetMate and SC Series watercraft. Major parts have been identified.

Check to be sure the rock grate blades are not bent or missing. Such a condition might allow a large rock to pass through and cause damage to the impeller.

Inspect the vanes in the pump case. File the vanes to a sharp edge on the forward end. Be sure to keep all the edges the same distance from the front of the bowl.

ASSEMBLING

The following procedures cover assembly of the complete jet pump -- all parts. The following procedures outline detailed instructions to assemble virtually all parts of jet pumps installed on the personal watercraft covered in this manual.

As the assembly work progresses, if a particular item was not removed or disassembled -- simply skip the steps involved for the part and proceed directly to the work required.

FIRST THESE WORDS

If working on a 550 Series mixed flow pump, proceed with Step 1.

If working on a 300 Series, 440 Series, 650 Series, or JetMate, proceed directly to Step 4.

Impeller Shaft Assembly
550 Series Mixed Flow

1- Obtain a suitable mandrel which contacts the impeller shaft bearing on its inner race and has an inside diameter larger then the impeller shaft. If the mandrel rests on the bearing cage or the outer race, the bearing will become distorted under the force of the arbor press.

Press the bearing onto the shaft until the inner race rests against the shoulder of the

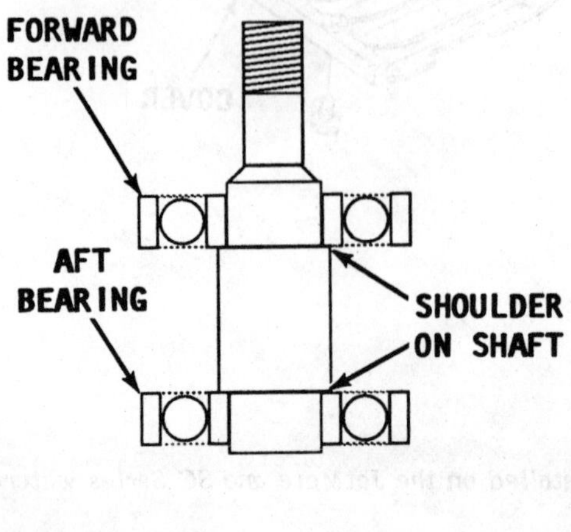

1

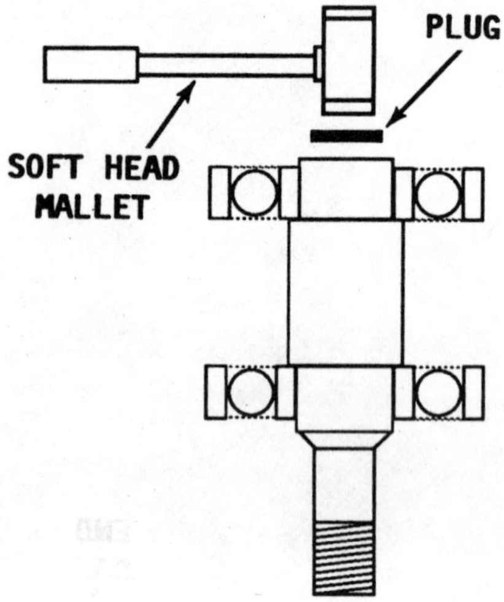

2

shaft. Install the bearing at the other end of the shaft in the same manner.

Pack both bearings with Shell Alvania EP1 or equivalent water resistant multipurpose lubricant.

2- Install a **NEW** plug into the threaded end of the impeller shaft, with the flat side facing down.

Guide Vane Casting Seal Installation

550 Series
Mixed Flow Pump

3- Pack the lips of both oil seals with the same lubricant as the bearings. Install the plain oil seal into the forward end of the casting with the seal lip facing **FORWARD** -- up -- toward the impeller -- when installed. The seal is correctly installed when the

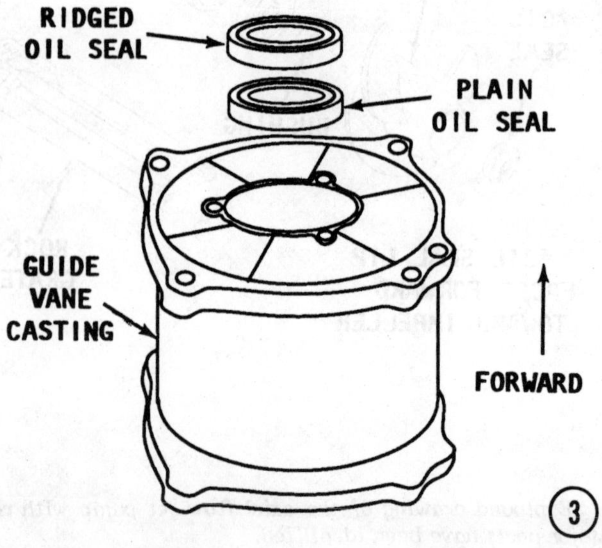

3

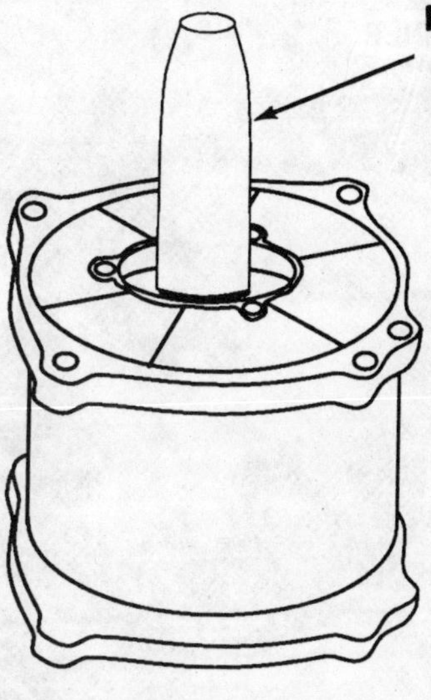

4

back side of the seal makes contact with an inner shoulder in the hub of the casting.

Install the ridged oil seal in the same manner -- until the back side of the seal rests against the first oil seal. The lip of the seal must face **FORWARD** -- toward the impeller -- after installation.

Impeller Shaft Installation
550 Series
Mixed Flow Pump

4- An oil seal protector **MUST** be used when installing the impeller shaft to avoid damage to the seal lips. If a small split occurs in the lip of one of these oil seals, water will find its way -- under pressure from the impeller -- past the two seals. This water will destroy the two bearings supporting the impeller shaft and the pump will seize.

Slide the protector over the end of the pump shaft and push the shaft into the guide vane casting from the aft end. Use a soft head mallet and drive the shaft into the casting until the lower bearing seats against a shoulder in the casting hub.

10-5 SHIMMING PROCEDURES 550 SERIES MIXED FLOW PUMPS

Using vernier calipers with a millimeter scale, measure the height of the shoulder on the end cap, as shown in the accompanying illustration -- dimension **"A"**. Now, measure the depth of the shoulder in the aft face of

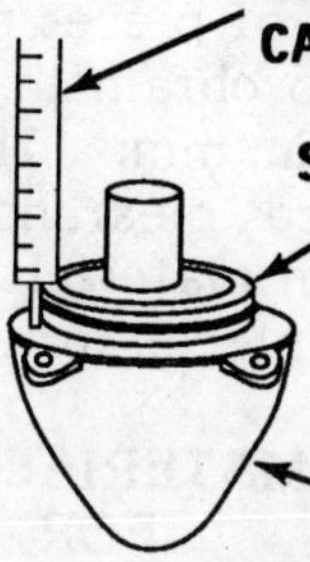

Simple line drawing to show how dimension "A" is determined as the height of the end cap shoulder.

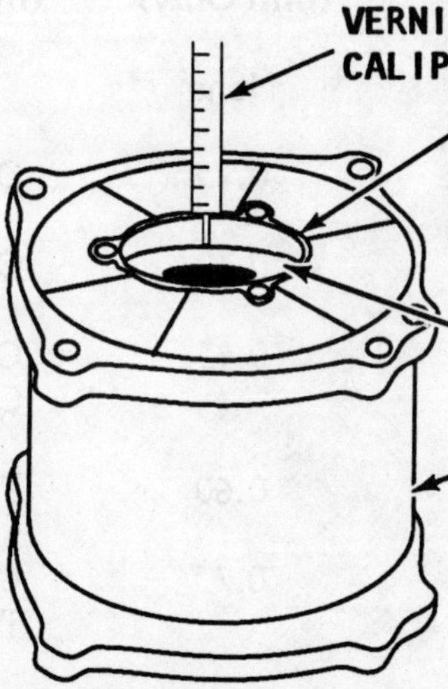

Line drawing to indicate how dimension "B" is obtained as the depth of the bearing, as mentioned in the text.

the guide vane casting hub -- from the outer bearing race to the mating surface of the end cap, as shown in the accompanying illustration -- dimension **"B"**.

Subtract the height of the end cap from the depth of the bearing to determine the clearance between the cap and the bearing, when the cap is installed.

B - A = Clearance

Example:

If the depth of the bearing -- **"B"** -- is 13mm, and the height of the cap is 12.5mm, the clearance will be as follows:

13.0 - 12.5 = 0.5mm

Compare the calculated clearance to the chart to determine the amount of shim material to be placed behind the impeller for the correct impeller-to-pump case clearance.

Shim material pieces are only available in two thicknesses -- 0.15mm and 0.30mm. A combination of pieces provide the required thickness to obtain the correct calculated clearance in mm. The following table lists calculated clearances and the total amount of shim material required, including the number of pieces for each thickness.

SHIM MATERIAL AMOUNT FOR CALCULATED CLEARANCE

Calculated Clearance (mm Only)	Total Shim Thickness (mm Only)	No. Pieces Ea. Size (mm Only)
0.10-0.25	None	None
0.26-0.40	0.15	One 0.15
0.41-0.55	0.30	One 0.30
0.56-0.70	0.45	One 0.15 One 0.30
0.71-0.85	0.60	Two 0.30
0.86-0.94	0.75	One 0.15 Two 0.30

In the given example, a clearance of 0.5mm falls in the 0.41-0.55mm range. Therefore, 0.30mm of shim material must be used.

Set aside the correct amount of shim material to be installed later -- Step 9, this section -- behind the impeller.

Pump Case Oil Seal and Bearing Installation 650 Series Including JetMate

5- Pack both bearings with Shell Alvania EP1 or equivalent water resistant multi-purpose lubricant. Install the aft pump bearing into the pump case hub using the bearing driver set. The bearing is correctly installed when the outer bearing race seats against a shoulder in the hub. Insert the spacer from the forward end of the pump case. Install the forward bearing using the same special tools. The bearing is correctly positioned inside the hub when the inner bearing race contacts the spacer.

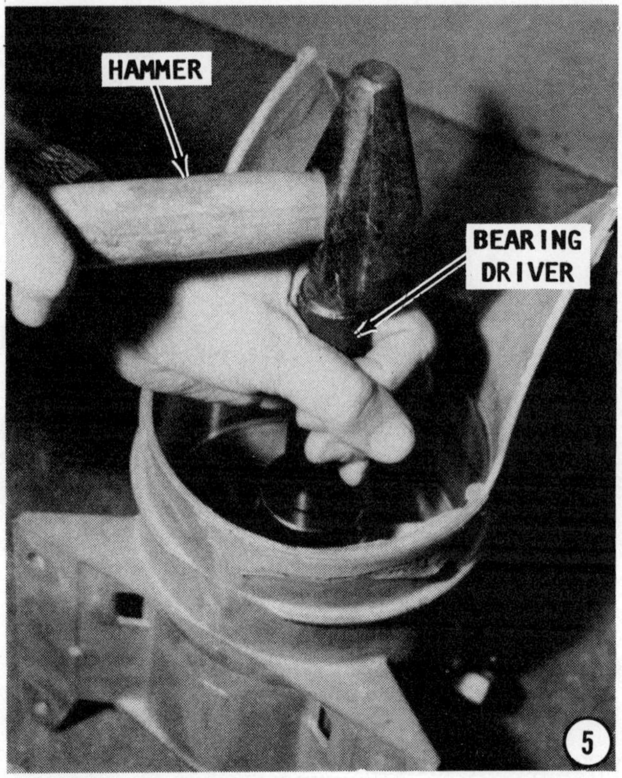

Pack the lips of the two oil seals with the same lubricant and install them into the forward face of the pump case. **BOTH** oil seals are installed with the lip facing **UP** -- forward -- when installed. The first seal is correctly installed when the back side seats against the bearing. The back side of the second seal must seat against the lip of the first.

6- Insert the short bushing into the forward end of the pump case hub.

Pump Case Oil Seal
and Bushing Installation
300 Series
and 440 Series

PLAN AHEAD

Place the bushing in a freezer for an hour or so to contract the metal and make installation easier.

Heat the pump case to about 200°F (93°C). The easiest method to heat the case is to boil the entire casting in water -- slowly increasing the temperature from room temperature to boiling. This heating **MUST** be done slowly to avoid cracking or distorting the casting. If a sufficient size container is not available, leave the casting out in the hot sun for as long as possible, and then pour boiling water over the hub.

7- Work **QUICKLY** -- to take advantage of the temperature differential. Position the cold bushing squarely over the hot pump case hub and drive in the bushing -- using the bushing/seal installation tool -- until the bushing seats against the shoulder inside the hub. Allow the case to cool to room temperature. Thoroughly dry the casting and installed bushing to remove all traces of water.

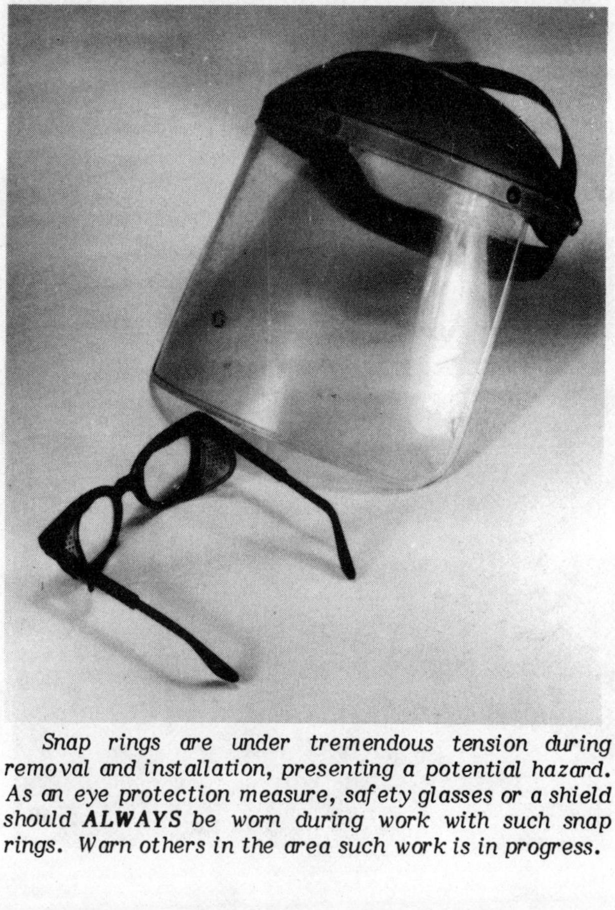

Snap rings are under tremendous tension during removal and installation, presenting a potential hazard. As an eye protection measure, safety glasses or a shield should ***ALWAYS*** *be worn during work with such snap rings. Warn others in the area such work is in progress.*

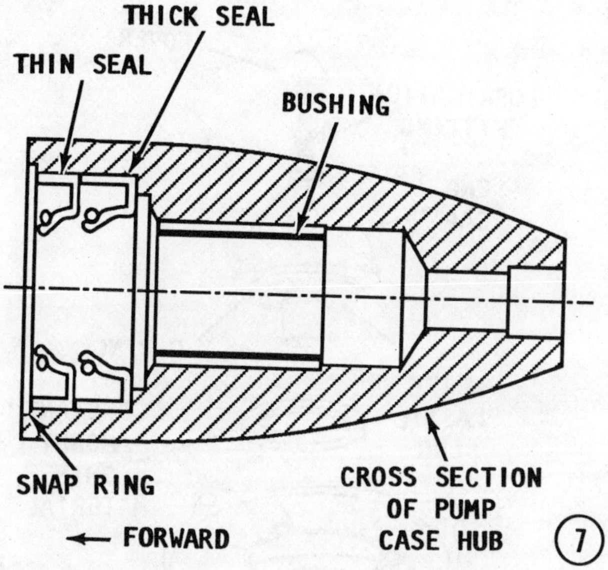

Pack the lips of the two oil seals with the same lubricant and install them into the forward face of the pump case. The thick oil seal is installed first, followed by the thin seal. Install both oil seals with seal lips facing **UP** -- forward when installed. The first seal is correctly installed when it seats against the shoulder inside the pump case. The second seal must seat against the first and not obstruct the snap ring groove.

SAFETY WORDS

The snap ring mentioned in the following step is under tremendous pressure while it is being installed. Therefore, wear a pair of safety glasses and exercise extra **CARE** while installing the snap ring to prevent injury to self or others in the area, if the ring should accidently "pop" out of the pliers.

Next, use a pair of internal snap ring pliers and install the snap ring into the groove above the thin oil seal.

Impeller Installation
550 Series
Mixed Flow Pump

8- Install the large White plastic plug into the end cap with the recessed side facing **OUTWARD.** Apply a coating of lubricant to a new **O**-ring, and then install the ring around the groove in the cap.

Apply a coating of Loctite Lock 'N Seal onto the threads of the attaching bolts.

Place the "constant" amount of shim material -- removed in Step 3 of disassemb-

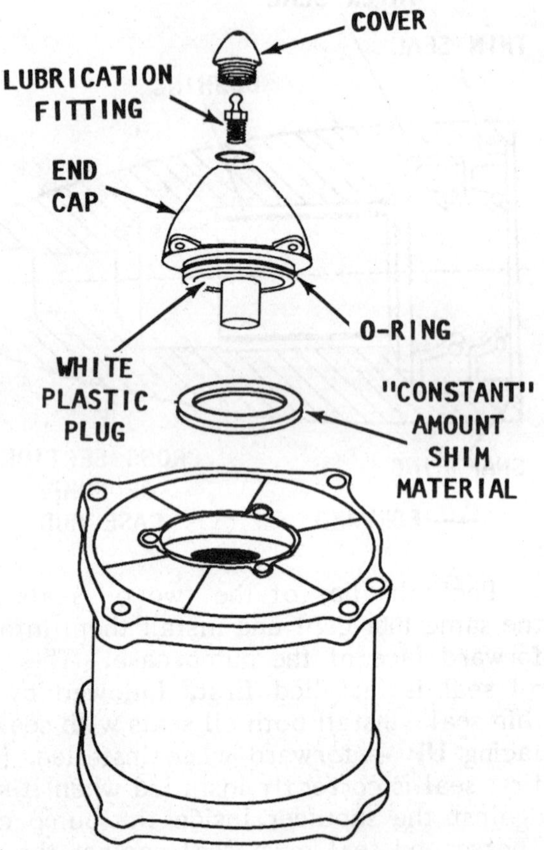

8

ly -- over the rear bearing in the hub. Install the cap over the guide vane casting hub. Tighten the bolts to a torque value of 52 in lb (6 Nm).

Next, install the lubrication fitting into the end of the end cap and tighten it securely. Install the lubrication fitting cover and new O-ring over the fitting. Tighten the cover with an allen wrench.

Using a grease gun with a cartridge filled with Shell Alvania EP1 or multi-purpose water resistant equivalent lubricant, pump the end cap full of lubricant until it oozes from the seal lip on the other side of the hub. Apply some of the excess lubricant onto the threads of the impeller shaft. Install the small O-ring onto the end of the impeller shaft.

9- Install the shim material required to obtain the correct impeller blade-to-pump case clearance onto the impeller shaft. This thickness of shim material was determined earler in section 10-5, on Page 10-24. Install the impeller.

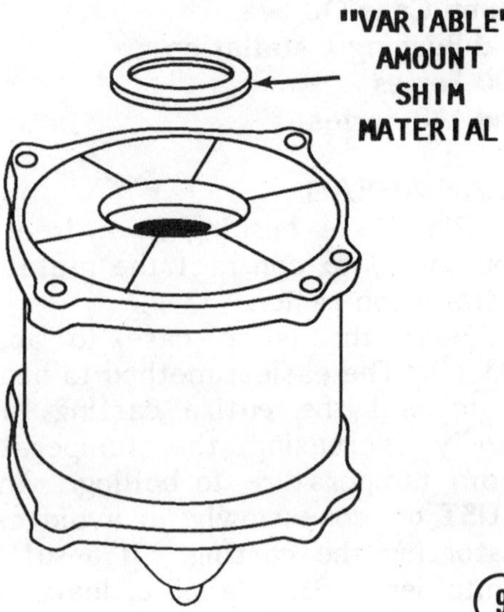

9

CRITICAL WORDS
550 SERIES
MANUFACTURED BEFORE 1991

The impeller has **LEFT-HAND** threads.

Units manufactured after 1991 have **RIGHT-HAND** threads, as do all the axial jet pumps.

Hold the impeller shaft with the special tool and tighten the impeller to a torque value of 14 ft lb (20 Nm).

A "crowsfoot" wrench the same size as the "cast" nut may be used to tighten the nut to the specified torque value. The crowsfoot wrench has a square hole to accommodate a torque wrench.

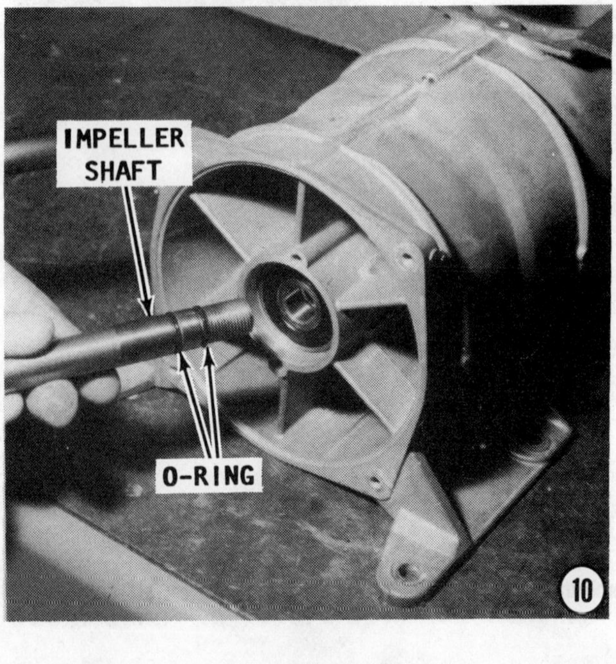

10

Impeller Shaft Installation
650 Series
Including JetMate

10- Install two new **O**-rings around the impeller shaft. Apply lubricant to both rings and the entire shaft, with the exception of the slotted nut.

Insert the shaft into the aft end of the pump case hub.

11- Thread the impeller onto the shaft.

12- Use the special tool inserted into the impeller splines to hold the impeller while the slotted nut is tightened. Tighten the slotted nut on the impeller shaft to a torque value of 72 ft lb (100 Nm).

If the small oil seal was removed, pack the lip of a new seal with lubricant and tap the seal into place with the seal lip facing **UP** -- forward when installed.

13- Pack the end cap with Shell Alvania EP1 lubricant, or equivalent. Apply a coating of the lubricant to the new **O**-ring around the cap. Apply a coating of Loctite Lock 'N Seal to the attaching bolts. Install the end cap over the hub. Secure it with the two bolts and washers. Tighten the bolts to a torque value of 52 in lb (6 Nm).

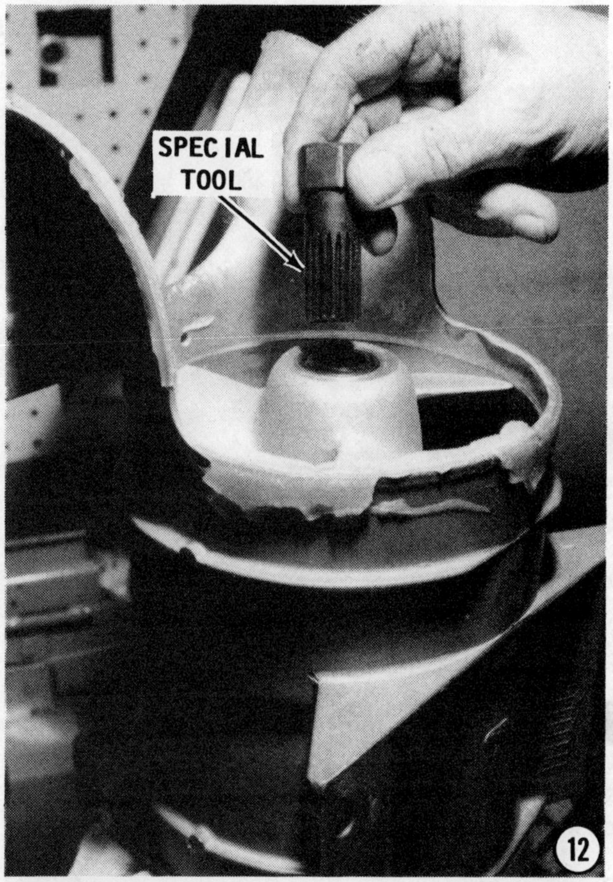

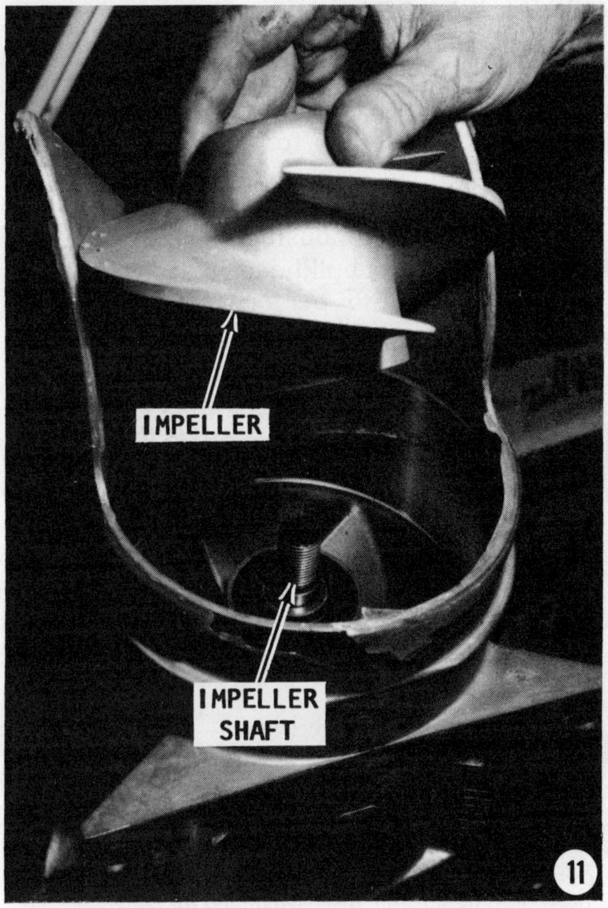

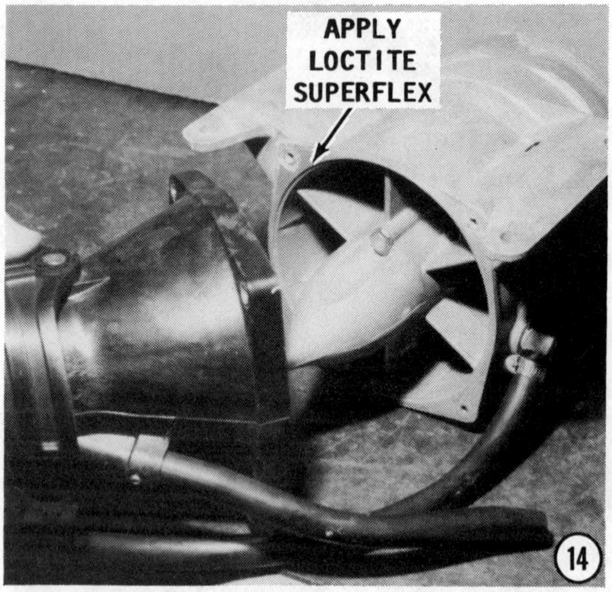

All Units
Except 550 Series
Mixed Flow Pump

14- Apply a light even coating of Loctite Superflex, or a good brand of silicone sealer, around the perimeter of the pump case and outlet nozzle mating surfaces.

15- Apply a coating of Loctite Stud N' Bearing Mount compound to the threads of the attaching bolts. Install the outlet nozzle, with steering nozzle attached, onto the pump case with the ball joint on the steering nozzle pointing **UP.** Install and tighten the securing bolts in a cross-tightening sequence to a torque value of 12 ft lb (16 Nm).

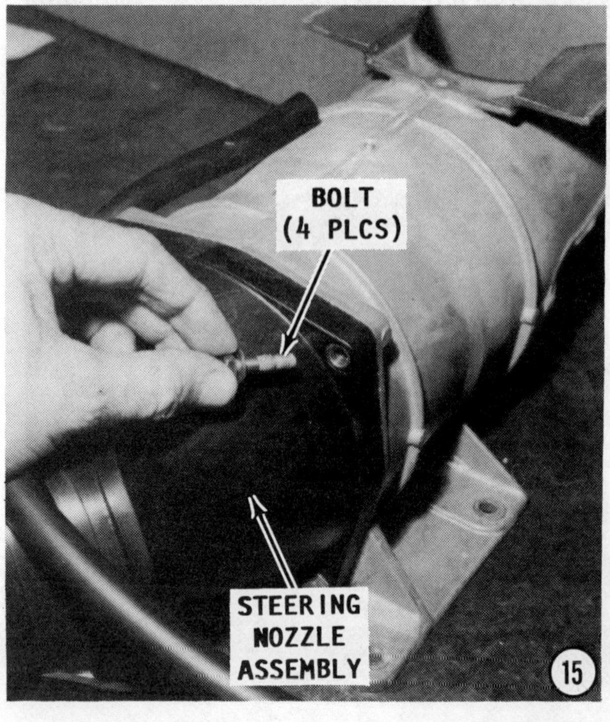

Check the action of the steering nozzle. The nozzle **MUST** swing smoothly from side to side.

550 Series
Mixed Flow Pump

Apply a light even coating of Loctite Superflex, or a good brand of silicone sealer, around the perimeter of the four mating surfaces -- the pump case, both sides of the guide vane casting, and the outlet nozzle. Check to be sure the two locating pins are in place at the two mating surfaces.

Apply a coating of Loctite Stud N' Bearing Mount compound to the threads of the four long attaching bolts. Hold the three pieces together -- with the locating pins indexed into the appropriate holes. All mating surfaces will come together in only one position -- the correct position -- with the locating pins properly indexed.

Install and tighten the securing bolts in a cross-tightening sequence to a torque value of 48 in lb (5.5 Nm).

Check the action of the steering nozzle. The nozzle **MUST** swing smoothly from side to side.

SPECIAL WORDS

Make one last check of the impeller blade-to-pump case clearance before the jet pump is installed into the craft.

Refer to Page 10-5 for specifications.

MORE SPECIAL WORDS

If the bearing housing, supporting the driveshaft at the bulkhead, needs service, this is the time to perform the necessary work. See Section 10-6.

Detailed service procedures for the bearing housing are presented in Section 10-6, beginning on Page 10-32.

INSTALLATION

1- If the engine has been removed and the driveshaft was pulled from the bearing housing, proceed as follows: Insert the driveshaft into the bearing housing. Insert a new rubber coupler into the crankshaft metal coupler and index the two metal couplers with the rubber coupler in the middle. Detailed comprehensive engine crankshaft-to-driveshaft alignment procedures are presented in the last section of the engine chapter -- Chapter 8.

Impeller Installation
300 Series and 440 Series
Axial Pump

Thread the impeller onto the driveshaft. Insert the pin in the adaptor into the hole in the driveshaft. Clamp the driveshaft holder tool around the adaptor and driveshaft. Index the pins of the impeller wrench into the holes in the impeller and tighten the impeller **VERY SECURELY.** The manufacturer does not provide a torque specification for this series axial pump.

2- Apply a coating of Loctite Stud N' Bearing Mount Compound to the threads of the attaching bolts. Bring the jet pump into position within the hull cavity and install the pump over the driveshaft.

3- Insert the stacks of shim material back into their original locations between the mounting ears of the pump case and the craft hull. Install the four allen head bolts and tighten them alternately and evenly to a torque value of 16 ft lb (22 Nm).

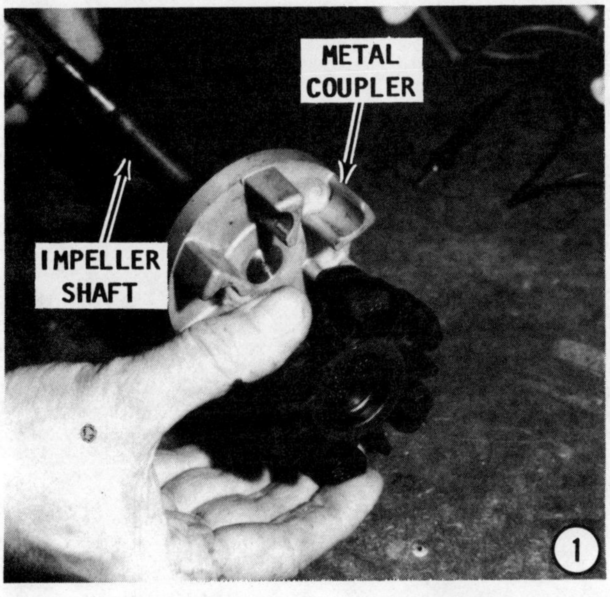

300 SERIES AND 440 SERIES
AXIAL PUMP

Install the lubrication fitting at the end of the end cap. This fitting was not installed until this time, to prevent hydraulic lock while the driveshaft was being installed.

4- Apply a solid bead of RTV Silicone around the perimeter of the pump case intake, as shown in the accompanying illustration. If there is an air gap between the pump case and hull, air will be sucked into

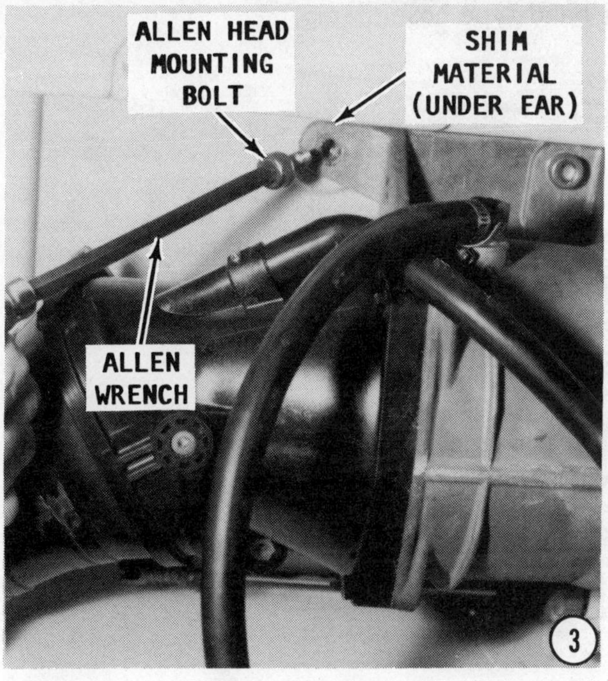

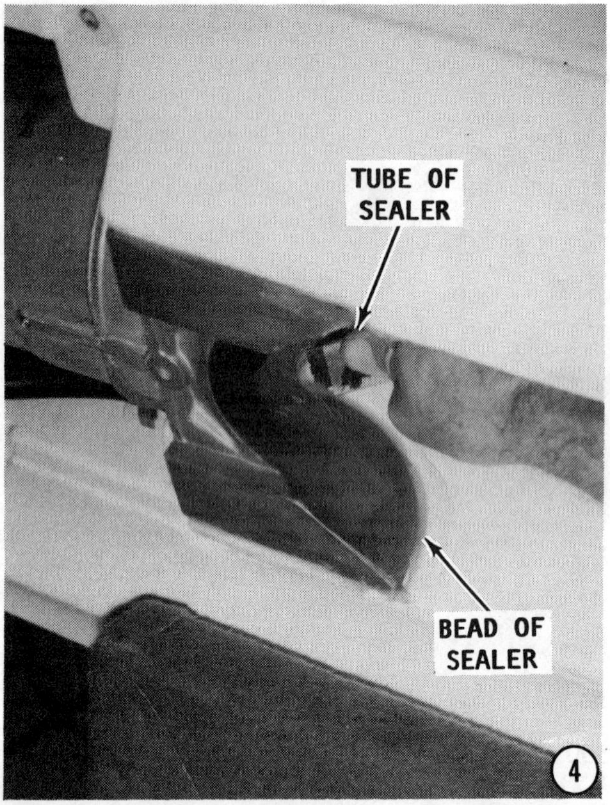

the pump and performance will be greatly reduced.

JETMATE AND SC SERIES

Pull back the reverse cable end and snap the cable onto the ball joint on the reverse bucket.

5- Connect the cooling hose to the fitting on the pump case. Secure the hose with a band type hose clamp.

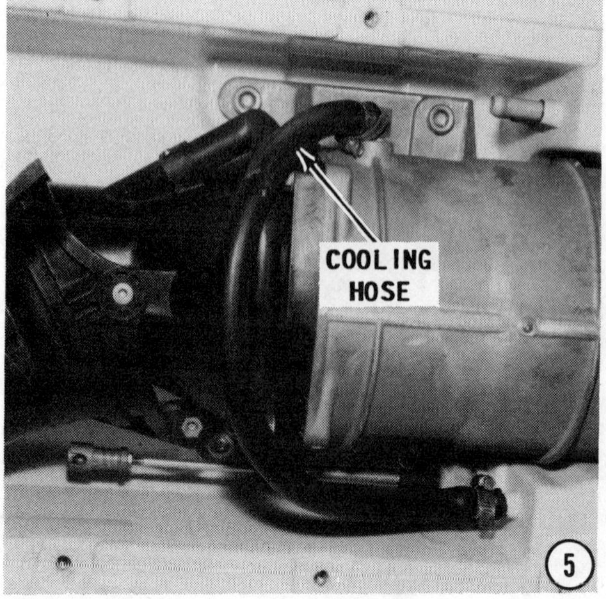

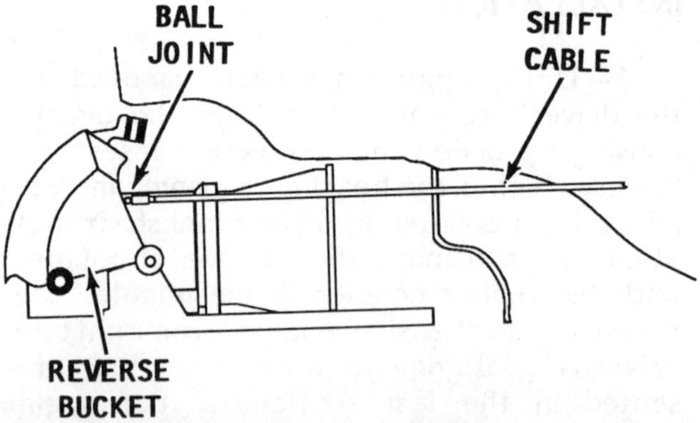

Simple line drawing to show the relationship of the reverse bucket, ball joint, and the shift cable on a JetMate or SC Series watercraft.

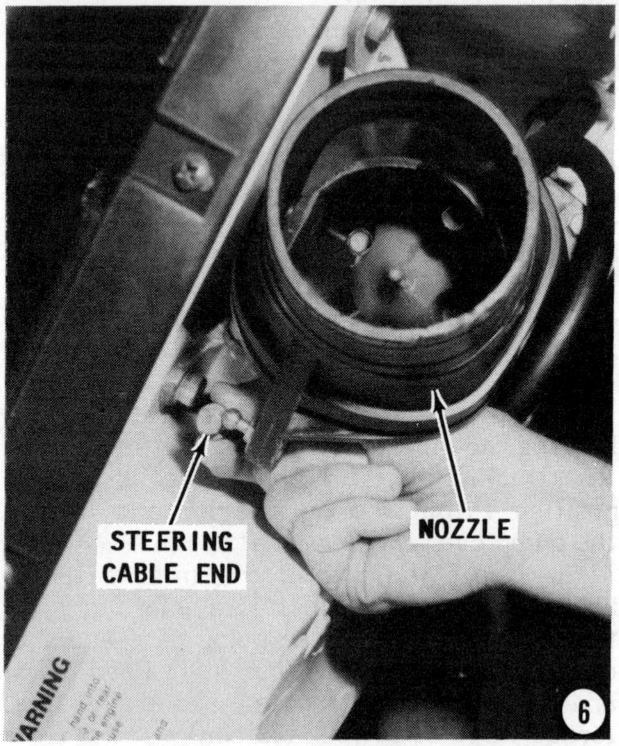

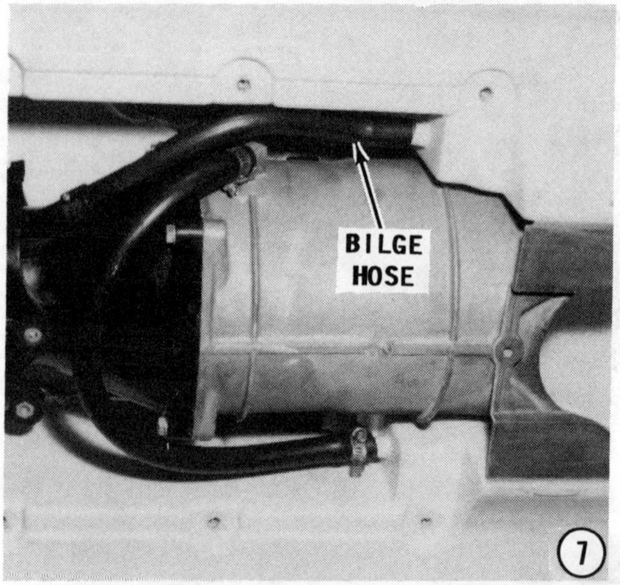

Tighten the hose clamp **SECURELY.** If the hose should blow off, the supply of cooling water to the engine would cease instantly and the engine would overheat.

6- Pull back the steering cable end and snap the cable end onto the steering ball joint on the steering arm of the nozzle.

7- Connect the bilge hose to the fitting on the hull. Secure the hose with a hose clamp.

8- Apply a coating of Loctite Stud N' Bearing Mount Compound Coat to the threads of the pump attaching bolts.

300 Series	Four bolts
440 Series	Four bolts
550 Series	Four bolts
JetMate	Four bolts
650 Series	Six bolts

9- Install the cover over the jet pump. Secure the cover with the attaching bolts and tighten them alternately and evenly to a torque value of 87 in lb (9.8 Nm).

300SX SERIES AXIAL PUMP

Connect the exhaust duct to the fitting on the hull. Tighten the hose clamp securely.

10- Apply a coating of Loctite Stud N' Bearing Mount Compound to the threads of the three rock grate attaching bolts.

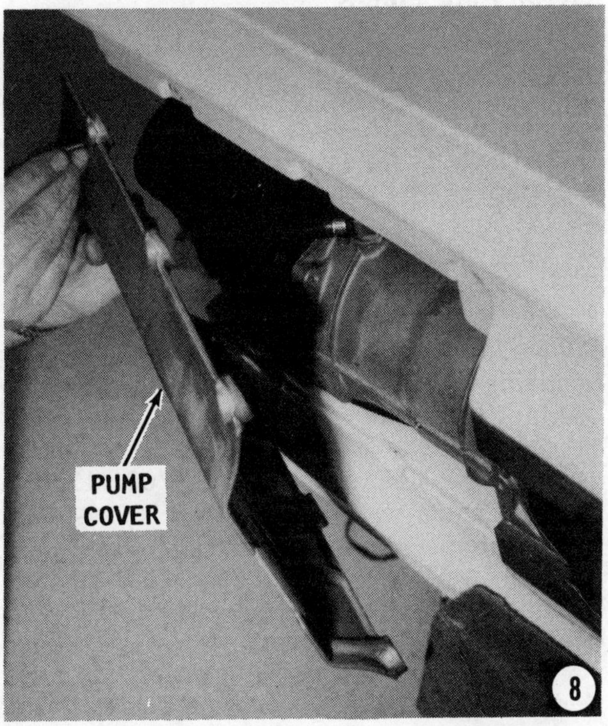

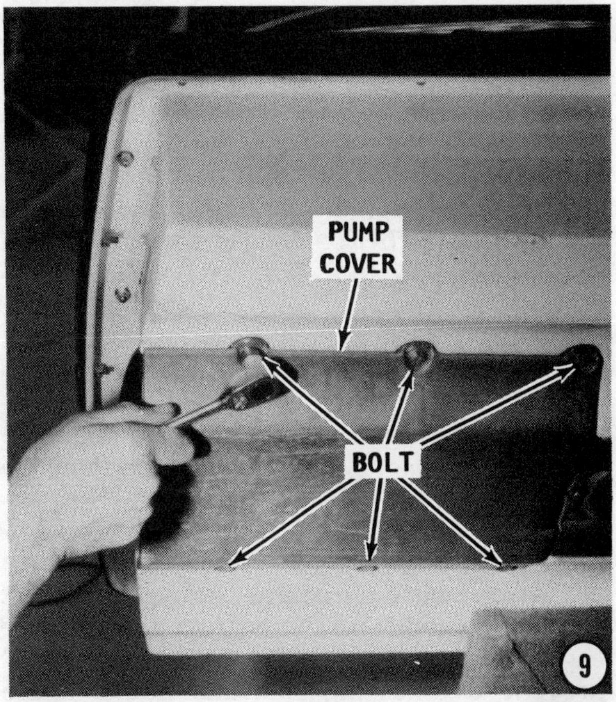

Install the rock grate and tighten the bolts to a torque value of 87 in lb (9.8 Nm).

Move the craft upright. Install the battery and connect the cables to the proper terminals. Install or fill the fuel tank.

If working on a 300 Series or 650 Series engine, fill the oil injection tank. The oil injection system **MUST** be bled while operating the engine on a premix as directed in Section 6-10 of Chapter 6.

Check Completed Work

Move the craft to a body of water or connect a garden hose to the engine cooling water supply fitting or flush fitting on the cylinder head.

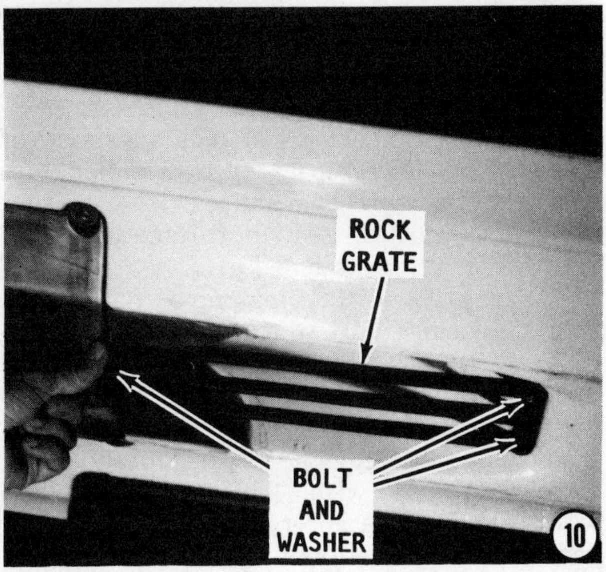

Start the engine and allow the rpm's to stabilize at idle speed **FOR JUST A FEW SECONDS,** and then turn the water on.

Adjust the water flow until a small trickle is discharged from the bypass outlet on the port side of the hull.

When the engine is to be shut down, turn the water off **FIRST** -- raise the aft portion of the hull -- **WHILE THE ENGINE IS OPERATING AT IDLE** -- "rev" the engine just a **COUPLE** times to clear water from the exhaust system -- and then shut it down.

NEVER allow the engine to operate without cooling water for more than 15 seconds.

CAUTION

Water must circulate through the jet pump -- to and from the engine, anytime the engine is operating. Circulating water will prevent overheating -- which could cause damage to moving engine parts and possible engine seizure.

NEVER, AGAIN, NEVER operate the engine at high speed with a flush device attached. The engine, operating at high speed with such a device attached, would **RUNAWAY** from lack of a load on the impeller, causing extensive damage.

JETMATE AND SC SERIES

The reverse gate adjustment is covered in Chapter 11, Section 11-3.

10-6 BEARING HOUSING SERVICE

The driveshaft bearing housing serves as a support for the driveshaft and prevents water from entering the engine compartment from the pump intake.

Older type bearing housings consisted of two ball bearings separated by a collar with double oil seals on each side.

The forward oil seals both faced forward and the aft oil seals faced aft.

A very few 650 Series pumps had only a single bearing, with the same oil seal arrangement.

Newer type bearing housings have bushing type bearings with no seals and are maintenance free. The older type housings were equipped with lubrication fittings, which the manufacturer recommended lubricating every 25 hours of craft operation.

Only the older type serviceable housings are covered in this section.

If a newer type bearing housing needs replacement, follow the service procedures in the section to remove the housing, and then skip the disassembly and assembly procedures -- move directly to the installation procedures to install the housing.

Normally, the bearing housing never requires service, unless the oil seals have failed and allowed water to enter the bearing cavity.

SPECIAL WORDS ON SPECIAL TOOLS

The following list of special tools are required in order to perform some of the necessary work on the bearing housing. Some tasks become extremely difficult, if not impossible, without the special tools or some type of equivalent. Manufacturers' part numbers are given. In specific steps of the procedures -- whenever possible -- a special effort has been made to give some type of equivalent or homemade tool which may be substituted. However, the cost of a special tool may be offset by money saved if an expensive part is not damaged.

SPECIAL TOOLS FOR BEARING HOUSING SERVICE

300 Series
440 Series
Axial Flow Pumps
AND
550 Series
Mixed Flow Pump

Coupling Holder	P/N 57001-1230
Driveshaft Holder	P/N W56019-003
Adapter	P/N 57001-1231
Oil Seal and Bearing Remover	P/N 57001-1058
Bearing Driver Set	P/N 57001-1129

650 Series
Including JetMate
Axial Flow Pump

Coupling Holder	P/N 57001-1230
Driveshaft Holder	P/N W56019-003
Adapter	P/N 57001-1231
Bearing Driver Set	P/N 57001-1129

Bearing Housing Removal

Remove the battery, the water box muffler, and the fuel tank from the craft.

Remove the three bolts securing the coupling cover to the bearing housing.

Remove the engine, see Chapter 8.

Remove the jet pump, see Section 10-3.

300 Series
440 Series
550 Series

Index the pin in the adaptor with the hole in the driveshaft. Clamp the driveshaft holder around the adaptor and driveshaft. Index the pins in the impeller wrench with the hole in the impeller and loosen the impeller from the shaft.

300SX Series

Loosen the hose clamp around the large connecting tube.

All Series

Remove the four bolts securing the bearing housing to the bulkhead or hull floor (300SX Series). There may be one or more pieces of shim material behind the mounting holes. Save the pieces of shim material and make a note of their locations. Identifying the shim material, at this time, will **ENSURE** correct shaft alignment at the pump.

The 300SX Series bearing housing is joined to the hull by the connecting tube -- no silicone sealant is used in this setup.

650 Series

The driveshaft may be pulled forward out of the bearing housing.

300 Series
440 Series
550 Series

Cut away any silicone sealant from the edges of the housing flange and then pull the bearing housing, with the driveshaft attached, from the bulkhead.

Clean off all traces of silicone sealer from the housing flange and the hull.

BEARING HOUSING DISASSEMBLING

300 Series
440 Series
550 Series

Older Models -- All with Lubrication Fitting Except 300SX Series

Refer to the accompanying exploded diagrams as an assist in disassembling.

Remove only the parts which will be replaced. If the bearings on the shaft are satisfactory, do not remove them from the shaft. If only the two sets of oil seals are to be replaced, the shaft may be pulled from the housing and the bearings left in place.

Insert the pin on the adapter into the hole in the driveshaft. Clamp the driveshaft holder tool around the adapter and driveshaft. Insert the coupling holder into the metal coupler and unscrew the coupler from the threaded end of the driveshaft. Remove the washer.

Remove the three bolts securing the end cover to the housing. If the cover is frozen in place, tap it lightly with a soft head mallet to free it from the housing. **AVOID** hitting the protruding brackets which support the coupling cover. These "brackets" are very fragile and may break off under the blow of a hammer. Remove the O-ring around the end cover and discard it, only if a replacement is available.

Use a slide hammer with a puller jaw attachment and pull out both oil seals from the end cover.

Reach inside the housing and pull out any pieces of shim material. If the oil seals are to be the only items to be replaced -- the original ball bearings are to be reused -- the exact thickness of shim material will be used again. If such is the case, **SAVE** the shim material.

Wrap the driveshaft in a shop towel about six inches away from the housing and clamp the wrapped portion in a vise equipped with soft jaws.

Bearing housing removed from a 650 Series watercraft.

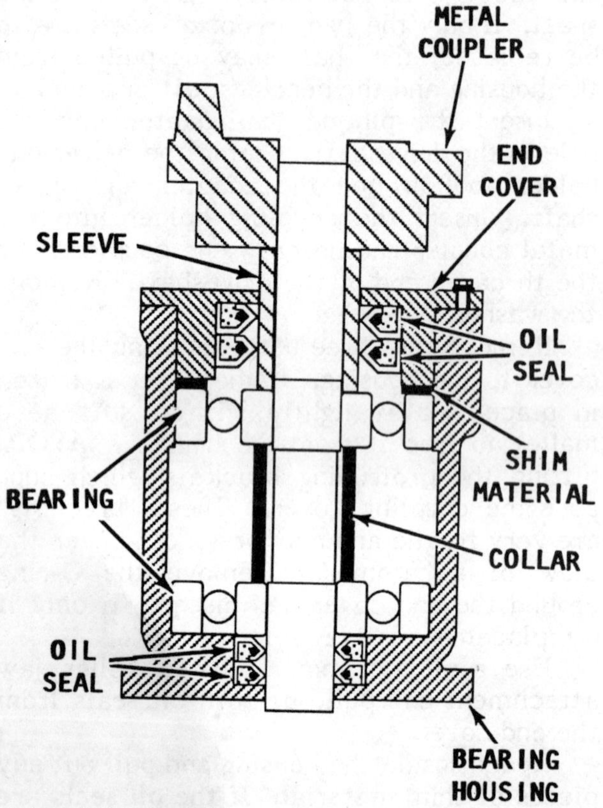

Cross-section line drawing of the bearing housing on a 300 Series, 440 Series, and 550 Series watercraft to show the relationship of the various parts mentioned in the text. This drawing is not valid for the 300SX bearing housing.

Using a soft head mallet, tap the bearing housing free of the shaft. Again, avoid striking the coupling cover brackets.

Use the slide hammer with a puller jaw attachment and reach into the housing to pull the second set of oil seals.

The sleeve, forward bearing, collar, and rear bearing must be pressed from the shaft. Make a setup in an arbor press to press these items from the shaft. Use a suitable mandrel which makes contact only with the inner bearing race. Do not allow the shaft to fall free to the floor.

Bearing Housing Disassembling 300SX Series

Refer to the accompanying exploded diagrams as an assist in disassembling.

Remove only the parts which will be replaced. Remove the inspection plate on the side of the housing. This may help to make a determination as to which parts need replacement.

If the bearing on the shaft is satisfactory, do not remove it from the shaft. If only the two sets of oil seals are to be replaced, the shaft may be pulled from the housing and the bearing left in place.

Insert the pin in the adaptor into the hole in the driveshaft. Clamp the driveshaft holder tool around the adaptor and shaft. Insert the coupling holder into the metal coupler and unscrew the coupler from the threaded end of the driveshaft.

Remove the breather bolt from the top of the housing.

SAFETY WORDS

The snap ring mentioned in the following step is under tremendous pressure when installed. Therefore, wear a pair of safety glasses and exercise extra **CARE** when removing the snap ring to prevent injury to self or others in the area.

Using a pair of internal snap ring pliers, remove the large snap ring from the housing.

Make a setup in an arbor press and press the driveshaft, with seals and bearings in place, from the housing.

Slide the four oil seals from the driveshaft. Remove the small external snap ring from the driveshaft and press the bearing free of the shaft.

If the two forward oil seals remained inside the housing, use a slide hammer with a puller jaw attachment and pull out both oil seals.

Bearing Housing and Driveshaft Disassembling All 650 Series — Including JetMate

Refer to the accompanying exploded diagrams as an assist in disassembling.

Remove only the parts which will be replaced.

If the bearing inside the housing is satisfactory, do not remove it. If only the two sets of oil seals are to be replaced, the seals may be pulled from the housing and the bearing left in place.

Insert the pin in the adaptor into the hole in the driveshaft. Clamp the driveshaft holder tool around the adaptor and shaft. Insert the coupling holder into the metal coupler and unscrew the coupler from the threaded end of the driveshaft. Remove the washer.

SAFETY WORDS

The snap ring mentioned in the following step is under tremendous pressure when in-

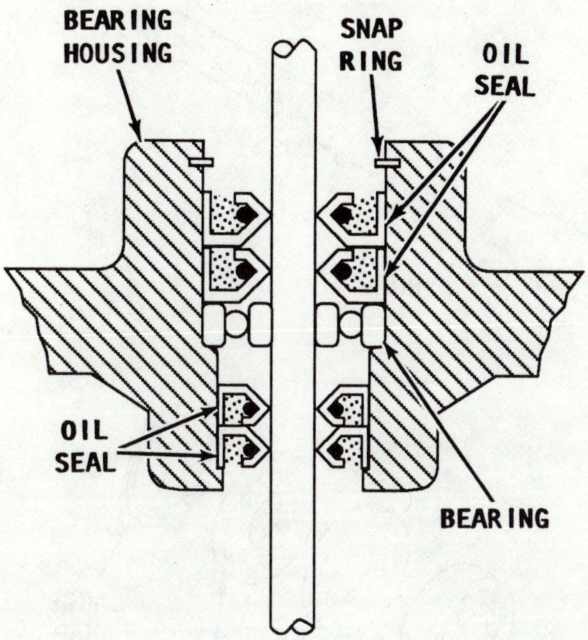

Cross-section line drawing of the bearing housing on a 650 Series, including the JetMate Series watercraft to show the relationship of the various parts mentioned in the text.

stalled. Therefore, wear a pair of safety glasses and exercise extra **CARE** when removing the snap ring to prevent injury to self or others in the area.

Using a pair of internal snap ring pliers, remove the snap ring at the forward face of the housing. Make a setup in an arbor press to remove the seals and bearing from the housing. With the forward face down over a wooden support, press on the top seal to remove the two large forward seals, the bearing, and finally the two small aft oil seals.

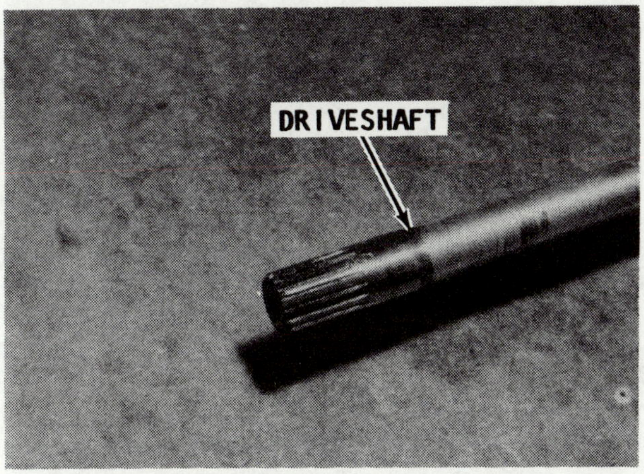

Splined end of a 650 Series driveshaft. The splines index with the splines in the impeller shaft. If the splines become excessively worn or damaged, impeller rotation will be seriously restricted.

CLEANING AND INSPECTING ALL SERIES

Place the driveshaft on a set of Vee blocks and install a dial indicator over the center of the shaft. Slowly rotate the shaft and observe the dial indicator. A shaft in satisfactory condition will have a runout less than 0.2mm. If the runout is more than 0.5mm, replace the shaft.

Next, position the dial indicator over the two bearing surfaces, and then the sleeve surface, if equipped. Measure the runout. A shaft in satisfactory condition will have runout less than 0.03mm. If the runout is more than 0.1mm, replace the driveshaft.

Rinse the bearings in a high flash point solvent and inspect them. Spin each bearing by hand -- never with compressed air -- and feel for rough spots. Inspect for water damage.

Bearing Housing Assembling
300 Series
440 Series
550 Series
All with Lubrication Fitting
Except 300SX Series

Pack the lips of all four **NEW** oil seals with Shell Alvania EP1 lubricant, or equivalent multi-purpose water resistant lubricant.

Using a suitable mandrel, install the two aft oil seals into the bearing housing with the seal lips facing **AFT.** Fill in the space between the two seals with lubricant.

Pack both bearings with the same lubricant. Using a suitable mandrel press the aft bearing onto the shaft. The bearing is correctly installed when the inner race contacts the shoulder on the shaft.

Next, press on the collar, until it makes contact with the aft bearing. Now, press on the forward bearing until it makes contact with the collar. Finally, press on the sleeve until it seats against the forward bearing.

Slide the bearing housing over the long end of the driveshaft. Wrap the shaft with a shop towel, as before, and then clamp it in a vise equipped with soft jaws. Using a soft head mallet **AND** avoiding the coupling cover brackets, gently tap the bearing housing over the bearings on the shaft. The bearing housing is correctly installed when the aft bearing is seated against the bottom of the housing.

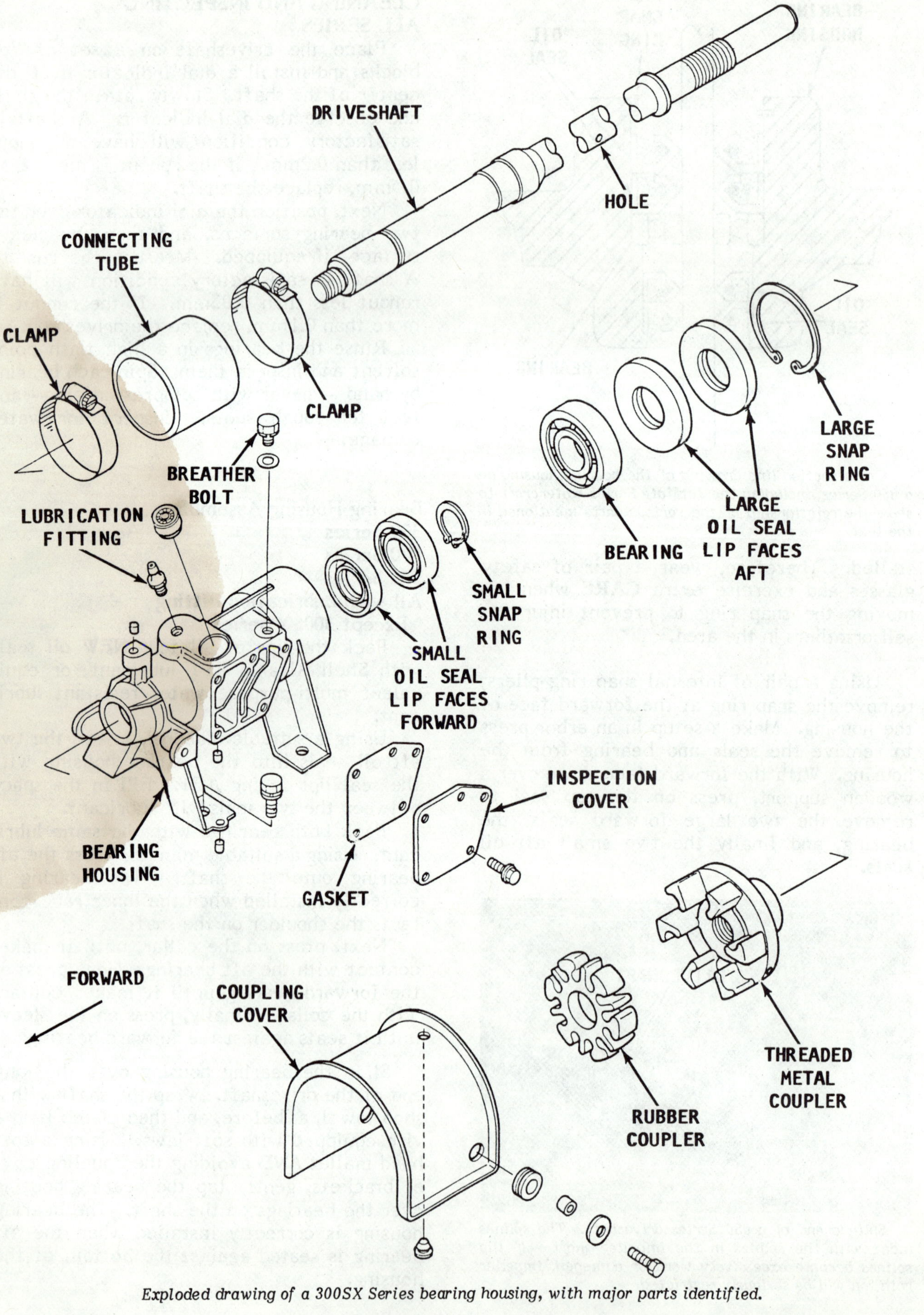

Exploded drawing of a 300SX Series bearing housing, with major parts identified.

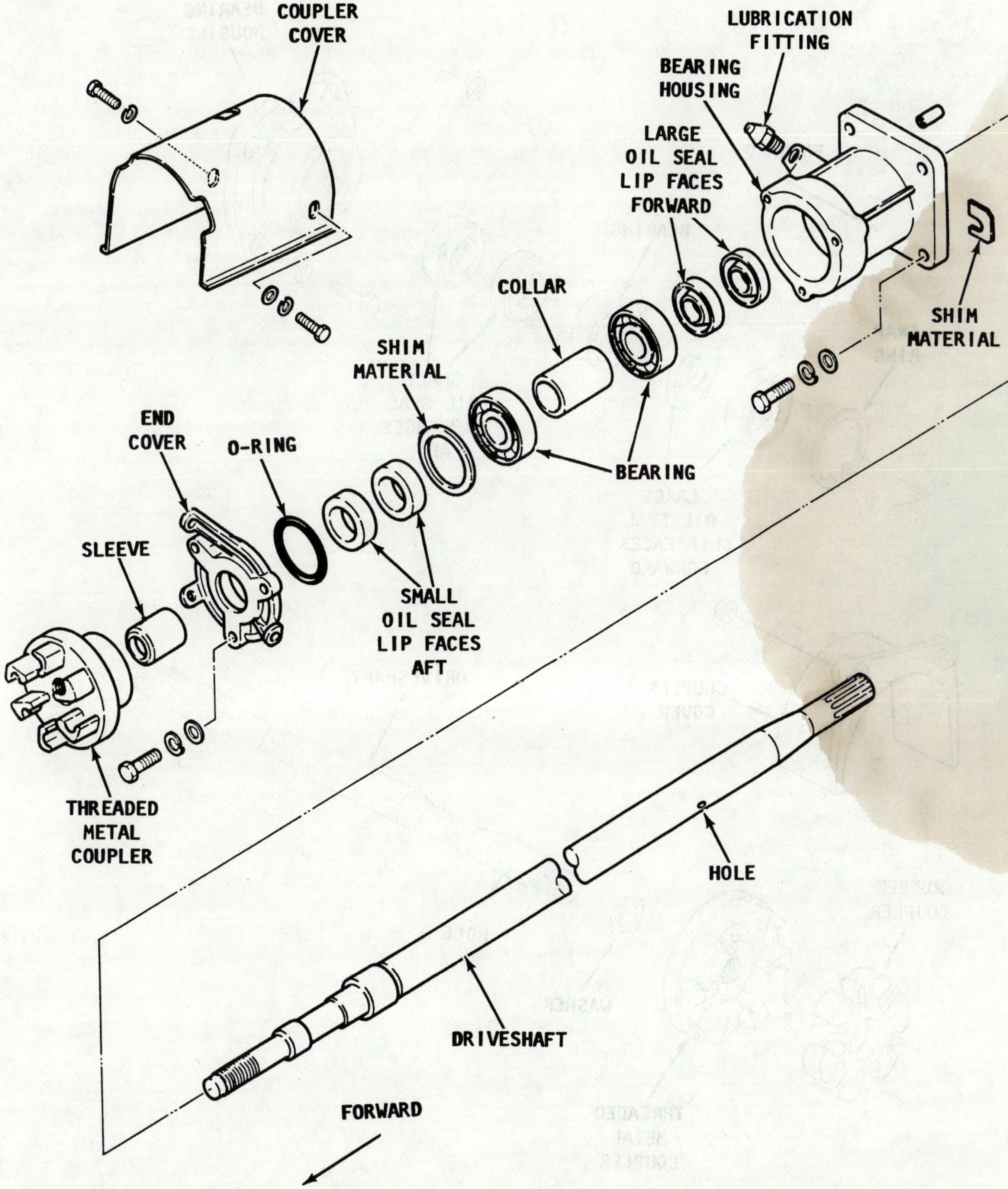

Exploded drawing of the bearing housing installed on the 300 Series (except the 300SX), 440 Series, and 550 Series watercraft. Major parts are identified.

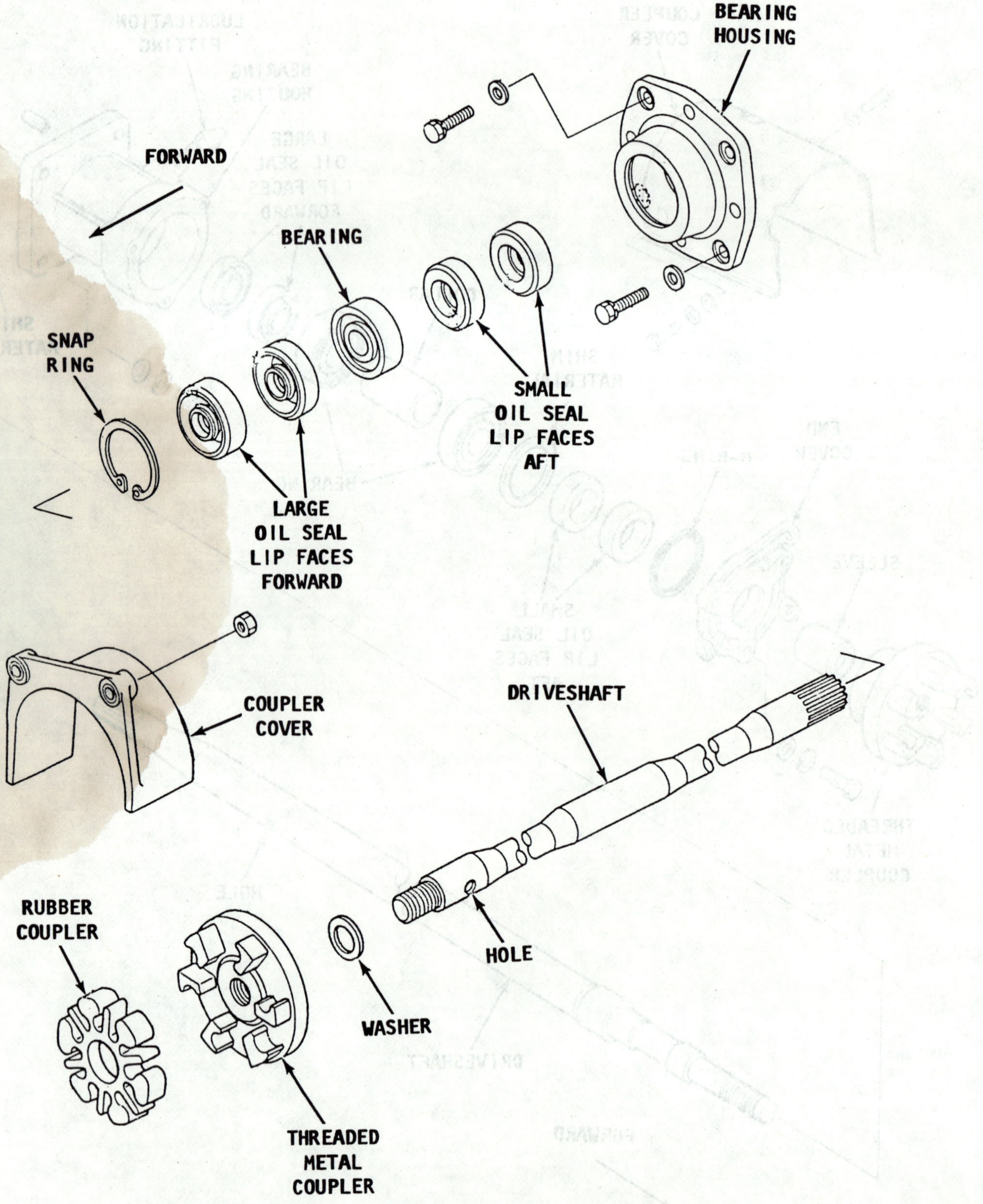

Exploded drawing of the bearing housing installed on the 650 Series, including the JetMate and SC Series. Major parts are identified.

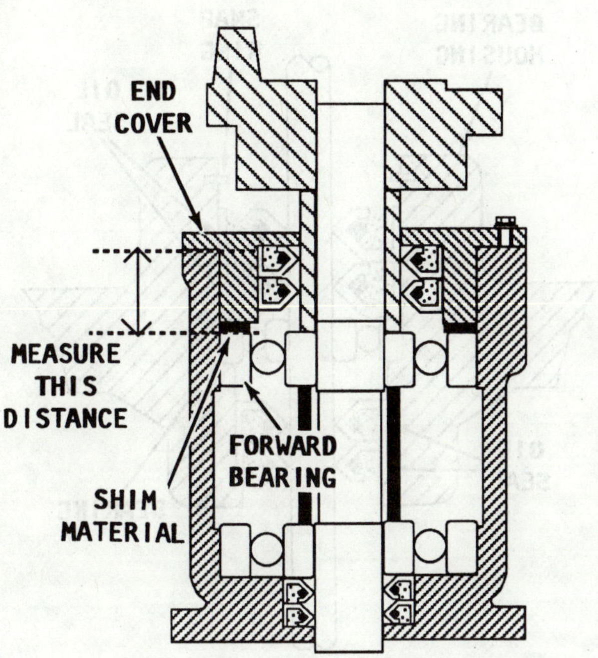

Cross-section line drawing of the bearing housing installed on the 300 Series, 440 Series and 550 Series (except 300SX), watercraft to illustrate placement of the shim material where the measurement mentioned in the text is to be taken.

If the original bearings were installed, the same shim material may be installed. If new bearing/s were used, measure the distance between the open end of the housing and the forward bearing.

If this distance is between 12.90 and 13.05mm, no shim material is used in assembling.

If the distance is between 13.06 and 13.20mm, use shim material P/N 92025-502.

If the distance is between 13.21 and 13.32mm, use shim material P/N 92025-503.

If the distance is greater than 13.32mm, the bearings, collar, spacer, or bearing housing have not seated correctly.

Place the correct amount of shim material in front of the forward bearing inside the bearing housing.

Using a suitable mandrel, install the two forward oil seals into the end cover with the seal lips facing **FORWARD.** Fill in the space between the two seals with lubricant.

Install a new O-ring around the cover. Apply a coating of the same lubricant to the O-ring.

Install the end cover over the bearing housing. Align the embossed word **"MADE"** with the lubrication fitting. This will position the coupling cover brackets correctly. Apply a coating of Loctite Lock N' Seal to the threads of the three attaching bolts. Install the plain washers, lockwashers, and bolts. Tighten the bolts to a torque value of 13.5 ft lb (19 Nm).

Place the washer over the shaft, and then thread the metal coupler onto the end of the driveshaft.

Clamp the driveshaft holder tool around the driveshaft. Insert the coupling holder into the metal coupler and tighten the coupler to a torque value of 72 ft lb (98 Nm).

Pump Shell Alvania EP1 lubricant into the lubrication fitting using a grease gun. The bearing housing is full when a slight amount of lubricant oozes from the seals. The bearing housing should be lubricated every 25 hours of craft operation.

Bearing Housing Assembly 300SX Series

Pack the lips of all four **NEW** oil seals with Shell Alvania EP1 lubricant, or equivalent multi-purpose water resistant lubricant.

Press a new bearing onto the driveshaft. The bearing is correctly positioned on the shaft when the forward face is next to the snap ring groove. Install the small external snap ring. Slide the two large aft oil seals onto the shaft with both seal lips facing **AFT.** Install both small forward oil seals into the bearing housing with the seal lips facing **FORWARD.**

Make a setup in an arbor press and press the driveshaft into the bearing housing. The shaft is correctly installed when the snap ring groove is visible above the last aft oil seal.

SAFETY WORDS

The snap ring mentioned in the following paragraph is under tremendous pressure while it is being installed. Therefore, wear a pair of safety glasses and exercise extra **CARE** when installing the snap ring to prevent injury to self or others in the area.

Install the large internal snap ring. Start the metal coupler onto the threaded end of the driveshaft. Insert the pin in the adaptor into the hole in the driveshaft. Clamp the driveshaft holder tool around the adaptor and shaft. Insert the coupling holder into the metal coupler and tighten the coupler onto the driveshaft.

Pump Shell Alvania EP1 lubricant into the lubrication fitting using a grease gun. The bearing housing is full when a slight amount of lubricant oozes from the breather hole. Install the breather bolt. The bearing housing should be lubricated every 25 hours of craft operation.

Bearing Housing and Driveshaft Assembly 650 Series Including JetMate

Pack the lips of all four **NEW** oil seals with Shell Alvania EP1 lubricant or equivalent multi-purpose water resistant lubricant.

Using a suitable mandrel, press the two small aft oil seals into the housing from the forward side, with both seal lips facing **DOWN** -- aft -- after installation.

Press the bearing into the housing until it makes contact with a shoulder inside the housing.

Next, press in the two large oil seals, with the seal lips facing **UP** -- forward -- after installation. The seals are correctly installed when the groove for the snap ring is visible.

SAFETY WORDS

The snap ring mentioned in the following paragraph is under tremendous pressure while it is being installed. Therefore, wear a pair of safety glasses and exercise extra **CARE** when installing the snap ring to prevent injury to self or others in the area.

Install the snap ring.

Insert the pin in the adaptor into the hole in the driveshaft. Clamp the driveshaft holder tool around the adaptor and shaft. Place the washer over the threaded end of the driveshaft. Thread the coupler onto the driveshaft. Insert the coupling holder into the metal coupler and tighten the coupler to the threaded end of the driveshaft to a torque value of 29 ft lb (39 Nm).

Bearing Housing Installation All Series

Check to be sure the hull surface and the bearing housing surface are free of any old sealant.

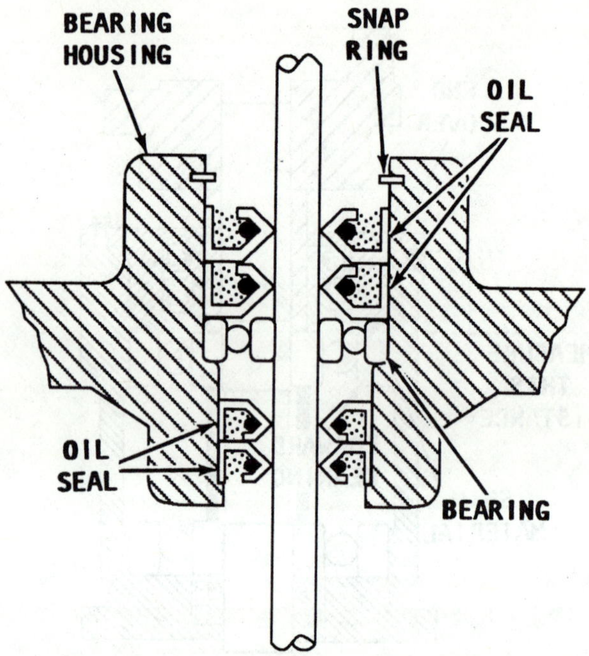

Cross-section of the bearing housing used on the 650 Series, including the JetMate. Note the direction of the oil seal lips during assembling.

Position any shim material removed in disassembly in the exact original location. The purpose of the shim material is to **ENSURE** correct driveshaft alignment at the pump.

Temporarily install the bearing housing -- without a silicone seal. Install and tighten the four bolts just **HAND** tight. Using a vernier caliper, measure the distance from the center of the driveshaft to the top wall of the hull cavity. This distance should be exactly 3.000" (77m). If the distance is not as specified, add or subtract shim material between the bearing housing and the hull. Equal amounts must be added or subtracted from the two top mounting corners or the two bottom mounting corners to keep the shaft in the same plane.

If working on a 300SX Series, the bearing housing is mounted on four feet to the hull floor. Therefore, if the distance is not as specified, add or subtract shim material between the bearing housing and the hull floor. Equal amounts must be added or subtracted from the two forward mounting feet or the two aft mounting feet to keep the shaft in the same plane.

Next, measure the distance between the center of the driveshaft and the two side walls of the hull cavity. This distance must

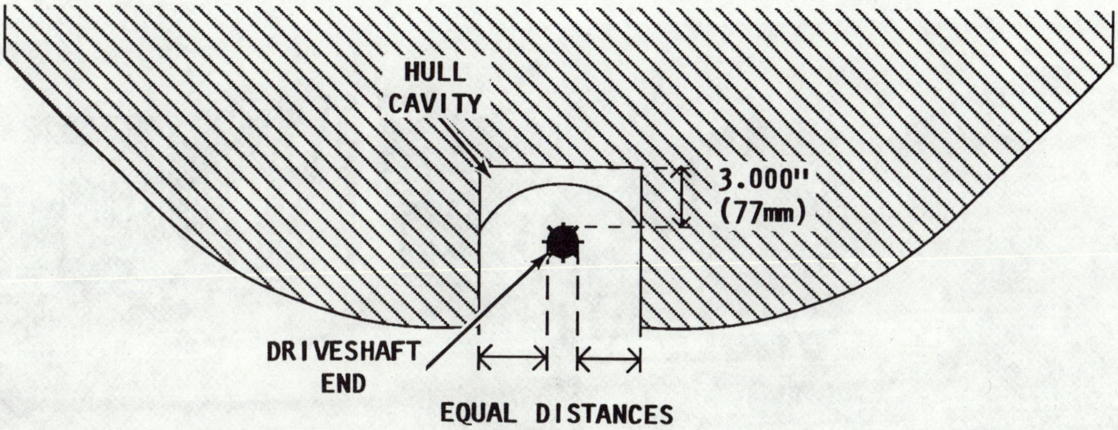

Simple line drawing depicting where the measurements mentioned in the text are to be taken to obtain proper engine/jet pump alignment.

be equal on both sides. If the two measurements are not equal, shim material must be added or subtracted, in equal amounts, to the two left mounting corners/feet or the two right mounting corners/feet, so as not to disturb the vertical setting.

Once the shimming is completed, take **TIME** and **CARE** to identify the four new stacks of shim material and remove the bearing housing from the hull.

If working on a 300SX Series bearing housing, tighten the hose clamp around the connecting tube. The tube seals the housing to the hull without the use of silicone sealant.

All Other Series

Apply a bead of Loctite Superflex around the perimeter of the mounting flange, looping around the four bolt holes. Install the bearing housing with the correct stack of shim material behind each corner. Apply a coating of Loctite Lock N' Seal to the attaching bolt threads. Install the plain washers, lockwashers and bolts. Tighten the bolts to a torque value of 13.5 ft lb (19 Nm).

650 Series Pump

Insert the driveshaft into the bearing housing and index the threads into the impeller.

Position the rubber coupler inside the driveshaft metal coupler.

300 Series
440 Series
550 Series

Start the impeller onto the driveshaft. Place the adaptor inside the driveshaft holder. Index the pins in the impeller wrench with the hole in the impeller and tighten the impeller securely to the shaft. The manufacturer does not provide a torque value for the impeller.

Install the jet pump, see Section 10-3.

Install the engine, see Chapter 8.

Install the fuel tank, the water box muffler and the battery.

Install the three bolts securing the coupling cover to the bearing housing and tighten them securely.

A new bearing housing in place ready for the jet pump and engine.

550

11 CONTROL ADJUSTMENTS

11-1 INTRODUCTION

Control adjustments on Kawasaki Personal Watercraft is limited to changing the length of the steering cable and the shift cable. Normally, adjustments are not necessary for extended periods of time, unless the craft has experienced an accident or the control has been disconnected to service other component/s.

Detailed procedures to adjust the steering cable and the shift cable are presented in this short chapter. Adjustment of the throttle cable and the choke cable is thoroughly covered in Chapter 6, **FUEL.**

11-2 STEERING CABLE ADJUSTMENT ALL MODELS EXCEPT JETMATE AND SC

If the steering nozzle is not positioned exactly in the center of the transom opening, with an equal clearance on both sides, the steering cable **MUST** be adjusted to prevent the craft from veering to port or starboard when the handle bar is positioned for "dead ahead".

Adjustment

1- Center the handle bar for a "dead ahead" course. Have an assistant prevent the handle bar from moving from this straight ahead position.

2- At the stern, measure the clearance, port and starboard, between the outside of the steering nozzle and the transom opening. Both measurements should be the same. If the measurements vary by more than 1/16", perform Steps 3 thru Step 7 to properly adjust the nozzle.

3- Raise the handle pole to the upright position. The steering cable is attached to a ball joint on the steering plate around the pole. Loosen the locknut on the steering cable.

4- Push the outer sleeve back from the ball joint and pry the cable end free.

5- With the handle bar still in the "dead ahead" position, use a tape measure to center the steering nozzle within the transom cavity.

6- Without disturbing the position of the handle bar or the position of the nozzle, rotate the cable end until the hole in the end aligns with the ball joint on the steering plate.

7- Push back the outer sleeve on the cable end, and then snap the cable end over the ball joint. Make another check of the steering nozzle position in relation to the handle still at "dead ahead". If the measurement, port and starboard has remained the same, tighten the locknut to hold the new adjustment.

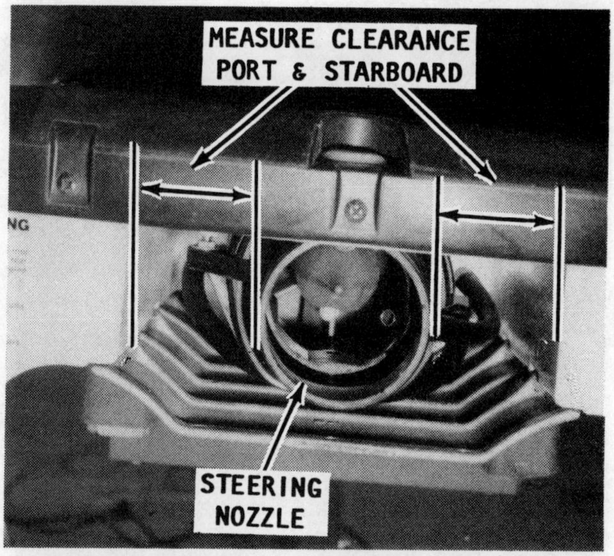

Close view of the steering nozzle to clearly indicate where the clearance measurements are to be taken when making a steering cable adjustment, on all models except the JetMate.

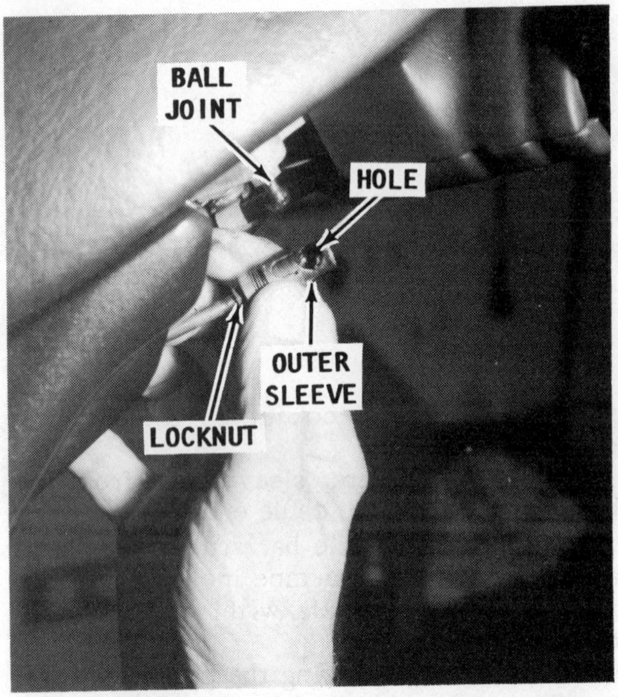

End fitting and ball joint for the steering cable on the 300 Series.

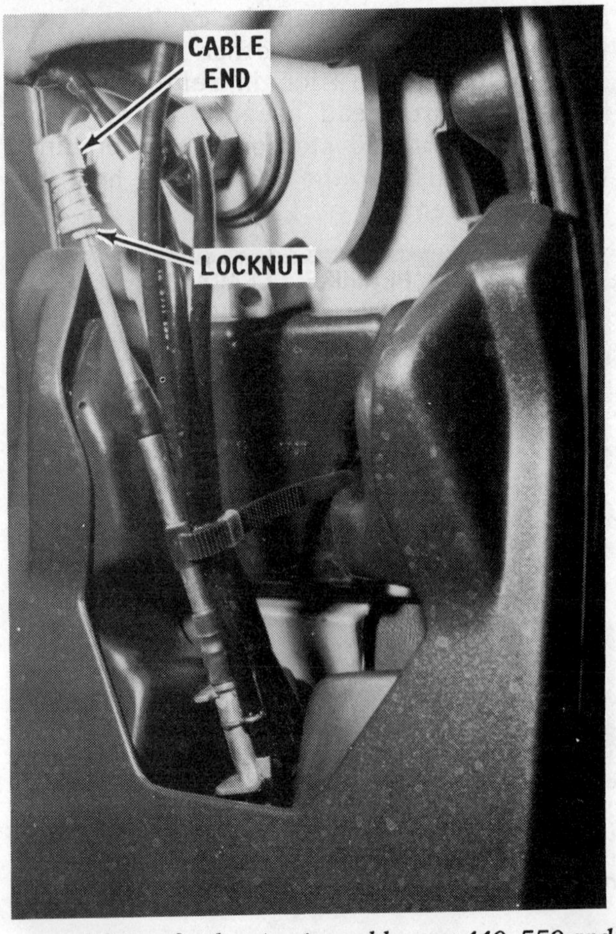

End fitting for the steering cable on a 440, 550 and 650 Series unit.

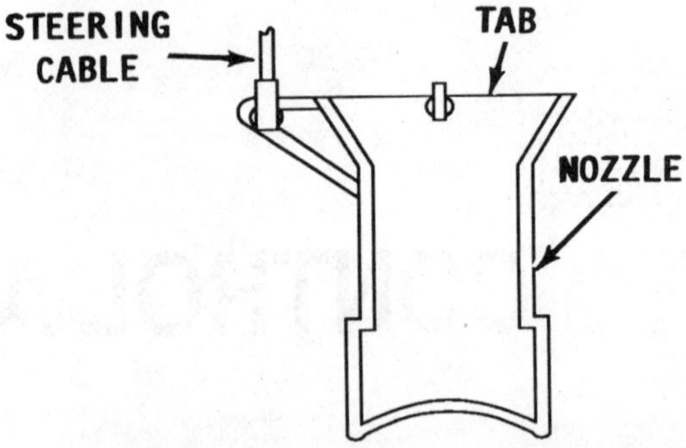

The JetMate and SC has a special balance tab centered inside the nozzle to reduce the force necessary to operate the "Joy Stick".

11-3 STEERING CABLE ADJUSTMENT JETMATE AND SC

General Information

The JetMate is maneuvered by a control lever ("Joy Stick") located on the craft centerline directly in front of the operator's seat. Through movement of the "Joy Stick", the operator is able to direct the jet pump thrust to move the bow port or starboard. The "Joy Stick" serves to house the trigger throttle and also the start and stop buttons.

A shift lever is located starboard of the operator's seat. This lever controls the reverse bucket mounted on the jet pump. The bucket can be lowered down over the output nozzle to direct the flow of water back under the craft and move the craft sternward.

The steering nozzle has a special steering tab to reduce the force needed to move the "Joy Stick".

Steering Cable Adjustment

1- Center the "Joy Stick" in the upright "dead ahead" position. Have an assistant

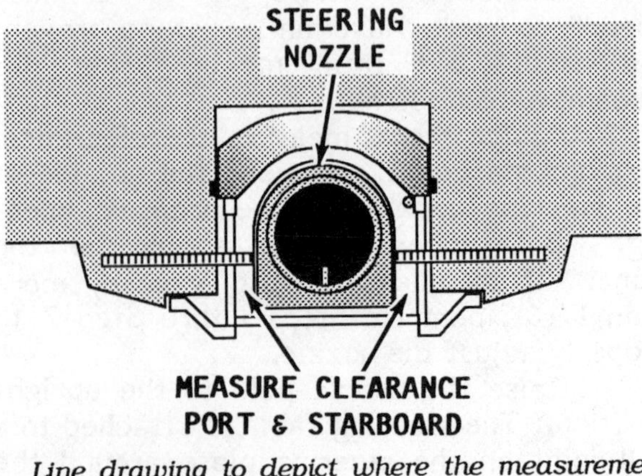

Line drawing to depict where the measurement between the nozzle and the edge of the transom opening is to be taken -- port and starboard -- when making a steering cable adjustment.

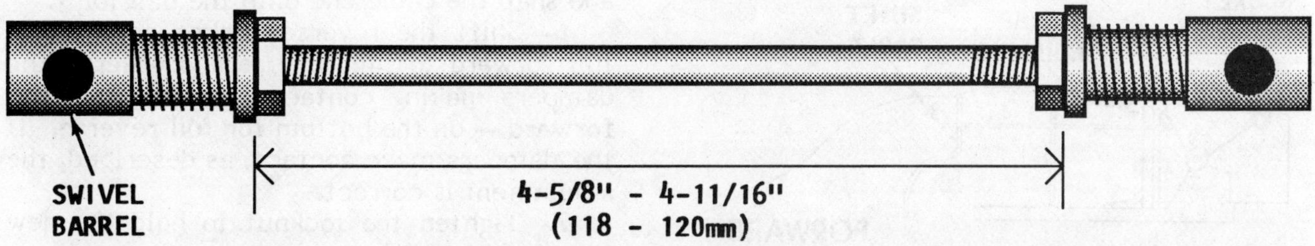

Line drawing of the steering rod for the JetMate and SC Series. The measurement -- described in the text and shown here -- between the two locknuts must be within 1/16" (1.59mm) of the distance indicated.

hold the "Joy Stick" to prevent it from inadvertently moving from the "dead ahead" position.

2- At the stern, measure the clearance between the outside of the steering nozzle and the transom opening. Take the measurement port and starboard. Both measurements must be the same. If they differ by more than 1/16", with the "Joy Stick" "dead ahead", the steering cable must be adjusted.

3- Remove the engine cover. The steering rod is located just above the steering cable. Measure the distance between the outer faces of the two locknuts on the rod. Normally, this distance would be 4-5/8" to 4-11/16" (118-120mm). Loosen the locknut at the steering shaft end of the steering rod. Push the outer sleeve back from the ball joint and pry the cable end free of the ball joint on the steering shaft arm.

4- Center the steering nozzle using a tape measure. Rotate the cable end on the steering rod until the hole in the cable end is aligned with the ball joint. Install the cable over the ball joint, and then tighten the locknut to hold this new adjustment.

5- A final check **MUST** be made with the craft in the water to ensure proper cable adjustment. Therefore, move the craft to a body of water or to a test tank.

6- Start the engine and allow it to warm to operating temperature.

CAUTION

Water must circulate through the jet pump -- to and from the engine, anytime the engine is operating. Circulating water will prevent overheating -- which could cause damage to moving engine parts and possible engine seizure.

7- With the engine operating, hold the handle bar in the "dead ahead" position; advance the throttle and observe if the bow tends to veer to port or starboard. If the bow remains unmoved, the adjustment is accurate -- adjustment task completed.

Reverse Cable Adjustment

The position of the reverse bucket determines the direction of thrust from water expelled from the nozzle. If the bucket is not adjusted properly forward motion of the craft will be seriously hampered because some of the expelled water will be forced back under the craft -- as for going astern.

More importantly, improper adjustment of the reverse bucket can cause the bucket to be unintentionally lowered -- throwing the craft into reverse -- suddenly raising the bow up out of the water -- causing possible injury to operator and/or passengers.

A shifting cable is attached between the shift lever arm ball joint to the ball joint on the reverse bucket. To check adjustment of this cable, proceed as follows:

1- Move the shift lever to the full **FORWARD** position. The rubber damper on top of the reverse bucket should make contact with the pump housing.

2- Move the shift lever aft to the full **REVERSE** position. The two rubber dampers on the lower edge of the reverse bucket should make contact with the pump cover, as indicated in the accompanying illustration.

If the three dampers fail to make contact, as described, the shift cable **MUST** be adjusted at the reverse bucket pivot barrel.

3- Determine if the bucket is adjusted too high or too low.

POSITION WORDS

If the two lower dampers fail to make contact, but the upper damper makes contact, as it should -- the bucket is adjusted too high -- the cable must be shortened.

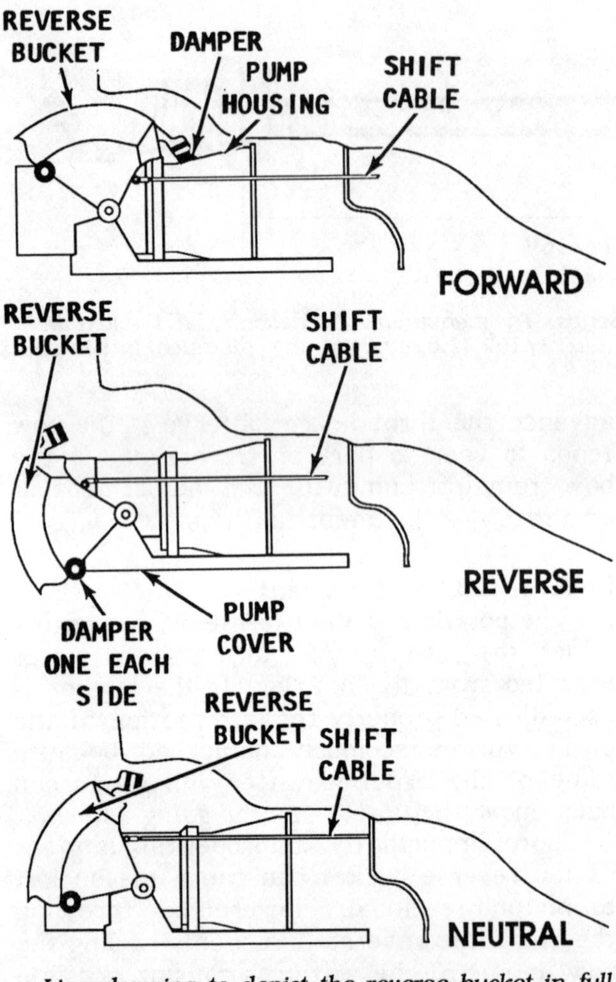

Line drawing to depict the reverse bucket in full ***FORWARD*** *(top), full* ***REVERSE*** *(center), and in the full* ***NEUTRAL*** *position (bottom).*

If the upper damper fails to make contact, but the lower dampers do, as they should -- the bucket is adjusted too low -- the shift cable must be lengthened.

4- Pry the cable end free of the ball joint on the reverse bucket and loosen the locknut. Rotate the cable end **COUNTER-CLOCKWISE** to lengthen the cable or **CLOCKWISE** to shorten the cable, as necessary.

5- Push the outer sleeve down the cable and snap the cable end onto the ball joint.

6- Shift the reverse bucket again from full forward to full reverse and observe the dampers making contact -- on top for full forward -- on the bottom for full reverse. If the dampers make contact, as described, the adjustment is correct.

7- Tighten the locknut to hold the new adjustment.

SPECIAL WORDS

A final check **MUST** be made with the craft in the water to ensure proper adjustment has been attained.

8- Move the craft to a body of water or to a test tank. Start the engine and allow it to warm to operating temperature.

CAUTION

Water must circulate through the jet pump -- to and from the engine, anytime the engine is operating. Circulating water will prevent overheating -- which could cause damage to moving engine parts and possible engine seizure.

9- With the craft in the water; with the engine operating; and the shift lever in the **NEUTRAL** position; the craft **MUST** not move forward or sternward at any engine rpm.

An adjustment to the steering cable end at the reverse bucket may be made immediately while the craft is in the water.

EXCEPTION

If the three dampers fail to make contact, as described, but the craft does not move forward or sternward in the water with the engine operating, such misadjustment can be considered acceptable.

APPENDIX

METRIC CONVERSION CHART

LINEAR

inches	X 25.4	= millimetres (mm)
feet	X 0.3048	= metres (m)
yards	X 0.9144	= metres (m)
miles	X 1.6093	= kilometres (km)
inches	X 2.54	= centimetres (cm)

AREA

inches^2	X 645.16	= millimetres^2 (mm^2)
inches^2	X 6.452	= centimetres^2 (cm^2)
feet^2	X 0.0929	= metres^2 (m^2)
yards^2	X 0.8361	= metres^2 (m^2)
acres	X 0.4047	= hectares (10^4 m^2) (ha)
miles^2	X 2.590	= kilometres^2 (km^2)

VOLUME

inches^3	X 16387	= millimetres^3 (mm^3)
inches^3	X 16.387	= centimetres^3 (cm^3)
inches^3	X 0.01639	= litres (l)
quarts	X 0.94635	= litres (l)
gallons	X 3.7854	= litres (l)
feet^3	X 28.317	= litres (l)
feet^3	X 0.02832	= metres^3 (m^3)
fluid oz	X 29.60	= millilitres (ml)
yards^3	X 0.7646	= metres^3 (m^3)

MASS

ounces (av)	X 28.35	= grams (g)
pounds (av)	X 0.4536	= kilograms (kg)
tons (2000 lb)	X 907.18	= kilograms (kg)
tons (2000 lb)	X 0.90718	= metric tons (t)

FORCE

ounces - f (av)	X 0.278	= newtons (N)
pounds - f (av)	X 4.448	= newtons (N)
kilograms - f	X 9.807	= newtons (N)

ACCELERATION

feet/sec^2	X 0.3048	= metres/sec^2 (m/S^2)
inches/sec^2	X 0.0254	= metres/sec^2 (m/s^2)

ENERGY OR WORK (watt-second - joule - newton-metre)

foot-pounds	X 1.3558	= joules (j)
calories	X 4.187	= joules (j)
Btu	X 1055	= joules (j)
watt-hours	X 3500	= joules (j)
kilowatt - hrs	X 3.600	= megajoules (MJ)

FUEL ECONOMY AND FUEL CONSUMPTION

miles/gal	X 0.42514	= kilometres/litre (km/l)

Note:
235.2/(mi/gal) = litres/100km
235.2/(litres/100 km) = mi/gal

LIGHT

footcandles	X 10.76	= lumens/metre^2 (lm/m^2)

PRESSURE OR STRESS (newton/sq metre - pascal)

inches HG (60° F)	X 3.377	= kilopascals (kPa)
pounds/sq in	X 6.895	= kilopascals (kPa)
inches H_2O (60° F)	X 0.2488	= kilopascals (kPa)
bars	X 100	= kilopascals (kPa)
pounds/sq ft	X 47.88	= pascals (Pa)

POWER

horsepower	X 0.746	= kilowatts (kW)
ft-lbf/min	X 0.0226	= watts (W)

TORQUE

pound-inches	X 0.11299	= newton-metres (N·m)
pound-feet	X 1.3558	= newton-metres (N·m)

VELOCITY

miles/hour	X 1.6093	= kilometres/hour (km/h)
feet/sec	X 0.3048	= metres/sec (m/s)
kilometres/hr	X 0.27778	= metres/sec (m/s)
miles/hour	X 0.4470	= metres/sec (m/s)

TEMPERATURE

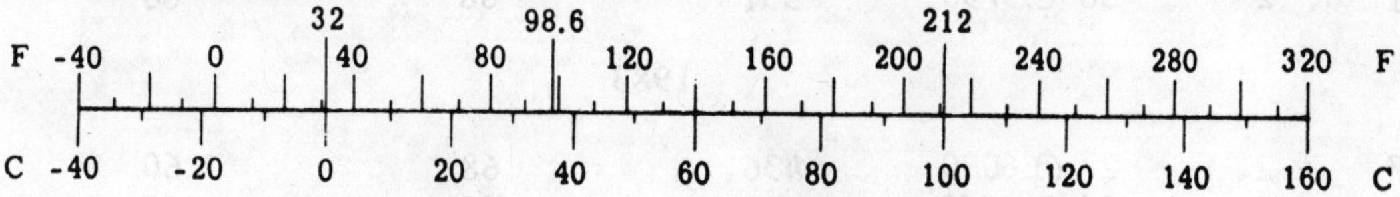

Celsius = 0.556 (°F -32) F = (1.8°C) +32

ENGINE SPECIFICATIONS AND

SERIES	CYL	HP AT RPM	DISPL. cc	BORE mm	STROKE mm	IDLE RPM IN WATER
1973				**1973**		
WSAA	2	26 @ 6000	398	65	60	1800
WSAB	2	26 @ 6000	398	65	60	1800
1974				**1974**		
WSAA	2	26 @ 6000	398	65	60	1800
1975				**1975**		
--	--	--	--	--	--	--
1976				**1976**		
JS 400-A3	2	26 @ 6000	398	65	60	1800
1977				**1977**		
JS 440-A1	2	27 @ 6000	436	68	60	1800
1978				**1978**		
JS 440-A1A	2	27 @ 6000	436	68	60	1800
JS 440-A2	2	27 @ 6000	436	68	60	1800
1979				**1979**		
JS 440-A3	2	27 @ 6000	436	68	60	1800
1980				**1980**		
JS 440-A4	2	27 @ 6000	436	68	60	1800
JS 440-A4A	2	27 @ 6000	436	68	60	1800
1981				**1981**		
JS 440-A5	2	27 @ 6000	436	68	60	1800
1982				**1982**		
JS 440-A6	2	27 @ 6000	436	68	60	1800
JS 550-A1	2	36 @ 5750	531	68	60	1500
1983				**1983**		
JS 440-A7	2	27 @ 6000	436	68	60	1800
JS 550-A2	2	36 @ 5750	531	68	60	1500

TUNE-UP ADJUSTMENTS

SPARK PLUG NGK	SPARK PLUG GAP in	TIMING (CRANK ANGLE AT 6000 RPM)	CARBURETOR TYPE	LOW SPEED SCREW	HIGH SPEED SCREW
1973			**1973**		
B7ES	0.020	25	Mikuni BN 38	3/4	1 1/8
B7ES	0.020	25	Mikuni BN 38	3/4	1 1/8
1974			**1974**		
B7ES	0.020	25	Mikuni BN 38	3/4	1 1/8
1975			**1975**		
--	--	--	--	--	--
1976			**1976**		
B7ES	0.030	25	Mikuni BN 38	3/4	1 1/8
1977			**1977**		
B7ES	0.030	25	Mikuni BN 38	5/8	3/4
1978			**1978**		
B7ES	0.030	25	Mikuni BN 38	5/8	3/4
B7ES	0.030	25	Mikuni BN 38	5/8	3/4
1979			**1979**		
B7ES	0.030	25	Mikuni BN 38	5/8	3/4
1980			**1980**		
B7ES	0.030	25	Mikuni BN 38	5/8	1
B7ES	0.030	25	Mikuni BN 38	5/8	1
1981			**1981**		
B7ES	0.030	25	Mikuni BN 38	1	5/8
1982			**1982**		
B7ES	0.030	25	Mikuni BN 38	1	5/8
B7ES	0.030	28	Mikuni BN 44	1 1/16	7/8
1983			**1983**		
B7ES	0.030	25	Mikuni BN 38	1	5/8
B7ES	0.030	28	Mikuni BN 44	1	5/8

ENGINE SPECIFICATIONS AND

SERIES	CYL	HP AT RPM	DISPL. cc	BORE mm	STROKE mm	IDLE RPM IN WATER
1984						
JS 440-A8	2	27 @ 6000	436	68	60	1800
JS 440-A8U	2	27 @ 6000	436	68	60	1800
JS 550-A3	2	36 @ 5750	530	75	60	1500
1985						
JS 440-A9	2	27 @ 6000	436	68	60	1800
JS 550-A4	2	36 @ 5750	530	75	60	1500
1986						
JS 300-B1	1	N/A	294	76	64.9	1700
JS 440-A10	2	27 @ 6000	436	68	60	1800
JS 550-A5	2	41 @ 6250	530	75	60	1500
JS 650-A1	2	52 @ 6000	635	76	70	1250
X-2	2	N/A	635	76	70	1250
1987						
JS 300-A1	1	30 @ 6000	294	76	64.9	1700
300SX	1	30 @ 6000	294	76	64.9	1700
JS 300-B2	1	N/A	294	76	64.9	1700
JS 440-A11	2	27 @ 6000	436	68	60	1800
JS 550-A6	2	41 @ 6250	530	75	60	1500
JF 650-A2	2	52 @ 6000	635	76	70	1250
X-2	2	N/A	635	76	70	1250
JS 650-A1	2	52 @ 6000	635	76	70	1250
650SX	2	N/A	635	76	70	1250
1988						
JS 300-A2	1	30 @ 6000	294	76	64.9	1700
300SX	1	30 @ 6000	294	76	64.9	1700
JS 300-B3	1	N/A	294	76	64.9	1700
JS 440-A12	2	27 @ 6000	436	68	60	1800
JS 550-A7	2	41 @ 6250	530	75	60	1500
JF 650-A3	2	52 @ 6000	635	76	70	1250
X-2	2	N/A	635	76	70	1250
JS 650-A2	2	52 @ 6000	635	76	70	1250
650SX	2	N/A	635	76	70	1250

TUNE-UP ADJUSTMENTS

SPARK PLUG NGK	SPARK PLUG GAP in	TIMING (CRANK ANGLE AT 6000 RPM)	CARBURETOR TYPE	LOW SPEED SCREW	HIGH SPEED SCREW
1984			**1984**		
B7ES	0.030	25	Mikuni BN 38	1	5/8
B7ES	0.030	25	Mikuni BN 38	1	5/8
B7ES	0.030	28	Mikuni BN 38	1	5/8
1985			**1985**		
B7ES	0.030	25	Mikuni BN 38	1	5/8
B7ES	0.030	28	Mikuni BN 38	1	5/8
1986			**1986**		
B7ES	0.030	18	Mikuni BN 34	5/8	1
B7ES	0.030	25	Mikuni BN 38	1	5/8
B8ES	0.030	21	Mikuni BN 44	1 1/16	15/16 *
B7ES	0.030	17	Keihin CDK 34	1	5/8
B7ES	0.030	17	Keihin CDK 34	1	5/8
1987			**1987**		
B7ES	0.030	18	Mikuni BN 34	1 9/16	1 1/4
B7ES	0.030	18	Mikuni BN 34	1 9/16	1 1/4
B7ES	0.030	18	Mikuni BN 34	1 3/8	1 7/8
B7ES	0.030	25	Mikuni BN 38	1	5/8
B8ES	0.030	21	Mikuni BN 44	1 1/16	15/16
B7ES	0.030	17	Keihin CDK 34	1	5/8
B7ES	0.030	17	Keihin CDK 34	1	5/8
B7ES	0.030	17	Keihin CDK 34	1 1/8	5/8
B7ES	0.030	17	Keihin CDK 34	1 1/8	5/8
1988			**1988**		
B7ES	0.030	18	Mikuni BN 34	1 11/16	3/8
B7ES	0.030	18	Mikuni BN 34	1 11/16	3/8
B7ES	0.030	18	Mikuni BN 34	1 3/4	1 3/8
B7ES	0.030	25	Mikuni BN 38	1	3/4
B8ES	0.030	21	Mikuni BN 44	1 1/16	15/16
B7ES	0.030	17	Keihin CDK 34	1 1/16	5/8
B7ES	0.030	17	Keihin CDK 34	1	5/8
B7ES	0.030	17	Keihin CDK 34	1 1/8	5/8
B7ES	0.030	17	Keihin CDK 34	1 1/8	5/8

* Pilot screw -- 1 5/16

ENGINE SPECIFICATIONS

SERIES	CYL	DISPL. cc	BORE mm	STROKE mm	IDLE RPM IN WATER
1989			**1989**		
JS 300-A3	1	294	76	64.9	1700
300SX	1	294	76	64.9	1700
JS 440-A13	2	436	68	60	1800
JS 550-A8	2	530	75	60	1500
JF 650-A4	2	635	76	70	1250
X-2	2	635	76	70	1250
JF 650-B1	2	635	76	70	1250
TS	2	635	76	70	1250
JS 650-A3	2	635	76	70	1250
650SX	2	635	76	70	1250
JB 650-A1 (JetMate)	2	635	76	70	1250
1990			**1990**		
JS 300-A4	1	294	76	64.9	1700
300SX	1	294	76	64.9	1700
JS 440-A14	2	436	68	60	1800
JS 550-B1	2	530	75	60	1500
550SX	2	530	75	60	1500
JF 650-A5	2	635	76	70	1250
X-2	2	635	76	70	1250
JF 650-B2	2	635	76	70	1250
TS	2	635	76	70	1250
JS 650-A4	2	635	76	70	1250
650SX	2	635	76	70	1250
JB 650-A2 (JetMate)	2	635	76	70	1250
1991			**1991**		
300SX	1	294	76	64.9	1700
JS 440-A15	2	436	68	60	1800
550SX	2	530	75	60	1500
650SX	2	635	76	70	1250
X-2	2	635	76	70	1250
TS	2	635	76	70	1250
SC	2	635	76	70	1250
JB 650-A3 (JetMate)	2	635	76	70	1250

ENGINE SPECIFICATIONS

SPARK PLUG NGK	SPARK PLUG GAP in	TIMING (CRANK ANGLE AT 6000 RPM)	CARBURETOR TYPE	LOW SPEED SCREW	HIGH SPEED SCREW
1989			**1989**		
B7ES	0.030	18	Mikuni BN 34	1 15/16	15/16
B7ES	0.030	18	Mikuni BN 34	1 3/4	1 3/8
B7ES	0.030	25	Mikuni BN 38	1	3/4
B7ES	0.030	21	Mikuni BN 44	1 1/16	15/16 *
B8ES	0.030	17	Keihin CDK 34	1 1/8	5/8
B7ES	0.030	17	Keihin CDK 34	1 1/8	5/8
B7ES	0.030	17	Keihin CDK 34	1 1/8	3/4
B7ES	0.030	17	Keihin CDK 34	1 1/8	5/8
B7ES	0.030	17	Keihin CDK 34	1 1/8	5/8
B7ES	0.030	17	Keihin CDK 34	1 1/8	5/8
BR7ES	0.030	17	Keihin CDK 34	1 1/2	1/2
1990			**1990**		
B7ES	0.030	18	Mikuni BN 34	1 15/16	15/16
B7ES	0.030	18	Mikuni BN 34	1 3/4	1 3/8
B7ES	0.030	25	Mikuni BN 38	7/8	3/4
B8ES	0.030	21	Mikuni BN 44	1 1/16	15/16
B7ES	0.030	21	Mikuni BN 44	1 1/16	15/16
B7ES	0.030	17	Keihin CDK 34	1 1/8	3/4
B7ES	0.030	17	Keihin CDK 34	1 1/8	3/4
B7ES	0.030	17	Keihin CDK 38	N/A	N/A
B7ES	0.030	17	Keihin CDK 34	1 1/8	3/4
B7ES	0.030	17	Keihin CDK 38	N/A	N/A
B7ES	0.030	17	Keihin CDK 34	1 1/8	5/8
BR7ES	0.030	17	Keihin CDK 34	1 1/8	3/8
1991			**1991**		
B7ES	0.030	18	Mikuni BN 34	1 3/4	1 3/8
B7ES	0.030	25	Mikuni BN 38	7/8	3/4
B8ES	0.030	21	Mikuni BN 38	N/A	N/A
B7ES	0.030	17	Mikuni BN 38	N/A	N/A
B7ES	0.030	17	Mikuni BN 38	N/A	N/A
B7ES	0.030	17	Mikuni BN 38	N/A	N/A
B7ES	0.030	17	Mikuni BN 38	N/A	N/A
BR7ES	0.030	17	Keihin CDK 34	1 1/8	3/8

* Pilot screw -- 1 5/16

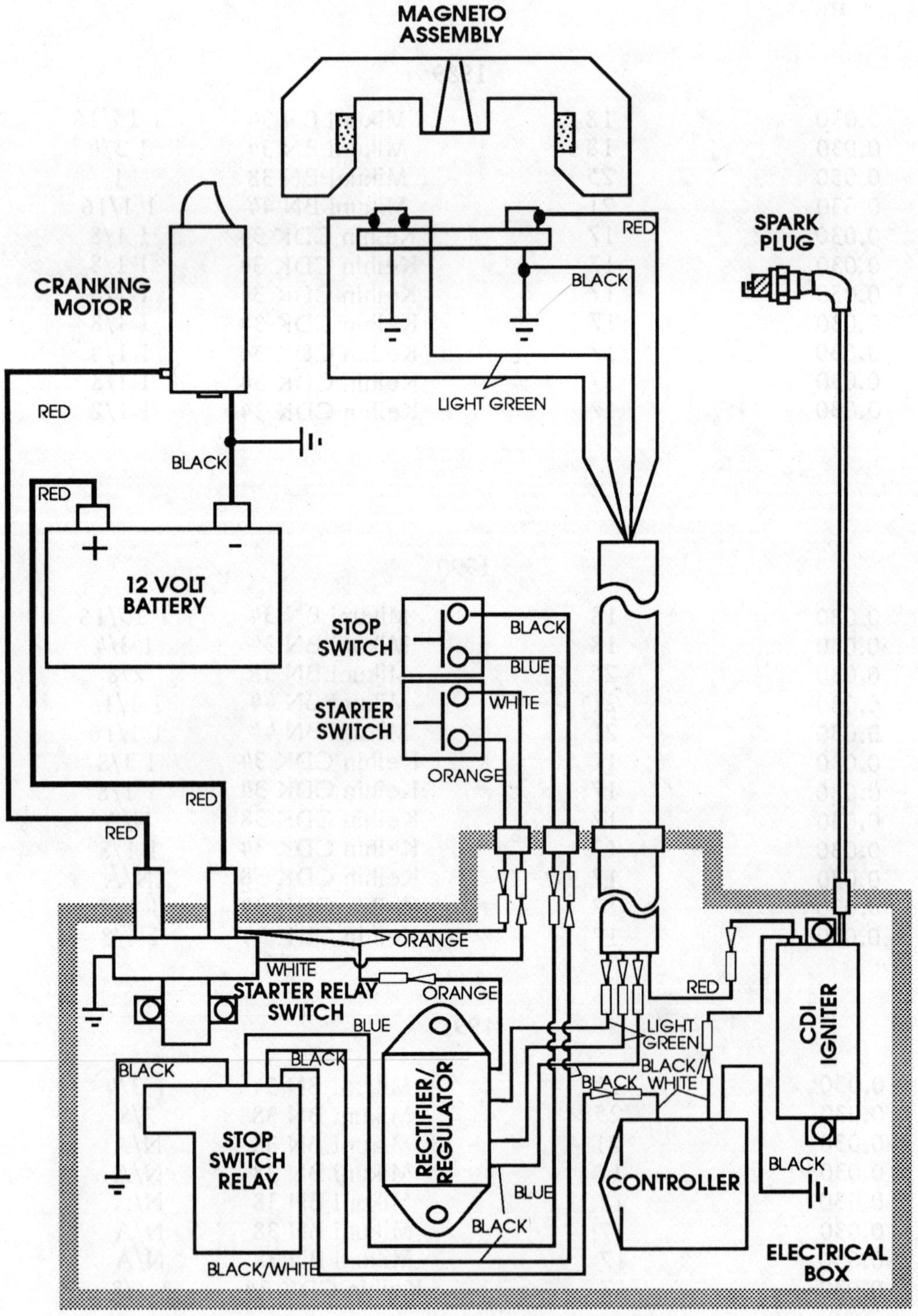

Wire Identification -- All 300 Series Engines.

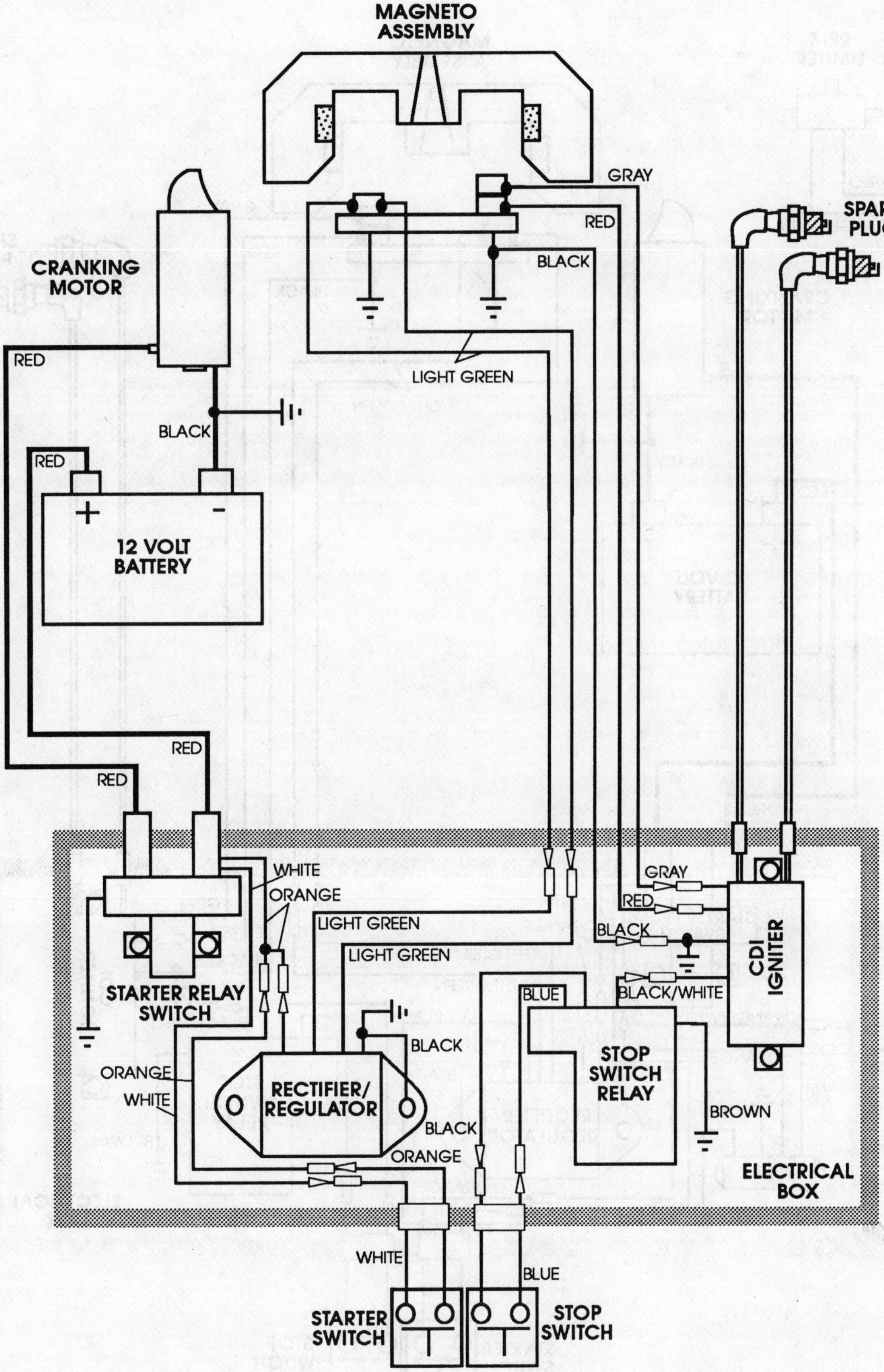

Wire Identification -- All 440 Series Engines and 550 Series Engines -- 1991 and on, without a mechanical fuel rev. limiter.

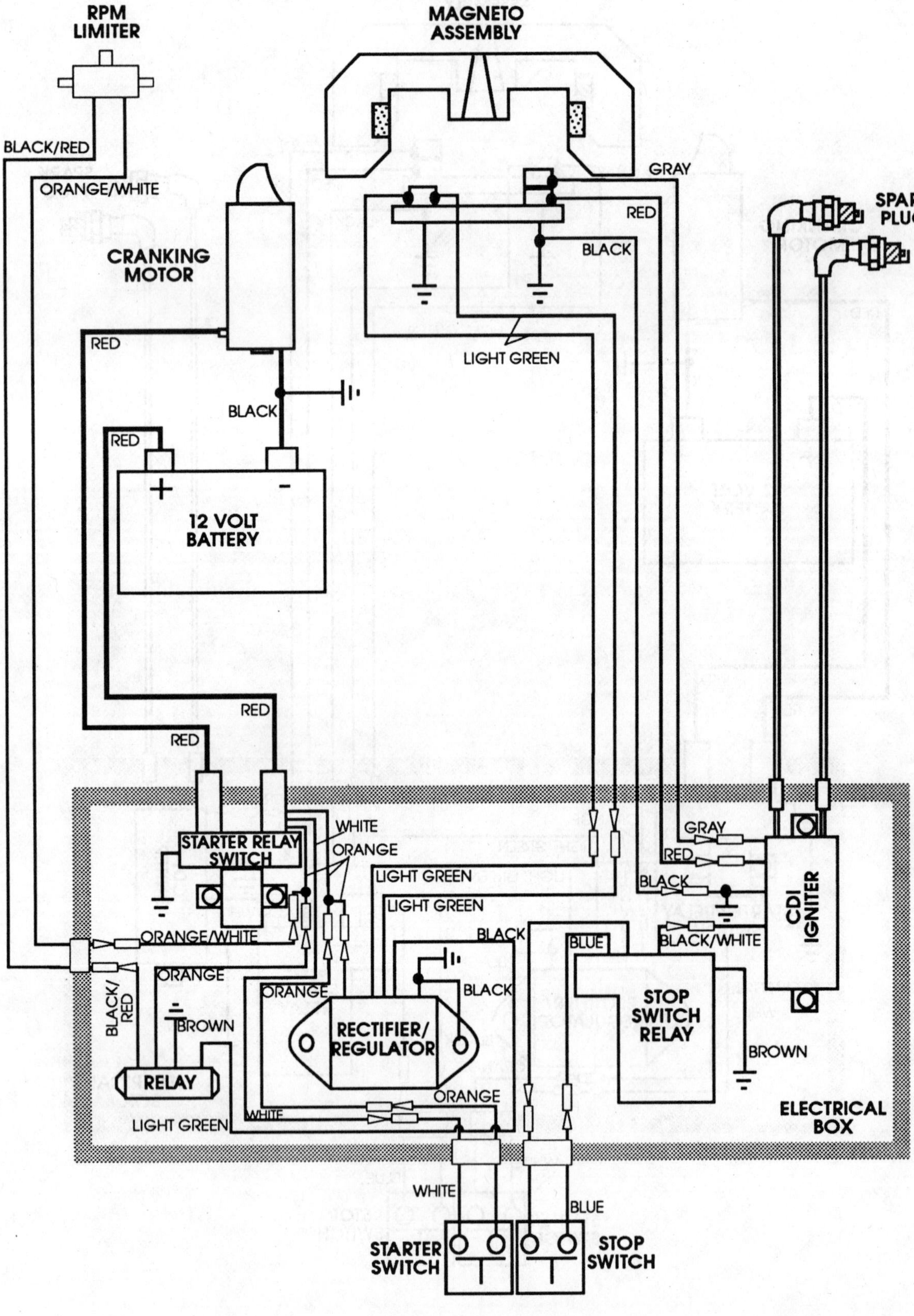

Wire Identification -- 550 Series Engines -- prior to1991, with a mechanical fuel rev. limiter.

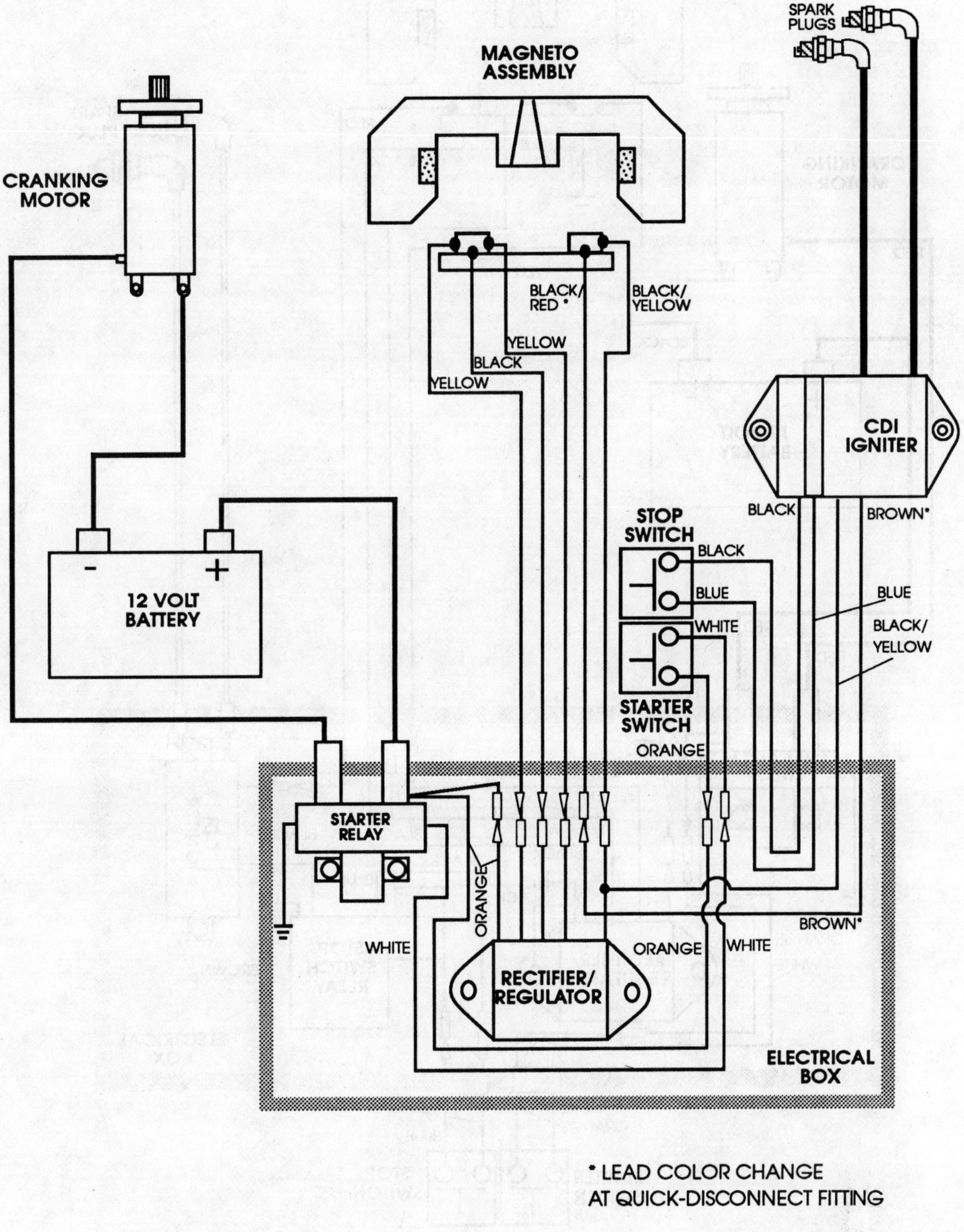

Wire Identification -- Typical 650 Series Engine -- except JF 650 Series Engine, JetMate and SC.

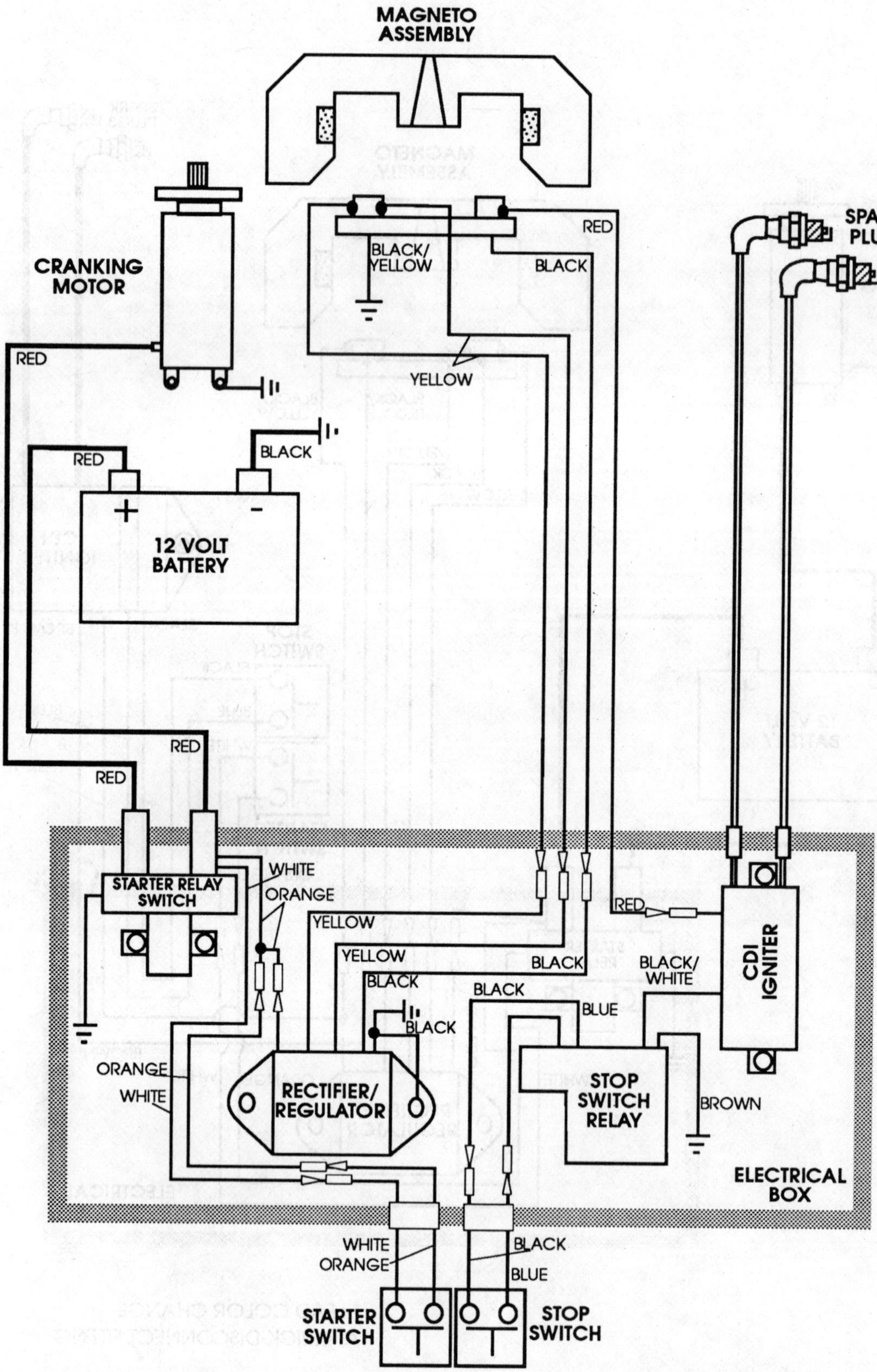

Wire Identification -- JF 650 Series Engine.

BK	BLACK	P	PINK
BL	BLUE	PU	PURPLE
BR	BROWN	R	RED
G	GREEN	W	WHITE
GY	GRAY	LG	LIGHT GREEN
O	ORANGE	Y	YELLOW

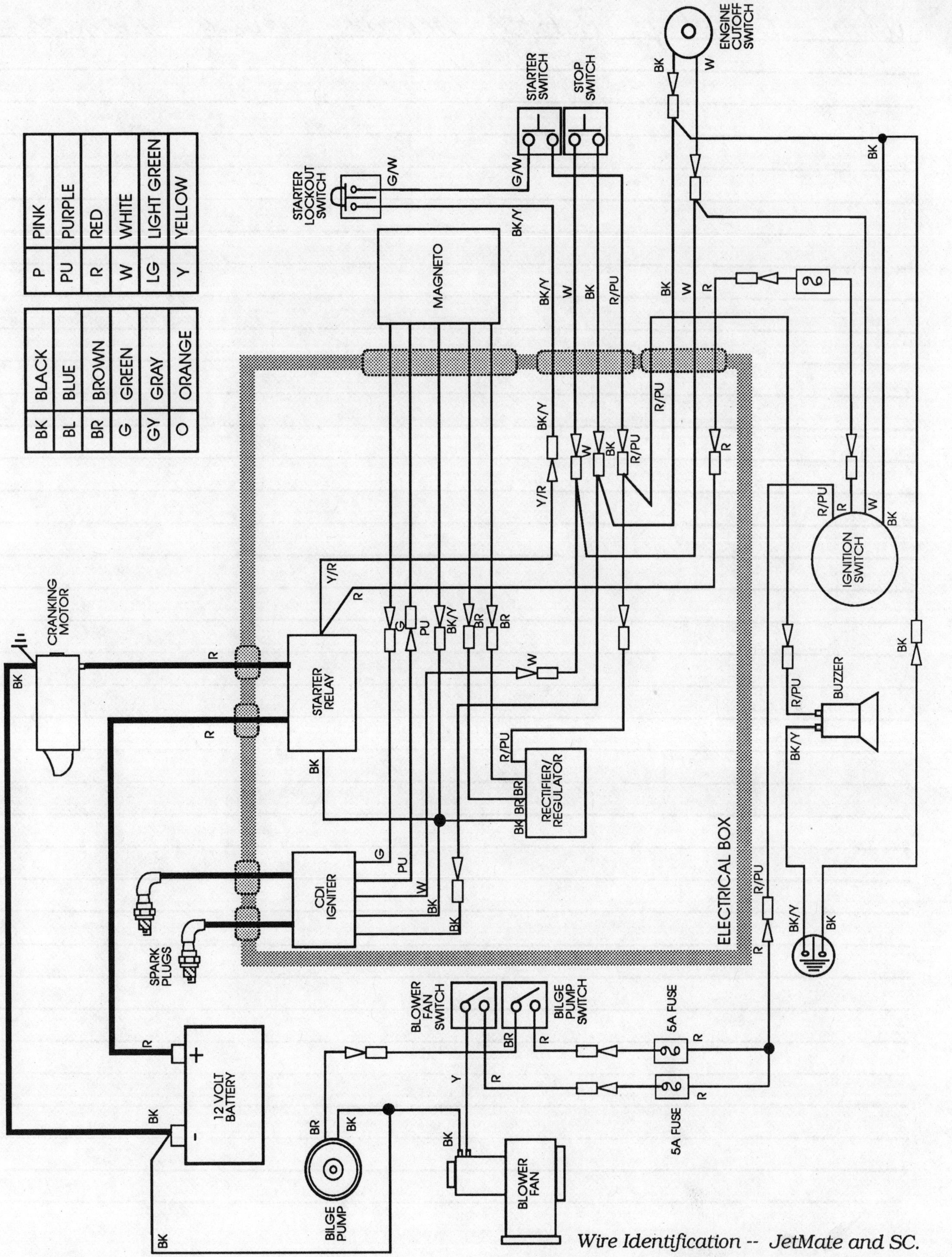

Wire Identification -- JetMate and SC.

NOTES & NUMBERS

When Installing Pistons ARROWS Towards Exhaust.

NOTES & NUMBERS

NOTES & NUMBERS

NOTES & NUMBERS

NOTES & NUMBERS

NOTES & NUMBERS

Other Seloc Marine Manuals

New titles are constantly being produced and the updating work on existing manuals never ceases.
All manuals contain complete detailed instructions, specifications, and wiring diagrams.